Biomaterials

Third Edition

Joon Park   R.S. Lakes

Biomaterials

An Introduction

Third Edition

 Springer

Joon Park
Biomedical and Mechanical Engineering
University of Iowa
Iowa City, IA 52242-1414
USA
joon-park@uiowa.edu

R.S. Lakes
Dept. Engineering Physics
University of Wisconsin
1500 Engineering Drive
Madison, WI 53792-3228
USA
lakes@engr.wisc.edu

ISBN 978-1-4419-2281-6   e-ISBN 978-0-387-37880-0

Printed on acid-free paper.

9 8 7 6 5 4 3 2 1

springer.com

PREFACE
TO THE THIRD EDITION

The field of biomaterials has grown tremendously since publication of the first edition in 1979 and the second edition in 1992. The purpose of this book is to provide for students a resource that includes current developments in the field. To that end, we have updated pertinent applications, incorporating the experience gained in clinical uses of materials. We have also introduced a chapter on tissue regeneration, emphasizing the use of materials as scaffolds to guide cell growth and differentiation in the new field of tissue engineering. We believe that a thorough knowledge of the basics is essential, and so we spend a great deal of time on the fundamentals of structure–properties relationships. The basic premise of the present edition is the same as that of the earlier ones: to describe the fundamentals of natural and man-made biomaterials. Much research is presently being done on the tissue-engineering aspects of biomaterials in the attempt to enlist nature to replace diseased or missing body parts; the biological aspects of biomaterials are crucial to this new area of study. Here, again, one needs to first understand the fundamentals, that is, cell–materials interactions, the effect of degradation of materials, and so on.

This book is intended as a general introduction to the use of artificial materials in the human body for the purposes of aiding healing, correcting deformities, and restoring lost function. It is an outgrowth of an undergraduate course for senior students in biomedical engineering, and is offered as a text to be used in such courses. Topics include biocompatibility, techniques to minimize corrosion or other degradation of implant materials, the principles of materials science as it relates to the use of materials in the body, and specific uses of materials in various tissues and organs. It is expected that the student will have successively completed elementary courses in the mechanics of deformable bodies and in anatomy and physiology, and preferably also an introductory course in materials science prior to undertaking a course in biomaterials.

Many quantitative examples are included as exercises for the student. We recognize that many of these involve unrealistic simplifications and are limited to simple mechanical or chemical aspects of the role of the implant. Many problems that may be used for midterm or final examinations are included in later chapters as a way of refreshing fundamentals learned in earlier chapters. We offer as an apology the fact that biomaterials engineering is still to a great extent an empirical discipline that is complicated by many unknowns associated with the human body. In recognition of that fact, we have endeavored to describe both the successes and failures in the use of materials in the human body. Many clinical statistics are included to illustrate the more realistic aspects of the success rate of implants in living subjects. Also included are many photographs and illustrations of implants and devices as an aid to visualization.

Any errors of commission or omission that have remained in spite of our efforts at correction are our responsibility alone.

We dedicate this book to Hyonsook (Danielle) Park and Diana Lakes for their patience and support during a lengthy undertaking.

Joon B. Park   Roderic S. Lakes
Iowa City, Iowa   *Madison, Wisconsin*

CONTENTS

4: Characterization of Materials — II: Electrical, Optical, X-Ray Absorption, Acoustic, Ultrasonic, etc.

5: Metallic Implant Materials

6: Ceramic Implant Materials

7: Polymeric Implant Materials

8: Composites as Biomaterials

9: Structure–Property Relationships of Biological Materials

10: Tissue Response to Implants

11: Soft Tissue Replacement — I: Sutures, Skin, and Maxillofacial Implants

12: Soft Tissue Replacement — II: Blood Interfacing Implants

Appendices

1

INTRODUCTION

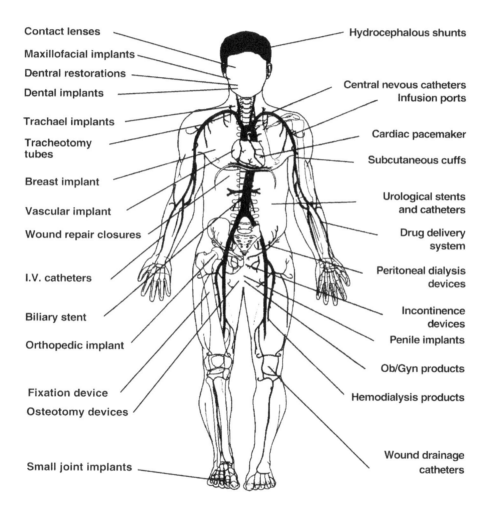

Contact lenses

Maxillofacial implants

Dentral restorations

Dental implants

Trachael implants

Tracheotomy tubes

Breast implant

Vascular implant

Wound repair closures

I.V. catheters

Biliary stent

Orthopedic implant

Fixation device

Osteotomy devices

Small joint implants

Hydrocephalous shunts

Central nevous catheters
Infusion ports

Cardiac pacemaker

Subcutaneous cuffs

Urological stents
and catheters

Drug delivery
system

Peritoneal dialysis
devices

Incontinence
devices

Penile implants

Ob/Gyn products

Hemodialysis products

Wound drainage
catheters

Illustrations of various implants and devices used to replace or enhance the function of diseased or missing tissues and organs. Adapted with permission from Hill (1998). Copyright © 1998, Wiley.

1.1. DEFINITION OF BIOMATERIALS

A *biomaterial* can be defined as any material used to make devices to replace a part or a function of the body in a safe, reliable, economic, and physiologically acceptable manner. Some people refer to materials of biological origin such as wood and bone as biomaterials, but in this book we refer to such materials as "biological materials." A variety of devices and materials is used in the treatment of disease or injury. Commonplace examples include sutures, tooth fillings, needles, catheters, bone plates, etc. A biomaterial is a synthetic material used to replace part of a living system or to function in intimate contact with living tissue. The Clemson University Advisory Board for Biomaterials has formally defined a biomaterial to be "a systemically and pharmacologically inert substance designed for implantation within or incorporation with living systems." These descriptions add to the many ways of looking at the same concept but expressing it in different ways. By contrast, a *biological material* is a material such as bone, skin, or artery produced by a biological system. Artificial materials that simply are in contact with the skin, such as hearing aids and wearable artificial limbs, are not included in our definition of biomaterials since the skin acts as a barrier with the external world.

Because the ultimate goal of using biomaterials is to improve human health by restoring the function of natural living tissues and organs in the body, it is essential to understand relationships among the properties, functions, and structures of biological materials. Thus, three aspects of study on the subject of biomaterials can be envisioned: biological materials, implant materials, and interaction between the two in the body.

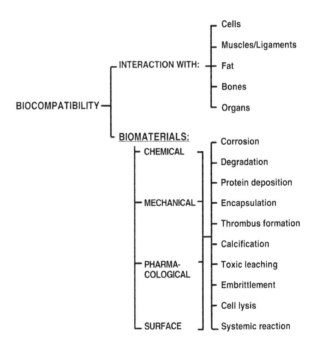

Figure 1-1. Schematic illustration of biocompatibility. Modified with permission from Hill (1998). Copyright © 1998, Wiley.

The success of a biomaterial or an implant is highly dependent on three major factors: the properties and biocompatibility of the implant (Figure 1-1), the health condition of the recipient, and the competency of the surgeon who implants and monitors its progress. It is easy to understand the requirements for an implant by examining the characteristics that a bone plate must satisfy for stabilizing a fractured femur after an accident. These are:

1. Acceptance of the plate to the tissue surface, i.e., biocompatibility (this is a broad term and includes points 2 and 3)
2. Pharmacological acceptability (nontoxic, nonallergenic, nonimmunogenic, noncarcinogenic, etc.)
3. Chemically inert and stable (no time-dependent degradation)
4. Adequate mechanical strength
5. Adequate fatigue life
6. Sound engineering design
7. Proper weight and density
8. Relatively inexpensive, reproducible, and easy to fabricate and process for large-scale production

Development of an understanding of the properties of materials that can meet these requirements is one of the goals of this book. The list in Table 1-1 illustrates some of the advantages, disadvantages, and applications of four groups of synthetic (manmade) materials used for implantation. Reconstituted (natural) materials such as collagen have been used for replacements (e.g., arterial wall, heart valve, and skin).

Table 1-1. Class of Materials Used in the Body

Materials	Advantages	Disadvantages	Examples
Polymers (nylon, silicone rubber, polyester, polytetrafuoroethylene, etc)	Resilient Easy to fabricate	Not strong Deforms with time May degrade	Sutures, blood vessels other soft tissues, sutures, hip socket, ear, nose
Metals (Ti and its alloys, Co–Cr alloys, Au, Ag stainless steels, etc.)	Strong, tough ductile	May corrode Dense Difficult to make	Joint replacements, dental root implants, pacer and suture wires, bone plates and screws
Ceramics (alumina zirconia, calcium phosphates including hydroxyapatite, carbon)	Very bio-compatible	Brittle Not resilient Weak in tension	Dental and orthopedic implants
Composites (carbon–carbon, wire- or fiber- reinforced bone cement)	Strong, tailor-made	Difficult to make	Bone cement, Dental resin

The materials to be used in vivo have to be approved by the FDA (United States Food and Drug Administration). If a proposed material is substantially equivalent to one used before the FDA legislation of 1976, then the FDA may approve its use on a Premarket Approval (PMA) basis. This process, justified by experience with a similar material, reduces the time and expense for the use of the proposed material. Otherwise, the material has to go through a series of "biocompatibility" tests. In general biocompatibility requirements include:

1. Acute systemic toxicity
2. Cytotoxicity
3. Hemolysis
4. Intravenous toxicity
5. Mutagenicity
6. Oral toxicity
7. Pyrogenicity
8. Sensitization

The guidelines on biocompatibility assessment are given in Table 1-2. The data and documentation requirements for all tests demonstrate the importance of good recordkeeping. It is also important to keep all documents created in the production of materials and devices to be used in vivo within the boundaries of Good Manufacturing Practices (GMP), requiring completely isolated clean rooms for production of implants and devices. The final products are usually sterilized after packaging. The packaged item is normally mass sterilized by γ-radiation or ETO (ethylene oxide gas).

Table 1-2. Guidance on Biocompatibility Assessment

A. Data required to assess suitability
 1. Material characterization. Identify the chemical structure of a material and any potential toxicological hazards. Residue levels. Degradation products. Cumulative effects of each process.
 2. Information on prior use. Documented proof of prior use, which would indicate the material(s) suitability.
 3. Toxicological data. Results of known biological tests that would aid in assessing potential reaction (adverse or not) during clinical use.
B. Supporting documents
 1. Details of application: shape, size, form, plus time in contact and use.
 2. Chemical breakdown of all materials involved in the product.
 3. A review of all toxicity data on those materials in direct contact with the body tissues.
 4. Prior use and details of effects.
 5. Toxicity tests [FDA* or ISO (International Standard Organization guides)]
 6. Final assessment of all information including toxicological significance.

*FDA internet address: http://www.fda.gov/cdrh/index.html.
CDRH (Center for Devices and Radiological Health of the FDA) administers medical devices.
Adapted with permission from Hill (1998). Copyright © 1998, Wiley.

Table 1-3 shows a series of criteria to be employed in developing new bone cement. The original bone cement was used by Dr. J. Charnley in total hip replacement fixation on the advice of Dr. D. Smith (a dentist) in the early 1960s. Cold curing acrylic was used in dentistry for many years, but this was the first time it was employed for such an application in orthopedics. Although this qualifies for PMA status under the FDA regulations, one still has to provide clinical data, proof of substantially the same or better performance than the previous bone cement, and chemical and physical performance in vitro and in vivo if one is trying to market a new or similar bone cement in the United States. More examples from the history of the development of biomaterials are given below in §1.3.

The surgical uses of implant materials are given in Table 1-4. One can classify biomaterials into permanent and transient, depending on the time intended to be in the body. Sometimes a temporary implant becomes permanent if one does not remove it, such as a bone plate after a fractured bone is completely healed.

Table 1-3. Criteria for Judgment and Registration of Bone Cements (McDermott, 1997) in the United States, as Specified by the Food and Drug Administration

Property	Parameter/test method	Standard/method (alternatives)
Chemical composition	Raw materials	NMR (if in the liquid phase), FTIR, HPLC/MS
	Added components	Ash
	Purity	ICP/MS, GC/FTIR/MS, titration
Molecular weight (MW)	Relative viscosity MW	Viscosimetry
		GPC (polystyrene standard)
Physical properties	Morphology	Light microscopy; SEM
	Porosity	Scanning acoustical microscopy, x-ray
	Aging due to water uptake	ISO 5833 (bending strength)
Handling properties	Doughing time	ISO 5833, ASTM F451
	Setting time	ISO 5833, ASTM F451
	Intrusion/viscosity	ISO 5833, ASTM F451
Polymerization	Maximum temperature	ISO 5833, ASTM F451
	Shrinkage	Density balance, pycnometer (ASTM D2566)
Degree of polymerization	Content of residual monomer	GC, HPLC/GPC, FTIR
	Release of residual monomer	CC, HPLC/GPC
Stability	Monomer stability (enforced)	ISO 5833, ASTM F451
	BPO content	Titration, FTIR
	Doughing/setting time	ISO 5833, ASTM F451
Modulus of elasticity	Four-point bending	ISO 5833
Compression modulus	Compression	ISO 5833
Tensile modulus	Tensile strength	ASTM D638
Fatigue	Tensile/compression fatigue; tensile/tensile fatigue	ASTM D638
	Four-point bending	Method of Dr. Soltesz, ASTM E399
Fracture toughness	Compact tension/notched bending strength	ASTM E399
Fatigue-crack propagation	Compact tension	ASTM E647
Static strength	ISO 5833	
Flexural strength	Four-point bending	ISO 5833
Compressive strength	Uniaxial compression	150 5833, ASTM F451
Tensile strength	Uniaxial tension	ASTM D638
Shear strength	Cement-cement shear; cement-implant shear	ASTM D732
Viscoelasticity	DMA/compressive creep	DMA/ASTM D2990
Shelf life		Mechanical properties of the hardened cement

See §7.3.4.2 of this text for many of the terms used in this table. Reprinted with permission from Kühn (2000). Copyright © 2000, Springer.

Another important area of study is that of the mechanics and dynamics of tissues and the resultant interactions between them. Generally, this study, known as *biomechanics,* is incorporated into the design and insertion of implants, as shown in Figure 1-2. More sophisticated analysis can be made using computer methods, such as FEM and FEA (finite-element modeling/analysis). These approaches help to design a better prosthesis or even custom make them for individual application.

Table 1-4. Surgical Uses of Biomaterials

Permanent implants

Muscular skeletal system — joints in upper (shoulder, elbow, wrist, finger) and lower (hip, knee, ankle, toe) extremities, permanently attached artificial limb

Cardiovascular system — heart (valve, wall, pacemaker, entire heart), arteries, veins

Respiratory system — larynx, trachea, and bronchus, chest wall, diaphragm, lungs, thoracic plombage

Digestive system — tooth fillings, esophagus, bile ducts, liver

Genitourinary system — kidney, ureter, urethra, bladder

Nervous system — dura, hydrocephalus shunt

Special senses — corneal and lens prosthesis, ear cochlear implant, carotid pacemaker

Other soft tissues — hernia repair sutures and mesh, tendons, visceral adhesion

Cosmetic implants — maxillofacial (nose, ear, maxilla, mandible, teeth), breast, eye, testes, penis, etc.

Transient implants

Extracorporeal assumption of organ function — heart, lung, kidney, liver, decompressive-drainage of hollow viscera-spaces, gastrointestinal (biliary),genitourinary, thoracic, peritoneal lavage, cardiac catheterization

External dressings and partial implants — temporary artificial skin, immersion fluids

Aids to diagnosis — catheters, probes

Orthopedic fixation devices — general (screws, hip pins, traction), bone plates (long bone, spinal, osteotomy), intertrochanteric (hip nail, nail-plate combination, threaded or unthreaded wires and pins), intramedullary (rods and pins), staples, sutures and surgical adhesives

Nanotechnology is a rapidly evolving field that involves material structures on a size scale typically 100 nm or less. New areas of biomaterials applications may develop using nanoscale materials or devices. For example, drug delivery methods have made use of a microsphere encapsulation technique. Nanotechnology may help in the design of drugs with more precise dosage, oriented to specific targets or with timed interactions. Nanotechnology may also help to reduce the size of diagnostic sensors and probes.

Transplantation of organs can restore some functions that cannot be carried out by artificial materials, or that are better done by a natural organ. For example, in the case of kidney failure many patients can expect to derive benefit from transplantation because an artificial kidney has many disadvantages, including high cost, immobility of the device, maintenance of the dialyzer, and illness due to imperfect filtration. The functions of the liver cannot be assumed by any artificial device or material. Liver transplants have extended the lives of people with liver failure. Organ transplants are widely performed, but their success has been hindered due to social, ethical, and immunological problems.

Since artificial materials are limited in the functions they can perform, and transplants are limited by the availability of organs and problems of immune compatibility, there is current interest in the regeneration or regrowth of diseased or damaged tissue. *Tissue engineering* refers to the growth of a new tissue using living cells guided by the structure of a substrate made of synthetic material. This substrate is called a *scaffold*. The scaffold materials are important since they must be compatible with the cells and guide their growth. Most scaffold materials are biodegradable or resorbable as the cells grow. Most scaffolds are made from natural or synthetic polymers, but for hard tissues such as bone and teeth ceramics such as calcium phosphate compounds can be utilized. The tissue is grown in vitro and implanted in vivo. There have been some clinical successes in repair of injuries to large areas of skin, or small defects in cartilage. The topic of tissue engineering, an area of current research activity, is discussed in Chapter 16.

It is imperative that we should know the fundamentals of materials before we can utilize them properly and efficiently. Meanwhile, we also have to know some fundamental properties

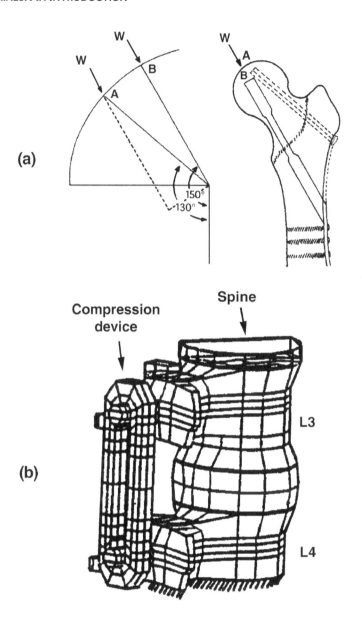

Figure 1-2. (**a**) Biomechanical analysis of femoral neck fracture fixation. Note that if the implant is positioned at 130°, rather than 150°, there will be a force component that will generate a bending moment at the nail-plate junction. The 150° implant is harder to insert and therefore not preferred by surgeons. Reprinted with permission from Massie (1964). Copyright © 1964, Charles C. Thomas. (**b**) Finite-element model (FEM) of spinal disc fusion. Reprinted with permission from Goel et al. (1991). Copyright © 1991, American Association of Neurological Surgeons.

and functions of tissues and organs. The interactions between tissues and organs with man-made materials have to be more fully elucidated. Fundamentals-based scientific knowledge can be a great help in exploring many avenues of biomaterials research and development.

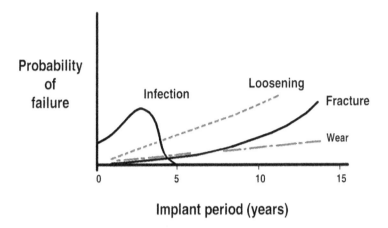

Figure 1-3. A schematic illustration of probability of failure versus implant period for hip joint replacements. Reprinted with permission from Dumbleton (1977). Copyright © 1977, Taylor & Francis.

1.2. PERFORMANCE OF BIOMATERIALS

The performance of an implant after insertion can be considered in terms of reliability. For example, there are four major factors contributing to the failure of hip joint replacements. These are fracture, wear, infection, and loosening of implants, as shown in Figure 1-3. If the probability of failure of a given system is assumed to be f, then the reliability, r, can be expressed as

$$r = 1 - f, \tag{1-1}$$

Total reliability r_t can be expressed in terms of the reliabilities of each contributing factor for failure:

$$r_t = r_1, r_2, ..., r_n, \tag{1-2}$$

where $r_1 = 1 - f_1$, $r_2 = 1 - f_2$, and so on.

Equation (1-2) implies that even though an implant has a perfect reliability of one (i.e., $r = 1$), if an infection occurs every time it is implanted then the total reliability of the operation is zero. Actually, the reliability of joint replacement procedures has greatly improved since they were first introduced.

The study of the relationships between the structure and physical properties of biological materials is as important as that of biomaterials, but traditionally this subject has not been treated fully in biologically oriented disciplines. This is due to the fact that in these disciplines workers are concerned with the biochemical aspects of function rather than the physical properties of "materials." In many cases one can study biological materials while ignoring the fact that they contain and are made from living cells. For example, in teeth the function is largely mechanical, so that one can focus on the mechanical properties of the natural materials. In other cases the functionality of the tissues or organs is so dynamic that it is meaningless to replace them with biomaterials, e.g., the spinal cord or brain.

1.3. BRIEF HISTORICAL BACKGROUND

Historically speaking, until Dr. J. Lister's aseptic surgical technique was developed in the 1860s, attempts to implant various metal devices such as wires and pins constructed of iron, gold, silver, platinum, etc. were largely unsuccessful due to infection after implantation. The aseptic technique in surgery has greatly reduced the incidence of infection. Many recent developments in implants have centered around repairing long bones and joints. Lane of England designed a fracture plate in the early 1900s using steel, as shown in Figure 1-4a. Sherman of Pittsburgh modified the Lane plate to reduce the stress concentration by eliminating sharp corners (Figure 1-4b). He used vanadium alloy steel for its toughness and ductility. Subsequently, Stellite® (Co–Cr-based alloy) was found to be the most inert material for implantation by Zierold in 1924. Soon 18-8 (18 w/o Cr, 8 w/o Ni) and 18-8sMo (2–4 w/o Mo) stainless steels were introduced for their corrosion resistance, with 18-8sMo being especially resistant to corrosion in saline solution. Later, another alloy (19 w/o Cr, 9 w/o Ni) named Vitallium® was introduced into medical practice. A noble metal, tantalum, was introduced in 1939, but its poor mechanical properties and difficulties in processing it from the ore made it unpopular in orthopedics, yet it found wide use in neurological and plastic surgery. During the post-Lister period, the various designs and materials could not be related specifically to the success or failure of an implant, and it became customary to remove any metal implant as soon as possible after its initial function was served.

(a)

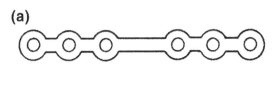

(b)

Figure 1-4. Early design of bone fracture plate: (a) Lane, (b) Sherman.

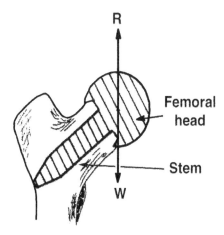

Figure 1-5. The Judet prosthesis for hip surface arthroplasty. Reprinted with permission from Williams and Roaf (1973). Copyright © 1973, W.B. Saunders.

Fracture repair of the femoral neck was not initiated until 1926, when Hey-Groves used carpenter's screws. Later, Smith-Petersen (1931) designed the first nail with protruding fins to prevent rotation of the femoral head. He used stainless steel but soon changed to Vitallium®. Thornton (1937) attached a metal plate to the distal end of the Smith-Petersen nail and secured it with screws for better support. Smith-Petersen later (1939) used an artificial cup over the femoral head in order to create new surfaces to substitute for the diseased joints. He used glass, Pyrex®, Bakelite®, and Vitallium®. The latter was found more biologically compatible, and 30–40% of patients gained usable joints. Similar mold arthroplastic surgeries were performed successfully by the Judet brothers of France, who used the first biomechanical designed prosthesis made of an acrylic (methylmethacrylate) polymer (Figure 1-5). The same type of acrylic polymer was also used for corneal replacement in the 1940s and 1950s due to its excellent properties of transparency and biocompatibility.

Due to the difficulty of surgical techniques and to material problems, cardiovascular implants were not attempted until the 1950s. Blood vessel implants were attempted with rigid tubes made of polyethylene, acrylic polymer, gold, silver, and aluminum, but these soon filled with clot. The major advancement in vascular implants was made by Voorhees, Jaretzta, and Blackmore (1952), when they used a cloth prosthesis made of Vinyon®N copolymer (polyvinyl chloride and polyacrylonitrile) and later experimented with nylon, Orlon®, Dacron®, Teflon®, and Ivalon®. Through the pores of the various cloths a pseudo- or neointima was formed by tissue ingrowth. This new lining was more compatible with blood than a solid synthetic surface, and it prevented further blood coagulation. Heart valve implantation was made possible only after the development of open-heart surgery in the mid-1950s. Starr and Edwards (1960) made the first commercially available heart valve, consisting of a silicone rubber ball poppet in a metal strut (Figure 1-6). Concomitantly, artificial heart and heart assist devices have been developed. Table 1-5 gives a brief summary of historical developments relating to implants.

Figure 1-6. An early model of the Starr-Edwards heart valve made of a silicone rubber ball and metal cage. Reprinted with permission from the Edwards Laboratories.

Table 1-5. Notable Developments Relating to Implants

Year	Investigator	Development
Late 18th–19th century		Various metal devices to fix fractures; wires and pins from Fe, Au, Ag, and Pt
1860–1870	J. Lister	Aseptic surgical techniques
1886	H. Hansmann	Ni-plated steel fracture plate
1893–1912	W.A. Lane	Steel screws and plates for fracture fixation
1909	A. Lambotte	Brass, Al, Ag, and Cu plate
1912	Sherman	Vanadium steel plate, first alloy developed exclusively for medical use
1924	A.A. Zierold	Stellite® (CoCrMo alloy), a better material than Cu, Zn, steels, Mg, Fe, Ag, Au, and Al alloy
1926	M.Z. Lange	18-8sMo (2–4% Mo) stainless steel for greater corrosion resistance than 18-8 stainless steel
1926	E.W. Hey-Groves	Used carpenter's screw for femoral neck fracture
1931	M.N. Smith-Petersen	Designed first femoral neck fracture fixation nail made originally from stainless steel, later changed to Vitallium®
1936	C.S. Venable, W.G. Stuck	Vitallium; 19 w/o Cr-9 w/o Ni stainless steel
1938	P. Wiles	First total hip replacement
1946	J. and R. Judet	First biomechanically designed hip prosthesis; first plastics used in joint replacement
1940s	M.J. Dorzee, A. Franceschetti	Acrylics for corneal replacement
1947	J. Cotton	Ti and its alloys
1952	A.B, Voorhees, A. Jaretzta, A.H. Blackmore	First blood vessel replacement made of cloth
1958	S. Furman, G. Robinson	First successful direct stimulation of heart
1958	J. Charnley	First use of acrylic bone cement in total hip replacements
1960	A. Starr, M.L. Edwards	Heart valve
1970s	W.J. Kolff	Experimental total heart replacement
1990s		Refined implants allowing bony ingrowth
1990s		Controversy over silicone mammary implants
2000s		Tissue engineering
2000s		Nanoscale materials

PROBLEMS

1-1. a. Determine the probability of failure of a hip joint arthroplasty after 15 and 30 years, assuming the following (t is in years).

   b. Which factor is the most important for the longevity of the arthroplasty?

Infection	$f_i = 0.05e^{-t}$
Loosening	$f_{lo} = 0.01e^{+0.15t}$
Fracture	$f_{fr} = 0.01e^{+0.01t}$
Wear	$f_w = 0.01e^{+0.1t}$
Surgical error	$f_{su} = 0.001$
Pain	$f_{pn} = 0.005$

1-2. Plot the individual failure versus time on a graph similar to that in Figure 1-1. Use any graphics software rather than spreadsheet software to achieve a high-quality graph. Also, plot the total success (r_t) versus time on the same graph.

1-3. How would the failure modes shown in Figure 1-1 differ if an obsolete material such as vanadium steel were used to make the hip joint implant (femoral stem)?

1-4. Discuss the feasibility and implications of replacing an entire arm.

1-5. Discuss the ethical problems associated with using fetal brain tissue for transplantation purposes to treat Parkinson's disease; or fetal bone marrow to treat leukemia.

1-6. Discuss the advantages and disadvantages of kidney transplantation as compared to the use of a dialysis machine.

1-7. Discuss the pros and cons of medical device litigation such as that associated with silicone breast implants in the United States. Be brief.

SYMBOLS/DEFINITIONS

Latin Letters

f: Probability of failure
r: Reliability or probability of success

Terms

Biomaterial: A synthetic material used to replace part of a living system or to function in intimate contact with living tissue. Also read the various definitions given by other authors in the text.

Biomechanics: The study of the mechanical laws relating to the movement or structure of living organisms.

Biological material: A material produced by a living organism.

Biocompatibility: Acceptance of an artificial implant by the surrounding tissues and by the body as a whole. The biomaterial must not be degraded by the body environment, and its presence must not harm tissues, organs, or systems. If the biomaterial is designed to be degraded, then the products of degradation should not harm the tissues and organs.

CDRH (Center for Devices and Radiological Health): Branch of the FDA that administers medical devices-related regulations.

Cytotoxicity: Toxic to living cells.

ETO (ethylene oxide gas, $(CH_2)_2O$): A flammable toxic gas used as a sterilization agent.

FDA (Food and Drug Administration): Government agency regulating testing, production, and marketing of food and drugs including medical devices within the United States.

FEM or FEA (finite-element modeling/analysis): Stress and strain analysis of a structural body using computer software. The object is divided into small elements that are amenable to analysis. Boundary conditions are applied and the distribution of stresses and strains calculated.

Gamma (γ)-radiation: The emission of energy as short electromagnetic waves that cause ionization. The radioactive isotope ^{60}Co is an effective source of the radiation. To be effective for sterilization, about 10^6 Gy (J/kg) is needed.

GMP (Good Manufacturing Practices): Medical devices are made in a clean room condition to prevent any contamination. Such practices are required by the FDA for manufacture of implants.

Hemolysis: Lysis (dissolution) of erythrocytes in blood with the release of hemoglobin.

ISO (International Standard Organization): ISO9000 is a set of standards related to medical devices necessary to maintain an efficient and quality system. A standard focuses on controlling organizations rather than specific requirements for final products. ASTM 13.01 focuses on specific products in the United States.

Microsphere: A microscopic hollow sphere, especially of a protein or synthetic polymer.

Mutagenicity: The capacity of a chemical or physical agent to cause permanent genetic alterations.

Nanotechnology: The branch of technology that deals with dimensions and tolerances of less than 100 nanometers — for example, manipulation of individual atoms and molecules.

PMA (Premarket Approval): Some medical devices can be approved by the FDA without extensive tests required by FDA through MDE (medical device exemptions) 510K (http://www.accessdata.fda.gov).

Pyrogenic: Caused or produced by combustion or the application of heat-inducing fever.

Sensitization: Making (an organism) abnormally sensitive to a foreign substance, such as a metal.

Systemic: Denoting the part of the circulatory system concerned with transportation of oxygen to and carbon dioxide from the body in general.

Tissue engineering: Generation of new tissue using living cells, optimally the patient's own cells, as building blocks, coupled with biodegradable materials as a scaffold.

BIBLIOGRAPHY

Encyclopedias and Handbooks

1. *Biomedical engineering handbook.* Boca Raton, FL: CRC Press
2. *Encyclopedia of medical engineering.* New York: J. Wiley
3. *Handbook of bioactive ceramics*, Vols. I and II. Boca Raton, FL: CRC Press
4. *Handbook of biomaterials evaluation.* New York: McMillan
5. *Handbook of materials for medical devices.* Materials Park, OH: ASM International

Journals on Biomaterials

1. Biomaterials
2. Bio-Medical Materials and Engineering
3. Journal of Biomedical Materials Research, Part A, and Part B: Applied Biomaterials

Meetings on Biomaterials

1. Society for Biomaterials, Annual
2. Orthopedic Research Society, part of American Association of Orthopedic Surgeons, Annual
3. American Society for Artificial Internal Organs, Annual

Standards

1. American Society for Testing and Materials, Annual Book of ASTM, Vol 13.01
2. International Standard Organization, ISO

Journals and Books

Bechtol CO, Ferguson AB, Liang PG. 1959. *Metals and engineering in bone and joint surgery.* London: Balliere, Tindall, & Cox.

Black J. 1981. *Biological performance of materials.* New York: Dekker.

Black J. 1992. *Biological performance of materials: fundamentals of biocompatibility.* New York: Dekker.

Bloch B, Hastings GW. 1972. *Plastics materials in surgery.* Springfield, IL: Thomas.

Block MS, Kent JN, Guerra LR, eds. 1997. *Implants in dentistry.* Philadelphia: W.B. Saunders.

Bokros JC, Atkins RJ, Shim HS, Atkins RJ, Haubold AD, Agarwal MK. 1976. Carbon in prosthetic devices. In *Petroleum derived carbons,* pp. 237–265. Ed ML Deviney, TM O'Grady. Washington, DC: American Chemical Society.

Boretos JW 1973. *Concise guide to biomedical polymers.* Springfield, IL: Thomas.

Boretos JW, Eden M, eds. 1984. *Contemporary biomaterials.* Park Ridge, NJ: Noyes.

Brånemark P-I, Hansson BO, Adell R, Breine U, Lidstrom J, Hallen O, Ohman A. 1977. *Osseous integrated implants in the treatment of the edentulous jaw, experience from a 10-year period.* Stockholm: Almqvist & Wiksell International.

Brown JHU, Jacobs JE, Stark L. 1971. *Biomedical engineering.* Philadelphia: Davis.

Brown PW, Constantz B, eds. 1994. *Hydroxyapatite and related materials.* Boca Raton, FL: CRC Press.

Bruck SD. 1974. *Blood compatible synthetic polymers: an introduction.* Springfield, IL: Thomas.

Bruck SD. 1980. *Properties of biomaterials in the physiological environment.* Boca Raton, FL: CRC Press.

Chandran KB. 1992. *Cardiovascular biomechanics.* New York: New York UP.

Charnley J. 1970. *Acrylic cement in orthopaedic surgery.* Edinburgh and London: Churchill/Livingstone.

Dardik H, ed. 1978. *Graft materials in vascular surgery.* Chicago: Year Book Medical Publishers.

de Groot K, ed. 1983. *Bioceramics of calcium phosphate.* Boca Raton, FL: CRC Press.

Ducheyne P, Van der Perre G, Aubert AE, eds. 1984. *Biomaterials and biomechanics.* Amsterdam: Elsevier Science.

Dumbleton JH. 1977. Elements of hip joint prosthesis reliability. *J Med Eng Technol* 1:341–346.

Dumbleton JH, Black J. 1975. *An introduction to orthopedic materials.* Springfield, IL: Thomas.

Edwards WS. 1965. *Plastic arterial grafts.* Springfield, IL: Thomas.

Gebelein CG, Koblitz FF, eds. 1980. *Biomedical and dental applications of polymers: polymer science and technology.* New York: Plenum.

Geesink RGT. 1993. Hydroxyapatite-coated hip implants: experimental studies. In *Hydroxyapatite-coatings in orthopedic surgery,* pp. 151–170. Ed RGT Geesink, MT Manley. New York: Raven Press.

Goel VK, Lim T-H, Gwon JK, Chen J-Y, Winterbottom JM, Park JB, Weinstein JN, Ahn J-Y. 1991. Effects of an internal fixation device: a comprehensive biomechanical investigation. *Spine* **16**:s155–s161.

Greco RS. 1994. *Implantation biology: the host response and biomedical devices.* Boca Raton, FL: CRC Press.

Guelcher SA, JO Hollinger. 2006. *An introduction to biomaterials.* Boca Raton, FL: CRC, Taylor & Francis

Hastings GW, Williams DF, eds. 1980. *Mechanical properties of biomaterials*, Part 3. New York: Wiley.

Helmus MN, ed. 2003. *Biomaterials in the design and reliability of medical devices.* New York: Springer.

Helsen JA, Breme HJ. 1998. *Metals as biomaterials.* Wiley Series in Biomaterials Science and Engineering. New York: Wiley.

Hench LL, ed. 1994. *Bioactive ceramics: theory and clinical applications.* Bioceramics, Vol. 7. Oxford: Pergamon/Elsevier Science.

Hench LL, Ethridge EC. 1982. *Biomaterials: an interfacial approach.* New York: Academic Press.

Hench LL, Jones JR, eds. 2005. *Biomaterials, artificial organs and tissue engineering.* Cambridge: Woodhead.

Hench LL, Wilson J, eds. 1993. *An introduction to bioceramics.* London: World Scientific.

Hill D. 1998. *Design engineering of biomaterials for medical devices.* New York: Wiley.

Homsy CA, Armeniades CD. 1972. *Biomaterials for skeletal and cardiovascular applications.* New York: Wiley Interscience.

Kawahara H, ed. 1989. *Oral implantology and biomaterials.* Amsterdam: Elsevier Science.

King PH, Fries RC. 2003. *Design of biomedical devices and systems.* New York: Dekker.

Kronenthal RL, Oser Z, eds. 1975. *Polymers in medicine and surgery.* New York, Plenum.

Kühn K-D. 2000. *Bone cements: up-to-date comparison of physical and chemical properties of commercial materials.* Berlin: Springer.

Lee H, Neville K. 1971. *Handbook of biomedical plastics.* Pasadena, CA: Pasadena Technology Press (Chapters 3–5 and 13).

Leinninger RI. 1972. *Polymers as surgical implants. CRC Crit Rev Bioeng* **2**:333–360.

Lendlein A, Kratz K, Kelch S. 2005. Smart implant materials. *Med Device Technol* **16**(3):12–14.

Levine SN. 1968a. Materials in biomedical engineering. *Ann NY Acad Sc* **146**:3–10.

Levine SN, ed. 1968b. *Polymers and tissue adhesives.* New York: New York Academy of Sciences.

Linkow LI. 1983. *Dental implants can make your life wonderful again!!* New York: Robert Speller & Sons.

Lipscomb IP, Nokes LDM. 1996. *The application of shape memory alloys in medicine.* Chippenham: Antony Rowe.

Lynch W. 1982. *Implants: reconstructing the human body.* New York: Van Nostrand Reinhold.

Massie WK. 1964. Fractures of the hip. *J Bone Joint Surg* **46A**:658–690.

McDermott B. 1997. *Preclinical testing of PMMA bone cement.* Draft FDA guidelines.

Mears DC. 1979. *Materials and orthopaedic surgery.* Baltimore: Williams & Wilkins.

Merritt K, Chang CC. 1991. Factors influencing bacterial adherence to biomaterials. *J Biomed Mater Res, Part B: Appl Biomat* **5**(3):185–203.

O'Brien WJ, ed. 2002. *Dental materials and their selection*, 3rd ed. Chicago: Quintessence Publishers.

Park JB. 1979. *Biomaterials: an introduction.* New York, Plenum.

Park JB. 1984. *Biomaterials science and engineering.* New York, Plenum.

Park JB, Bronzino JD. 2002. *Biomaterials: principles and applications.* Boca Raton: CRC Press.

Park JB, Lakes RS. 1992. *Biomaterials: an introduction*, 2d ed. New York: Plenum.

Peppas NA, Langer R. 2004. Origins and development of biomedical engineering within chemical engineering. *AIChE J* **50**(3):536–546.

Ratner, 2004. BD. *Biomaterials science: an introduction to materials in medicine.* Amsterdam: Elsevier Academic.

Ratner BD, Bryant SJ. 2004. Biomaterials: where we have been and where we are going. *Annu Rev Biomed Eng* **6**:41–75.

Rubin LR, ed. 1983. *Biomaterials in reconstructive surgery.* St. Louis: Mosby.

Schaldach M, Hohmann D, eds. 1976. *Advances in artificial hip and knee joint technology.* Berlin: Springer-Verlag.

Schnitman PA, Schulman LB, eds. 1978. *Dental implants: benefits and risk.* NIH Consensus Development Conference Statement. Bethesda, MD: US Department of Health and Human Services.

Sharma CP, Szycher M, eds. 1991. *Blood compatible materials and devices.* Lancaster, PA: Technomics Publishing.

Stallforth H, Revell P, eds. 2000. *Materials for medical engineering,* EUROMAT 99, Vol. 2. New York: Wiley-VCH.

Stark L, Agarwal G. 1969. *Biomaterials.* New York: Plenum.

Syrett BC, A. Acharya, eds. 1979. *Corrosion and degradation of implant materials.* Philadelphia: American Society for Testing and Materials.

Szycher M, ed. 1991. *High performance biomaterials.* Lancaster, PA: Technomics Publishing.

Szycher M, Robinson WJ, eds. 1993. *Synthetic biomedical polymers, concepts and applications.* Lancaster, PA: Technomics Publishing.

van Loon J, Mars P. 1997. Biocompatibility: the latest developments. *Med Device Technol* **8**(10):20–4.

Vogler EA. 1998. Structure and reactivity of water at biomaterial surfaces. *Adv Colloid Interface Sci* **74**:69–117.

von Recum AF, Laberge M. 1995. Educational goals for biomaterials science and engineering: prospective view. *J Appl Biomaterials* **6**(2):137–144.

Webster JG, ed. 1988. *Encyclopedia of medical devices and instrumentation.* New York: J. Wiley.

Webster JG. 2006. *Encyclopedia of medical devices and instrumentation.* New York: Wiley Interscience.

Wesolowski SA, Martinez A, McMahon JD. 1966. *Use of artificial materials in surgery.* Chicago: Year Book Medical Publishers.

Williams D. 2003. Revisiting the definition of biocompatibility. *Med Device Technol* **14**(8):10–3.

Williams D. 2005. Environmentally smart polymers. *Med Device Technol* **16**(4):9–10, 13.

Williams DF, ed. 1976. *Compatibility of implant materials.* London: Sector.

Williams DF, ed. 1981a. *Fundamental aspects of biocompatibility.* Boca Raton, FL: CRC Press.

Williams DF, ed. 1981b. *Systemic aspects of blood compatibility.* Boca Raton, FL: CRC Press.

Williams DF, ed. 1982. *Biocompatibility in clinical practice.* Boca Raton, FL: CRC Press.

Williams DF, ed. 1987. Definitions in biomaterials: progress in biomedical engineering. Amsterdam: Elsevier.

Williams DF, Roaf R. 1973. *Implants in surgery.* Philadelphia: W.B. Saunders.

Worthington P, Brånemark P-I. 1992. *Advanced osseointegration surgery: application in the maxillofacial region.* Chicago: Quintessence Publishers.

Worthington P, Lang BR, Rubenstein JE, eds. 2003. *Osseointegration in dentistry.* Chicago: Quintessence Publishers.

Yamamuro T, Hench LL, Wilson J, eds. 1990. *Handbook of bioactive ceramics,* Vols. I and II. Boca Raton, FL: CRC Press.

2

THE STRUCTURE OF SOLIDS

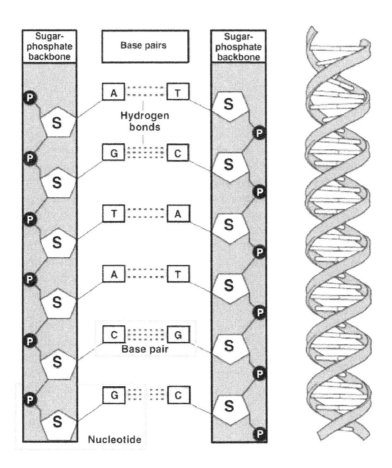

The structure of deoxyribonucleic acid (DNA) was discovered by James Watson and Francis Crick in 1953. DNA was determined to be a right-handed double helix based on x-ray crystallographic data provided to Watson and Crick by Maurice Wilkins and Rosalind Franklin. DNA is composed of repeating subunits called nucleotides. Nucleotides are further composed of a phosphate group, a sugar, and a nitrogenous base. Four different bases are commonly found in DNA: adenine (A), guanine (G), cytosine (C), and thymine (T). In their common structural configurations, A and T form two hydrogen bonds while C and G form three hydrogen bonds. Because of the specificity of base pairing, the two strands of DNA are said to be complementary. This characteristic makes DNA unique and capable of transmitting genetic information. Reprinted courtesy of the National Genomic Research Institute, National Institutes of Health.

The properties of a material are determined by its structure and chemical composition. Since chemical behavior depends ultimately upon the internal structural arrangement of the atoms, all material properties may be attributed to structure. Structure occurs on many levels of scale. These scales may be somewhat arbitrarily defined as the atomic or molecular (0.1–1 nm), nanoscale or ultrastructural (1 nm–1 μm), microstructural (1 μm–1 mm), and macrostructural (>1 mm). In pure elements, alloys, ceramics, and in polymers, the major structural features are on the atomic/molecular scale. Polycrystalline materials such as cast metals consist of grains that may be quite large; however, the boundaries between the grains are atomic scale features.

The first five sections of this chapter deal with the atomic/molecular structural aspects of materials. These sections constitute a review for those readers who have had some exposure to materials science. The final section deals with larger-scale structure associated with composite and cellular materials.

This and the next two chapters give a brief review on the background of the materials science and engineering mostly pertinent to the subsequent chapters. Any reader who cannot comprehend chapters 2–4 should review or study the basic materials science and engineering texts such as those given in the Bibliography.

2.1. ATOMIC BONDING

All solids are made up of atoms held together by the interaction of the outermost (*valence*) electrons. The valence electrons can move freely in the solid but can only exist in certain stable patterns within the confines of the solid. The nature of the patterns varies according to the *ionic, metallic,* or *covalent* bonding. In metallic bonds the electrons are *loosely* held to the ions, which makes the bond nondirectional. Therefore, in many metals it is easy for *plastic deformations* to occur (i.e., the ions can rearrange themselves permanently to the applied external forces). The ionic bonds are formed by exchanging electrons between metallic and nonmetallic atoms. The metallic atoms, such as Na, donate electrons, becoming positive ions (Na^+), while the nonmetallic atoms (e.g., Cl) receive electrons, becoming negative ions (Cl^-). The valence electrons are much more likely to be found in the space around the negative ions than the positive ions, thus making the bonds very directional. The ionic solid structures are limited in their atomic arrangement due to the strong repulsive forces of like ions. Therefore, the positive ions are surrounded by negative ions, and vice versa. The covalent bonds are formed when atoms share the valence electrons to satisfy their partially filled electronic orbitals. The greater the overlap of the valence orbitals or shells, the stronger the bonds become, but bond strength is limited by the strong repulsive forces between nuclei. Covalent bonds are also highly directional and strong, as can be attested by diamond, which is the hardest material known.

In addition to the primary bonds there are *secondary bonds*, which can be a major factor in contributing to material properties. Two major secondary bonds are the hydrogen and van der Waals bonds. The *hydrogen* bonds can arise when the hydrogen atom is covalently bonded to an electronegative atom so that it becomes a positive ion. The electrostatic force between them can be substantial since the hydrogen ion is quite small and can approach the negative ion very closely. The *van der Waals* forces arise when electrons are not distributed equally among ions that can form dipoles. The dipole–dipole interactions do not give rise to directional bonds and the effect is over a short distance. These bonds are much weaker than the hydrogen bonds, as given in Table 2-1.

Table 2-1. Strength of Different Chemical Bonds as Reflected
in Their Heat of Vaporization

Bond type	Substance	Heat of vaporization (kJ/mol)
van der Waals	N_2	13
Hydrogen	Phenol	31
	HF	47
Metallic	Na	180
	Fe	652
Ionic	NaCl	1062
	MgO	1880
Covalent	Diamond	1180
	SiO_2	2810

Reprinted with permission from Harris and Bunsell (1977). Copyright ©
1977, Longman.

Although we have categorized the bonds as discussed above, the real materials may show some combination of bonding characteristics. For example, silicon atoms share electrons covalently but a fraction of the electrons can be freed and permit limited conductivity (semiconductivity). Thus silicon has covalent as well as some metallic bonding characteristics, as shown in Figure 2-1.

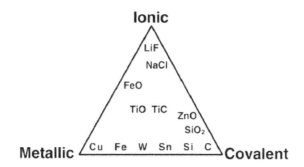

Figure 2-1. Some materials exhibit nearly ideal covalent, metallic, or ionic bonding, but most materials exhibit a hybrid of different bond types.

2.2. CRYSTAL STRUCTURE

2.2.1. Atoms of the Same Size

Crystals contain an orderly arrangement of atoms or molecules. A crystal need not be transparent. Diamond, salt crystals, and quartz are indeed transparent. Also, an ordinary piece of metal such as steel, aluminum, or brass contains many crystals in contact: it is polycrystalline. The crystals are ordinarily not visible since each one has nearly the same appearance. The polycrystalline nature of a metal can be revealed by etching the surface with an acid, which selectively dissolves the surface of crystals that have different orientations. Such etching may occur unintentionally in old brass door knobs and handles, in which the acid secreted by the hands of many people reveals the crystals, which can exceed 1 mm in size.

The arrangement of atoms in a crystal can be treated as an arrangement of hard spheres in view of their maintenance of characteristic equilibrium distances (bond length). Measurement of this distance is done by using x-rays, which have short wavelengths — of the order of one Angstrom (1 Å = 10^{-10} m), approaching the atomic radius. When the atoms are arranged in a regular array the structure can be represented by a *unit cell*, which has a characteristic dimension, the *lattice constant*, *a*, as shown in Figure 2-2. If this atomic structure is extended into three dimensions, the corresponding crystal structure will be cubic. This is a simple cubic space lattice, which is one of the three types of cubic crystals.

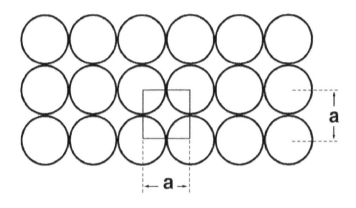

Figure 2-2. Stacking of hard balls (atoms) in simple cubic structure (*a* is the lattice spacing).

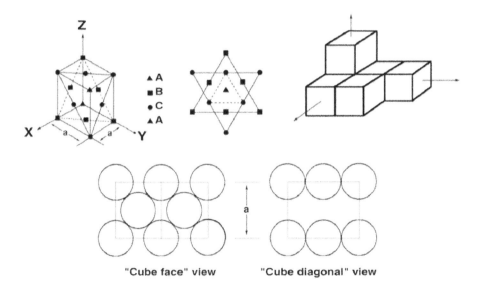

"Cube face" view   "Cube diagonal" view

Figure 2-3. Face-centered cubic structure. Note the arrangement of atomic planes.

The *face-centered cubic* (fcc) structure is another cubic crystal, as shown in Figure 2-3. This structure is called *close packed* [actually, it should be called the closest packed] in three dimensions. Because each atom touches 12 neighbors [hence the *coordination number (CN)* = 12] rather than six as in simple cubic, it results in a most efficiently packed structure. The *hexagonal close-packed* (hcp) structure is arranged by repeating layers of every other plane, that is, the atoms in the third layer occupy sites directly over the atoms in the first layer, as shown in Figure 2-4. This can be represented as ABAB..., while the fcc structure can be represented by three layers of planes ABCABC... Both hcp and fcc have the same packing efficiency (74%); both structures have the most efficient packed planes of atoms with the same coordination number.

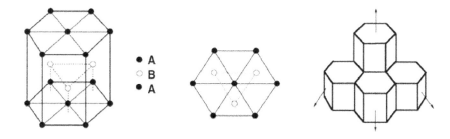

Figure 2-4. Hexagonal close-packed structure. Compare the arrangement of atomic planes with fcc structure.

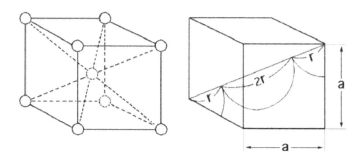

Figure 2-5. Body-centered cubic unit cell. Note that each unit cell has two atoms and $4r = a/\sqrt{3}$.

Another cubic structure is the body-centered cubic (bcc) structure, in which an atom is located in the center of the cube, as shown in Figure 2-5. This structure has a lower packing efficiency (68%) than the fcc structure. Other crystal structures include orthorhombic, in which the unit cell is a rectangular parallelepiped with unequal sides; hexagonal, with hexagonal prisms as unit cells; monoclinic, in which the unit cell is an oblique parallelepiped with one oblique angle and unequal sides; and triclinic, in which the unit cell has unequal sides and all oblique angles. Some examples of the crystal structures of real materials are presented in Table 2-2.

Table 2-2. Examples of Crystal Structures

Material	Crystal structure
Cr	bcc
Co	hcp (below 460°C)
	hcp (above 460°C)
Fe	bcc (below 912°C), ferrite (α)
	fcc (912–1394°C), austenite (γ)
	bcc (above 1394°C), delta iron (δ)
Mo	bcc
Ni	fcc
Ti	hcp (below 900°C)
	bcc (above 900°C)
Rock salt (NaCl)	fcc
Alumina (Al_2O_3)	hcp
Polyethylene	orthorhombic
Polyisoprene	orthorhombic

Determination of crystal structures is actually made by an x-ray instrument where mono-chromatic waves are diffracted from the planes of atoms, as shown in Figure 2-6. The diffracted x-rays are detected by film or recorded, as shown in Figure 2-7 for iron (α-Fe). Figure 2-8 shows a schematic diagram of an x-ray diffractometer. The crystallographic directions and planes are expressed by Miller indices where (x,y,z) indicates the location of a point, [hkl] indicates the direction from (0,0,0) to (x,y,z) in whole numbers. For example, [111] indicates the direction from the origin to the point (1,1,1), a diagonal direction. A family of directions is designated as <hkl>. A crystallographic plane is designated by (hkl), and a family of planes are designated by {hkl}.

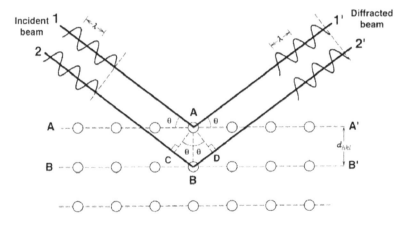

Figure 2-6. X-ray diffraction by planes of atoms (AA' and BB'). $\lambda = BC + BD = 2AB \sin \theta = 2d_{hkl} \sin \theta$. Modified with permission from Callister (2000). Copyright © 2000, Wiley.

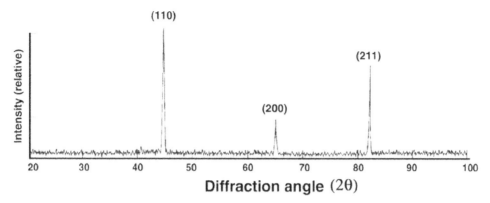

Figure 2-7. X-ray diffraction pattern of polycrystalline α-Fe. Reprinted with permission from Callister (2000). Copyright © 2000, Wiley.

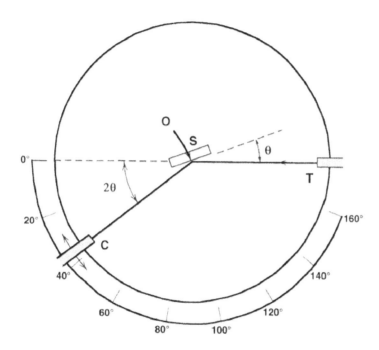

Figure 2-8. Schematic diagram of an x-ray diffractometer; T: x-ray source; S: specimen; C: detector; and O: axis around which the specimen and detector rotate. Reprinted with permission from Callister (2000). Copyright © 2000, Wiley.

Example 2.1

Iron (Fe) has a bcc structure at room temperature with atomic radius of 1.24 Å. Calculate its density (atomic weight of Fe is 55.85 g/mol).

Answer

From Figure 2-5,

$$a = \frac{4r}{\sqrt{3}},$$

and the density (ρ) is given by

$$\rho = \frac{\text{weight/unit cell}}{\text{volume/unit cell}}$$

$$= \frac{2 \text{ atoms/u.c.} \times 55.85 \text{ g/mol}}{(4 \times 1.24 / \sqrt{3} \times 10^{-24} \text{ cm})^3 / \text{u.c.} \ 6 \times 10^{23} \text{ atoms/mol}}$$

$$= \underline{7.87 \text{ g/cm}^3}.$$

Example 2.2

Determine the Miller indices for the plane shown:

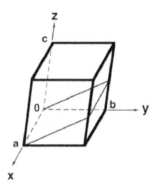

Answer

Usually five steps are used to determine the Miller indices for a plane.

	x	y	z
Intercepts	∞	$-b$	$c/2$
Intercepts in lattice spaces	∞	-1	$1/2$
Reciprocal	0	-1	2
Reduction (not necessary)			
Enclosure		$(0\bar{1}2)$	

Note that the -1 is expressed as $\bar{1}$.

Example 2.3

Determine the d_{110} of α-Fe based on the diffraction pattern depicted in Figure 2-7. Fe has an atomic radius of 0.1241 nm (Appendix IV). The relationship between space $dhkl$ and lattice space a for a cubic unit cell is given as $a = d_{hkl}(h^2 + k^2 + l^2)^{1/2}$.

Answer

For a BCC unit cell, $4r = \sqrt{3}a$, $a = 0.1241 \times 4 / \sqrt{3} = 0.2866$ nm.

2.2.2. Atoms of Different Size

We seldom use pure materials for implants. Most of the materials used for implants are made of more than two elements. When two or more different sizes of atoms are mixed together in a solid, two factors must be considered: (1) the type of site and (2) the number of sites occupied. Consider the stability of the structure shown in Figure 2-9. In Figure 2-9a,b the interstitial atoms touch the larger atoms, and hence they are stable; but they are not stable in 2-9c. At some critical value the interstitial atom will fit the space between six atoms (only four atoms are shown in two dimensions), which will yield the maximum interaction between atoms and consequently the most stable structure results. Thus, at a certain radius ratio of the host and interstitial atoms the arrangement will be most stable. Figure 2-10 gives the maximum radius ratios for given coordination numbers. Note that these radius ratios are determined solely by geometric considerations.

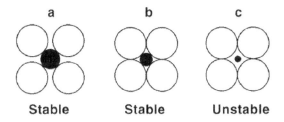

Figure 2-9. Possible arrangements of interstitial atoms as given in Figure 2-10. The critical radius for (**b**) is given by $r + R = \sqrt{2}R$; hence $r/R = 0.414$.

Example 2.4

Calculate the minimum radius ratios for a coordination number of 6.

Answer

For coordination number 6, from Figure 2-10 looking down:

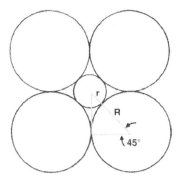

Two-dimensional representation of a structure with coordination number 6.

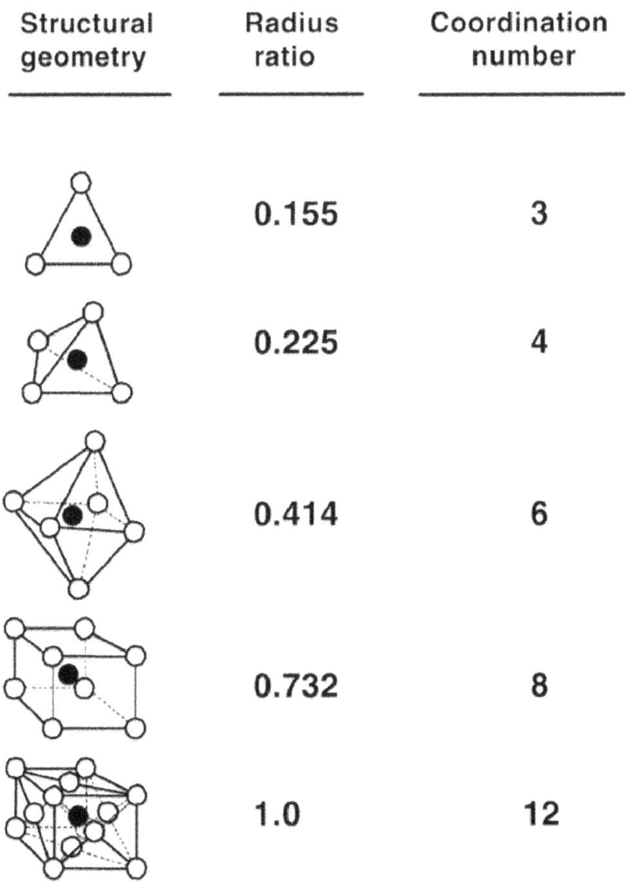

Structural geometry	Radius ratio	Coordination number
	0.155	3
	0.225	4
	0.414	6
	0.732	8
	1.0	12

Figure 2-10. Minimum radius ratios and coordination numbers.

From the diagram,

$$\cos 45° = \frac{R}{R+r}, \quad \frac{1}{\sqrt{2}} = \frac{R}{R+r},$$

$$\sqrt{2}R = R+r, \quad \frac{r}{R} = \sqrt{2} - 1 = 0.414.$$

2.3. IMPERFECTIONS IN CRYSTALLINE STRUCTURES

Imperfections in crystalline solids are sometimes called defects, and they play a major role in determining their physical properties. Point defects commonly appear as lattice vacancies and substitutional or interstitial atoms, as shown in Figure 2-11. The interstitial or substitutional atoms are sometimes called alloying elements if placed intentionally, and impurities if they are unintentional.

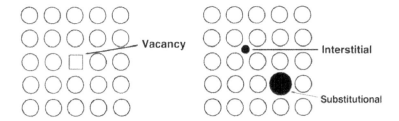

Figure 2-11. Point defects in the form of vacancies and interstitials.

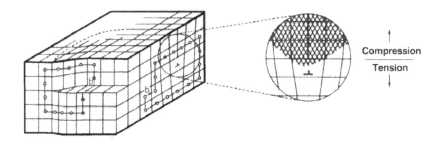

Figure 2-12. Line defects. The displacement is perpendicular to the edge dislocation but parallel to the screw dislocation (left-hand side). The unit length *b* has magnitude (one lattice space) and direction and is called a Burger's vector.

The line defects, called dislocations, are created when an extra plane of atoms is displaced or dislocated out of its regular lattice space registry (Figure 2-12). The line defects or dislocations will lower the strength of a solid crystal enormously since it takes much less energy to move or deform a whole plane of atoms one atomic distance at a time rather than all at once. This is analogous to moving a carpet or a heavy refrigerator on the floor. The carpet cannot be pulled easily if one tries to move it all at once, but if one folds it and propagates the fold until the fold reaches the other end, it can be moved without too much force. Similarly, the refrigerator can be moved easily if one puts one or two logs under it. However, if one uses too many logs (analogous to many dislocations), it becomes harder to move the refrigerator. Correspondingly, if a lot of dislocations are introduced in a solid the strength increases considerably. The reason is that the dislocations become entangled with each other, impeding their movement. The blacksmith practices this principle when he heats a horseshoe to red hot and hammers it. Hammering introduces dislocations. He has to repeat the heating and hammering process in order to increase the number of dislocations without breaking the horseshoe.

Planar defects exist at the grain boundaries. Grain boundaries are created when two or more crystals are mismatched at the boundaries. This occurs during crystallization. Within each grain all the atoms are in a lattice of one specific orientation. Other grains have the same crystal lattice but different orientations, creating a region of mismatch. The grain boundary is less dense than the bulk, hence most diffusion of gas or liquid takes place along the grain boundaries. Grain boundaries can be seen by polishing and subsequent etching of a "polycrystalline" material. This is due to the fact that the grain boundary atoms possess higher energy than the bulk, resulting in a more chemically reactive site at the boundary. Figure 2-13 shows a polished surface of a metal implant. The size of grains plays an important role in determining the physical properties of a material. In general a fine-grained structure is stronger than a coarse one for a given material at a low recrystallization temperature since the former contains more grain boundaries, which in turn interfere with the movement of atoms during deformation, resulting in a stronger material.

Figure 2-13. Midsection of a femoral component of a hip joint implant that show grains (Co–Cr alloy). Notice the size distribution along the stem and from the core to the surface.

2.4. LONG-CHAIN MOLECULAR COMPOUNDS (POLYMERS)

Polymers have very long-chain molecules that are formed by covalent bonding along the backbone chain. The long chains are held together either by secondary bonding forces — such as van der Waals and hydrogen bonds — or primary covalent bonding forces through crosslinks between chains. The long chains are very flexible and are easily tangled. In addition, each chain can have side groups, branches, and copolymeric chains or blocks that can also interfere with the long-range ordering of chains. For example, paraffin wax has the same chemical formula as polyethylene $[(CH_2CH_2)_n]$ but will crystallize almost completely because of its much shorter chain lengths. However, when the chains become extremely long [from 40 to 50 repeating units $[-CH_2CH_2-]$ to several thousands as in linear polyethylene] they cannot be crystallized completely (up to 80–90% crystallization is possible). Also, branched polyethylene, in which side chains are attached to the main backbone chain at positions where a hydrogen atom normally occupies, will not crystallize easily due to the steric hindrance of side chains, resulting in a more noncrystalline structure. The partially crystallized structure is called semicrystalline which is the most commonly occurring structure for linear polymers. The semicrystalline structure is represented by disordered noncrystalline regions and ordered crystalline regions, which may contain folded chains, as depicted in Figure 2-14.

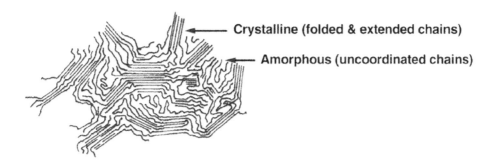

Crystalline (folded & extended chains)

Amorphous (uncoordinated chains)

Figure 2-14. Fringed (micelle) model of linear polymer with semicrystalline structure.

The degree of polymerization (*DP*) is one of the most important parameters in determining polymer properties. It is defined as the average number of mers or repeating units per molecule, i.e., chain. Each chain may have a small or large number of mers depending on the condition of polymerization. In addition, the length of each chain may be different. Therefore, we deal with the average degree of polymerization or average molecular weight (*MW*). The relationship between molecular weight and degree of polymerization can be expressed as

$$MW = DP \times MW \text{ of mer (or repeating unit)}, \tag{2-1}$$

The average molecular weight can be calculated according to the weight fraction (W_i) in each molecular weight fraction (MW_i);

$$M_w = \frac{\sum W_i \cdot MW_i}{\sum W_i} = \sum W_i \, MW_i, \tag{2-2}$$

since $\sum W_i = 1$. This is the average molecular weight.

As the molecular chains become longer by the process of polymerization, their relative mobility decreases. Chain mobility is also related to the physical properties of the final polymer. Generally, the higher the molecular weight, the lesser the mobility of chains, which results in higher strength and greater thermal stability. The polymer chains can be arranged in three ways: linear, branched, and a crosslinked or three-dimensional network, as shown in Figure 2-15. Linear polymers such as polyvinyls, polyamides, and polyesters are much easier to crystallize than the crosslinked or branched polymers. However, they cannot be crystallized 100% as with metals. Instead they become semicrystalline polymers. The arrangement of chains in crystalline regions is believed to be a combination of folded and extended chains. The chain folds, which are seemingly more difficult to form, are necessary to explain observed single crystal structures in which the crystal thickness is too small to accommodate the length of the chain as determined by electron and x-ray diffraction studies. Figure 2-16 shows the two-dimensional representation of chain arrangements. The classical "fringed micelle" model (Figure 2-14) in which the amorphous and crystalline regions coexist has been modified to include chain folds in the crystalline regions.

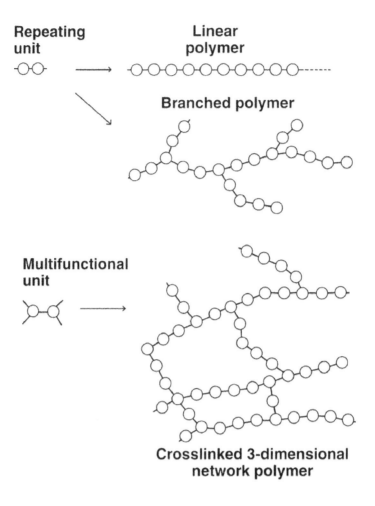

Figure 2-15. Types of polymer chains.

Such crosslinked or three-dimensional network polymers as (poly)phenolformaldehyde cannot be crystallized at all and become noncrystalline, amorphous polymers.

Vinyl polymers have a repeating unit, $-CH_2-CHX-$, where X is a monovalent side group. There are three possible arrangements of side groups X: (1) atactic, (2) isotactic, and (3) syndiotactic, as shown in Figure 2-17. In atactic arrangements the side groups are randomly distributed, while in syndiotactic and isotactic arrangements they are either in alternating positions or on one side of the main chain. If side groups are small as in polyethylene (X=H) and the chains are linear, the polymer crystallizes easily. However, if the side groups are large as in polyvinyl chloride (X=Cl) and polystyrene (X=C_6H_6, benzene ring) and are randomly distributed along the chains (atactic), then a noncrystalline structure will be formed. The isotactic and syndiotactic polymers usually crystallize even when the side groups are large. Note that polyethylene does not have tacticity since it has symmetric side groups.

Copolymerization, in which two or more homopolymers (repeating units of one type throughout its structure) are chemically combined, always disrupts the regularity of polymer chains, thus promoting the formation of noncrystalline structure (Figure 2-18). Plasticizers may be added to a polymer to achieve greater compliance. Plasticizers contain small molecules that facilitate the movement of the long-chain molecules of the polymer itself. Plasticizers also prevent crystallization by keeping the molecular chains separated from one another. This results in a noncrystalline version of a polymer that normally crystallizes. An example is celluloid, which is made of normally crystalline nitrocellulose plasticized with camphor. Plasticizers are also used to make rigid noncrystalline polymers like polyvinylchloride (PVC) into a more flexible solid (a good example is Tygon® tubing).

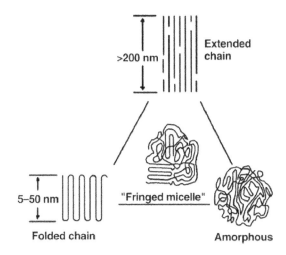

Figure 2-16. Two-dimensional representation of polymer solid.

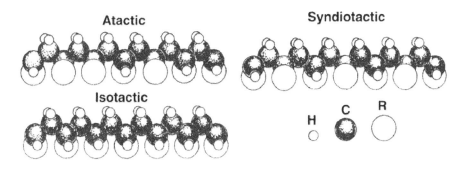

Figure 2-17. Tacticity of vinyl polymers.

Elastomers or rubbers are polymers that exhibit low stiffness and high stretchability at room temperature. They snap back to their original dimensions when the load is released. The elastomers are noncrystalline polymers that have an intermediate structure consisting of long-chain molecules in three-dimensional networks (see the next section for more details). The chains also have "kinks" or "bends" in them that straighten when a load is applied. For example, the chains of cis-polyisoprene (natural rubber) are bent at the double bond due to the methyl group interfering with the neighboring hydrogen in the repeating unit [$-CH_2-$ $C(CH_3)=CH-CH_2-$]. If the methyl group is on the opposite side of the hydrogen, then it becomes trans-polyisoprene, which will crystallize due to the absence of the steric hindrance present in the cis form. The resulting polymer is a very rigid solid called gutta percha, which is not an elastomer.

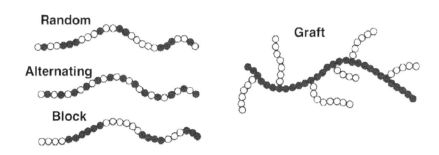

Figure 2-18. Possible arrangements of copolymers.

Below the glass transition temperature (T_g; second-order transition temperature between viscous liquid and solid) natural rubber loses its compliance and becomes a glass-like material. Therefore, to be flexible, all elastomers should have T_g well below room temperature. What makes the elastomers not behave like liquids above T_g is in fact due to the crosslinks between chains that act as pinning points. Without crosslinks the polymer would deform permanently. An example is latex, which behaves as a viscous liquid. Latex can be crosslinked with sulfur (vulcanization) by breaking double bonds (C=C) and forming C—S–S—C bonds between the chains. The more crosslinks are introduced, the more rigid the structure becomes. If all the chains are crosslinked together, the material will become a three-dimensional rigid polymer.

2.5. SUPERCOOLED AND NETWORK SOLIDS

Some solids such as window glass do not have a regular crystalline structure. Solids with such an atomic structure are called amorphous or noncrystalline materials. They are usually super-cooled from the liquid state and thus retain a liquid-like molecular structure. Consequently, the density is always less than that of the crystalline state of the same material, indicating inclusion of some voids (free volume, Figure 2-19). Due to the metastable state of the structure, the amorphous material tends to slowly crystallize with time. The process may be so slow (as in window glass) that crystallization is never seen over humanly accessible time scales. Amorphous materials are also more brittle and less strong than their crystalline counterparts. It is very difficult to make metals amorphous since the metal atoms are extremely mobile in the liquid state, and crystallization to a solid is abrupt. However, one can make amorphous metal from liquid metal by cooling the melt sufficiently rapidly. This is done by splat cooling in which a thin layer of liquid is rapidly placed in contact with a highly heat-conductive metal plate at low temperature. This process causes the amorphous structure of the liquid to be preserved in the solid, since there is insufficient time for crystallization. Polymers can be easily made amorphous because of the relatively sluggish mobility of their molecules.

The network structure of a solid results in a three-dimensional, amorphous structure since the restrictions on the bonds and rigidity of subunits prevent them from crystallizing. Common network structure materials are phenolformaldehyde (Bakelite®) polymer and silica (SiO_2) glass, as shown in Figure 2-20. The network structure of phenolformaldehyde is formed by crosslinking through the phenol rings, while the Si–O tetrahedra are joined corner to corner via oxygen atoms. The three-dimensional network solids do not flow at high temperatures.

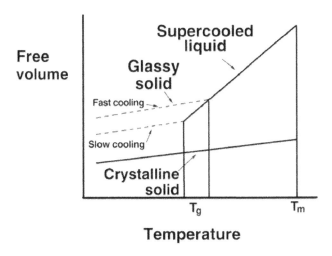

Figure 2-19. Change of volume versus temperature of a solid. The glass transition temperature (T_g) depends on the rate of cooling; below T_g the material behaves as a solid like a window glass.

Example 2.3

Calculate the free volume of 100 g of iodine supercooled from the liquid state, which has a density of 4.8 g/cm³. Assume the density of amorphous iodine is 4.3 g/cm³ and the crystalline density is 4.93 g/cm³.

Answer

The fraction of supercooled iodine can be calculated by extrapolation. We consider the density to be proportional to crystallinity and calculate the slope of the line:

$$\frac{4.93 - 4.8}{4.93 - 4.3} = 0.21.$$

The weight of the supercooled liquid is 0.21 × 100 g = 21 g; therefore, the total free volume is

$$\left(\frac{1}{4.3} \frac{cm^3}{g} - \frac{1}{4.93} \frac{cm^3}{g} \right) \times 21 \text{ g} = \underline{0.65 \text{ cm}^3}.$$

Upon complete crystallization, the volume of the iodine will decrease by 0.65 cm³.

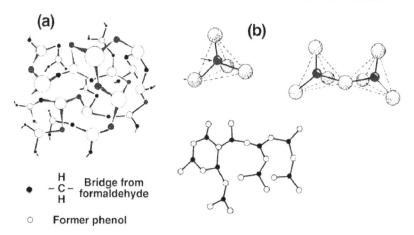

Figure 2-20. Network structure of noncrystalline solids: (**a**) phenolformaldehyde (Bakelite®); (**b**) silica glass structure. Subunit SiO_4 is a tetrahedron with a silicon atom at its center.

2.6. COMPOSITE MATERIAL STRUCTURE

Composite materials are those which consist of two or more distinct parts. The term "composite" is usually reserved for those materials in which the distinct phases are separated on a scale larger than the atomic, and in which properties such as the elastic modulus are significantly altered in comparison with those of a homogeneous material. Accordingly, bone and fiberglass are viewed as composite materials, but alloys such as brass or metals such as steel with carbide particles are not. Although many engineering materials, including biomaterials, are not composites, virtually all natural biological materials are composites.

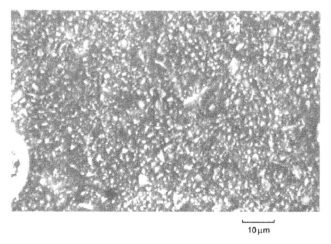

Figure 2-21. Dental composite. The particles are silica (SiO_2) and the matrix is polymeric. Reprinted with permission from Boyer (1989). Courtesy of the author.

The properties of a composite material depend upon the shape of the heterogeneities (second-phase material), upon the volume fraction occupied by them, and upon the stiffness and integrity of the interface between constituents. The shape of the heterogeneities in a composite material is classified as follows. Possible inclusion shapes are the particle, with no long dimension; the fiber, with one long dimension; and the platelet or lamina, with two long dimensions. Cellular solids are those in which the "inclusions" are voids, filled with air or liquid. In the context of biomaterials, it is necessary to distinguish the above cells, which are structural, from biological cells, which occur only in living organisms.

Examples of composite material structures are as follows. The dental composite filling material shown in Figure 2-21 has a particulate structure. This material is packed into the tooth cavity while still soft, and the resin is polymerized in situ. The silica particles serve to provide hardness and wear resistance superior to that of the resin alone. A typical fibrous solid is shown in Figure 2-22. The fibers serve to stiffen and strengthen the polymeric matrix. In this example, pullout of fibers during fracture absorbs mechanical energy, conferring toughness on the material. Fibers have been added to the polymeric parts of total joint replacement prostheses, in an attempt to improve the mechanical properties. Figure 2-23 shows a laminated structure in which each lamina is fibrous, while Figure 2-24 shows representative synthetic cellular materials.

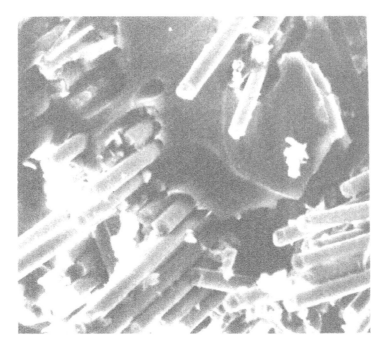

Figure 2-22. Glass-fiber-epoxy composite: fracture surface showing fiber pullout. Reprinted with permission from Agarwal and Broutman (1980). Copyright © 1980, Wiley.

Figure 2-23. Structure of a cross-ply laminate. Reprinted with permission from Agarwal and Broutman (1980). Copyright © 1980, Wiley.

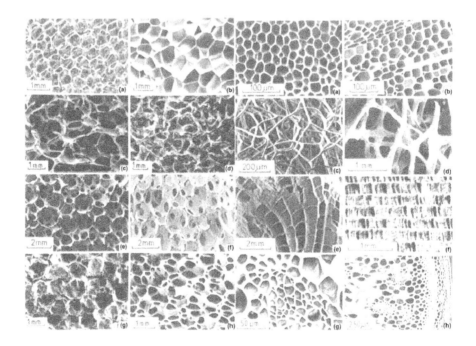

Figure 2-24. Synthetic (**a**) and natural (**b**) cellular materials. Reprinted with permission from Gibson and Ashby (1988). Copyright © 1988, Pergamon Press.

The properties of composite materials depend very much upon structure, as they do in homogeneous materials. Composites differ in that the engineer has considerable control over the larger-scale structure, and hence over the desired properties. The relevant structure–property relations will be developed in Chapter 8.

PROBLEMS

2-1.   Identify the type of bond (ionic or covalent) in the following compounds. (a) ammonia (NH_3), (b) salt (NaCl), (c) carbon tetrachloride (CCl_4), (d) hydrogen peroxide (H_2O_2), (e) ozone (O_3), (f) ethylene ($CH_2=CH_2$), (g) water (H_2O), (h) magnesium oxide (MgO), and (i) diamond (C).

2-2.   Calculate the number of atoms present per cm^3 for alumina (Al_2O_3), which has a density of 3.8 g/cm^3.

2-3.   Calculate the diameters of the smallest cations that have a 6- and 8-fold coordination with O^{2-} ions (see Table 6-1 and Figure 2-10).

2-4.   Steel contains carbon, which forms iron carbide (Fe_3C). Determine the weight percentage of carbon in Fe_3C.

2-5.   A metal alloy of 92.5 w/o Ag and 7.5 w/o Cu is called sterling silver. Determine the atomic percentage of silver.

2-6.   Titanium (Ti) is bcc above 882°C. The atomic radius increases 2% when the bcc structure changes to hcp during cooling. Determine the percentage volume change. Hint: there will be a change in the atomic packing factor.

2-7.   The molecular weight of polymers can be either a number or weight average (M_n or M_w) which are defined as

$$M_n = \frac{\sum (N_i M_i)}{\sum N_i} \quad \text{and} \quad M_w = \frac{\sum (W_i M_i)}{\sum W_i} \,,$$

where N_i is the number of molecules with molecular weight M_i, M_i is the molecular weight of the ith species, and W_i is the weight fraction of molecules of the ith species. Show that

$$M_n = \frac{1}{\sum (W_i / M_i)}.$$

2-8.   Calculate the weight of an iron atom. The density of iron is 7.87 g/cm^3. Avogadro's number is 6.02×10^{23}. How many iron atoms are contained in a cubic centimeter?

2-9.   A polyethylene is made of the following weight distributions (see chart).

   a.   Calculate M_n, M_w, and M_w/M_n (polydispersity).
   b.   Plot W_i versus M_i and also M_w and M_n in the same plot.
   c.   Why is M_w always greater than M_n?

W_i (grams)	10	20	30	30	10
M_i (kg/mol)	10	20	30	40	50

2-10. Ultrahigh-molecular-weight polyethylene is used to make the acetabular cup of a hip joint prosthesis. If the average molecular weight is 2.8 mg/mol,

   a. Calculate the number of repeating units (monomer, $-CH_2-CH_2-$) in an average chain molecule.

   b. Calculate the length of the stretched chain. Due to the tetrahedral structure of the carbon, a C–C–C bond will make an 108° angle, resulting in a 0.126-nm bond length projected horizontally even though the C–C bond length is 0.154 nm.

   c. Calculate M_n if the polydispersity is 1.5.

SYMBOLS/DEFINITIONS

Greek letters

ρ: Density, mass per unit volume.

Latin letters

a: Lattice constant, which is the spacing of atoms in a crystal lattice. For example, in cubic crystal systems all the lattice constants are the same, i.e., $a = b = c$, so that the atomic spacing is the same in each principal direction.

bcc: Body-centered cubic lattice; one atom is positioned in the center of the cubic unit cell.

CN: Coordination number; number of atoms touching an atom in the middle.

fcc: Face-centered cubic lattice; one atom is positioned in each face of the cubic unit cell.

hcp: Hexagonal close-packed lattice; twelve atoms surround and touch a central atom of the same species resulting in the hexagonal prism symmetry.

M_n: Average molecular weight of a polymer (number).

M_w: Average molecular weight of a polymer (weight).

r, R: Radius of an atom.

T_g: Glass transition temperature at which solidification without crystallization takes place from a viscous liquid.

Words

Composites: Materials obtained by combining two or more materials at a scale larger than the atomic/molecular, e.g., nanoscale, microscale, or macroscale, taking advantage of salient features of each material. An example is (high-strength) fiber-reinforced epoxy resin.

Copolymers: Polymers made from two or more homopolymers; can be obtained by grafting, block, alternating, or random attachment of the other polymer segment.

Covalent bonding: Bonding of atoms or molecules by sharing valence electrons.

Elastomers: Rubbery materials. The restoring force comes from uncoiling or unkinking of coiled or kinked molecular chains. They can be highly stretched.

Free volume: The difference in volume occupied by the crystalline state (minimum) and non-crystalline state of a material for a given temperature and a pressure.

Glass transition temperature (T_g): see T_g.

Hydrogen bonding: A secondary bonding through dipole interactions in which a hydrogen ion is one of the dipoles.

Ionic bonding: Bonding of atoms or molecules through electrostatic interaction of positive and negative ions.

Imperfections: Defects created in a perfect (crystalline) structure by vacancy, interstitial, and substitutional atoms by the introduction of an extra plane of atoms (dislocations) or by mismatching at the crystals during solidification (grain boundaries).

Lattice constant: The spacing of atoms in a crystal lattice. For example, in cubic crystal systems all the lattice constants are the same, i.e., $a = b = c$, so that the atomic spacing is the same in each principal direction.

Metallic bonding: Bonding of atoms or molecules through loosely bound valence electrons around positive metal ions.

Minimum radius ratio (r/R): Ratio between the radius of a smaller atom to be fitted into the space among the larger atom's radius based on geometric consideration.

Packing efficiency: The (atomic) volume per unit volume (or space).

Plasticizer: Substance made of small molecules, mixed with (amorphous) polymers to make the chains slide more easily past each other, making the polymer less rigid.

Polycrystalline: Structure of a material that is an aggregate of single crystals (grains).

Semiconductivity: Electrical conductivity of a material lies between that of a conductor and an insulator. There is an energy band gap of the order of 0.1 eV (Sn) to 6 eV (C, diamond), which governs the conductivity and its dependence on temperature.

Semicrystalline solid: Solid that contains both crystalline and noncrystalline regions; for example, in polymers due to their long-chain molecules.

Steric hindrance: Geometrical interference that restrains movement of molecular groups such as side chains and main chains of a polymer.

Tacticity: Arrangement of asymmetrical side groups along the backbone chain of polymers; groups could be distributed at random (atactic), on one side (isotactic), or alternating (syndiotactic).

Unit cell: Smallest repeating unit of a space lattice representing the whole crystal structure.

Valence electrons: Outermost (shell) electrons of an atom.

van der Waals bonding: Secondary bonding arising through the fluctuating dipole–dipole interactions.

Vinyl polymers: Thermoplastic linear polymers synthesized by free radical polymerization of vinyl monomers having a common structure of $CH_2=CHR$.

Vulcanization: Crosslinking of rubbers by sulfur.

BIBLIOGRAPHY

Agarwal BD, Broutman LJ. 1980. *Analysis and performance of fiber composites*. New York: Wiley.

Ashby MF, Jones DRH. 1988. *Engineering materials*, Vol. 1: *An introduction to properties, applications and design*. Oxford: Pergamon Press.

Ashby MF, Jones DRH. 1988. *Engineering materials*, Vol. 2: *An introduction to microstructures, processing and design*. Oxford: Pergamon Press.

Askeland DR, Phule PP. 2002. *The science and engineering of materials*. Boston, PWS-Kent.

Boyer, D. 1989. Personal communication. The University of Iowa, Iowa City.

Callister Jr WD. 2000. *Materials science and engineering: an introduction*. New York: Wiley.

Chang YL, Lew D, Park JB, Keller JC. 1999. Biomechanical and morphometric analysis of hydroxyapatite-coated implants with varying crystallinity. *J Oral Maxillofac Surg* **57**(9):1096–1108.

Cottrell AH. 1967. The nature of metals. *Sci Am* **217**(3).

Flinn RA, Trojan PK. 1994. *Engineering materials and their applications*. Boston: Houghton Mifflin.

Gibson LJ, Ashby MF. 1988. *Cellular solids*. Oxford: Pergamon.

Harris B, Bunsell AR. 1977. *Structure and properties of engineering materials*. London: Longman.

Hayden WH, Moffatt WG, Wulff J. 1965. *The structure and properties of materials*, Vol. 3. New York: Wiley.

LeGeros JP, LeGeros RZ, Burgess A, Edwards B, Zitelli J. 1994. X-ray diffraction method for the quantitative characterization of calcium phosphate coatings. In *Characterization and performance of calcium phosphate coatings for implants*, pp. 33–42. Ed E Horowitz, JE Parr. Philadelphia: American Society for Testing Materials.

LeGeros RZ, LeGeros JP, Kim Y, Kijkowska R, Zheng R, Wong JL. 1995. Calcium phosphates in plasma-sprayed HA coatings. *Ceramic Trans* **48**:173–189.

Pauling, L. 1960. *The nature of the chemical bonding*. Ithaca, NY: Cornell UP.

Shackelford JF. 2004. *Introduction to materials science for engineers*. Upper Saddle River, NJ: Pearson/Prentice Hall.

Van Vlack LH. 1989. *Elements of materials science*. Reading, MA: Addison-Wesley.

3

CHARACTERIZATION OF MATERIALS — I

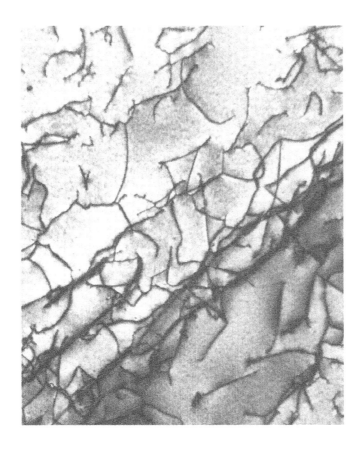

A transmission electron micrograph of a titanium alloy in which dark lines are dislocations. Courtesy of M.R. Plichta, Michigan Technological University. Reprinted with permission from Callister (2000). Copyright © 2000, Wiley.

The characterization of materials is an important step to be taken before utilizing the materials for any purpose. Depending on the purpose one can subject the material to mechanical, thermal, chemical, optical, electrical, and other characterizations to make sure that the material under consideration can function without failure for the life of the final product. We will consider only mechanical, thermal, and surface properties in this chapter, while in the next chapter we will study electrical, optical, and diffusive properties.

3.1. MECHANICAL PROPERTIES

Among the most important properties for the application of materials in medicine and dentistry are the mechanical properties. We will study the fundamental mechanical properties that will be used in later chapters.

3.1.1. Stress–Strain Behavior

For a material that undergoes a mechanical deformation, the *stress* is defined as a force per unit area, which is usually expressed in Newtons per square meter (Pascal, Pa) or pounds force per square inch (psi):

$$\text{Stress } (\sigma) = \frac{\text{force}}{\text{cross-sectional area}} \left[\frac{N}{m^2}\right] \text{ or } \left[\frac{lbf}{in^2}\right]. \tag{3-1}$$

A load (or force) can be applied upon a material in *tension, compression,* and *shear* or any combination of these forces (or stresses). Tensile stresses are generated in response to loads (forces) that pull an object apart (Figure 3-1a), while compressive stresses squeeze it together (Figure 3-1b). Shear stresses resist loads that deform or separate by sliding layers of molecules past each other on one or more planes (Figure 3-1c). The shear stresses can also be found in uniaxial tension or compression since the applied stress produces the maximum shear stress on planes at 45° to the direction of loading (Figure 3-1d).

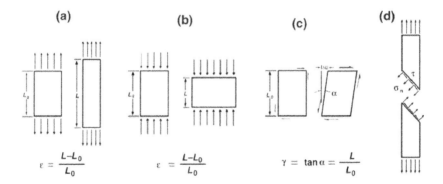

Figure 3-1. Three different modes of deformation: (**a**) tension, (**b**) compression, (**c**) shear, and (**d**) shear in tension. The shear stresses can be produced by tension or compression as in (**d**).

The deformation of an object in response to an applied load is called *strain*:

$$\text{Strain } (\sigma) = \frac{\text{deformed length} - \text{original length}}{\text{original length}} \left[\frac{\text{m}}{\text{m}}\right] \text{ or } \left[\frac{\text{in}}{\text{in}}\right]. \tag{3-2}$$

It is also possible to denote strain by the stretch ratio, i.e., deformed length/original length. The deformations associated with different types of stresses are called tensile, compressive, and shear strain (cf. Figure 3-1).

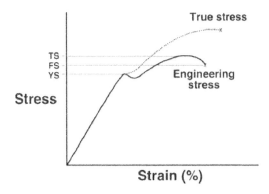

Figure 3-2. Stress–strain behavior of an idealized material.

If the stress–strain behavior is plotted on a graph, a curve that represents a continuous response of the material toward the imposed force can be obtained, as shown in Figure 3-2. The stress–strain curve of a solid sometimes can be demarcated by the yield point (σ_y or YP) into elastic and plastic regions. In the elastic region, the strain ε increases in direct proportion to the applied stress σ (Hooke's law):

$$\sigma = E\varepsilon : \text{stress} = (\text{initial slope}) \times (\text{strain}). \tag{3-3}$$

The slope (E) or proportionality constant of the tensile/compressive stress–strain curve is called *Young's modulus* or the *modulus of elasticity*. It is the value of the increment of stress over the increment of strain. The stiffer a material is, the higher the value of E and the more difficult it is to deform. Similar analysis can be performed for deformation by shear, in which the *shear modulus* (G) is defined as the initial slope of the curve of shear stress versus shear strain. The unit for the modulus is the same as that of stress since strain is dimensionless. The shear modulus of an *isotropic* material is related to its Young's modulus by

$$E = 2G(1+\nu), \tag{3-4}$$

in which ν is the *Poisson's ratio* of the material. Poisson's ratio is defined as the negative ratio of the transverse strain to the longitudinal strain for tensile or compressive loading of a bar. Poisson's ratio is close to 1/3 for common stiff materials, and is slightly less than 1/2 for rubbery materials and for soft biological tissues. For example, stretch a rubber band by 10% of its original length and the cross-sectional dimensions will decrease by about 5%.

In the *plastic region*, strain changes are no longer proportional to the applied stress. Further, when the applied stress is removed, the material will not return to its original shape but will be permanently deformed which is called a *plastic deformation*. Figure 3-3 is a schematic illustration of what will happen on the atomic scale when a material is deformed. Note that the individual atoms are distorted and stretched while part of the strain is accounted for by a limited movement of atoms past one another. When the load is released before atoms can slide over other atoms, the atoms will go back to their original positions, making it an elastic deformation. When a material is deformed plastically the atoms are moved past each other in such a way they will have new neighbors, and when the load is released they can no longer go back to their original positions.

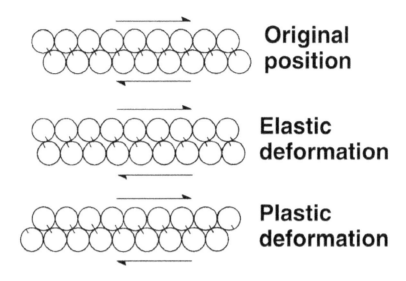

Figure 3-3. Schematic diagrams of a two-dimensional atomic model after elastic and plastic deformation.

Referring back to Figure 3-2, a peak stress can be seen that is often followed by an apparent decrease until a point is reached where the material ruptures. The peak stress is known as the *tensile* or *ultimate tensile strength* (TS); the stress where failure occurs is called the *failure* or *fracture strength* (FS).

In many materials such as stainless steels, definite yield points occur. This point is characterized by temporarily increasing strain without a further increase in stress. Sometimes it is difficult to decipher the yield point since the deviation from linear behavior may be obscured by noise in the data. Therefore, an *offset* (usually 0.2 to 1.0%) yield point is used in lieu of the original yield point. Specifically, the offset-based yield point is the stress at which a 0.2% (or 1%) residual strain occurs after removal of the load.

Thus far we have been examining the *engineering stress–strain* curves, which differ from the *true stress–strain* curves since the former curve is obtained by assuming a constant cross-sectional area over which the load is acting from the initial loading until final rupture. This assumption is not correct, which accounts for the peak seen at the ultimate tensile stress. For example, as a specimen is loaded under tension, sometimes *necking* (Figure 3-4) occurs, which reduces the area over which the load is acting. If additional measurements are made of the changes in cross-sectional area that occur, and the true area is used in the calculations, then a dotted curve like that in Figure 3-2 is obtained.

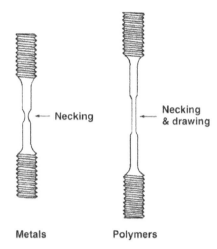

Metals   Polymers

Figure 3-4. Deformation characteristics of metals and plastics under stress. Note that metals rupture without further elongation after necking occurs; by contrast, in plastics the necked region undergoes further deformation, called drawing.

Example 3.1

The following data were obtained for an aluminum alloy. A standard tensile test specimen with a 2-inch gauge length and 0.505 inch diameter was used:

a. Plot the engineering stress–strain curve.
b. Determine the Young's modulus, yield strength (0.2% offset), and tensile strength.
c. Determine the engineering and true fracture strength. The diameter of the broken pieces was 0.4 in.

Load (kilo-lbf)	Gauge length (in)
2	2.002
4	2.004
8	2.008
10	2.010
12	2.011
13	2.014
14	2.020
16	2.050
16 (maximum)	2.099
15.6 (fracture)	2.134

Answer

a. See the plot from the calculations based on cross-sectional area = π (diameter)^2/4 = 0.2 in^2).

Stress (ksi)	Strain (%)
20	0.2
40	0.4
50	0.5
60	0.6
65	0.7
70	1.0
80	2.5
80	5.0
78	6.7

b.   Young's modulus = 40 ksi/0.004 = <u>1 × 10^7 psi (69 GPa)</u>
0.2% yield stress: from graph: <u>62,000 psi (428 MPa)</u>
Tensile strength: <u>80,000 psi (560 MPa)</u>

c.   Engineering fracture strength =15,600 lbs/0.2 in^2 = <u>78,000 psi (538 MPa)</u>
True fracture strength = 15,600 lbs/π(0.2 in)^2 = <u>124,140 psi (856 MPa)</u>

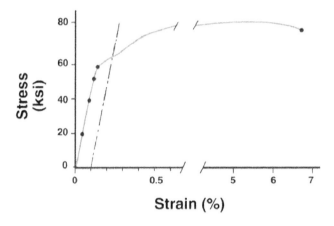

Stress versus strain for an aluminum alloy.

3.1.2. Mechanical Failure

3.1.2.a. Static failure

Mechanical failure usually occurs by fracture. The fracture of a material can be characterized by the amount of energy per unit volume required to produce the failure. The quantity is called *toughness* and can be expressed in terms of stress and strain:

$$\text{toughness} = \int_{\varepsilon_o}^{\varepsilon_f} \sigma \, d\varepsilon = \int_{l_0}^{l_f} \sigma \frac{dl}{l}. \tag{3-5}$$

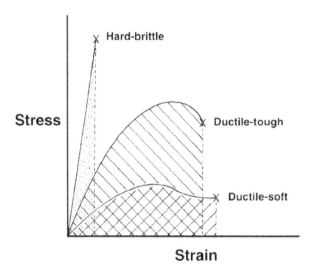

Figure 3-5. Stress–strain curves of different types of materials. The areas underneath the curves are the measure of toughness.

Expressed another way, toughness is the summation of stress times the normalized distance over which it acts (strain) taken in small increments. The area under the stress–strain curve provides a simple method of estimating toughness, as shown in Figure 3-5.

A material that can withstand high stresses and can undergo considerable plastic deformation (*ductile-tough* material) is tougher than one that resists high stresses but has no capacity for deformation (*hard-brittle* material) or one with a high capacity for deformation but can only withstand relatively low stresses (*ductile-soft* or plastic material).

Brittle materials exhibit fracture strengths far below the theoretical strength predicted based on known atomic bond strengths. Moreover, there is much variation in strength from specimen to specimen, so that the practical strength is difficult to predict. These facts, along with the comparative weakness of ceramics in tension, are the major reasons why ceramic and glassy materials are not used extensively for implantation despite their excellent compatibility with tissues.

The comparative weakness of brittle materials is explained as follows. The stress on a brittle material is not uniformly distributed over the entire cross-sectional area if a crack or flaw is present, as shown in Figure 3-6. If the crack is a narrow elliptic hole in a specimen subjected to a tensile stress, the maximum stress (σ_{max}) acting at the ends of the hole is given by

$$\sigma_{max} = \sigma_{app}\left(1+\frac{2a}{b}\right), \tag{3-6}$$

where σ_{app} is the applied (or nominal) tensile stress experienced away from the crack. One can rearrange Eq. (3-6):

$$\frac{\sigma_{max}}{\sigma_{app}} = \left(1+\frac{2a}{b}\right), \tag{3-7}$$

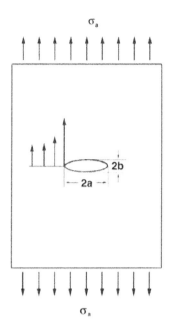

Figure 3-6. An elliptic microcrack inclusion in a brittle material.

where the ratio $\sigma_{max}/\sigma_{app}$ is called the *stress concentration factor* (scf), which can be substantial if ratio a/b is high, i.e., a sharp crack. If the crack tip has a radius of curvature r ($= b^2/a$), then Eq. (3-6) can be rewritten:

$$\sigma_{max} = \sigma_{app}\left[1+2\left(\frac{a}{b}\right)^{1/2}\right]. \tag{3-8}$$

Since $a \gg r$ for a crack,

$$\sigma_{max} \cong 2\sigma_{app}\left(\frac{a}{r}\right)^{1/2}. \tag{3-9}$$

Equation (3-9) indicates that the stress concentration becomes very large for a sharp crack tip as well as for long cracks. Thus, the propagation of a sharp crack can be blunted if one increases the crack tip radius. For example, progression of a crack in a large glass window can be halted or slowed by drilling a hole in the glass at the crack tip.

Griffith proposed in 1920 an energy approach to fracture. The elastic energy stored in a test specimen of unit thickness is

$$\sigma \times \varepsilon = \pi a^2 \sigma\left(\frac{\sigma}{E}\right) = \frac{\pi(a\sigma)^2}{E}. \tag{3-10}$$

Observe that the elastic energy for a brittle material is twice the area under the stress–strain curve. The elastic energy is used to create two new surfaces as the crack propagates. The surface energy, $4\gamma a$ (γ is the surface energy) should be smaller than the elastic energy for the crack to grow. Thus, the incremental changes of both energies for the crack to grow can be written as

$$\frac{d}{da}\left(\frac{\pi}{(a\sigma)^2}\right) = \frac{d}{da}(4\gamma a).$$

(3-11)

Hence,

$$\sigma = \sigma_f \sqrt{\frac{2\gamma E}{\pi a}}.$$

(3-12)

Since for a given material E and γ are constants,

$$\sigma_f = \frac{K}{\sqrt{\pi a}}.$$

(3-13)

In this case, K has the units of psi $\sqrt{\text{in}}$ or MPa $\sqrt{\text{m}}$ and is proportional to the energy required for fracture. K is also called *fracture toughness*. Stress concentrations also occur in ductile materials, but their effect is usually not as serious as in brittle ones since local yielding that occurs in the region of peak stress will effectively blunt the crack and alleviate the stress concentration.

Example 3.2

Estimate the size of the surface flaw in a glass whose modulus of elasticity and surface energy are 70 GPa and 800 erg/cm^2, respectively. Assume that the glass breaks at a tensile stress of 100 MPa.

Answer

From Eq. (3-12), and keeping in mind the transformation from cgs to SI units,

$$a = \frac{2\gamma E}{\pi \sigma_f^2}$$

$$= \frac{2 \times 800 \text{ dyne/cm} \times 70 \text{ GPa}}{\pi (100 \text{ MPa})^2}$$

$$= \underline{3.6 \ \mu\text{m}}.$$

[Note that if the crack is on the surface its length is a; if it is inside the specimen it is $2a$. Remember that 1 dyne = 10^{-5} N, 1 cm = 10^{-2} m, and 1 erg = 1 dyne cm.]

3.1.2.b. Dynamic fatigue failure

When a material is subjected to a constant or a repeated load below the fracture stress, it can fail after some time. This is called static or dynamic (cyclic) fatigue respectively. The effect of cyclic stresses (Figure 3-7) is to initiate microcracks at centers of stress concentration within the material or on the surface, resulting in the growth and propagation of cracks, leading to failure. The rate of crack growth can be plotted in a log–log scale versus time, as shown in Figure 3-8. The most significant portion of the curve is the crack propagation stage, which can be estimated as follows:

$$\frac{da}{dN} = A(\Delta K)^m, \tag{3-14}$$

where a, N, and ΔK are the crack length, number of cycles, and range of stress intensity factor (cf. Eq. (3-13)), $\Delta K = \Delta \sigma \sqrt{\pi a}$. A and m are the intercept and slope of the linear portion of the curve. This is called the Paris equation.

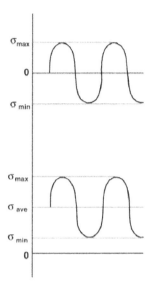

Figure 3-7. Cyclic stresses. σ_{min} and σ_{max} are the maximum and minimum values of the cyclic stresses. The range of stresses $\Delta \sigma = \sigma_{min} - \sigma_{max}$ and average stress $\sigma_{ave} = (\sigma_{max} + \sigma_{min})/2$. The top curve is fluctuating, and the bottom curve is for reversed cyclic loading.

Another method of testing the fatigue properties is to monitor the number of cycles to failure at various stress levels, as shown in Figure 3-9. This test requires a large number of specimens compared with the crack propagation test. The *endurance limit* is the stress below which the material will not fail in fatigue no matter how many cycles are applied. Normally 10^7 cycles is considered as a representative limit for normal fatigue failure. Not all materials exhibit an endurance limit. Since implants are often flexed many times during a patient's life, the fatigue properties of materials are very important in implant design.

3.1.2.c. Friction and wear failure

Wear properties of an implant material are important, especially for various joint replacements. Wear cannot be discussed without some understanding of friction between two materials. When two solid materials contact, they touch only at the tips of their highest asperities (microscopic protuberances). Therefore, the real contact area is much smaller than the apparent surface area. It is found that the true area of contact increases with applied load (P) for ductile materials. Ductile materials can be pressure welded due to the formation of plastic junctions, as shown in Figure 3-10. The plastic junctions are the main source of an adhesive friction when two materials are sliding over each other with or without a lubricating film. The

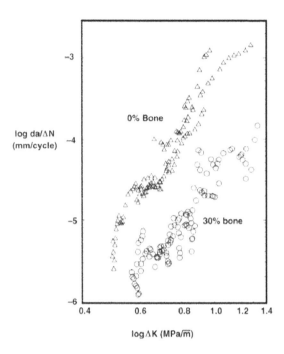

Figure 3-8. Log *da/dN* versus log ΔK for polymethylmethacrylate bone cement. Reprinted with permission from Liu et al. (1987). Copyright © 1987, Wiley.

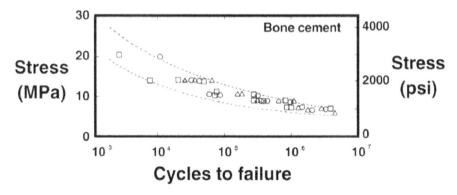

Figure 3-9. Stress versus log *N* (number of cycles) of PMMA bone cement. The Simplex P bone cement test specimens were fabricated at pressures between 5 and 50 psi and tested in air at 22°C. Reprinted with permission from Freitag and Cannon (1977). Copyright © 1977, Wiley.

resistance to the shear failure of the plastic junction results in a frictional force. Therefore, the sliding force *F* will be simply proportional to the shear yield strength *k* of the junctions and the contact area *A*:

$$F = Ak. \qquad (3\text{-}15)$$

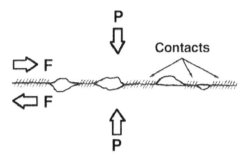

Figure 3-10. Schematic representation of two surfaces under pressure. Plastic junctions are formed when ductile materials are pressed together between asperities.

Since for ductile materials the area of contact increases with P,

$$P = HA, \tag{3-16}$$

where H is the penetration hardness or yield pressure. If we combine Eqs. (3-15) and (3-16), the coefficient of sliding friction μ can be obtained:

$$\mu = F/P = k/H. \tag{3-17}$$

This equation implies that the friction coefficient is merely the ratio of two plastic strength parameters of the weaker material and is independent of the contact area, load, sliding speed including surface roughness, and geometry. Figure 3-11 shows an example of the effect of contact area changes on the friction coefficient. Note that there are no significant variations in friction coefficient with wide variations of the contact area.

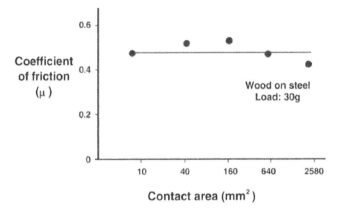

Figure 3-11. Friction coefficient versus sliding contact area, wood on steel. Modified with permission from McClintock and Argon (1966). Copyright © 1966, Addison-Wesley.

Wear is surface damage or material removal from surfaces in contact and in motion with respect to each other. Material may be damaged at the surface, broken loose as a wear particle, or it may be transferred from one surface to another. The rate of wear tends to increase with the applied load P and, for metals in contact, to decrease with the hardness H of the surface that is worn. An empirical relation for the wear volume ΔV of wear debris as it depends on the distance x moved by the sliding surfaces is as follows:

$$V = k(Px/H), \tag{3-18}$$

with k as a dimensionless *wear coefficient* or *constant* that depends on the materials in contact and the presence or absence of lubrication. The hardness H can be considered as the yield strength of the material being worn. Indeed, the above equation can be derived via a model in which asperities (microscopic protuberances) experience yield and failure during the wear process.

If metals in contact are similar, wear is higher than if they are dissimilar. Specifically, the lower the mutual solid solubility of the metals, the lower the wear. Typical wear coefficients for various conditions are listed in Table 3-1.

Table 3-1. Typical Values of Wear Coefficients k (in 10^{-6}) for Various Conditions

Condition	Similar metals	Dissimilar metals	Nonmetal–metal
No lubrication	1500	15–500	1.5
Poor lubrication	300	3–100	1.5
Average lubrication	30	0.3–10	0.3
Excellent lubrication	1	0.03–0.3	0.03

In the context of biomaterials, dissimilar metals are problematic in view of the fact that corrosion occurs when dissimilar metals are placed in an aqueous, saline environment such as the body. Moreover, wear of metals in a corrosive environment is exacerbated.

Polymer–metal interfaces can exhibit relatively low friction and wear, but low friction does not guarantee low wear. For example, Teflon® (PTFE, polytetrafluoroethylene) offers a low coefficient of friction, but it undergoes severe wear. Teflon® was tried in early implants but was not successful due to excess wear. Currently, UHMWPE (ultrahigh-molecular-weight polyethylene) is favored for use in joint replacement implants. Polymers cannot withstand as much contact stress as metals, but they are adequate in this regard for use in joint replacements.

Wear debris in hip replacement implants can cause tissue reactions, including proliferation of local fibroblast-like cells and activated macrophages. While PMMA and UHMWPE are inert in the bulk, small particles of these materials act as cellular irritants. Tissue reaction to wear debris is a contributing factor to bone resorption and implant loosening, a major cause of joint implant failure. In joint replacement implants, metal-on-metal designs offer reduced wear rates in comparison with metal-on-polymer ones, and are therefore considered an alternative. However, since the new designs and better congruent head and socket designs are relatively new, only early and midterm clinical results are available, and no long-term results. Elevated levels of metal ions due to wear debris have been observed in patients with these implants, but it is not yet known if there is a health risk. Diamond coatings have been explored in the laboratory in an effort to achieve wear-free surfaces. Alumina ceramic surfaces have been used clinically to achieve low wear.

3.1.3. Viscoelasticity

3.1.3.a. Viscoelastic material behavior

Viscoelastic materials are those for which the relationship between stress and strain depends on time. In such materials the stiffness will depend on the *rate of application* of the load. In addition, mechanical energy is dissipated by conversion to heat in the deformation of viscoelastic materials. All materials exhibit some viscoelastic response. In metals such as steel or aluminum at room temperature, as well as in quartz, the response at small deformation is almost purely elastic. Metals can behave plastically at large deformation, but ideally plastic deformation is independent of time. Also, plastic deformation occurs only if a threshold stress is exceeded. By contrast, materials such as synthetic polymers, wood, and human tissue display significant viscoelastic effects, and these effects occur at small or large stress.

3.1.3.b. Characterization of viscoelastic materials

Creep is a slow, progressive deformation of a material under constant stress. Suppose the history of stress σ as it depends on time t is a step function beginning at time zero:

$$\sigma(t) = \sigma_0 H(t), \tag{3-19}$$

where $H(t)$ is the unit Heaviside step function defined as zero for t less than zero, and one for t equal to zero. The step function is normally defined to be 1/2 at zero. The strain $\varepsilon(t)$ will increase, as shown in Figure 3-12. The ratio

$$J(t) = \frac{\varepsilon(t)}{\sigma_0} \tag{3-20}$$

is called the creep compliance. In linear materials, it is independent of stress level. If the load is released at a later time t_s, the strain will exhibit recovery, as shown in Figure 3-12. For linear materials we may invoke the *Boltzmann superposition principle*, which states that the effect of a compound cause is the sum of the effects of individual causes. The stress may then be written as a superposition of a step up followed by a step down:

$$\sigma(t) = [H(t) - H(t - t_1)], \tag{3-21}$$

so the strain is, assuming superposition,

$$\varepsilon(t) = \sigma_0 [J(t) - J(t - t_1)]. \tag{3-22}$$

The strain may or may not recover to zero, depending on the material.

Stress relaxation is the gradual decrease of stress when the material is held at constant extension. If we suppose the strain history to be a step function beginning at time zero — $\varepsilon(t) = \varepsilon_0 H(t)$ — the stress $\sigma(t)$ will decrease as shown in Figure 3-13. The ratio

$$E(t) = \frac{\sigma(t)}{\varepsilon_0} \tag{3-23}$$

is called the *relaxation modulus*. In linear materials it is independent of strain level.

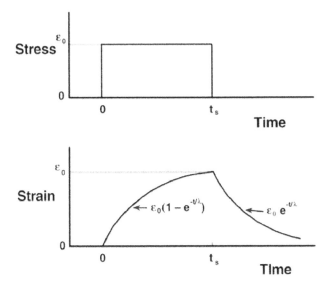

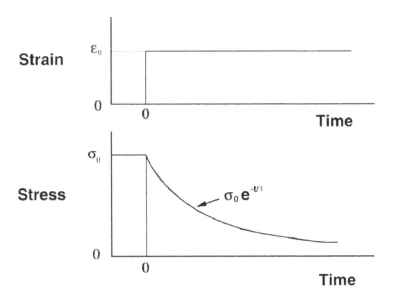

Figure 3-12. Creep and creep recovery of an idealized viscoelastic material.

Figure 3-13. Stress relaxation of an idealized viscoelastic material.

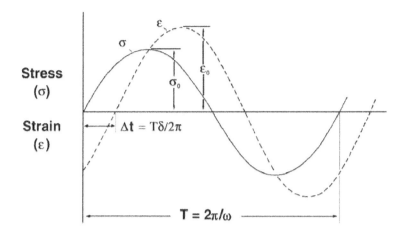

Figure 3-14. Loss angle δ for oscillatory loading of viscoelastic material.

If a sinusoidally varying stress,

$$\sigma(t) = \sigma_0 \sin(2\pi ft), \tag{3-24}$$

of frequency f is applied to a linearly viscoelastic material, the strain

$$\varepsilon(t) = \varepsilon_0 \sin(2\pi ft - \delta) \tag{3-25}$$

will also be sinusoidal in time but will lag the stress by a phase angle δ, as shown in Figure 3-14. The loss angle δ is a measure of the viscoelastic damping of the material. Both the loss angle and the dynamic stiffness E (= σ_0/ε_0) depend on frequency. The tangent of the loss angle is referred to as the *loss tangent*, tan δ.

For a given material, viscoelastic damping, creep, and stress relaxation are not independent. For example, the creep and relaxation functions are related by a convolution; the dynamic properties are related to the creep or relaxation behavior by Fourier transformation; and the loss angle and dynamic stiffness are related by the Kramers-Kronig relations. We remark that the larger the value of the loss tangent, the more rapidly the dynamic stiffness changes with frequency, and the more rapidly the relaxation modulus and creep compliance change with time. An approximate correspondence between the frequency scale, f, for dynamic behavior and the time scale, t, for creep and relaxation is the relation

$$t = \frac{1}{2\pi f}. \tag{3-26}$$

Furthermore, the viscoelastic properties of a material in tension need not follow the same time or frequency dependence as those in shear.

3.1.3.c. Prediction of the response

Viscoelastic materials may be used under conditions of complex loading. It is possible to predict the response of such materials based on the material properties discussed above. The

Boltzmann superposition principle for linear materials is applied to decompose the load history into a series of differential creep and recovery episodes. Summing the effects of these, one obtains the Boltzmann superposition integral:

$$\varepsilon(t) = \int_0^t J(t-\tau)\frac{d\sigma(\tau)}{d\tau}d\tau, \tag{3-27}$$

or conversely,

$$\sigma(t) = \int_0^t E(t-\tau)\frac{d\varepsilon(\tau)}{d\tau}d\tau, \tag{3-28}$$

in which τ is a time variable of integration. Consequently, if the response of a material to step stress or strain has been determined experimentally, the response to *any* load history can be found for the purpose of analysis or design.

3.1.3.d. Mechanical models

Simple mechanical models may be considered as an aid for visualizing viscoelastic response. The models consist of springs, which are purely elastic, and dashpots, which are purely viscous. In a spring, the stress is proportional to the strain:

$$\sigma = k\varepsilon \tag{3-29}$$

(cf. Eq. 3-3). The constant of proportionality k is the spring constant. By contrast, in a dashpot the stress is proportional to the time derivative of the strain (the strain rate):

$$\sigma = \eta\frac{d\varepsilon}{dt}. \tag{3-30}$$

The constant of proportionality η is a viscosity. For example, the Kelvin (Voigt) model consists of a spring in parallel with a dashpot. The strain is the same in both elements, but the stress in the Kelvin model is the sum of the spring and dashpot stresses. After some reductions, one may obtain a differential form of the stress–strain relation:

$$\eta\frac{d\varepsilon}{dt} + k\varepsilon = \sigma. \tag{3-31}$$

The solution of this differential equation for a step stress input gives the creep compliance:

$$J(t) = \frac{[1 - e^{\frac{-kt}{\eta}}]}{k}. \tag{3-32}$$

Real materials, both synthetic and natural, exhibit behavior that cannot be described by simple mechanical models. In a two- or three-element model, the creep or relaxation is completed within about one logarithmic decade [a factor of ten in time] while in real materials the creep or relaxation is distributed over many decades.

Although the simple Eq. (3-3) can describe the elastic behavior of many materials at low strains, as represented by a spring in Figure 3-15, it cannot be used to characterize the polymers and tissues that are some of the major concerns of this book. The fluid-like behavior of a material (such as water or oil) can be described in terms of (shear) stress and (shear) strain as in the elastic solids, but the proportionality constant (viscosity, η) is derived from the relationship given in Eq. (3-30). It is noted that the stress and strain are shear in this case rather than tensile or compressive, although the same symbols are used to avoid complications.

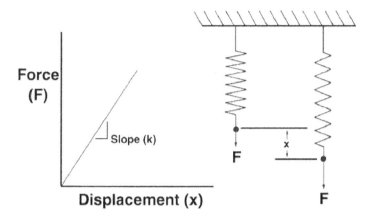

Figure 3-15. Force versus displacement of a spring.

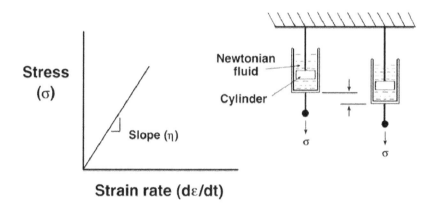

Figure 3-16. Stress-versus-strain rate of a dashpot.

A mechanical analog (dashpot) can be used to model the viscous behavior of Eq. (3-30), as shown in Figure 3-16. An automobile shock-absorbing cylinder that contains oil as the damping fluid has a similar construction. By examining Eq. (3-30) one can see that the stress is *time*-dependent, i.e. if the deformation is accomplished in very short time ($t \approx 0$) then the stress becomes infinite. On the other hand, if the deformation is achieved slowly ($t \Rightarrow \infty$) the stress approaches zero regardless of the viscosity value.

Real materials that have both elastic and viscous aspects to their behavior are known as *viscoelastic* materials. Simple equations (3-29) and (3-30), when combined together as if the material is made of springs and dashpots, can describe, in principle, the viscoelastic behavior of an idealized material. The stress–strain behavior of a spring and dashpot combination can be represented as shown in Figure 3-17. If the spring and dashpot are arranged in series and parallel they are called Maxwell and Voigt (or Kelvin) models, respectively. Remember that Eq. (3-29) does not involve time, implying the spring acts instantaneously when stressed. Hence, if the Maxwell model is stressed suddenly the spring reacts instantaneously, while the dashpot cannot react immediately since the piston of the dashpot cannot move abruptly due to the infinite stress that would be required by the surrounding fluid, following Eq. (3-30). However, if we hold the Maxwell model after instantaneous deformation, the dashpot will react due to retraction of the spring, and this will take time (t = finite). The foregoing description can be expressed concisely by a simple mathematical formulation. The deformation response to stress by the Maxwell model is the sum of deformations, since the displacements are additive:

$$\varepsilon_{total} = \varepsilon_{spring} + \varepsilon_{dashpot}. \tag{3-33}$$

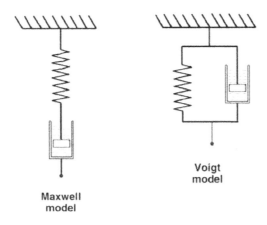

Voigt model

Maxwell model

Figure 3-17. Two-element viscoelastic models.

Differentiating both sides and using Eqs. (3-29) and (3-30) will result in Eq. (3-31). Replacing the spring constant k with Young's modulus E to express the relationship in the context of material rather than structural properties, we obtain

$$\frac{d\varepsilon}{dt} = \frac{1}{E}\frac{d\sigma}{dt} + \frac{\sigma}{\eta}. \tag{3-34}$$

Equation (3-34) can be applied easily to a simple mechanical test such as stress relaxation in which the specimen is strained instantaneously and the relaxation of the load is monitored while the specimen is held at constant length, as shown in Figure 3-18. Thus, the strain rate becomes zero ($d\varepsilon/dt = 0$) and at $t = 0$ and $\sigma = \sigma_0$, in σ_0 = constant; hence,

$$\frac{\sigma}{\sigma_0} = \exp\left(-\frac{E}{\eta}t\right). \tag{3-35}$$

Constant η/E has dimension of time and is defined as the *relaxation time* τ, and Eq. (3-35) becomes

$$\sigma = \sigma_o e^{\frac{-t}{\tau}} = \frac{\sigma_o}{e^{\frac{t}{\tau}}}. \tag{3-36}$$

Examining Eq. (3-36), one can see that if the relaxation time is short in comparison to the present time t, then the stress σ at a given time becomes small. On the other hand, if the relaxation time is long, then the stress σ is nearly the same as the original stress σ_o, as shown in Figure 3-18.

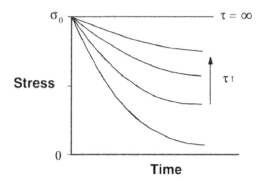

Figure 3-18. Relaxation curve for Maxwell model with different values of relaxation time (τ).

Example 3.3

A stress of 1 MPa was required to stretch a 2-cm aorta strip to 2.3 cm. After an hour in the same stretched position, the strip exerted a stress of 0.75 MPa. Assume the mechanical property of the aorta did not vary appreciably during the experiment.

a. What is the relaxation time, assuming a simple exponential decay model?
b. What stress would be exerted by the aorta strip in the same stretched position after five hours?

Answer

a. From Eq. (3-32):

$$\sigma = \sigma_o e^{-t/\tau}, \qquad \frac{0.75}{1} = \exp\left(-\frac{1}{\tau}\right).$$

Therefore, $\tau = \underline{3.48 \text{ hour}}$

b. Substituting the relaxation time

$$\sigma = 1 \ \exp\left(-\frac{5}{3.48}\right) = \underline{0.24 \text{ MPa}}.$$

In comparison, window glass has a distribution of relaxation times including very large relaxation times of many years; minimal stress relaxation occurs over short times. By contrast, water and oil have short relaxation times. Thus, when stressed their shape changes immediately to relieve the applied stress (instantaneous stress relaxation).

Similar analysis can be made with the Voigt model (Figure 3-17, right). In this case the strain of the spring and dashpot are equal since both are assumed to deform together, in parallel:

$$\varepsilon_{total} = \varepsilon_{spring} = \varepsilon_{dashpot} = \varepsilon. \tag{3-37}$$

The total stress is the summation of that in the spring and dashpot:

$$\sigma_{total} = \sigma_{spring} = \sigma_{dashpot} = \sigma. \tag{3-38}$$

The reason is that the forces in this parallel system are additive.

Substituting Eqs. (3-29) and (3-30) into (3-38), with E again substituted for k:

$$\sigma = E\varepsilon + \eta\frac{d\varepsilon}{dt}. \tag{3-39}$$

The creep behavior of the Voigt model is given by a compliance $J(t) = \varepsilon(t)/\sigma$. In terms of retardation time, λ, the creep behavior can be expressed as

$$J(t) = \frac{1}{E}(1 - e^{-t/\lambda}).$$

If a stress is applied and maintained for a long time, and if the stress is removed, then the strain recovers with time in a way that can be derived from Eq. (3-39):

$$\varepsilon(t) = \varepsilon_0 \exp\left(-\frac{E}{\eta}t\right), \tag{3-40}$$

where ε_0 is the strain at the time of stress removal. The constant η/E is termed the *retardation time* λ for this *creep recovery* process. Since the strain is being recovered from the original strain ε_0, Eq. (3-40) can be rewritten as

$$\varepsilon_{recovery} = \varepsilon_0 - \varepsilon_0 \exp\left(-\frac{t}{\lambda}\right) = \varepsilon_0\left[1 - \exp\left(-\frac{t}{\lambda}\right)\right]. \tag{3-41}$$

Therefore, the pattern of recovery of a creep strain follows the same pattern as the original creep, provided the material is linearly viscoelastic (stress and strain proportional for each time or frequency). Figure 3-19 shows the creep response of a Voigt model with varying retardation times.

Example 3.4

A piece of polyethylene is stretched to 20% of its length. When the stress was released it recovered 50% of its strain after one hour at room temperature.

a. What is the retardation time assuming a single exponential model?
b. What is the amount of strain recovered after 5 hours at room temperature?

Answer

a.   From Eq. (3-41):

$$\varepsilon = \varepsilon_0[1 - \exp(-t/\lambda)]], \quad \frac{\varepsilon}{\varepsilon_0} = 0.5 = 1 - \exp\left(-\frac{1}{\lambda}\right).$$

Therefore, λ = 1.443 hour

b. ε = 0.2 [1 – exp(–5/1.443) = 0.194 [or 19.4%, which is 96.9% recovery of the strain since the original strain is 20%].

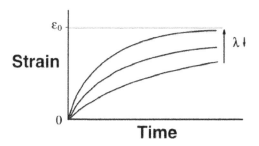

Figure 3-19. Retarded elastic deformation (creep and recovery) curves for Voigt model with various retardation time (λ).

3.1.3.e. Behavior of viscoelastic materials

The viscoelastic behavior of various materials is shown in Figure 3-20. The dynamic stiffness (Figure 3-20a) and loss tangent (Figure 3-20b) are shown since they are the properties most easy to conceptualize. Observe that the scales are logarithmic. The very low frequencies correspond to very long times in creep: $t = 1/2\pi f$. The behavior of a three-element spring–dashpot model is also shown in Figure 3-20b for comparison. Observe that the glass-to-rubber transition in polymers is associated with a large peak in the loss tangent in the frequency domain. Crosslinked polymers exhibit a nonzero limit to the stiffness at low frequency; the stiffness of uncrosslinked polymers tends to become zero at sufficiently low frequency or sufficiently long time. Such materials are viscoelastic liquids. Some such materials may superficially appear solid; the excess creep becomes apparent only after months or years. Blood is also a viscoelastic liquid.

3.1.3.f. Applications

There are a variety of consequences of viscoelastic behaviors that influence the application of viscoelastic materials. For example, in those applications for which a steady-state stress is applied, the creep behavior is of greatest importance. The expected service life of implant materials may be very long; consequently, attention to the long-term creep behavior is in order. Blood vessels experience a steady-state internal pressure that gives rise to circumferential stress, so creep in blood vessel materials is important. Creep also occurs in the polyethylene

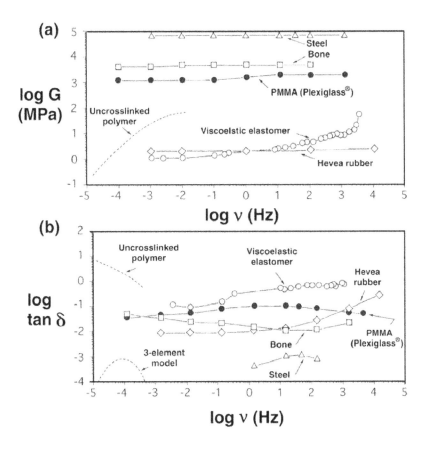

Figure 3-20. Viscoelastic behavior of various materials: (**a**) stiffness, (**b**) loss.

socket component of the total hip replacement. This creep gives rise to an indentation of the socket component by the ball joint. Wear and loosening are usually more important than creep in these implants. Stress relaxation is relevant to such situations as the loosening of screws that have been tightened to a specified extension. Bone screws are an example. In applications involving vibration, the width of the vibration amplitude-versus-frequency curve is proportional to the loss tangent. The maximum vibration amplitude at resonance is inversely proportional to the loss tangent. Vibrations damp out more rapidly for higher loss tangent values. Viscoelastic materials may therefore be used to reduce vibration in machinery or to protect the body from vibration or from mechanical shock. At higher frequency, sound or ultrasound waves are attenuated in proportion to the loss tangent. Diagnostic ultrasound is therefore absorbed in tissues, and will not penetrate far into a body composed of highly viscoelastic materials.

3.2. THERMAL PROPERTIES

The most familiar thermal properties are the melting and freezing (solidification) temperatures. These are phase transformations that occur at specific temperatures. These transformation temperatures depend on the bond energy, e.g., the higher the bonding strength, the higher the melting temperature. If the material is made of different elements or compounds, then it may have a range of melting or solidification temperatures, that is, the liquid coexists with solid over a range of temperatures, unlike a pure material.

The thermal energy spent on converting one gram of material from solid to liquid is called the heat of fusion. The unit is Joules per gram, where one Joule is equivalent to one Newton meter. The heat of fusion is closely related to the melting temperature (T_m), i.e., the higher is T_m, the higher the heat of fusion, although there are many exceptions (Table 3-2).

Table 3-2. Thermal Properties of Materials

Substance	Melting temp. (°C)	Specific heat (J/g)	Heat of fusion (J/g)	Thermal conductivity (W/mK)	Linear thermal expansion coeff. ($\times 10^{-6}$/°C)
Mercury	−38.87	0.138	12.7	68	60.6
Gold	1,063	0.13	67	297	14.4
Silver	960.5	0.2345	108.9	421	19.2
Copper	1,083	0.385	205.2	384	16.8
Platinum	1,773	0.134	113	70	–
Enamel	–	0.75	–	0.82	11.4
Dentine	–	1.17	–	0.59	8.3
Acrylic	70*	1.465	–	0.2	81.0
Water	0	4.187	334.9 (ice)	–	–
Paraffin	52	2.889	146.5	–	–
Beeswax	62	–	175.8	0.4	350
Alcohol	−117	2.29	104.7	–	–
Glycerin	18	2.428	75.4	–	–
Amalgam	480	–	–	23	22.1–28
Porcelain	–	1.09	–	1	4.1

*Softening temperature (T_g).

The thermal energy spent on changing the temperature of a material by 1°C per unit mass is called *specific heat*. Traditionally, water is usually chosen as a standard substance, and 1 calorie is the heat required to raise 1 gram of water from 15 to 16°C, but now the standard unit of energy including energy associated with heat is the Joule. Thus, the specific heat is in units of J/g. (1 calorie is equivalent to 4.187 J. The calorie used to represent food or metabolic energy is actually a kilocalorie, or 1000 calories.)

The change in length Δl for a unit length l_0 per unit temperature is called the *linear coefficient of expansion* (α), which can be expressed as

$$\alpha = \frac{\Delta l}{l_0 \Delta T}. \tag{3-42}$$

The thermal expansion may depend on the direction in a single crystal or composite, and it may depend on temperature. If the material is homogeneous and isotropic, then the *volumetric thermal expansion coefficient* (V_{exp}) can be approximated:

$$V_{exp} \approx 3\alpha . \tag{3-43}$$

Another important thermal property is the *thermal conductivity*, which is defined as the amount of heat passed for a given time, thickness, and area of material. The unit is Watt/mK, where one Watt is equivalent to one J per second. Generally, the thermal conductivity of metals is much higher than ceramics and polymers due to the free electrons in metals that act as energy conductors.

Example 3.5

In order to fill a cavity, a cylindrical hole with $r = 2$ mm is made in a molar tooth, the remaining thickness of which is 1 mm. The length of the hole is $L = 4$ mm. Consider that it is filled with amalgam; then consider it to be filled with acrylic resin. Assume the temperature variation is 50°C. The modulus of elasticity of the amalgam and resin are $E_{amalgam} = 20$ GPa and $E_{resin} = 2.5$ GPa, respectively. Moreover, $E_{dentin} = 14$ GPa, $E_{enamel} = 48$ GPa. The thermal expansions are $\alpha_{amalgam} = 25 \times 10^{-6}/°C$, $\alpha_{resin} = 81 \times 10^{-6}/°C$, $\alpha_{enamel} = 11 \times 10^{-6}/°C$. Assume that Poisson's ratio for all the materials is $\nu = 1/3$.

a. Calculate the volume changes for the fillings.
b. Calculate, based upon an elementary one-dimensional model and neglecting any remaining dentin, the force developed between the enamel and fillers.
c. Calculate the force between the filling and the remaining enamel tooth structure, considered as a cylindrical shell in a more realistic two-dimensional model, considering the filling as a cylinder. Also determine the stress at the interface and the stress in the enamel and discuss the results.

Answer

a. Since the volume expansion coefficient can be defined as in Eq. (3-39),

$$\frac{\Delta V}{V_0 \Delta T} = 3\alpha;$$

therefore, $\Delta V = V_0 \times 3\alpha \times \Delta T$

The net volume change after the filling will be

$$\Delta V_{amalgam} = \pi (1 \text{ mm})^2 \times 4 \text{ mm} \times 3(25 - 8.3) \times 10^{-6} \times 50 = \underline{0.03 \text{ mm}^3},$$

$$\Delta V_{resin} = \pi (1 \text{ mm})^2 \times 4 \text{ mm} \times 3(81 - 8.3) \times 10^{-6} \times 50 = \underline{0.14 \text{ mm}^3}.$$

b. First, it is necessary to choose a simple model for the thermal mismatch. If we have two dissimilar rods end to end, then they expand freely and do not generate mismatch stress. If they are constrained at the ends, they generate thermal strains even if there is no mismatch in thermal expansion. Two dissimilar bars or plates bonded together over their long surfaces will bend when heated, but this geometry is not representative of a tooth filling.

Consider the elementary analysis for a cylindrical pressure vessel of radius r and thickness t containing a pressure P. The circumferential stress is $\sigma = Pr/t$. This solution may be used to analyze the tooth modeled as a thin outer ring (the remaining tooth structure) bonded to an inner ring representing the filling. The strain is $\varepsilon = \sigma/E + \alpha\Delta T$, with σ as stress, E as Young's modulus, α as thermal expansion, and ΔT as temperature change. The strain in the tooth and the strain in the filling are the same since they are modeled as bonded layers. So,

$$\sigma_f / E_f + \alpha_f \Delta T = \sigma_t / E_t + \alpha_t \Delta T.$$

Here f represents the filling and t represents the tooth. Rearranging,

$$\sigma_t / E_t - \sigma_f / E_f = (\alpha_f - \alpha_t) \Delta T.$$

But the contact pressures P are equal but the filling and tooth stresses are opposite in sign. So

$$\sigma_t = (\alpha_f - \alpha_t) \Delta T / [1 / E_f + 1 / E_t].$$

But $\sigma = Pr/t$. The force is $F = PA$, with $A = 2\pi rL$ as the contact area. So

$$F = (t / r) 2\pi rL (\alpha_f - \alpha_t) \Delta T / [1 / E_f + 1 / E_t].$$

Substituting values given above for an amalgam filling, $F = 248$ N.

In reality, the filling is solid, not hollow. A more realistic analysis is given in the following.

c.   Consider the remaining tooth structure to be a hollow cylindrical shell of thickness t, length L, and radius R. The filling occupies the inside of this shell.

Calculate the strain in the filling ε_f, in terms of the stresses σ in different directions, Poisson's ratio ν, the coefficient of thermal expansion α, and the temperature change ΔT.

First, write the three-dimensional constitutive equation relating stress, strain, and temperature change:

$$\varepsilon_{fx} = [\sigma_x - \nu(\sigma_y + \sigma_z)] / E + \alpha \Delta T.$$

Similarly, in the y direction,

$$\varepsilon_{fy} = [\sigma_y - \nu(\sigma_x + \sigma_z)] / E + \alpha \Delta T.$$

Let us neglect stress in the z direction. Moreover, the stresses should be equal in the x and y directions in view of the cylindrical geometry assumed. So the radial and tangential strain in the filling (subscript f) are given by:

$$\varepsilon_f = \sigma(1 - \nu_f) / E_f + \alpha_f \Delta T.$$

Since Poisson's ratio is about 1/3, neglect of the elementary three-dimensional aspect of this problem would generate an error of about 30%. This would perhaps be acceptable in view of the other simplifications involved in this problem.

The surrounding tooth (represented by subscript t) may be thought of as a thin-walled pressure vessel in this approximation, so the tangential stress is much less than the radial stress. The tangential strain is given by

$$\varepsilon_t = \sigma_t / E_t + \alpha_t \Delta T.$$

If no gap opens up between the filling and tooth (since the filling expands more than the tooth), the tangential strains will be equal, so

$$\sigma_f(1-\nu_f)/E_f + \alpha_f/E_t + \alpha_t \Delta T.$$

Now consider the radial force F. Stress is force per area, so for the filling we consider equal radial and tangential stresses, so

$$\sigma_f = F/2\pi RL.$$

For the tooth, represented as a thin cylindrical shell with pressure P from within (radial stress), the tangential stress is given by

$$\sigma_t = PR/t = [R/t][F/2\pi RL] = F/2\pi Lt.$$

Substituting the stresses above, simplifying, and recognizing if the filling is in compression, the shell of tooth structure is in tension:

$$F = [2\pi(\alpha_f - \alpha_t)\Delta TL]/[(1/E_t t) + ((1-\nu_f)/ER)].$$

Observe that the force between filling and tooth is a result of the *difference* in thermal expansion. Of course, the actual geometry of a filled tooth is more complex than we have assumed here. A more accurate result would be obtained by finite-element analysis involving the true geometry.

We obtain, after substituting appropriate values, forces of 547 N for a resin filling and 459 N for an amalgam filling. Moreover, we calculate the tangential tensile stress in the enamel with an amalgam filling to be 18 MPa. This is in the neighborhood of the fracture stress of enamel. Ordinarily, the dentist would avoid having such a thin layer of tooth structure remaining. Moreover, the filling is placed at a temperature between room temperature (20°C) and body temperature (37°C). Even though a range of 50°C can occur in the mouth, a portion of that range is below ambient, as in the eating of ice cream. Cooling of the filling can result in a gap forming between the filling and tooth, and leakage can occur in that gap. Leakage of oral fluids can give rise to tooth sensitivity or further decay of the tooth.

3.3. PHASE DIAGRAMS

When two or more metallic elements are melted and cooled they form an intermetallic compound or a *solid solution* or, more commonly, a mixture thereof. Such combinations are called "alloys." The alloys can exist as either a single phase or a blend of multiple phases depending on temperature and composition. A *phase* is defined as a physically homogeneous part of a material system. Thus, a liquid and gas are both single phase, but there can be more than one phase for a solid, such as fcc iron and bcc iron, depending on pressure and temperature. Among multiphase metals, steels are iron-based alloys containing various amounts of a carbide (usually Fe_3C) phase. In this case, the carbon atoms occupy the interstitial sites of the iron atoms (cf. Figure 2-9); this is called an *interstitial* solid solution. Most metal atoms are too large to exist in the interstitial sites. If the two metal atoms are roughly the same size, have the same bonding tendencies, and tend to crystallize in the same types of crystal structure, then a *substitutional* solid solution may form. This structure is composed of a random mixture of two different atoms, as shown in Figure 3-21. Unless the elements are very similar in properties, such a solution will exhibit a limited solubility, i.e., as more substitutional atoms are added into the matrix, the lattice will be more and more distorted until phase separation occurs at the solubility limit. In some systems, such as Cu–Ni, as shown in Figure 3-21, complete solid solubility exists.

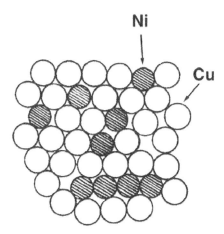

Ni

Cu

Figure 3-21. Substitutional solid solution of a Cu–Ni system.

The phase diagram is constructed first by preparing known compositions of Cu–Ni, and then melting and cooling them under thermal equilibrium. During the cooling cycle one has to determine at what temperatures the first solid phase (α) appears and all of the liquid disappears. These points will determine the *liquidus* and *solidus* line in the phase diagram. From this phase diagram one can determine the types of phase and amount of each element present for a given composition and temperature. Thus, if we cool a 40 w/o Ni–60 w/o Cu liquid solution, from Figure 3-22;

Temperature (°C)	Phase (relative amount)	Composition of each phase
above 1270	liquid (all)	40 Ni – 60 Cu
1250	liquid (63%)	33 Ni – 67 Cu
	α (37%)	52 Ni – 48 Cu
1220	liquid (5%)	26 Ni – 74 Cu
	α (95%)	43 Ni – 57 Cu
below 1210	α (all)	40 Ni – 60 Cu

The relative amount of each phase present at a given temperature and composition is determined by the *lever rule* after making a horizontal isothermal (tie) line at the temperature of interest. Let C_A and C_B be the composition of element A (Ni) and B (Cu) in the two-phase region met by the tie line (say 1240°C) with the same composition given above (40 w/o Ni = C_A); then the amount of liquid (L) phase can be calculated as follows:

$$\frac{L}{\alpha + L} = \frac{C_l - C_A}{C_l - C_\alpha} = \frac{52 - 40}{52 - 33} = \frac{12}{19} = \underline{0.63}, \qquad (3\text{-}42)$$

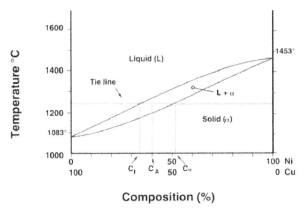

Figure 3-22. The Cu–Ni phase diagram: an example of complete solid solubility. See text for an explanation of the tie line.

where C_A is the original composition of element A. This principle can be applied in more complicated systems such as Ag–Cu (eutectic) or Fe–C (eutectic + eutectoid) systems, as shown in Figures 3-23 and 3-24. The eutectic and eutectoid reactions are defined as

$$L_2 \Leftrightarrow S_1 + S_3 \text{ [eutectic]},$$

(3-45)

$$S_2 \Leftrightarrow S_1 + S_3 \text{ [eutectoid]},$$

where L refers to liquid, S refers to solid phases, and the numbers indicate the relative amount of phases. There is a relatively larger amount of one of the components at those temperatures. For example, the copper content increases from 8.8% (S_1 or α), 28.1% (L), and 92% (S_3 or β) for the Cu–Ag alloy at 779.4°C, as can be deduced from Figure 3-23. Note that the last liquid will disappear at eutectic temperature and composition (cf. Figure 3-23).

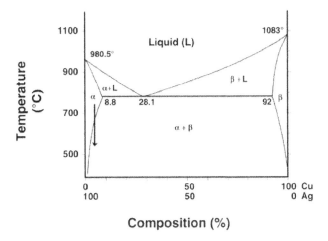

Figure 3-23. The Cu–Ag phase diagram: the dotted vertical line indicates precipitation hardening by quenching of the α phase.

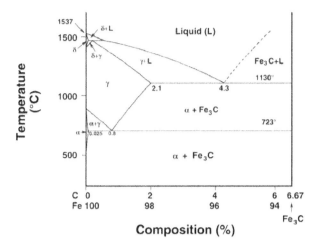

Figure 3-24. Fe–C phase diagram.

Example 3.6

Copper and silver metal are mixed thoroughly in powder form, in proportions 80 w/o Cu and 20 w/o Ag and heated to well above the melting temperature of the alloy. The liquid metal is then cooled and allowed to reach thermodynamic equilibrium. Give the composition of each phase and the relative amount of each phase:

a. At 1,000°C.   b. At 780°C.   c. At 700°C.

Answer

From Figure 3-23:

a.   At 1,000°C, all liquid (80 w/o Cu + 20 w/o Ag).
b.   At 780°C, β (92 w/o Cu + 8 w/o Ag) 81 w/o.
        L (28.1 w/o Cu + 71.9 w/o Ag) 19 w/o.

$$\frac{\beta}{\beta+L} = \frac{28.1-80}{28.1-92} = \underline{0.81}.$$

c.   At 700°C, a (6 w/o Cu + 94 w/o Ag); 15 w/o.
        b (93 w/o Cu + 7 w/o Ag); 85 w/o.

$$\frac{\alpha}{\alpha+\beta} = \frac{93-80}{93-6} = \frac{13}{87} = \underline{0.15}.$$

3-4. STRENGTHENING BY HEAT TREATMENTS

3.4.1. Metals

One of the strengthening processes is precipitation (or age) hardening of alloys by heat treatments. This is accomplished by rapidly cooling (quenching) a solid solution of decreasing solubility, as shown in Figure 3-23 along the dotted vertical line. If quenching is done properly, there will not be enough time for the second phase (β) to form. Hence, a quasi-thermal equilibrium exists, but depending on the amount of thermal energy (related to temperature) and time, the second phase (β) will form (precipitation). If the β phase particles are small and uniformly dispersed throughout the matrix, their presence can increase the strength greatly. It is important that they be dispersed within a grain as well as at grain boundaries, so that the dislocations can be impeded during the deformation process, as in the case of cold-working.

In relation to the precipitation process, the diffusionless *martensitic* transformation process is another mechanism of strengthening steel and other alloys. When fcc iron or steel is quenched from the austenitic temperature range (γ phase in Figure 3-24), there is no time for carbon and other alloying elements to form $\alpha + \underline{C}$ phases. Almost all the carbon atoms must diffuse to form carbide ($\underline{C}$) as well as the carbide formers (Cr, Mo, and V), which should concentrate in the carbide, whereas ferrite formers (Ni and Si) must diffuse into the ferrite (α). These reactions take time at low temperature (below 400°C). Since the fcc structure of austenite is not in equilibrium, a driving force develops and at low enough temperatures this driving force becomes sufficient to force transformation by shear. The resulting structure is a *tetragonal martensite* instead of the body-centered cubic ferrite. The martensite is extremely hard because it is non-cubic (has fewer slip systems than bcc structure) and the interstitially entrapped carbon prevents slip. Martensite is the hardest iron-rich phase material but is extremely brittle. Hence, tempering by heating (600°C) and slow cooling is necessary to make the material tough and strong. Martensite crystal structures are also of interest in the context of shape memory materials such as nickel–titanium alloys:

$$\text{Martensite} \xrightarrow{\text{tempering}} \alpha + \text{Iron carbide (C; Fe}_3\text{C)}. \tag{3-46}$$

3.4.2. Ceramics and Glasses

As mentioned earlier, ceramics and glasses are hard and brittle due to their *non-yielding* character during deformation, which in turn is due to their bonding characteristics. Because of this brittleness they are subject to stress concentration effect at the microcracks present in the material when in tensile deformation mode (cf. §3.1.2). Therefore, if we want to increase their strength we can employ means to overcome these problems by (1) introducing surface compression that has to be overcome before the net stress becomes tensile, and (2) reducing stress concentration by minimizing sharp cracks on the surface and in the bulk.

We can accomplish surface compression by thermal treatment (quenching from high to low temperature and surface crystallization) or by chemical treatment (ion exchange). In the latter process, a small ion such as Na$^+$ is exchanged with a larger ion such as K$^+$. Reduction of the number of microcracks for glasses can be accomplished by a simple fire-polishing process that will also concomitantly introduce surface compression, making the glass very strong (this can increase its strength up to 200-fold). Strong glass can also be achieved by drawing the glass into fibers. The fibers have good surface quality, and they are smaller in size than the typical distance between defects such as microcracks. Therefore, such fibers can be essentially defect-free and are very strong. Such fibers are used in composite materials.

3.4.3. Polymers and Elastomers

Polymers and elastomers cannot be heat treated to increase their strength in general. The strengths of these materials are sensitive to the chemical composition, side group, branching, molecular weight, polydispersity, and degree of crosslinking. Considerable strength increase can be achieved by substituting the main chain repeating unit of polyamide (nylon) with a benzene ring. The resulting polymer is called aramid (Kevlar®). This fiber is stronger than piano wire if compared in terms of specific strength (see §7.3.1). Drawing and annealing indeed involve a thermal process but usually are used to increase the strength of fibers or thin sheets. We will consider this topic in chapter 8.

3.5. SURFACE PROPERTIES AND ADHESION

Surface properties are important since all implants interface with the tissues at their surfaces. The surface property is directly related to the bulk property since the surface is the discontinuous boundary between different phases. If ice is being melted, then there are two surfaces created between three phases: liquid (water), gas (air and water vapor), and solid (ice).

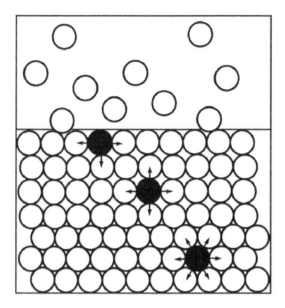

Figure 3-25. Two-dimensional representation of a surface. Surface molecules are not subject to balanced forces from surrounding atoms in the solid, and are therefore more reactive.

The *surface tension* develops near the phase boundaries since the equilibrium bonding arrangements are disrupted, leading to an excess energy that will minimize the surface area, as shown in Figure 3-25. Other means of minimizing the surface energy is to attract foreign materials (*adsorption*) and bonding with adsorbent (*chemisorption*). The surface free energy (dG) can be expressed as,

$$dG = dw - \gamma dA, \tag{3-47}$$

where w is work done on the surface area change dA, and γ is the surface energy of the material.

The conventional units used to describe surfaces are dynes per cm or ergs per cm^2 for surface energy (or tension), but these units are exactly the same since one erg is one dyne cm. The SI unit is N/m, as given in Table 3-3.

Table 3-3. Surface Tension of Materials

Substances	Temperature (°C)	Surface tension (N/m)
Mercury	20	0.465
Lead	327	0.452
Zinc	419	0.785
Copper	1131	1.103
Gold	1120	1.128

1 N/m = 10^3 ergs/cm^2 = 10^3 dynes/cm

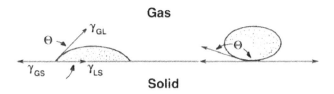

Figure 3-26. Wetting and non-wetting of a liquid on the flat surface of a solid. Note the contact angle.

If a liquid is dropped on a solid surface, then the liquid droplet will spread or make a spherical globule, as shown in Figure 3-26. At equilibrium the sum of surface tensions (γ_{GS}, γ_{LS}, and γ_{GL}) among the three phases (gas, liquid, and solid) in the solid plane should be zero since the liquid is free to move until force equilibrium is established. Therefore,

$$\gamma_{GS} - \gamma_{LS} - \gamma_{GL} \cos\theta = 0, \quad \cos\theta = \frac{\gamma_{GS} - \gamma_{LS}}{\gamma_{GL}}, \quad (3\text{-}48)$$

where θ is called the contact angle. The wetting characteristic can be described as

$$\theta = 0 \quad \text{(complete wetting)},$$
$$0 < \theta < 90° \quad \text{(partial wetting)}, \quad (3\text{-}49)$$
$$\theta > 90° \quad \text{(non-wetting)}.$$

Note that Eq. (3-48) gives only ratios rather than absolute values of surface tension. Some values of contact angle are given in Table 3-4.

The lowest surface tension of a liquid (γ_{GL}) in contact with a solid surface with a contact angle (θ) greater than zero degrees is termed *critical surface tension* (γ_c). The critical surface tension can normally be used as the surface tension of a solid, as given in Table 3-5, for some polymers. The critical surface tension can be measured by measuring contact angles with various liquids of known surface energy. Extrapolating the curve to a zero contact angle would be equivalent to the surface energy of the substrate material. The plot is called a *Zisman plot*, as given in Figure 3-27.

Table 3-4. Contact Angle Values

Liquid	Substrate	Contact angle(°)
Methylene iodine	Soda-lime glass	29
(CH_2I_2)	Fused quartz	33
Water	Paraffin wax	107
Mercury	Soda-lime glass	140

Table 3-5. Critical Surface Tension of Polymers

Polymer	γ_c (dyne/cm)
Polyhexamethylene adipamide, nylon 66	46
Polyethylene terephthalate	43
Poly(6-amino caproic acid), nylon 6	42
Polyvinyl chloride	39
Polyvinyl alcohol	37
Polymethylmethacrylate	33–44
Polyethylene	31
Polystyrene	30–35
Polydimethyl siloxane	24
Poly(tetrafluroethylene)	18.5

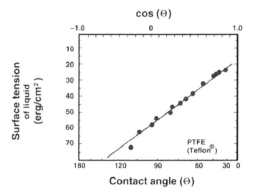

Figure 3-27. Zisman plot of contact angles: the PTFE substrate with various liquids (Baier, 1980).

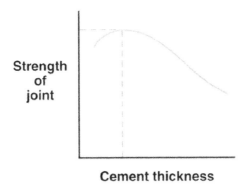

Figure 3-28. Variation of the strength of a joint versus the thickness of the cement between adherends.

When two surfaces are bonded together it is called adhesion if the two materials are different and cohesion if the same. All surfaces cemented with a cementing agent are bonded by adhesion; hence, the cementing agent is an adhesive. For the maximum interfacial strength the thickness of the adhesive layer must be optimal, as shown in Figure 3-28.

In dental and medical applications the adhesives should be considered a temporary remedy since the tissues are living, replacing the old cells with new ones, and thus destroying the initial bonding. This problem led to the development of porous implants, which allows tissues to grow into the interstices (pores), creating a viable, interlocking system between implants and tissues.

PROBLEMS

3-1.   Which of the following materials will be best described by the three stress–strain curves of Figure 3-5?

  a. Ceramics and glasses.
  b. Plastics (polymers) such as polyethylene.
  c. Glassy polymers such as Plexiglas® (polymethylmethacrylate).
  d. Soft tissues such as skin, blood vessel walls, etc.
  e. Hard mineral tissues of bone and teeth.
  f. Copper.
  g. Rubber band.

3-2.   Poisson's ratio (ν) is defined by the following expression:

$$\nu = -\varepsilon_x / \varepsilon_z,$$

where ε is strain; the load is applied in the z direction in simple tension or compression.

Silicone rubber has a Poisson's ratio of 0.4. Consider a silicone rubber drain hose for a surgical procedure; the hose has an outer diameter of 4 mm and a wall thickness of 0.5 mm. How much does the lumen (interior space) of the catheter constrict if the hose is stretched by 20%?

3-3.　The following data were obtained using a stainless steel tensile specimen:

Stress (MPa)	Strain (%)	Stress (MPa)	Strain (%)
98	0.06	700	0.50
160	0.10	770	0.60
280	0.16	830	0.70
350	0.20	870	0.80
500	0.30	920	0.90
620	0.40	930	1.00

a. Plot the stress–strain curve.

b. Determine the modulus of elasticity, 0.2% offset yield strength, fracture strength and the toughness.

3-4.　A piece of suture is tested for its stress relaxation properties after cutting a 3-cm long sample with a diameter of 1 mm. The initial force recorded after stretching 0.1 cm between grips was 5 Newtons. Assume the suture material will behave as if it has one relaxation time. The gauge length was 1 cm.

a. Calculate the initial stress.

b. Calculate the initial strain.

c. Calculate the modulus of elasticity of the suture if the initial stretching can be considered as linear and elastic.

d. Calculate the relaxation time if the force recorded after 10 hours is 4 Newtons.

3-5.　The following data were obtained using unknown liquids and a polyethylene sheet.

Liquid	γ (erg/cm^2)	Contact angle (°)
A	75	96
B	40	63
C	30	15
Polyethylene	29	–

a. Plot the interfacial surface tension versus contact angle.

b. Obtain a linear relationship between interfacial surface tension and contact angle.

c. What conclusions can you draw regarding γ_{GL} and γ_{SL}?

3-6.　Surface properties change after a material is implanted inside the body.

a. Explain how properties will be changed as a result of adsorption of protein, or as a result of diffusion of water into the material.

b. What methods should be used to understand the interaction between the tissue and implant. Can you use the data obtained from in vitro surface experiment in vivo?

3-7.　From the data given in Figure 3-9:

a. Estimate the endurance limit of bone cement.

    b. If a tooth implant is made of this material with a 4-mm diameter and is cylindrical in shape, how long will it last? Assume the maximum force of chewing is 100 N in compression.

3-8.   From the data given in Table 3-2:

    a. Plot melting temperature (T_m) versus specific heat, heat of fusion, thermal conductivity, and linear thermal expansion coefficient.

    b. What conclusions can you draw?

3-9.   Show that the maximum (resolved) shear stress operates on a 45° angle with respect to the force (stress) being applied. Hint: the shear stress $\tau = F_s/A_s$ can be resolved into a component upon a slip direction which has an angle of λ and a force normal to the slip plane. The slip plane has an angle of ϕ; then $\tau = \sigma \cos \lambda \cos \phi$, where σ is the applied stress on the cross-sectional area A.

3-10.   Prove the lever rule.

3-11.   A 70 w/o Cu and 30 w/o Ag alloy was made and cooled from liquid. From the Cu–Ag phase diagram (Figure 3-23).

    a. At what temperature does the solid phase start to appear?

    b. What phases exist at 780°C?

    c. What are the compositions of the phases in b?

    d. What is the weight of Cu in the solid phase and liquid phase in b if we started with 100 g of alloy?

    e. Can this alloy be solution hardened?

3-12.   From the following general stress–strain curves, answer the following. Observe the axis labels carefully.

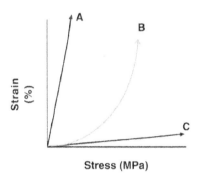

    a. Which curve represents the lowest Young's modulus material?

    b. Which curve represents the toughest material?

    c. Which curve represents alumina (Al_2O_3)?

    d. Which curve represents aluminum?

    e. Which curve represents polyethylene $[-(-CH_2CH_2-)_n-]$?

    f. Which curve represents material with the lowest Poisson's ratio?

3-13. The plastic was stretched from 100.00 cm to 120.00 cm and released at room tem-
        perature. The length changes with time were obtained as the following. Assume the
        Voigt model can be applied for this mechanical behavior.

Time (hr)	Length (cm)
0	120.0
1	110.0
3	103.0
6	100.4
20	100.1

a. What is the retardation time(hours)?

b. What is the percent strain at 15 minutes based on the original length of the sample?

c. What is the amount of strain (%) recovered after 15 minutes?

d. If the test was made at body temperature, would the recovery be faster, slower, or the
    same compared to the room temperature test.

SYMBOLS/DEFINITIONS

Greek Letters

α: Linear thermal expansion coefficient, amount of length change per unit length per unit tem-
    perature.

δ: Loss angle, the phase angle between stress and strain in oscillatory loading of viscoelastic
    materials; it is zero in an elastic material.

ε: Strain, change in length per unit length.

γ: Surface energy. See definitions of words.

γ_c: Critical surface tension that is the lowest surface energy a solid exhibits.

η: Viscosity, measure of the flow characteristic of a material [Pa $*$ s or Poise].

λ: Retardation time at which the strain level is reduced to its original value by $1/e$ at constant
    stress.

ν: Poisson's ratio — the ratio of lateral contraction to longitudinal extension in simple tension.

θ: Contact angle which is formed between liquid and solid substrate due to the partial or non
    wetting nature of the surface.

σ: Stress (tensile, compressive, or shear), force per unit cross-sectional area.

κ: Yield stress of the softer material between two sliding materials that can wear.

τ: Relaxation time at which the stress level is reduced to its original value by $1/ie$ at constant
    strain.

μ: Friction coefficient.

Latin Letters

a: Surface crack length of an elliptic crack ($2a$ if the crack is in the bulk).

b: One half-width of an elliptic crack.

$\underline{C}$: Iron carbide (Fe_xC).

E: Young's modulus or modulus of elasticity; slope of the stress–strain curve in the elastic portion.

FS: Failure or fracture strength of a material.

$H(t)$: Heaviside step function; 0 for $t < 0$, 1 for $t > 0$.

$J(t)$: Creep compliance, ratio of time-varying strain to constant stress, $\varepsilon(t)/\sigma_0$.

K: Fracture toughness derived from the micromechanics of crack propagation. Units are [MPa $\sqrt{m}$ or   psi $\sqrt{in}$]

r: Crack tip radius.

scf: Stress concentration factor, the ratio of the stress at a tip of a crack or hole to nominal stress away from the tip.

TS: maximum or ultimate tensile strength of a material.

ΔV: wear volume.

YS: yield point or stress beyond which the material will be permanently deformed.

Words

Adhesion: Joining of two different materials.

Adsorption: Physical attachment of foreign material (usually gas) on a surface.

Boltzmann superposition principle: Effect of a compound cause is the sum of the effects of individual causes. This is the statement of linearity for viscoelastic materials.

Chemisorption: Chemical attachment of foreign material (usually gas) on a surface.

Cohesion: Joining of identical materials.

Contact angle (θ): Angle made when a drop of liquid spreads over a solid surface. It is governed by the balance of surface tension at various interfaces.

Convolution: Type of integral equation used in the analysis of viscoelastic materials.

Creep: Increase of strain with time in viscoelastic materials under constant stress.

Endurance limit: Stress level below which the material will not fail by cyclic fatigue loading no matter how many cycles (a practical limit is often chosen as 10^7 cycles).

Engineering stress: Stress calculated based on its original cross-sectional area.

Fourier transform: Integral transform equation involving sinusoids. In the context of viscoelastic materials, Fourier transforms relate the time and frequency domains.

Griffith theory: Energy approach to the fracture of a (brittle) material in which the fracture strength in tension is inversely proportional to the square root of theh crack length. The intrinsic properties of surface energy and Young's modulus are directly related to the strength.

Heaviside step function: Unit step function; has a value of zero for arguments less than zero, and one for arguments greater than zero. It is used in analysis of viscoelastic materials.

Hooke's law: Stress is linearly proportional to strain. Most materials follow this law at low strains.

Isotropic: Properties of material are the same in every direction. Materials such as steel and glassy polymers are usually isotropic, but composite materials and biological materials are not.

Kramers-Kronig relations: Relationships between the compliance and loss functions of frequency in viscoelastic materials.

Limited solubility: Results when only a maximum amount of solute material can be dissolved in a solution (solid).

Martensite: Iron carbon alloy (steel) obtained by quenching from austenite (γ); has a body-centered tetragonal crystal structure. Other similar crystal structures such as those occurring in shape memory alloys (e.g., nickel–titanium) are also called martensite.

Maxwell model: Mechanical analog model consisting of a spring and a dashpot in series for describing viscoelastic material properties.

Necking: Unstable irreversible flow of material locally during tensile deformation, resulting in a neck-like shape.

Phase: Having the same (atomic or micro) structure and properties throughout.

Relaxation modulus: Stress relaxation of viscoelastic materials is the decrease in stress that occurs under constant strain. The relaxation modulus is the ratio of stress to strain during stress relaxation.

Solid solution: Solid phase that contains more than two elements that are mixed uniformly everywhere in the phase.

Surface tension (surface energy): Amount of free energy exhibited at the surface of a material.

Tempering: Toughening of martensite by heat treatment; the structure becomes more stable by converting to ferrite (α) and carbide microstructure.

Thermal conductivity: Amount of heat (thermal energy) passed for a given thickness, time, and cross-sectional area of a material.

Toughness: amount of energy expended before its fracture or failure.

True stress: Stress calculated based on a specimen's true cross-sectional area.

Voigt or Kelvin model: Mechanical analog model describing material properties by arranging a spring and a dashpot in parallel.

Yield point: Point of the stress–strain curve where transition takes place from elastic to plastic deformation, i.e., the curve deviates from initial linear portion.

BIBLIOGRAPHY

Ashby MF, Jones DRH. 1988. *Engineering materials*, Vol. 1: *An introduction to properties, applications and design*. Amsterdam: Elsevier.

Baier RE. 1980. *Guidelines for physicochemical characterization of biomaterials*. Devices and Technology Branch, National Heart, Lung and Blood Institute. NIH Publication No. 80-2186.

Bhushan B. 1999. *Principles and applications of tribology* (Chapter 8). New York: Wiley.

Bonow RO, Carabello B, de Leon AC, Edmunds LH Jr, Fedderly BJ, Freed MD, Gaasch WH, McKay CR, Nishimura RA, O'Gara PT, O'Rourke RA, Rahimtoola SH, Ritchie JL, Cheitlin MD, Eagle KA, Gardner TJ, Garson A Jr, Gibbons RJ, Russell RO, Ryan TJ, Smith SC Jr. 1998. ACC/AHA task force report: ACC/AHA guidelines for the management of patients with valvular heart disease. *J Am Coll Cardiol* **32**(5):1486–1588. Full text available at http://en.wikipedia.org/wiki/Mitral_stenosis.

Callister Jr WD. 2000. *Materials science and engineering: an introduction*. New York: Wiley.

Cottrell AH. 1981. *The mechanical properties of matter*. Huntington, NY: Krieger.

Freitag TA, Cannon SL. 1977. Fracture characteristics of acrylic bone cement, II: fatigue. *J Biomed Mater Res* **11**:609–624.

Griffith, 1920. AA. The phenomena of rupture and flow in solids. *Philos Trans R Soc London Ser A* **221**:163–198.

Hayden WH, Moffatt WG, Wulff J. 1965. *The structure and properties of materials*, Vol. 3. New York: Wiley.

Liu YK, Park JB, Njus GO, Stienstra D. 1987. Bone-particle-impregnated bone cement: an in vitro study. *J Biomed Mater Res* **21**:247–261.

McClintock FA, Argon AS. 1966. *Mechanical behavior of materials*. Reading, MA: Addison-Wesley.

Paris P, Erdogan F. 1963. A critical analysis of crack propagation. *J Basic Eng* **85**:528–535.

Santavirta S. 2003. Compatibility of the totally replaced hip: reduction of wear by amorphous diamond coating. *Acta Orthop Scandinavica Suppl* **74**(310):1–19.

Van Vlack LH. 1970. *Materials science for engineers*. Reading, MA: Addison-Wesley.

Van Vlack LH. 1982. *Materials for engineering: concepts and applications*. Reading, MA: Addison-Wesley.

Van Vlack LH. 1989. *Elements of materials science*. Reading, MA: Addison-Wesley.

4

CHARACTERIZATION OF MATERIALS — II:
ELECTRICAL, OPTICAL, X-RAY ABSORPTION, ACOUSTIC, ULTRASONIC, ETC.

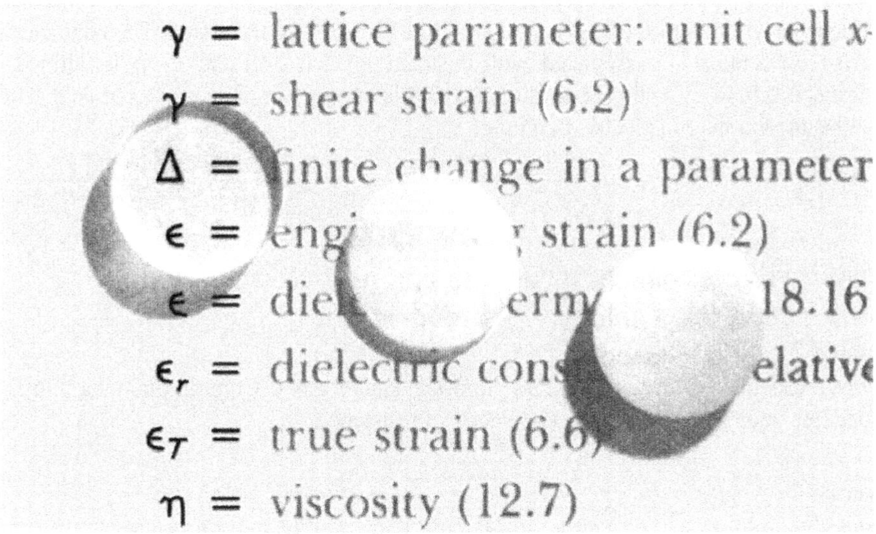

γ = lattice parameter: unit cell x

γ = shear strain (6.2)

Δ = finite change in a parameter

ϵ = eng... strain (6.2)

ϵ = diel... erm... 18.16

ϵ_r = dielectric cons... elative

ϵ_T = true strain (6.6)

η = viscosity (12.7)

The light transmittance of three aluminum oxide specimens. From left to right: single-crystal material (sapphire), which is transparent; a polycrystalline and fully dense (nonporous) material, which is translucent; and a polycrystalline material that contains approximately 5% porosity, which is opaque. Specimen preparation, P.A. Lessing; photography by J. Telford. Reprinted with permission from Callister (2000). Copyright © 2000, Wiley.

In addition to the mechanical, thermal, and surface properties of materials, other physical properties could be important in particular applications of biomaterials. Properties considered in this chapter include electrical, optical, absorption of x-rays, acoustic, ultrasonic, density, porosity, and diffusion.

4.1. ELECTRICAL PROPERTIES

The electrical properties of materials are important in such applications as pacemakers and stimulators, as well as in piezoelectric implants to stimulate bone growth.

Electrical resistance, R, is defined as the ratio between the potential difference (voltage) V applied to the object and the current i that flows through:

$$R = \frac{V}{i}. \tag{4-1}$$

If potential difference V is measured in volts (V) and current i in amperes (A), resistance R is in ohms, denoted by a capital Greek omega (Ω). *Ohm's law* states that voltage is proportional to the current in a conductor, so that resistance R is independent of voltage. Metals obey Ohm's law if the temperature does not change much but semiconductors do not. The resistance of an object depends upon both the material of which it is made and the shape. The characteristic of *resistivity*, by contrast, is associated with the material itself. Resistivity ρ_e is defined as the ratio of electric field E to current density J, which is current per cross-sectional area. The electric field is the gradient in electric potential:

$$\rho_e = \frac{E}{J}. \tag{4-2}$$

The unit of resistivity is ohm-meter (Ω-m).

Example 4.1

Consider a pacemaker wire of circular cylindrical shape, $d = 0.1$ mm in diameter and $L = 100$ mm long, made of gold. Determine the electrical resistance.

Answer

In view of the uniform cross-section, the electric field, and the current density will be uniform over the length of the wire (L). The electric field is the gradient in potential, so

$$E = V / L.$$

The current density is the current per unit area, or

$$J = i / A = 4i / \pi d^2.$$

The resistance, therefore, is

$$R = V / i = 4EL / J \pi d^2 = 4L\rho_e / \pi d^2.$$

For gold, $\rho_e = 2.35 \times 10^{-8}$ ohm-m, so $R = 0.3$ ohm. This is much smaller than the resistance of the surrounding tissue, so it is not likely to be a problem. Single-strand pacemaker wire would be vulnerable to mechanical fatigue due to flexure from a beating heart, so multi-strand wires are used.

The electrical resistivity of materials varies over many orders of magnitude. Insulators, or materials with very high resistivity, are used to isolate electrical equipment, including implantable devices such as pacemakers and other stimulators, from body tissues. Polymers and ceramics tend to be good insulators. The electrical resistivities of representative materials are given in Table 4-1.

Table 4-1. Electrical Resistivity of Various Materials

Material	Resistivity (Ω-m)
UHMWPE	$>5 \times 10^{14}$
PMMA	10^{14}
Al_2O_3	$10^9 - 10^{12}$
Zr_2O (3% Y_2O_3)	10^{10}
SiO_2	10^{10}
Bone (wet, longitudinal)	46
Muscle (wet, longitudinal)	2
Physiological saline	0.7
Stainless steel	7.3×10^{-7}
Platinum	10^{-7}
Gold	2.35×10^{-8}
Copper	1.7×10^{-8}
Silver	1.6×10^{-8}

Piezoelectricity is a coupling between mechanical deformation and electrical polarization of a material. Specifically, mechanical stress results in electric polarization, the direct effect; and an applied electric field causes strain, the converse effect. The piezoelectric constitutive equations are as follows:

$$D_i = \sum_{jk} d_{ijk}\sigma_{jk} + \sum_j K_{ij}E_j + p_i\Delta T, \qquad (4\text{-}3)$$

$$\varepsilon_{ij} = \sum_{kl} S_{ijkl}\sigma_{kl} + \sum_k d_{kij}E_k + \alpha_{ij}\Delta T, \qquad (4\text{-}4)$$

in which D is the electric displacement, σ is the stress, d is the piezoelectric sensitivity tensor, K is the dielectric permittivity, E is the electric field, p is the pyroelectric coefficient, T is the temperature, ε is the strain, S is the elastic compliance, and α is the thermal expansion. Only materials with sufficient asymmetry exhibit piezoelectricity or pyroelectricity and consequently have d and p coefficients not equal to zero. The physical origin of piezoelectricity lies in the presence of asymmetric charged groups in the material, as shown in Figure 4-1. As the material is deformed, the charges move with respect to each other so that change in dipole moment occurs. Figure 4-1 also shows the polarization that results from stress via several d coefficients. As for elastic compliance S, the first term in Eq. (4-4) represents Hooke's law, as discussed in the previous chapter. All piezoelectric materials are anisotropic, and the S coefficients represent different compliances for different directions of loading. Young's modulus E is the inverse of compliance S in the direction considered.

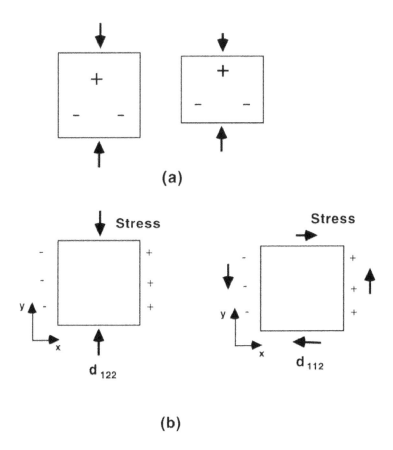

Figure 4-1. Piezoelectric materials. (**a**) Physical origin of piezoelectricity: charge separation in asymmetric unit cell under deformation. (**b**) Piezoelectric d coefficients from Eqs. (4-3) and (4-4). **Left**: polarization in the $x(1)$ direction due to compressive stress in the $y(22)$ direction. **Right**: polarization in the x direction due to shear stress in the $xy(12)$ direction.

Fukada and Yasuda first demonstrated that dry bone is piezoelectric in the classic sense. The piezoelectric properties of bone are of interest in view of their hypothesized role in bone remodeling. Wet collagen, however, does not exhibit a piezoelectric response. Piezoelectric effects occur in the kilohertz range, well above the range of physiologically significant frequencies. Both the dielectric and piezoelectric properties of bone depend strongly upon frequency. The magnitude of the piezoelectric sensitivity coefficients of bone depends on frequency, on direction of load, and on relative humidity. Values up to 0.7 pC/N (= 10–12 C/N) have been observed in bone, to be compared with 0.7 and 2.3 pC/N for different directions in quartz, and 600 pC/N in some piezoelectric ceramics. It is, however, uncertain whether bone is piezoelectric in the classic sense at the relatively low frequencies that dominate in the normal loading of bone. *Streaming potentials* can result in stress-generated potentials at relatively low frequencies even in the presence of dielectric relaxation, but this process is as yet poorly understood. The electrical potentials observed in transient deformation of wet bone in vivo may be mostly due to streaming potentials.

Compact bone also exhibits a permanent electric polarization as well as *pyroelectricity*, which is a change in polarization with temperature. These phenomena are attributed to the polar structure of the collagen molecule; these molecules are oriented in bone. The orientation of permanent polarization has been mapped in various bones and has been correlated with developmental events.

The electrical properties of bone are relevant not only as a hypothesized feedback mechanism for bone remodeling but also in the context of external electrical stimulation of bone to aid its healing and repair.

Example 4.2

Suggest an application of artificial piezoelectric materials in the body.

Answer

In some cases natural growth or repair of bone may be inadequate. Inclusion of an active piezoelectric element in a bone plate or joint replacement will generate electrical signals in vivo that will stimulate the growth of bone. This approach has been used on a research basis.

Example 4.3

A piezoelectric stimulator one square centimeter in cross-sectional area and one millimeter thick is incorporated in a composite bone plate. It experiences 1% of the stress seen in a healthy leg bone during walking (8 MPa). The material is a lead titanate zirconate ceramic for which the relevant piezoelectric coefficient is 100 pC/N and the dielectric constant 1,000. Determine the peak voltage produced by the device. For the purpose of calculation, neglect the leakage of charge through the conductive pathways in bone.

Answer

The charge density may be calculated as follows. The charge density q/A is the piezoelectric coefficient times the stress; the stress is 0.01×8 MPa as given above:

$$q/A = 100 \text{ pC/N} [0.01][8 \times 10^6 \text{ N/m}^2] = [10^{-10} \text{ C/N}][8 \times 10^4 \text{ N/m}^2] = 8 \text{ } \mu\text{C/m}^2.$$

Under the assumptions given, the implant behaves as a capacitor of capacitance C, for which the charge q is $q = CV$, in which V is the voltage. $V = q/[k\varepsilon_0 A/d]$, with k as the dielectric constant, ε_0 as the permittivity of space, A as the cross-sectional area, and d the thickness. Using the charge density from above,

$$V = [8 \text{ } \mu\text{C/m}^2][1 \text{ mm}]/[1,000 \times 8.85 \times 10^{-12} \text{ F/m}^2] = \underline{0.9 \text{ Volts}}.$$

We have neglected the parallel capacitance and leakage conductance of the surrounding tissue. As for the parallel capacitance, the dielectric constant of muscle, for example, is about 10^5 at 100 Hz and 10^8 at 0.01 Hz. The actual stimulating voltage will therefore be considerably less than determined above and will moreover decay rapidly with time, following each step in walking as a result of the conductivity of the surrounding bone and muscle.

4.2. OPTICAL PROPERTIES

The optical properties of materials are relevant to their performance when used in the eye, as well as to cosmetic aspects of dental materials. A ray of light incident upon a transparent material will be partly reflected and partly transmitted. The transmitted ray is bent or *refracted* by the material. It is observed experimentally (and can be deduced from Maxwell's equations) that the incident ray, normal to the surface, and refracted ray all lie in the same plane, and the angle of incidence equals the angle of reflection. The angle of the refracted ray depends upon a material property known as the *refractive index*, usually denoted by n. The refractive index is defined as the ratio of the speed of light in a vacuum to the speed of light in the medium. The relationship between the angles of incidence and of refraction is given by *Snell's law*:

$$n_1 \sin \theta_1 = n_2 \sin \theta_2, \qquad (4\text{-}5)$$

in which θ_1 is the angle of the incident ray with respect to the normal to the surface, n_1 is the refractive index of the medium containing the incident ray, and θ_2 is the angle of the refracted ray to the normal to the surface of the material, which has refractive index n_2, as shown in Figure 4-2. Some representative indices of refraction are given in Table 4-2 for yellow-orange light at a wavelength of 589 nm.

Table 4-2. Refractive Index of Some Materials

Material	Refractive index
Vacuum	1.0
Air	1.0003
Water	1.33
Human aqueous humor	1.336
Human vitreous humor	1.338
Human cornea	1.376
Human lens	1.42
HEMA hydrogel, wet	1.44
PMMA	1.49
Polyethylene(film)	1.5
Crown glass	1.52
Flint glass	1.66
Diamond	2.41

In ophthalmologic biomaterials, transparent materials find application in lenses. Refraction of light by a convex lens is shown in Figure 4-3. The focal length of such a lens is defined as the distance from the lens to the image plane when parallel rays of light (from far away) impinge upon the lens. The focal length f of a simple, thin lens (in air or vacuum) depends on its refractive index n and the surface radii of curvature, r_1 and r_2, as follows:

$$\frac{1}{f} = (n-1)\left(\frac{1}{r_1} - \frac{1}{r_2}\right). \qquad (4\text{-}6)$$

This is the "lens maker's equation" and is derivable from Snell's law. The sign convention for the radii of curvature is that they are considered positive if the center of curvature is on the right side of the lens. Consequently, both surfaces of a double convex lens contribute positive power. In biomedical applications, the lens is likely to have one or more interfaces with tissue fluid or the tissues of the eye. The optical power P of lenses is usually expressed in diopters (D) in the ophthalmic setting:

$$P\ (\text{D}) = \frac{1}{f\ (\text{meters})}, \qquad (4\text{-}7)$$

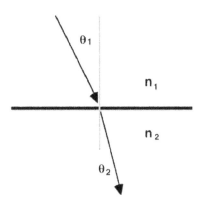

Figure 4-2. Snell's law for refracted light.

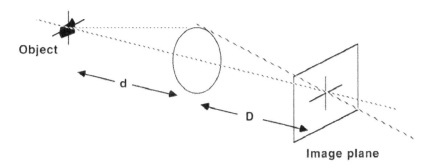

Figure 4-3. Refraction of light by a convex lens.

so that a converging lens with a focal length of 100 mm has a power of +10 D. Diverging lenses have negative focal lengths, hence negative optical power. A typical, normal human eye has an optical power of about 60 D, of which about 43 D is associated with the curvature of the cornea. The eye measures about 22 mm from cornea to retina.

Example 4.4

Consider a polycarbonate intraocular lens of diameter $d = 8.3$ mm, thickness $t = 2.4$ mm, anterior radius of curvature 17.8 mm, and posterior radius of curvature 10.7 mm. Determine the optical power and focal length of this lens in air and in water.

Answer

Use the thin lens Eq. (4-6). In air, assume $n = 1$ for air and take $n = 1.59$ for polycarbonate from Table 4-2.

$$P = (1.59 - 1)[1/0.0178 \text{ m} - 1/-0.0107 \text{ m}] = 88.2 \text{ m}^{-1} = \underline{88.2 \text{ D}},$$

so

$$f = 1/P = \underline{13.6 \text{ mm}} \text{ in air.}$$

Water has an index of refraction $n = 1.33$, and so

$$P = (1.59 - 1.33)[1/0.0178 \text{ m} - 1/-0.0107 \text{ m}] = 39 \text{ m}^{-1} = \underline{39 \text{ D}},$$

so

$$f = 1/P = \underline{25.6 \text{ mm}} \text{ in water.}$$

We remark that the normal lens of the eye has an optical power of about 19 D.

Transparent biomaterials are used to make contact lenses and intraocular lenses. PMMA is the material of choice for intraocular lenses and for hard contact lenses. Its main disadvantage as a contact lens material is that its permeability to oxygen is low, so that the cornea, which receives its oxygen by diffusion from the air, suffers hypoxia. Other materials are used in ocular applications. Representative material properties are given in Table 4-3. Polycarbonate is an amorphous thermoplastic that has been used in the manufacture of contact lenses, particularly those of strong power. The comparatively high refractive index of polycarbonate permits a thinner, lighter lens to be made. Soft contact lenses are commonly made of the water-absorbing hydrogel material poly-HEMA [hydroxyethyl methacrylate], which is highly permeable to oxygen.

Table 4-3. Physical Properties of Some Transparent Materials

Material	Density (g/cm^3)	Refractive index	Young's modulus (MPa)	Tensile strength (MPa)
PMMA	1.19	1.49	2800	55
Silicone rubber	0.99–1.5	1.43	6	2.4–6.9
Silicone rubber, contact lens	1.09	1.43	6	1.4
Polycarbonate	1.2	1.59	2200	60

4.3. X-RAY ABSORPTION

The ability of materials to absorb x-rays is of interest in the context of the visibility of an implanted object in diagnostic radiographs. X-rays are electromagnetic waves similar to light except that their wavelength is much shorter and their energy much higher. The index of refraction for x-rays in matter of any kind is very nearly unity. Consequently, x-rays are neither bent nor reflected to any appreciable extent as they interact with matter. The contrast that gives rise to an image in an x-ray film comes about from differences in materials' ability to absorb x-rays. The absorption follows *Beer's law*:

$$I = I_0 e^{-\alpha x}, \tag{4-8}$$

in which I is the intensity at a depth x, and α is the absorption coefficient. Absorption of x-rays in matter is governed by the *photoelectric effect*, in which the incident x-ray photon is absorbed by one of the bound electrons in an atom of material, and the electron is ejected; and by the *Compton effect*, in which the x-ray photon is scattered from a free or weakly bound electron in the material. The absorption due to the photoelectric effect process is proportional to the fifth power of the atomic number N, and increases with the wavelength (λ) (hence decreases with energy) in the x-ray energy range 100 to 350 keV:

$$\alpha = N^5 \lambda^{7/2}. \tag{4-9}$$

At energies below 20 keV, a resonant effect occurs in which absorption becomes very strong when the x-ray energy is equal to the binding energy of the electrons in the inner "K" shell of the atoms of the absorber material. Clinical x-ray diagnostic equipment operates at tube voltages of from 20 to 200 kV. The emitted x-rays are at energies (in electron volts) equal to or less than the tube voltage. Most radiological techniques involve tube voltages between 60 and 100 kV, for which absorption by the photoelectric effect and the Compton effect are comparably important. Since the x-ray energy is

$$E = h\nu = \frac{hc}{\lambda}, \tag{4-10}$$

in which h is Planck's constant and c is the speed of light, the x-rays have wavelengths from 100 pm (0.1 nm) at 10 keV to 5 pm at 200 keV. These wavelengths are much smaller than those of visible light — 400 to 700 nm.

Table 4-4. Mass Absorption Coefficient for X-Rays in Various Materials

Material	Atomic no.	Density ρ (g/cm^3)	Specific absorption coefficient μ/ρ (cm^2/g)
Al	13	2.70	48.7
P	15	1.82	73
Ca	20	1.55	172
Cr	24	7.19	259
Fe	26	7.87	324
Co	27	8.9	354
Pb	82	11.34	241

For CuKα x-rays, wavelength λ =1.54 Å or 0.154 nm.

It is clear that the heavier elements absorb x-rays strongly, as shown in Table 4-4. Heavy metals such as lead are commonly used to shield x-ray equipment. Human soft tissue contains a great deal of the lighter elements (hydrogen, carbon, and oxygen) and is consequently relatively transparent to x-rays. Bone, by virtue of its calcium and phosphorus content, absorbs more strongly and therefore shows up well in x-ray images. Metallic implants absorb strongly and also are highly visible in x-ray images. Polymers, by contrast, are relatively transparent to x-rays. Barium sulfate is incorporated in bone cement to make it visible in diagnostic x-ray images.

4.4. ACOUSTIC AND ULTRASONIC PROPERTIES

The acoustic and ultrasonic properties of biomaterials are relevant in the context of their importance in diagnostic ultrasound images. Important properties are the acoustic velocity v, the acoustic attenuation α, and the material density ρ. The relation for the attenuation of ultrasound is identical to that for x-rays — Eq. (4-8).

Signals for ultrasonic imaging devices are generated by *reflection* of the waves from interfaces in the body, by contrast to diagnostic x-rays, in which differences in attenuation are exploited to create an image. The amplitude reflection coefficient associated with an interface between material 1, containing the incident wave, and material 2, is given by

$$R_A = \frac{Z_2 - Z_1}{Z_2 + Z_1}. \tag{4-11}$$

The acoustic impedance Z is defined as

$$Z = \rho v, \tag{4-12}$$

in which the acoustic velocity v is proportional to the square root of the stiffness (the modulus C_{1111} for longitudinal waves, or the shear modulus for shear waves) divided by the material mass density. By contrast to x-rays, the atomic number has no bearing upon ultrasound. The reflection of electromagnetic waves such as light is similarly determined from the index of refraction, which is also defined in terms of wave velocity.

Table 4-5. Acoustic Properties of Some Materials

Material	Velocity v(m/s)	Impedance Z (kRayl)	Attenuation coeff. α (dB/cm)
Air (at STP)	330	0.04	12
Water	1480	148	0.002
Fat	1450	138	0.63
Blood	1570	161	0.18
Kidney	1560	162	1.0
Soft tissues(avg)	1540	163	0.7
Liver	1550	165	0.94
Muscle	1580	170	1.3–3.3
Bone	4080	780	15
PMMA	2670	320	–
UHMWPE	2000	194	–
Ti6Al4V	4955	2225	–
Stainless steel	5800	4576	–
Barium titanate	4460	2408	–

kRayl = 10^4 kg/m^2/s

Comparative acoustic properties of some relevant materials are given in Table 4-5. Attenuations are given for a frequency of 1 MHz, at the lower end of the range used clinically. The ultrasonic attenuation α (per unit length) is directly related to the viscoelastic loss tangent (tan δ) discussed in Chapter 3; the attenuation therefore depends on frequency. The relation is

$$\alpha = \left(\frac{2\pi v}{v}\right) \tan \frac{\delta}{2}, \tag{4-13}$$

in which v is the frequency of the wave and v is its velocity. As for biomaterials, we observe that polymeric implants, which do not show up well in x-ray images, will usually be highly visible in a diagnostic ultrasound scan as a result of the differences in acoustic properties in comparison with natural tissues.

4.5. DENSITY AND POROSITY

The density ρ of a material is defined as the ratio of mass to volume for a sample of the material:

$$\rho = \frac{m}{v}. \tag{4-14}$$

A biomaterial that replaces an equivalent volume of tissue may have a different weight, as a result of the differences in density. In some applications this can be problematic. Densities of representative materials are given in Table 4-6.

Table 4-6. Density of Some Materials

Material	Density(g/cm^3)
Air (at STP)	0.0013
Fat	0.94
Polyethylene (UHMWPE)	0.94
Water	1.0
Soft tissue	1.01–1.06
Rubber	1.1–1.2
Silicone rubber	0.99–1.50
PMMA	1.19
Compact bone	1.8–2.1
Glass	2.4-2.8
Aluminum	2.8
Diamond	3.5
Titanium	4.5
Stainless steel	7.93
Wrought CoCr	9.2
Gold	19.3

Porous materials are used in a variety of biomedical applications, including implants and filters for extracorporeal devices such as heart-lung machines. In other applications such as bone plates, porosity may be an undesirable characteristic since pores concentrate stress and decrease mechanical strength. Perhaps the most important physical quantity associated with porous materials is the *solid volume fraction* (V_s). The *porosity*, often expressed as a percent figure, is given by

$$\text{Porosity} = 1 - V_s. \tag{4-15}$$

It is also noted that there are three measurements of volume, i.e., true, apparent, and total(bulk)volume:

$$\text{True volume} = \text{total (bulk) volume} - \text{total pore volume}, \tag{4-16}$$

$$\text{Apparent volume} = \text{total (bulk) volume} - \text{open pore volume}, \tag{4-17}$$

$$\text{Total pore volume} = \text{open pore volume} + \text{closed pore volume}. \tag{4-18}$$

We can also have three types of densities corresponding to the three definitions of the volume.

Pore size is important in situations in which tissue ingrowth is to be encouraged, or if the permeability of the porous material is of interest. Porous materials may be characterized by a single pore size, or may exhibit a distribution of pore sizes. Porosity and pore size can be measured in a variety of ways. If the density of the parent solid is known, a measurement of the apparent density of a block of material suffices to determine the porosity. Mercury intrusion porosimetry is a more precise method that delivers a measurement of the porosity and the pore size distribution. In this method, mercury is forced into the pores under a known pressure and the relationship between pressure and mercury volume is determined. Since the mercury has a high surface tension and does not wet most materials, higher pressures are required to force the mercury into progressively smaller pores. If a single pore size predominates, it can be measured by optical or electron microscopy.

Example 4.5

A polyurethane open cell foam is to be used for a lining for an artificial leg. The solid volume fraction is 4%, the pores are 0.5 mm in diameter and the solid polyurethane has a density of 0.9 g/cm^3. Determine the porosity and apparent density of the material, and the weight of a sheet 200 mm by 200 mm by 1 cm thick.

Answer

$$\text{Porosity } P = 1 - V_s, \text{ so } P = 1 - 0.04 = 0.96, \text{ or } \underline{96\% \text{ porosity}}.$$

$$\text{Apparent density} = \rho_{app} = V_s \rho_{true} = 0.04[0.9 \text{ g/cm}^3] = \underline{0.036 \text{ g/cm}^3}.$$

$$\text{Mass} = V\rho_{app} = 20 \times 20 \times 1 \text{ (cm}^3\text{) } 0.036 \text{ g/cm}^3 = \underline{14.4 \text{ g}}, \text{ so weight} = \underline{0.032 \text{ lb}}.$$

The weight of the foam lining is much less than that of an artificial leg. In some cases, the effect of implant density is important. For example, intraocular lenses made of PMMA are denser than the soft tissues of the eye, as shown in Table 4-6. During rapid eye movement, the lens is accelerated by the structures to which it is attached, e.g., the iris. By contrast, the natural eye lens is essentially neutrally buoyant. Damage to these eye structures by intraocular lenses has been observed clinically. Lenses of a silicone rubber composition of lower density are currently under investigation; these offer the added benefit of being softer, and hence easier to insert through a small incision. Another example of the density of implants is augmentation mammoplasty, in which the added weight of the implant can be problematical.

4.6. DIFFUSION PROPERTIES

The diffusion properties of materials are important in applications in which transport of biologically significant constituents is required. Examples include transport of oxygen and carbon dioxide from the atmosphere to the blood in the artificial lung component of the heart-lung machine and transport of oxygen to the cornea through contact lenses. The diffusion equation that governs the motion of dissolved materials under a gradient of concentration C is given by

$$\frac{\partial C}{\partial t} = D\nabla^2 C, \tag{4-19}$$

in which D is the diffusion coefficient, and ∇^2 is the Laplacian, which in one dimension reduces to $\partial^2 C/\partial x^2$. The driving force for material transport may be a pressure gradient rather than a concentration gradient. Moreover, the geometry in many biomedical applications may be approximated by a thin film. In that case, the volumetric flux F (in units of volume per unit time) across a layer of area A is given by

$$F = KA\,DP, \tag{4-20}$$

in which ΔP is the pressure difference across the layer, and K is the permeability coefficient. Representative permeabilities for oxygen transport are given in Table 4-7. Permeabilities for other gases are in general different. Carbon dioxide, for example, diffuses through these materials from two to five times more rapidly than oxygen.

Table 4-7. Gas Permeability of Various Materials

Material	Permeability to O_2	Applications
Silicone rubber	50	Contact lens, lung
Polyalkylsulfone	6	Lung
Polyethylenecellulose- perfluorobutyrate	5	Lung
Teflon film	1.1	Lung
Poly(HEMA)	0.69	Contact lens
PMMA	0.0077	Contact lens

Units: $[cm^3/sec][(cm\ thick/cm^2)(cm\ Hg \times 10^{-9})]$

When high oxygen transport is desired, a material with a large permeability coefficient should be chosen, if all other aspects of the materials under question are comparable. In the case of contact lenses, poly-HEMA (poly hydroxyethyl methacrylate) lenses are commonly used for soft lenses, even though the permeability is lower than that of silicone rubber. The permeability of poly-HEMA is adequate for the oxygenation of the cornea, and it is chosen for other reasons, such as ease of manufacture. As for membrane materials for oxygenators in heart-lung machines, oxygen transport depends on membrane thickness as well as on permeability. Minimum thickness is dictated by membrane strength, so a strong material with a lower permeability may result in the highest oxygen flux.

PROBLEMS

4-1. The crystalline lens of a normal human eye has radii of curvature 10.2 and 6 mm, is convex, and has a thickness of 2.4 mm. The refractive index varies with position but may be taken as 1.386. Determine the optical power and focal length, recognizing that in vivo the lens is immersed in ocular fluid. We remark that "crystalline" in this context is an anatomical term that means transparent; it does not mean the lens has a regular atomic arrangement. The lens is actually fibrous.

4-2. Barium compounds and iodine compounds are used to enhance contrast in diagnostic radiology. Why?

4-3. Show that a contact lens which fits the cornea exactly will provide the same optical correction whether the lens is in fact on the cornea or whether it is separated by an infinitesimal air gap. *Hint*: Consider two lenses of different indices of refraction and one common curvature. Determine the focal length for two cases: lenses in contact and lenses separated by a layer of air.

4-4.   Consider a hemispherical silicone augmentation mammoplasty 130 mm in diameter. Determine the weight of two such mammoplasties. Compare with a corresponding volume of natural tissue. Discuss the implications.

4-5.   Determine the reflection coefficient for ultrasonic waves crossing a muscle-to-bone interface. Some of the needed material properties may be found elsewhere in the book. Discuss the implications for ultrasonic imaging through bone. Discuss the use of other kinds of waves for imaging through bone.

4-6.   Calculate the voltage generated in bone by mechanical deformation associated with walking. Use the given piezoelectric coefficient for bone, and find any other needed material properties elsewhere in the book.

4-7.   It is often difficult to remove the bone cement from leg bones from which a hip joint prosthesis must be removed for revision arthroplasty. An experimental technique for this purpose is extracorporeal shock wave lithotripsy (ESWL), a method that was originally developed to shatter kidney stones by intense sonic shock pulses without surgery. Calculate the relative sound intensity of a wave after passing through 5 cm of muscle and 2 cm of bone, as a percent of the initial intensity.

4-8.   Discuss the advantages and disadvantages of the extracorporeal shock wave lithotripsy method described in Problem 4-7.

SYMBOLS/DEFINITIONS

Greek Letters

α: Attenuation coefficient for ultrasound or for x-rays.

λ: Wavelength of light, x-rays, or ultrasound.

ν: Frequency of a wave.

ρ: Density.

ρ_e: Electrical resistivity.

∇^2: Laplacian operator, which in one dimension reduces to $\partial^2/\partial x^2$.

Latin Letters

c: Speed of light.
C: Concentration.
dB: Decibel — a logarithmic unit of ratios, often applied to sound pressure levels.
D: Diopter, see below.
E: Young's modulus, the ratio of stress to strain for simple tension or compression.
f: Focal length of a lens.
F: Volumetric flux: volume per unit time.
h: Planck's constant, associated with the quantum nature of light.

K: Permeability coefficient.

m: Mass.

n: Refractive index.

N: Atomic number.

R: Electrical resistance. Also, amplitude reflection coefficient.

S: Elastic compliance. For anisotropic materials the compliance tensor describes the strain-to-stress ratio for different directions. For example, $S_{1111} = 1/E_1$.

v: Ultrasonic wave speed.

V: Voltage.

V_s: Solid volume fraction of porous material.

Z: Acoustic impedance, equal to density times sound velocity.

Words

Attenuation: Absorption of sound or ultrasound waves.

Compton effect: Mechanism for scattering of x-rays by free or weakly bound electrons in matter.

Diopter: Optical power of lens, or degree of ray divergence of a bundle of light rays. The power in diopters is the inverse of the focal length measured in meters.

Piezoelectricity: Electrical polarization of a material in response to mechanical stress.

Permeability: Ability of a material to pass a gaseous or ionic species, under a pressure gradient or a concentration gradient.

Photoelectric effect: Mechanism for the extinction of x-rays by the x-ray photon knocking bound electrons out of atoms.

Pyroelectricity: Electrical polarization of a material in response to temperature change.

Streaming potential: Electric potential developed when charged particles such as ions flow through a tube or porous medium such as bone; potential may be attributed to the imbalance created by washing away the electric double layer .

Maxwell's equations: Set of four field equations that govern electricity, magnetism, radio wave propagation, and the behavior of light.

Refractive index: Ratio of the speed of light in a vacuum to the speed of light in a material. It is a measure of the ability of a material to refract [bend] a beam of light.

Snell's law: Governing equation for the angle of incident and refracted light rays.

Tensor: A mathematical expression with well-defined transformation properties under changes in coordinates. A vector is a tensor of rank one; a scalar is a tensor of rank zero. Stress and strain are examples of tensors of rank two.

BIBLIOGRAPHY

Arya AP. 1966. *Fundamentals of nuclear physics*. Boston: Allyn and Bacon.

Baier RE. 1980. *Guidelines for physicochemical characterization of biomaterials*. Devices and Technology Branch, National Heart, Lung and Blood Institute. NIH Publication No. 80-2186.

Baier N, Lowther G. 1977. *Contact lens correction*. London: Butterworths.

Balter S. 1988. X-ray equipment design. In *Encyclopedia of medical devices and instrumentation*, Ed JG Webster. Boston: Wiley.

Bur AJ. 1976. Measurements of the dynamic piezoelectric properties of bone as a function of temperature and humidity. *J Biomech* **9**:495–507.

Cima LG, Ron ES. 1991. *Tissue-inducing biomaterials*. Materials Research Society Symposium, Boston.

Cooney DO. 1976. *Biomedical engineering principles*. New York: Marcel Dekker.

Feng J, Yuan H, Zhang X. 1997. Promotion of osteogenesis by a piezoelectric biological ceramic. *Biomaterials* **18**(23):1531–1534.

Fukada E, Yasuda I. 1957. On the piezoelectric effect of bone. *J Physiol Soc Jap* **12**:1158.

Geddes LA, Baker LE. 1989. *Principles of applied biomedical instrumentation*. New York: Wiley.

Janshoff A, Galla HJ, Steinem C. 2000. Piezoelectric mass-sensing devices as biosensors: an alternative to optical biosensors? *Angew Chem, Int Ed* **39**(22):4004–4032.

Kohles SS, Bowers JR, Vailas AC, Vanderby Jr R. 1997. Ultrasonic wave velocity measurement in small polymeric and cortical bone specimens. *J Biomech Eng* **119**(3):232–236.

Kronenthal RL, Oser Z, eds. 1975. *Polymers in medicine and surgery*. New York: Plenum.

Maniglia AJ, Proops DW. 2001. *Implantable electronic otologic devices: state of the art*. Philadelphia: W.B. Saunders.

Moss SC. 2002. *Advanced biomaterials: characterization, tissue engineering, and complexity*. Materials Research Society Symposium. Boston: Materials Research Society.

Nye JF. 1976. *Physical properties of crystals*. Oxford: Oxford UP.

O'Neal MR. 1988. Contact lenses. In *Encyclopedia of medical devices and instrumentation*, Ed JG Webster. New York: Wiley.

Schnur JM, Peckerar MC, Stratton HM. 1992. *Synthetic microstructures in biological research*. International Conference on Synthetic Microstructures in Biological Research. New York: Plenum.

Schoeff LE, Williams RH. 1993. *Principles of laboratory instruments*. St. Louis: Mosby.

Spector M, Miller M, Bealso N. 1988. Porous materials. In *Encyclopedia of medical devices and instrumentation*, Ed JG Webster. New York: Wiley.

Wang YH, Grenabo L, Hedelin H, Pettersson S, Wikholm G, Zachrisson BF. 1993. Analysis of stone fragility *in vitro* and *in vivo* with piezoelectric shock waves using the EDAP LT-01. *J Urol* **149**(4):699–702.

Webster JG. 2004. *Bioinstrumentation*. New York: Wiley.

Webster JG. 2005. *Encyclopedia of medical devices and instrumentation*. New York: Wiley-Interscience.

Webster JG, Clark JW. 1998. *Medical instrumentation: application and design*. New York, Wiley.

Wells PNT, ed. 1982. *Scientific basis of medical imaging*. New York: Churchill Livingstone.

5

METALLIC IMPLANT MATERIALS

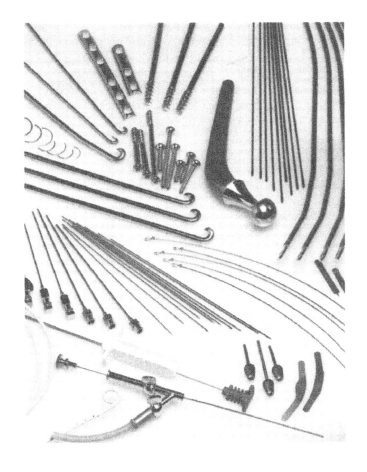

Examples of medical devices made of steels. Reprinted with permission from Hill (1998). Copyright © 1998, Wiley.

Metals have been used in various forms as implants. The first metal developed specifically for human use was "Sherman Vanadium Steel," which was used to manufacture bone fracture plates and screws. Most metals used for manufacturing implants (e.g., Fe, Cr, Co, Ni, Ti, Ta, Mo, and W) can be tolerated by the body in minute amounts. Sometimes those metallic elements, in naturally occurring forms, are essential in cell functions (Fe), synthesis of a vitamin B_{12} (Co), and crosslinking of elastin in the aorta (Cu) but cannot be tolerated in large amounts in the body. The biocompatibility of implant metals is of considerable concern because they can corrode in the hostile body environment. The consequences of corrosion include loss of material, which will weaken the implant, and probably more important, that the corrosion products escape into tissue, resulting in undesirable effects. In this chapter we study the composition–structure–property relationship of metals and alloys used for implant fabrications.

5.1. STAINLESS STEELS

The first stainless steel used for implant materials was 18-8 (type 302 in modern classification), which is stronger than the steel and more resistant to corrosion. Vanadium steel is no longer used in implants since its corrosion resistance is inadequate, as discussed in §5.6. Later 18-8sMo stainless steel was introduced. It contains molybdenum to improve corrosion resistance in salt water. This alloy became known as type 316 stainless steel. In the 1950s the carbon content of 316 stainless steel was reduced from 0.08 w/o to 0.03 w/o maximum for better corrosion resistance to chloride solution, and it became known as 316L.

Table 5-1. Compositions of 316L Stainless Steel
Surgical Implants (ASTM, 2000)

Element	Composition (w/o)
Carbon	0.030 max
Manganese	2.00 max
Phosphorus	0.025 max
Sulfur	0.010 max
Silicon	0.75 max
Chromium	17.00–19.00
Nickel	13.00–15.00
Molybdenum	2.25–3.00
Nitrogen	0.10 max
Copper	0.50 max
Fe	Balance

Stainless steels (F138 and F139 of ASTM).

5.1.1. Types and Composition of Stainless Steels

Chromium is a major component of corrosion-resistant stainless steel. The minimum effective concentration of chromium is 11 w/o. The chromium is a reactive element, but it and its alloys can be passivated to give excellent corrosion resistance.

The *austenitic stainless steels* — especially types 316 and 316L — are most widely used for implants. These are *not* hardenable by heat-treatment but can be hardened by cold-working. This group of stainless steels are nonmagnetic and possess better corrosion resistance than any others. The inclusion of molybdenum enhances resistance to pitting corrosion in salt water.

The ASTM (American Society of Testing and Materials) recommends type 316L rather than 316 for implant fabrication. The specifications for 316 and 316L stainless steels are given in Table 5-1.

The nickel serves to stabilize the austenitic phase at room temperature and, in addition, to enhance corrosion resistance. The austenitic phase stability can be influenced by both the Ni and Cr contents, as shown in Figure 5-1 for 0.10 w/o carbon stainless steels.

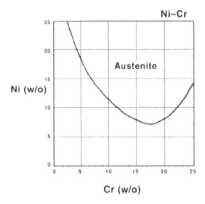

Figure 5-1. Effect of Ni and Cr contents on the austenitic phase of stainless steels containing 0.1 w/o C. Reprinted with permission from Keating (1956). Copyright © 1956, Butterworths.

5.1.2. Properties of Stainless Steel

Table 5-2 gives the mechanical properties of 316 and 316L stainless steels. As can be noted, a wide range of properties exists depending on the heat-treatment (to obtain softer materials) or cold-working (for greater strength and hardness). The designer must consequently be careful when selecting materials of this type. Even type 316L stainless steels may corrode inside the body under certain circumstances, such as in a highly stressed and oxygen-depleted region. They are, however, suitable to use in *temporary* devices such as fracture plates, screws, and hip nails.

5.1.3. Manufacturing of Implants Using Stainless Steel

The austenitic stainless steels work-harden very rapidly, as shown in Figure 5-2, and therefore cannot be cold-worked without intermediate heat-treatments. The heat-treatments, however, should not induce formation of chromium carbide (CCr_4) in grain boundaries, which might deplete Cr and C in the grains, causing corrosion. For the same reason, austenitic stainless steel implants are not usually welded.

Distortion of components by heat-treatments can occur, but this problem can be solved easily by controlling the uniformity of heating. Another undesirable effect of heat-treatment is the formation of surface oxide scales that have to be removed either chemically (acid) or mechanically (sandblasting). After the scales are removed the surface of the component is polished to a mirror or matte finish. The surface is then cleaned, degreased, and passivated in nitric acid (ASTM Standard F86). The component is washed and cleaned again before packaging and sterilizing.

Table 5-2. Mechanical Properties of Stainless Steel Surgical Implants (ASTM, 2000)

Condition	Ultimate tensile strength, min ksi (MPa)	Yield strength (0.2% offset), min, ksi (MPa)	Elongation 2 in. (50.8 mm) min,%	Rockwell hardness max
Bar and wire (F138)				
Annealed	71 (490)	27.5 (190)	40	–
Cold worked	125 (860)	100 (690)	12	–
Extra hard	196 (1350)	–	–	–
Cold drawn	125 (860)	–	–	–
	150 (1035)	–	5	–
Sheet and strip (F139)				
Annealed	71 (490)	27.5 (190)	40	95HRB
Cold worked	125 (860)	100 (690)	10	–

Stainless steels (F138 and F139 of ASTM).
1 ksi = 1,000 psi, 1 psi = 6,895 Pa

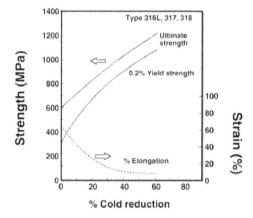

Figure 5-2. Effect of cold-work on the yield and ultimate tensile strength of 18-8 stainless steel. Reprinted with permission from ASM (1978). Copyright © 1978, American Society for Metals.

Example 5-1

Calculate the amount of volume change when iron is oxidized to FeO (ρ = 5.95 g/cm^3). The density of Fe is 7.787 g/cm^3.

Answer

Since the molecular weight of Fe is 55.85 g/mol,

$$\frac{55.87 \text{ g/mol}}{7.7 \text{ g/cm}^3} = 7.1 \text{ cm}^3/\text{mol} .$$

The molecular weight of FeO is 71.85 g/mol; hence,

$$\frac{71.85 \text{ g/mol}}{5.95 \text{ g/cm}^3} = 12.08 \text{ cm}^3/\text{mol} .$$

Therefore, $\Delta V = (12.08 - 7.1)/7.1 = \underline{0.7}$ (70% volume increase by oxidation). This increase in volume by oxidation causes the oxides formed to be very porous. The oxide layer may flake off as a result of the volume mismatch. Continued oxidation can then take place via further diffusion of oxygen to the underlying metal.

5.2. Co-BASED ALLOYS

These materials are usually referred to as cobalt-chromium alloys. There are basically two types: one is the CoCrMo alloy, which is usually used to *cast* a product, and the other is CoNiCrMo alloy, which is usually *wrought* by (hot) *forging*. The castable CoCrMo alloy has been in use for many decades in dentistry and in making artificial joints. The wrought CoNiCrMo alloy has been used for making the stems of prostheses for heavily loaded joints (such as the knee and hip).

5.2.1. Types and Composition of Co-Based Alloys

ASTM lists four types of Co-based alloys that are recommended for surgical implant applications: (1) cast CoCrMo alloy (F75), (2) wrought CoCrWNi alloy (F90), (3) wrought CoNiCrMo alloy (F562) and wrought CoNiCrMoWFe alloy (F563). Their chemical compositions are summarized in Table 5-3 except F563. At the present time only two of the four alloys are used extensively in implant fabrications — the castable CoCrMo and the wrought CoNiCrMo alloy. As can be noted from Table 5-3, these compositions are quite different from each other.

Table 5-3. Chemical Compositions of Co-Based Alloys (ASTM, 2000)

Element	Co28Cr6Mo (F75) Castable		Co20Cr15W10Ni (F90) Wrought		Co28Cr6Mo (F1537) Wrought		Co35Ni20Cr10Mo (F562)	
	Min.	Max.	Min.	Max.	Min.	Max.	Min.	Max.
Cr	27.0	30.00	19.00	21.00	26.0	30.0	19.0	21.0
Mo	5.0	7.00	–	–	5.0	7.0	9.0	10.5
Ni	–	2.5	9.00	11.00	–	1.0	33.0	37.0
Fe	–	0.75	–	3.00	–	0.75	9.0	10.5
C	–	0.35	0.05	0.15	–	0.35	–	0.025
Si	–	1.00	–	1.00	–	1.0	–	0.15
Mn	–	1.00	–	2.00	–	1.0	–	0.15
W	–	0.20	14.00	16.00	–	–		
P	–	0.020	–	0.040	–	–	–	0.015
S	–	0.010	–	0.030	–	–	–	0.010
N	–	0.25	–	–	–	0.25	–	–
Al	–	0.30	–	–	–	–	–	–
Bo	–	0.01	–	–	–	–	–	0.015
Ti							–	1.0
Co				Balance				

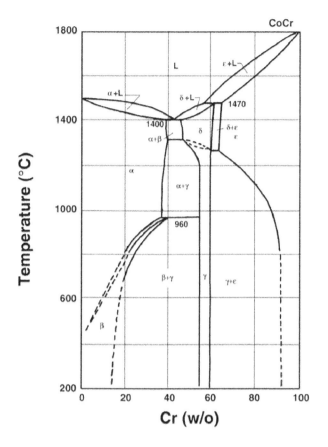

Figure 5-3. Phase diagram of Co–Cr. Reprinted with permission from Smithells (1976). Copyright © 1976, Butterworths.

5.2.2. Properties of Co-Based Alloys

The two basic elements of Co-based alloys form a solid solution of up to 65 w/o Co and the remainder is Cr, as shown in Figure 5-3. Molybdenum is added to produce finer grains, which results in higher strength after casting or forging.

One of the most promising wrought Co-based alloys is the CoNiCrMo alloy originally called MP35N (Standard Pressed Steel Co.), which contains approximately 35 w/o Co and Ni each. The alloy has a high degree of corrosion resistance to seawater (containing chloride ions) under stress. Cold-working can increase the strength of the alloy considerably, as shown in Figure 5-4. However, there is a considerable difficulty of cold-working, especially when making large devices such as hip joint stems. Only hot-forging can be used to fabricate an implant with the alloy.

The abrasive wear properties of the wrought CoNiCrMo alloy is similar to the cast Co-CrMo alloy (about 0.14 mm/year in a joint simulation test). Such a wear rate would be unacceptable in an actual prosthesis. However, the former is not recommended for the bearing surfaces of any joint prosthesis because of its poor frictional properties with itself or other materials. The superior fatigue and ultimate tensile strength of the wrought CoNiCrMo alloy make it very suitable for applications that require a long service life without fracture or stress fatigue. Such is the case for the stems of the hip joint prostheses. This advantage is more appreciated when the implant has to be replaced with another one since it is quite difficult to remove the failed piece of implant embedded deep in the femoral medullary canal. Furthermore, the revision arthroplasty is usually inferior to the original in terms of its function due to poorer fixation of the implant.

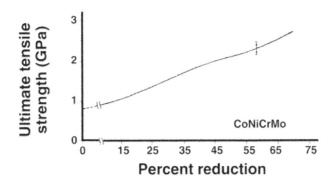

Figure 5-4. Relationship between ultimate tensile strength and the amount of cold-work for CoNiCrMo alloy. Reprinted with permission from Devine and Wulff (1975). Copyright © 1975, Wiley.

Table 5-4 shows the mechanical properties required of Co-based alloys. As is the case with other alloys, the increased strength is accompanied by decreased ductility. Both the cast and wrought alloys have excellent corrosion resistance.

Experimental determination of the rate of nickel release from the CoNiCrMo alloy and 316L stainless steel in 37°C Ringer's solution showed an interesting result. Although the cobalt alloy had more initial release of nickel ions into the solution, the rate of release was about the same (3×10^{-10} g/cm^2/day) for both alloys, as shown in Figure 5-5. This is rather surprising since the nickel content of the CoNiCrMo alloy is about three times that of the 316L stainless steel.

The modulus of elasticity for the cobalt-based alloys ranges from 220 to 234 GPa. These values are higher than the moduli of other materials such as stainless steels. Cold-work and heat-treatment procedures have little effect on the elastic modulus but substantial effects on strength and toughness. Differences in elastic modulus may have some implications for different load transfer modes to the bone, although it is not established clearly what is the effect of the increased modulus. Whenever one intends to reduce the stiffness of the implants one could substitute Ti-alloys, which have a modulus and density about half that of the Co-alloys (see §5.3).

5.2.3. Manufacturing Implants Using Co-Based Alloys

The CoCrMo alloy is particularly susceptible to the work-hardening so that the normal fabrication procedure used with other metals cannot be employed. Instead the alloy is cast by a lost wax (or investment casting) method that involves the following steps (Figure 5-6):

1. A wax pattern of the desired component is made.
2. The pattern is coated with a refractory material, first by a thin coating with a slurry (suspension of silica in ethyl silicate solution) followed by complete investing after drying.
3. The wax is melted out in a furnace (100–150°C).
4. The mold is heated to a high temperature, burning out any traces of wax or gas-forming materials.
5. Molten alloy is poured with gravitational or centrifugal force. The mold temperature is about 800–1000°C and the alloy is at 1350–1400°C.

Table 5-4. Mechanical Property Requirements of Co-Based Alloys (ASTM, 2000)

Condition	Ultimate tensile strength min ksi (MPa)	Yield strength (0.2% offset) min, ksi (MPa)	Fatigue strength[a] ksi (MPa)	Elongation min (%)	Reduction of area min (%)
Co28Cr6Mo (F75)					
As cast	95 (655)	65 (450)	45 (310)	8	8
Co20Cr15W10Ni (F90)					
Annealed	125 (860)	45 (310)	–	30	–
Co28Cr6Mo (F1537)					
Annealed[b]	130 (897)	75 (517)	–	20	20
Hot worked	145 (1000)	101 (700)	–	12	12
Warm worked	170 (1172)	120 (827)	–	12	12
Co35Ni20Cr10Mo (F562)					
Annealed[b]	115 (793)	35(241)	49.3(340)	50.0	65.0
	145 (1000)	65(448)			
Cold worked,aged[c]	260 (1793) min	230 (1586) min	–	8.0	35.0

1 ksi = 1,000 psi, 1 psi = 6,895 Pa
[a] Reprinted with permission from Semlitsch (1980). Copyright 1980 © Institute of Mechanical Engineers.
[b] 1–2 hrs at 1050 ± 15°C air or water quenched to room temperature.
[c] Cold worked 50% and aged 540–640 ± 15°C for 4 hrs, then air cooled.

Controlling the mold temperature will have an effect on the grain size of the final cast; coarse ones are formed at higher temperatures, which will decrease the strength. However, high processing temperatures will result in larger carbide precipitates with greater distances between them, resulting in a less brittle material. Again there is a complementary (tradeoff) relationship between strength and toughness.

Example 5-2

Calculate the number of Co atoms released during a year from the femoral head of a hip joint prosthesis made of CoCrMo alloy. Assume that the wear rate is 0.14 mm/yr and that all of the atoms become ionized.

Answer

Assume a nominal diameter of the prosthetic femoral head of 28 mm. The surface area is

$$A = 4\pi(1.4 \text{ cm})^2 = 24.63 \text{ cm}^2.$$

Half of this area is in contact with the socket portion of the joint. Therefore, the volume of wear material is $1/2 \times 24.63$ cm$^2 \times 0.014$ cm/yr = 0.172 cm^3/yr:

$$\frac{\text{atoms}}{\text{yr}} = \frac{0.65 \times 0.172 \text{ cm}^3/\text{yr} \times 8.83 \text{ g/cm}^3 \times 6.02 \times 10^{23} \text{ atoms/mol}}{58.93 \text{ g/mol}}.$$

Since the density of Co is 8.83 g/cm^3, and the atomic weight is 58.93, and the alloy is about 65% cobalt,

$$\text{atoms/yr} = \underline{1.0 \times 10^{22} \text{ atoms/yr}}, \quad \text{or} \quad \underline{3.2 \times 10^{14} \text{ atoms per second}}.$$

Wear usually gives rise to particulate debris rather than full dissolution of the metal. The above number should therefore be regarded as an upper bound on the number of dissolved atoms. The clinical importance is that people may develop an allergic reaction to the dissolved metal. It is more likely that the wear particles will be corroded or oxidized heavily since the relative amount of surface compared to the metal volume increases tremendously by fragmentation into particles.

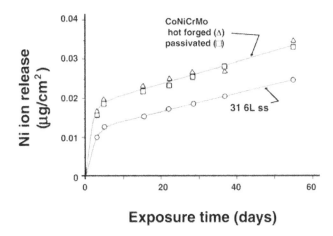

Exposure time (days)

Figure 5-5. Nickel ion release versus time for hot-forged and passivated CoNiCrMo and 316L stainless steel in 37°C Ringer's solution. Reprinted with permission from *Biophase Implant Material*, Technical Information Publ. No. 3846, Richards Manufacturing Company, Memphis, TN. Copyright © 1980, Richards Manufacturing Company.

5.3. Ti AND Ti-BASED ALLOYS

Attempts to use titanium for implant fabrication date to the late 1930s. It was found that titanium was tolerated in cat femurs, as was stainless steel and Vitallium® (CoCrMo alloy). The lightness of titanium (4.5 g/cm^3 compared to 7.9 g/cm^3 for 316 stainless steel, 8.3 g/cm^3 for cast CoCrMo, and 9.2 g/cm^3 for wrought CoNiCrMo alloys) and good mechanochemical properties are salient features for implant application.

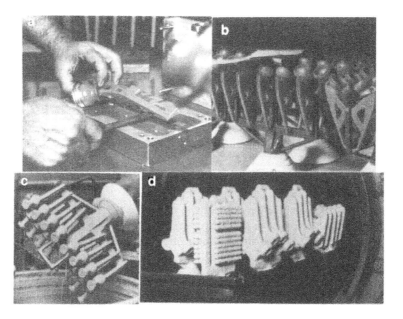

Figure 5-6. Lost wax casting of femoral joint prosthesis. (**a**) Injection of wax into a brass mold. (**b**) Wax patterns assembled for a ceramic coating (note the hollow part of the femoral head). (**c**) Application of ceramic coating. (**d**) A hot pressure chamber retrieves the wax, leaving behind a ceramic coating. (**e**) Pouring molten metals into the preheated ceramic mold. Reprinted courtesy of Howmedica Inc., Rutherford, NJ.

5.3.1. Compositions of Ti and Ti-Based Alloys

There are four grades of unalloyed titanium for implant applications as given in Table 5-5. The impurity contents distinguish them; oxygen, iron and nitrogen should be controlled carefully. Oxygen in particular has a great influence on ductility and strength. The chemical compositions of some titanium alloys are given in Table 5-6.

One titanium alloy (Ti6Al4V) is widely used to manufacture implants, and its chemical requirements are given in Table 5-6. The main alloying elements of the alloy are aluminum (5.5–6.5 w/o) and vanadium (3.5–4.5 w/o). There are some other wrought Ti alloys being used for implant fabrication (see Table 5-7 for their chemical compositions).

Table 5-5. Chemical Compositions of Pure Titanium (F67; ASTM, 2000)

Element	Grade 1	Grade 2	Grade 3	Grade 4
N	0.03	0.03	0.05	0.05
C	0.10	0.10	0.10	0.10
H	0.015	0.015	0.015	0.015
Fe	0.20	0.30	0.30	0.50
O	0.18	0.25	0.35	0.40
Ti		Balance		

All are in maximum % allowed.

Table 5-6. Chemical Compositions of TI6Al4V Alloys (ASTM, 2000)

Element	Wrought, forging (F136, F620)	Casting (F1108)	Coating (F1580)
N	0.05	0.05	0.05
C	0.08	0.10	0.08
H	0.012	0.015	0.015
Fe	0.25	0.30	0.30
O	0.13	0.20	0.20
Cu	–	–	0.10
Sn	–	–	0.10
Al	5.5–6.50	5.5–6.75	5.50–6.75
V	3.5–4.5	3.5–4.5	3.50–4.50
Ti		Balance	

Another wrought Ti6Al4V alloy (F1472) is very similar to F136 alloy. All are in maximum % allowed.

Table 5-7. Chemical Compositions of Wrought Ti Alloys (ASTM, 2000)

Element	Wrought Ti8Al7Nb (F1295)	Wrought Ti13Nb13Zr (F1713)	Wrought Ti121Mo6Zr2Fe (F1813)
N	0.05	0.05	0.05
C	0.08	0.08	0.05
H	0.009	0.012	0.020
Fe	0.25	0.25	1.5–2.5
O	0.20	0.15	0.08–0.28
Ta	0.50	–	–
Al	5.5–6.50	–	5.50–6.75
Zr	–	12.5–14.0	5.0–7.0
Nb	6.50–7.50	12.5–14.0	–
Mo	–	–	10.0–13.0
Ti		Balance	

All are in maximum % allowed.

5.3.2. Structure and Properties of Ti and Ti Alloys

Titanium is an allotropic material that exists as a hexagonal close-packed structure (α-Ti) up to 882°C and a body-centered cubic structure (β-Ti) above that temperature. The addition of alloying elements to titanium enables it to have a wide range of properties:

1. Aluminum tends to stabilize the α phase, that is, increase the transformation temperature from α to β phase (Figure 5-7).
2. Vanadium stabilizes the β phase by lowering the temperature of the transformation from α to β.

The α alloys have single-phase microstructure (Figure 5-8a), which promotes good weldability. The stabilizing effect of the high aluminum content of these groups of alloys makes for excellent strength characteristics and oxidation resistance at high temperature (300–600°C). These alloys cannot be heat-treated for strengthening since they are single phased since the precipitation of the second or third phase increases the strength by precipitation hardening process.

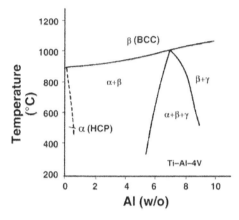

Figure 5-7. Part of phase diagram of Ti-Al-V at 4 w/o V. Reprinted with permission from Smith and Hughes (1966). Copyright © 1966, Institute of Mechanical Engineers.

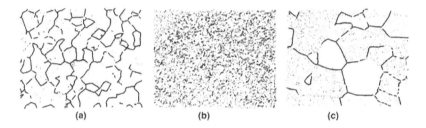

Figure 5-8. Microstructure of Ti alloys (all are 500×): (**a**) annealed α-alloy; (**b**) Ti6Al4V, α–β alloy, annealed; (c) β-alloy, annealed. Reprinted with permission from Hille (1966). Copyright © 1966, The Minerals, Metals, and Materials Society.

The addition of controlled amounts of β-stabilizers causes the higher strength β phase to persist below the transformation temperature, which results in the two-phase system. As discussed in §3.4, the precipitates of β phase will appear by heat-treatment in the solid solution temperature and subsequent quenching, followed by aging at a somewhat lower temperature. The aging cycle causes the precipitation of some fine β particles from the metastable β, imparting a structure that is stronger than the annealed α–β structure (Figure 5-8b).

The higher percentage of β-stabilizing elements (13 w/o V in Ti13V11Cr3Al alloy) results in a microstructure that is substantially β, which can be strengthened by heat-treatment (Figure 5-8c).

The mechanical properties of the commercially pure titanium, Ti6Al4V, and other alloys are given in Tables 5-8, 5-9, and 5-10 respectively. The modulus of elasticity of these materials is about 110 GPa, which is half the value of Co-based alloys. From Table 5-8 one can see that the higher impurity content leads to higher strength and reduced ductility. The strength of the Ti alloys is similar to 316 stainless steel or the Co-based alloys, as given in Tables 5-9 and 5-10. When compared by specific strength (strength per density), the titanium alloy excels any other implant materials, as shown in Figure 5-9. Titanium, nevertheless, has poor shear strength, making it less desirable for bone screws, plates, and similar applications. It also tends to gall or seize when in sliding contact with itself or another metal.

Table 5-8. Mechanical Properties of Pure Titanium (F67, 1992) (ASTM, 2000)

Properties	Grade 1	Grade 2	Grade 3	Grade 4
Tensile strength ksi (MPa)	35 (240)	50 (345)	65 (450)	80 (550)
Yield strength (0.2% offset) ksi (MPa)	25 (170)	40 (275)	55 (380)	70 (485)
Elongation (%)	24	20	18	15
Reduction of area(%)	30	30	25	25

1 ksi = 1,000 psi, 1 psi = 6,895 Pa

Table 5-9. Mechanical Properties of Ti6Al4V (ASTM, 2000)

Properties	Wrought (F136)	Casting (F1108)
Tensile strength ksi (MPa)	125 (860)	125 (860)
Yield strength (0.2% offset) ksi (MPa)	115 (795)	110 (758)
Elongation (%)	10 min	8 min
Reduction of area (%)	20 min	14 min

1 ksi = 1,000 psi, 1 psi = 6,895 Pa. Forging Ti6Al4V alloy (F620) does not have any mechanical properties requirement.

Table 5-10. Mechanical Properties of Wrought Ti Alloys (ASTM, 2000)

Alloys		Tensile strength ksi (MPa)	Yield strength (0.2% offset) ksi (MPa)	Elongation (%)	Reduction of area (%)
Wrought Ti8Al7Nb (F1295)		130.5 (900)	116 (800)	10	25
Wrought Ti13Nb13Zr (F1713)	Aged	125 (860)	105 (725)	8	15
	Annealed	80 (550)	50 (345)	15	30
	Unannealed	80 (550)	50 (345)	8	15
Wrought Ti121Mo6Zr2Fe (F1813)		135 (931.5)	130 (897)	12	30

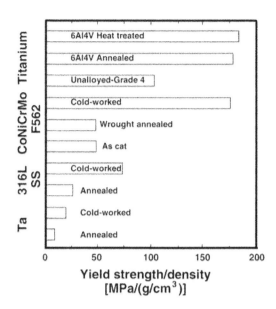

Figure 5-9. Yield strength-to-density ratio of some implant materials. Reprinted with permission from Hille (1966). Copyright © 1966, The Minerals, Metals, and Materials Society.

Titanium derives its resistance to corrosion by the formation of a solid oxide layer. Under in vivo conditions the oxide (TiO_2) is the only stable reaction product. The oxide layer forms a thin adherent film and passivates the material. Corrosion resistance mechanisms are discussed further in §5.6.

5.3.3. Manufacture of Implants

Titanium is very reactive at high temperature and burns readily in the presence of oxygen. It therefore requires an inert atmosphere for high-temperature processing or is processed by vacuum melting. Oxygen diffuses readily in titanium, and the dissolved oxygen embrittles the metal. As a result any hot-working or forging operation should be carried out below 925°C. Machining at room temperature is not the solution to all the problems since the material also tends to gall or seize the cutting tools. Very sharp tools with slow speeds and large feeds are used to minimize this effect. Electrochemical machining is an attractive means.

Example 5-3

The allotropic phase changes in titanium at 882°C is from hcp structure (a = 2.95 Å, C = 4.683 Å) to bcc structure (a = 3.32 Å). Calculate the volume change in cm³/g by heating above the transformation temperature. The density is 4.54 g/cm³ and the atomic weight 47.9 g.

Answer

$$\text{Area of hexagon} = a \times a \times \sqrt{\frac{3}{2}} \times \frac{1}{2} \times 6 = \frac{3\sqrt{3}}{2} \times a^2 \text{ Å}^2,$$

$$V_{hcp} = \frac{3\sqrt{3}}{2} a^2 c = 106.3 \text{ Å}^3, \text{ which contains 6 atoms,}$$

$$V_{bcc} = a^3 36.59 \text{ Å}^3, \text{ which contains 2 atoms,}$$

$$\text{hcp: 0.2227 cm}^3/\text{g}, \quad \text{bcc: 0.23 cm}^3/\text{g}.$$

The difference is $\underline{0.0073 \text{ cm}^3/\text{g}}$.

5.4. DENTAL METALS

5.4.1. Dental Amalgam

An *amalgam* is an alloy in which one of the component metals is mercury. The rationale for using amalgam as a tooth filling material is that since mercury is a liquid at room temperature it can react with other metals such as silver and tin and form a plastic mass that can be packed into the cavity, and which hardens (sets) with time. To fill a cavity the dentist mixes solid alloy, supplied in particulate form, with mercury in a mechanical triturator. The resulting material is readily deformable and is then packed into the prepared cavity. The solid alloy is composed of at least 65% silver, and not more than 29% tin, 6% copper, 2% zinc, and 3% mercury. The reaction during setting is thought to be

$$\gamma + \text{Hg} \Leftrightarrow \gamma + \gamma_1 + \gamma_2, \tag{5-1}$$

in which the γ phase is Ag_3Sn, the γ_1 phase is Ag_2Hg_3, and the γ_2 phase is Sn_7Hg, as shown in Figure 5-10.

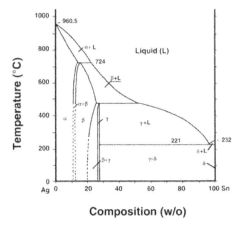

Figure 5-10. Phase diagram of Ag and Sn.

The phase diagram for the Ag–Sn–Hg system shows that over a wide compositional range all three phases are present. Dental amalgams typically contain 45 to 55% mercury, 35 to 45% silver, and about 15% tin, when fully set. The strength of the restoration increases during the setting process, so that the amalgam has attained one quarter of its final strength after one hour, and almost all of its final strength after one day.

5.4.2. Gold

Gold and gold alloys are useful metals in dentistry as a result of their durability, stability, and corrosion resistance. Gold fillings are introduced by two methods: casting and malleting. *Cast* restorations are made by taking a wax impression of the prepared cavity, making a mold from this impression in a material such as gypsum silica, which tolerates high temperature, and casting molten gold in the mold. The patient is given a temporary filling for the intervening time. Gold *alloys* are used for cast restorations, since they have mechanical properties superior to those of pure gold. Corrosion resistance is retained in these alloys provided they contain 75 w/o or more of gold and other noble metals. Copper, alloyed with gold, significantly increases its strength. Platinum also improves strength, but no more than about 4% can be added, or the melting point of the alloy is elevated excessively. Silver compensates for the color of copper. A small amount of zinc may be added to lower the melting point and to scavenge oxides formed during melting. Gold alloys of different composition are available. Softer alloys containing more than 83% gold are used for inlays, which are not subjected to much stress. Harder alloys containing less gold are chosen for crowns and cusps, which are more heavily stressed.

Malleted restorations are built up in the cavity from layers of *pure* gold foil. The foils are degassed before use, and the layers are welded together by pressure at room temperature. In this type of welding the metal layers are joined by thermal diffusion of atoms from one layer to another. Since intimate contact is required in this procedure, it is particularly important to avoid contamination. The pure gold is relatively soft, so this type of restoration is limited to areas not subjected to much stress.

5.4.3. Nickel-Titanium Alloys

The nickel-titanium alloys show an unusual property in that after the metal is deformed they can snap back to their previous shape following heating. This phenomenon is called the *shape memory effect*. The shape memory effect (SME) of Ni-Ti alloy was first observed by Buehler and Wiley at the U.S. Naval Ordnance Laboratory (NOL). The equiatomic Ni-Ti alloy (Nitinol®) exhibits an exceptional SME near room temperature: if it is plastically deformed below the transformation temperature, it reverts back to its original shape as the temperature is raised. The SME can be generally related to a diffusionless martensitic phase transformation that is also thermoelastic in nature, the thermoelasticity being attributed to ordering in the parent and martensitic phases. The thermoelastic martensitic transformation exhibits the following general characteristics:

1. Martensite formation can be initiated by cooling the material below M_s, defined as the temperature at which the martensitic transformation begins. Martensite formation can also be initiated by applying a mechanical stress at a temperature above M_s.
2. M_s and A_s (temperature at which the reverse austenitic transformation begins upon heating) temperatures can be increased by applying stresses below the yield point; the increase is proportional to the applied stress.
3. The material is more resilient than most metals.
4. The transformation is reversible.

Shape memory alloys are used in orthodontic dental arch wires. They also are used in arterial blood vessel stents, and may be used in vena cava filters, intracranial aneurysm clips, and orthopedic implants. On a more speculative level, they might find use in contractile artificial muscles for an artificial heart.

In order to develop such devices it is necessary to understand fully the mechanical and thermal behavior associated with the martensitic phase transformation. A widely known Ni-Ti alloy is 55-Nitinol (55 weight % (w/o) or 50 atomic % (a/o) Ni), which has a single phase and "mechanical memory" plus other properties — for example, high acoustic damping, direct conversion of heat energy into mechanical energy, good fatigue properties, and low temperature ductility. Deviation from the 55-Nitinol (near stoichiometric Ni-Ti) in the Ni-rich direction yields a second group of alloys that are also completely nonmagnetic but differ from 55-Nitinol in their capability of being thermally hardened to higher hardness levels. Shape recovery capability decreases and heat-treatability increases rapidly as Ni content approaches 60 w/o. Both 55- and 60-Nitinols have relatively low moduli of elasticity and can be tougher and more resilient than stainless steel, Ni-Cr, or Co-Cr based alloys.

The efficiency of 55-Nitinol shape recovery can be controlled by changing the final annealing temperatures during preparation of the alloy device. For the most efficient recovery, the shape is fixed by constraining the specimen in a desired configuration and heating to between 482 and 510°C. If the annealed wire is deformed at a temperature below the shape recovery temperature, shape recovery will occur upon heating, provided the deformation has not exceeded crystallographic strain limits (~8% strain in tension). The Ni-Ti alloys also exhibit good biocompatibility and corrosion resistance in vivo.

The mechanical properties of Ni-Ti alloys are especially sensitive to the stoichiometry of composition (typical composition given in Table 5-11) and to the individual thermal and mechanical history. Although much is known about the processing, mechanical behavior, and properties relating to the shape memory effect, considerably less is known about the thermomechanical and physical metallurgy of the alloy.

Table 5-11. Chemical Composition of
Ni–Ti Alloy Wire

Element	Composition (w/o)
Ni	54.01
Co	0.64
Cr	0.76
Mn	0.64
Fe	0.66
Ti	balance

The differential scanning calorimeter (DSC) is capable of measuring the heat capacity of materials as a function of temperature. Figure 5-11 shows a typical DSC plot for Ni-Ti alloy and identifies several relevant parameters. The Ni-Ti alloys generally display two peaks that are associated with the martensitic phase transformation temperature in cooling (M_s and M_f) and heating (austenitic phase transformation temperature, A_s and A_f), where subscripts s and f indicate the starting and finishing of transformations respectively. The area under the specific heat-versus-temperature curve can be used to calculate the amount of thermal energy used for the phase transformation.

Typical curves for the bending moment versus bend angle are shown in Figure 5-12. Ni-Ti alloy wire specimens were tested at 0°C and room temperature. As can be seen from the curves, the samples deformed at room temperature recovered almost completely to their original states, indicating that the transformation temperature is close to room temperature. From these curves the elastic stiffnesses of the alloy were calculated using the straight portions of the curves (given in Table 5-12). The results also indicate that the elastic modulus is higher at higher temperature.

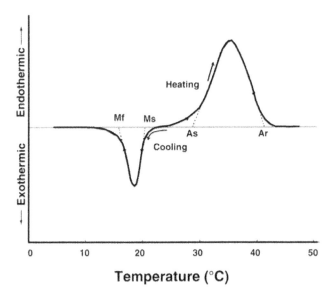

Figure 5-11. Typical DSC curve of specific heat (thermal energy) versus temperature. Reprinted with permission from Lee et al. (1988). Copyright © 1988, Wiley.

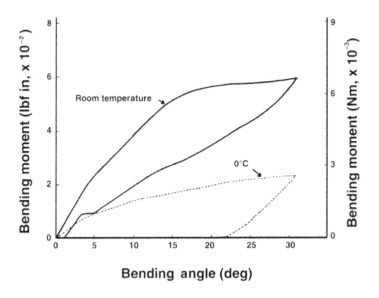

Figure 5-12. Bending moment versus bending angle curves for an NiTi specimen at 0°C and room temperature. Reprinted with permission from Lee et al. (1988). Copyright © 1988, Wiley.

Table 5-12. Elastic Stiffness of Ni–Ti Alloy Wire

Test temperature (°C)	Elastic stiffness $(E_s)^a$	
	lbf in/degree, $\times 10^{-3}$	N m/degree, $\times 10^{-4}$
0	1.4	1.58
Room temp.	4.3	4.86

$^a E_s = \Delta I_b / \Delta\omega$ [bending moment/angle of bending],

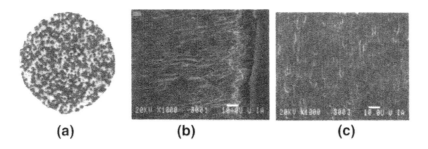

(a)   (b)   (c)

Figure 5-13. Microstructure of an NiTi specimen at room temperature. (**a**) Optical micrograph of the cross-section (100×) showing nonmetallic inclusions (black particulates) and NiTi matrix (white background). (**b**) Scanning electron micrograph of unbent specimen at a longitudinal section (1000×) showing elongated pores in the wire drawing direction. (**c**) Scanning electron micrograph of bent specimen at a longitudinal section (1000×). Note the martensites formed at about 45° to the wire drawing direction (top to bottom in this photo). Reprinted with permission from Lee et al. (1988). Copyright © 1988, Wiley.

Figure 5-13 shows the typical microstructure of an Ni-Ti specimen at room temperature. The optical microscopic picture of the cross-section of the wire shown in Figure 5-13a illustrates the evenly dispersed nonmetallic inclusions in the Ni-Ti matrix. The inclusions are presumed to be primarily titanium carbonitrides, with a few nickel-titanium oxides. Figures 5-13b,c show SEM photomicrographs of the unbent and bent portions of specimens, respectively, at room temperature. From these photomicrographs one can see that long pores are aligned with the longitudinal direction of the specimen. The long pores appear to be created by the polishing and chemical etching steps used to prepare the specimen. However, Figure 5-13c shows transformed martensitic structure near the surface in contrast to the undeformed structure shown in Figure 5-13b, which does not show martensitic structure at room temperature.

5.5. OTHER METALS

Several other metals have been used for a variety of specialized implant applications. Tantalum has been subjected to animal implant studies and has been shown very biocompatible due to the thin oxide layer formed that prevents further oxygen penetration as is the case with titanium. Due to its relatively poor mechanical properties (Table 5-13) and its high density (16.6 g/cm^3), it is restricted to a fe   re sutures used by plastic surgeons and neurosurgeons. Radioactive tantalum (Ta182) has been used to treat head and neck tumors. Arterial stents made from a single woven tantalum filament help open occluded vessels. Also, porous tantalum (72–81% porosity) has been tested as a bone graft substitute for the femoral head, as shown in Figure 5-14.

Table 5-13. Mechanical Properties of Tantalum (F560) (ASTM, 2000)

Properties	Fully annealed	Cold worked
Tensile strength ksi (MPa)	30 (207)	75 (517)
Yield strength (0.2% offset) ksi (MPa)	20(138)	50 (345)
Elongation (%)	20	2
Young's modulus ksi (GPa)	–	27550 (190)

1 ksi = 1,000 psi, 1 psi = 6,895 Pa. Flat mill products.

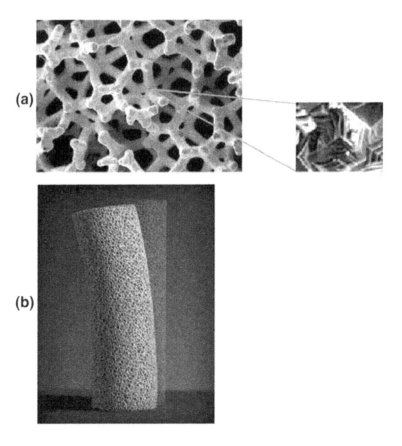

Figure 5-14. SEM picture of porous Ta to be used for hard tissue ingrowth for implant fixation (**a**) and the porous rod being bent, showing its flexibility (**b**). Reprinted courtesy of Zimmer Inc. (http://www.zimmer.com/ctl?template=CP&op=global&action=1&id=33).

NiCu and CoPd alloys are of interest in cancer treatment since their magnetic properties enable them to be heated by an oscillating magnetic field. In this form of cancer treatment, referred to as *hyperthermia*, a "seed" (1 cm long, 1 mm diameter) is implanted in the tumor, then subjected to induction heating to kill tumor cells without harming adjacent tissue. This method is used for prostate cancer. Induction heating is by a Helmholtz coil at radio frequency (RF). These alloys, however, have a *Curie temperature* above which ferromagnetic behavior becomes paramagnetic and induction heating ceases. Therefore, one could design self-regulating thermal seeds that can be heated whenever necessary from outside of the body. The maximum temperature is the same as the Curie temperature. The Curie temperatures of Ni and Co can be manipulated mainly by varying the amount of alloying element and to a lesser degree by annealing the drawn wires. The Pd–6.15%Co and Ni28Cu give the Curie temperature of about 60°C, as shown in Figure 5-15.

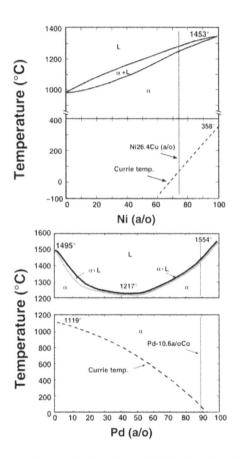

Figure 5-15. Binary phase diagram for Ni–Cu and Pd–Co. Reprinted with permission from Paulus (1994). Copyright © 1994, University of Iowa Press.

Platinum and other noble metals in the platinum group (Pd, Rh, and Ir) are extremely corrosion resistant but have poor mechanical properties. They are mainly used as alloys for electrodes such as pacemaker tips because of their high resistance to corrosion and low threshold potential.

5.6. CORROSION OF METALLIC IMPLANTS

Corrosion is the unwanted chemical reaction of a metal with its environment, resulting in continued degradation to oxides, hydroxides, or other compounds. Tissue fluid in the human body contains water, dissolved oxygen, proteins, and various ions such as chloride and hydroxide. As a result, the human body presents a very aggressive environment to metals used for implantation. Corrosion resistance of a metallic implant material is consequently an important aspect of its biocompatibility.

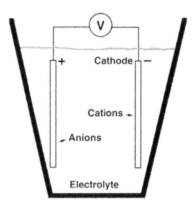

Figure 5-16. Electrochemical cell.

5.6.1. Electrochemical Aspects

The lowest free energy state of many metals in an oxygenated and hydrated environment is that of the *oxide*. Corrosion occurs when metal atoms become ionized and go into solution, or combine with oxygen or other species in solution to form a compound that flakes off or dissolves. The body environment is very aggressive in terms of corrosion since it is not only aqueous but also contains chloride ions and proteins. A variety of chemical reactions occur when a metal is exposed to an aqueous environment, as shown in Figure 5-16. The electrolyte, which contains ions in solution, serves to complete the electric circuit. In the human body, the required ions are plentiful in body fluids. Anions are negative ions that migrate toward the anode, and cations are positive ions that migrate toward the cathode. The electrical component (V) in Figure 5-16 can be a voltmeter with which to measure the potential produced; or it can be a battery, in which case the cell is an electroplating cell; or it can be tissue resistance, in which case the electrochemical cell is an unwanted corrosion cell for a biomaterial in the body. In the body an external electrical driving source may be present in the form of a cardiac pacemaker, or an electrode used to stimulate bone growth. At the anode, or positive electrode, the metal oxidizes. The following reactions involving a metal M may occur:

$$M \Leftrightarrow M^{+n} + ne^-. \tag{5-2}$$

At the cathode, or negative electrode, the following reduction reactions are important:

$$M^{+n} + ne^- \Leftrightarrow M, \tag{5-3}$$

$$M^{++} + OH^- + e^- \Leftrightarrow MOH, \tag{5-4}$$

$$2H_3O^+ + 2e^- \Leftrightarrow H_2O + 2H_2O, \tag{5-5}$$

$$\tfrac{1}{2}O_2 + H_2O + 2e^- \Leftrightarrow 2OH^-. \tag{5-6}$$

Consider as an example the corrosion of a metal such as iron. Metallic iron goes into solution in ionized form as follows:

$$Fe + 2e^- + 2H_2O \Leftrightarrow Fe^{++} + H_2O + 2OH^-. \tag{5-7}$$

In the presence of oxygen, rust may be formed in the following reactions:

$$4Fe^{++} + O_2 + 2H_2O \Leftrightarrow 4Fe^{+++} + 4OH^-, \tag{5-8}$$

and

$$4Fe^{+++} + 12OH^- \Leftrightarrow 4Fe(OH)_3 \downarrow. \tag{5-9}$$

If less oxygen is available, Fe_3O_4 (magnetite) may precipitate rather than ferric hydroxide.

The tendency of metals to corrode is expressed most simply in the standard electrochemical series of *Nernst potentials*, shown in Table 5-14. These potentials are obtained in electrochemical measurements in which one electrode is a standard hydrogen electrode formed by bubbling hydrogen through a layer of finely divided platinum black. The potential of this reference electrode is defined to be zero. Noble metals are those that have a potential higher than that of a standard hydrogen electrode; base metals have lower potentials.

Table 5-14. Standard Electrochemical Series

Reaction	ΔE^0 (volts)
Li $\Leftrightarrow$ Li$^+$	−3.045
Na $\Leftrightarrow$ Na$^+$	−2.714
Al $\Leftrightarrow$ Al^{+++}	−1.66
Ti $\Leftrightarrow$ Ti^{+++}	−1.63
Fe $\Leftrightarrow$ Fe^{++}	−0.44
H$_2$ $\Leftrightarrow$ 2H$^+$	0.000
Ag $\Leftrightarrow$ Ag$^+$	+0.799
Au $\Leftrightarrow$ Au$^+$	+1.68

If two dissimilar metals are present in the same environment, the one that is most negative in the *galvanic series* will become the anode, and bimetallic (or galvanic) corrosion will occur. Galvanic corrosion can be much more rapid than the corrosion of a single metal. Consequently, implantation of dissimilar metals (mixed metals) is to be avoided. Galvanic action can also result in corrosion within a single metal, if there is inhomogeneity in the metal or in its environment, as shown in Figure 5-17.

The potential difference E actually observed depends on the concentration of metal ions in solution according to the Nernst equation:

$$E = E_0 - (RT / nF)\ln(M^{n+}), \tag{5-10}$$

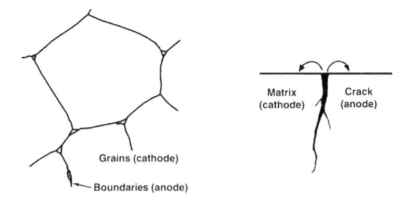

Figure 5-17. Micro-corrosion cells. **Left**: grain boundaries are anodic with respect to the grain interior. **Right**: crevice corrosion due to oxygen-deficient zone in the metal's environment.

in which R is the gas constant, E_0 is the standard electrochemical potential, T is the absolute temperature, F is Faraday's constant (96,487 C/mole), and n is the number of moles of ions.

Gold, platinum, and silver are noble metals. They resist corrosion and other chemical reactions. The order of nobility, which refers to the rank ordering of the electrochemical series, observed in actual practice may differ from that predicted thermodynamically. The reasons are that some metals become covered with a *passivating* film of reaction products that protects the metal from further attack. The dissolution reaction may be strongly irreversible, so that a potential barrier must be overcome. In this case corrosion may be inhibited even though it remains energetically favorable. Finally, the corrosion reactions may proceed slowly: the kinetics are not determined by the thermodynamics.

5.6.2. Pourbaix Diagrams in Corrosion

A Pourbaix diagram is a plot of regions of corrosion, passivity, and immunity as they depend on electrode potential and pH. Pourbaix diagrams are derived from the Nernst equation and from the solubility of the degradation products and the equilibrium constants of the reaction. For the sake of definition, the *corrosion region* is set arbitrarily at a concentration of greater than 10^{-6} gram atom per liter (molar) or more of metal in the solution at equilibrium. This corresponds to about 0.06 mg/liter for metals such as iron and copper, and 0.03 mg/liter for aluminum. *Immunity* is defined as equilibrium between metal and its ions at less than 10^{-6} molar. In the region of immunity, corrosion is energetically impossible. Immunity is also referred to as cathodic protection. In the passivation domain, the stable solid constituent is an oxide, hydroxide, a hydride, or a salt of the metal. *Passivity* is defined as equilibrium between a metal and its reaction products (oxides, hydroxides, etc.) at a concentration of 10^{-6} molar or less. This situation is useful if reaction products are adherent. In the biomaterials setting, passivity may or may not be adequate: disruption of a passive layer may cause an increase in corrosion. The equilibrium state may not occur if reaction products are removed by tissue fluid. Materials differ in their propensity to reestablish a passive layer that has been damaged. This layer of material may protect the underlying metal if it is firmly adherent and nonporous; in that case, further corrosion is prevented. Passivation can also result from a concentration polarization due to a buildup of ions near the electrodes. This is not likely to occur in the body since the ions are continually replenished. Cathodic depolarization reactions can aid in the passivation of a metal by virtue of an energy barrier that hinders the kinetics. Equations (5-5) and (5-6) are examples.

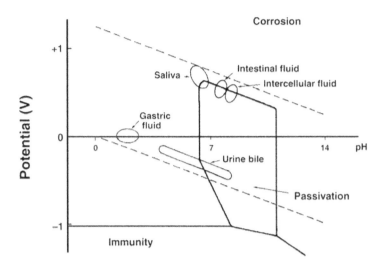

Figure 5-18. Pourbaix diagram for chromium, showing regions associated with various body fluids. Reprinted with permission from Dumbleton and Black (1975). Copyright © 1975, Charles C. Thomas Publisher.

In the diagrams shown in Figures 5-18 to 5-21, there are two diagonal lines. The top ("oxygen") line represents the upper limit of the stability of water and is associated with oxygen-rich solutions or electrolytes near oxidizing materials. In the region above this line, oxygen is evolved according to $2H_2O \rightarrow O_2 + 4H^+ + 4e^-$. In the human body, saliva, intracellular fluid, and interstitial fluid occupy regions near the oxygen line, since they are saturated with oxygen. The lower ("hydrogen") diagonal line represents the lower limit of the stability of water. Hydrogen gas is evolved according to Eq. (5-5). Aqueous corrosion occurs in the region between these diagonal lines on the Pourbaix diagram. In the human body, urine, bile, the lower gastrointestinal tract, and secretions of ductless glands occupy a region somewhat above the hydrogen line.

The significance of Pourbaix diagrams is as follows. Different parts of the body have different pH values and oxygen concentrations. Consequently, a metal that performs well (is immune or passive) in one part of the body may suffer an unacceptable amount of corrosion in another part. Moreover, pH can change dramatically in tissue that has been injured or infected. In particular, normal tissue fluid has a pH of about 7.4, but in a wound it can be as low as 3.5, and in an infected wound the pH can increase to 9.0.

Pourbaix diagrams are useful but do not tell the whole story; there are some limitations. Diagrams are made considering equilibrium among metal, water, and reaction products. The presence of other ions (e.g., chloride) may result in very different behavior, and large molecules in the body may also change the situation. Prediction of "passivity" may in some cases be optimistic, since reaction rates are not considered.

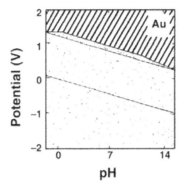

Figure 5-19. Pourbaix diagram for an immune metal: gold. Reprinted with permission from Pourbaix (1974). Copyright © 1974, Pergamon Press.

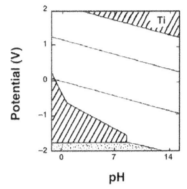

Figure 5-20. Pourbaix diagram for a passive metal: titanium. Reprinted with permission from Pourbaix (1974). Copyright © 1974, Pergamon Press.

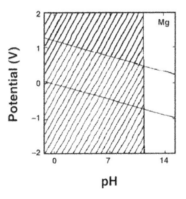

Figure 5-21. Pourbaix diagram for magnesium. Reprinted with permission from Pourbaix (1974). Copyright © 1974, Pergamon Press.

5.6.3. Rates of Corrosion and Polarization Curves

The regions in the Pourbaix diagram specify whether corrosion will take place, but they do not determine the rate. The rate, expressed as an electric current density (current per unit area) depends upon electrode potential, as shown in the polarization curves of Figure 5-22. From such curves it is possible to calculate the number of ions per unit time liberated into the tissue, as well as the depth of metal removed by corrosion in a given time. An alternative experiment is one in which the weight loss of a specimen of metal due to corrosion is measured as a function of time.

The rate of corrosion also depends on the presence of synergistic factors, such as those of mechanical origin. For example, in corrosion fatigue, repetitive deformation of a metal in a corrosive environment results in acceleration of both the corrosion and the fatigue microdamage. Since the body environment involves both repeated mechanical loading and a chemically aggressive environment, fatigue testing of implant materials should always be performed under physiological environmental conditions: under Ringer's solution at body temperature. In *fretting corrosion*, rubbing of one part on another disrupts the passivation layer, resulting in accelerated corrosion. In *pitting*, the corrosion rate is accelerated in a local region. Stainless steel is vulnerable to pitting. Localized corrosion can occur if there is inhomogeneity in the metal or in the environment. *Grain boundaries* in the metal may be susceptible to initiation of corrosion as a result of their higher energy level. *Crevices* are also vulnerable to corrosion, since the chemical environment in the crevice may differ from that in the surrounding medium. The area of contact between a screw and a bone plate, for example, can suffer crevice corrosion.

Example 5-4

Consider a restoration made of dental amalgam. Suppose that the corrosion current density is 100 μA/cm². Is this a reasonable value? This material is an alloy; assume the density is 11 g/cm³ and that the mean atomic weight is 150 amu.

    a.   How many univalent ions will be released per year from a 3.16 × 3.16 mm restoration surface?

    b.   If the corrosion is uniform, how many millimeters of depth will be lost per year? Discuss your answer.

Answer

    a.   Ion flow:

$$J = \left(10^{-4} \frac{A}{cm^2}\right)\left(\frac{Coul}{sec \times 1.6 \times 10^{-19} \ Coul}\right),$$

$$(0.1 \ cm^2)(3.15 \times 10^7 \ sec/yr) = \underline{1.97 \times 10^{21} \ e^-/yr}$$

    b.   Depth loss:

$$L = J\left(\frac{1 \ ion}{e^-} \frac{150 \ g}{mol}\right)\left(\frac{1}{6.02 \times 10^{23} \ ions/mol}\right),$$

$$\left(\frac{1}{11 \ g/cm^3}\right)\left(\frac{1}{0.1 \ cm^2}\right) = \underline{0.45 \ cm/yr.}$$

(Depth loss is far in excess of what is known to occur in dental amalgams. However, the value of 100 µA/cm^2 is a conservative estimate of the corrosion current density from the polarization curve in Figure 5-22. Surface changes that depend upon time may occur in vivo. Such changes may cause the corrosion rate to slow.)

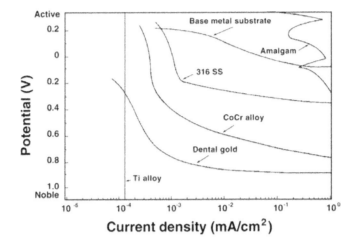

Figure 5-22. Potential–current density curves for some biomaterials. Reprinted with permission from Greener et al. (1972). Copyright © 1972, Williams & Wilkins.

5.6.4. Corrosion of Available Metals

Choice of a metal for implantation should take into account the corrosion properties discussed above. Metals in current use as biomaterials include gold, cobalt chromium alloys, type 316 stainless steel, titanium, nickel-titanium alloys, and silver-mercury amalgam. This section deals with the corrosion aspects of these metals. Their composition and physical properties are discussed in §5.4.

The noble metals are immune to corrosion and would be ideal materials if corrosion resistance were the only concern. Gold is widely used in dental restorations, and in that setting it offers superior performance and longevity. Gold is not, however, used in orthopaedic applications as a result of its high density, insufficient strength, and high cost.

Titanium is a base metal in the context of the electrochemical series; however, it forms a robust passivating layer and, as shown in Figure 5-20, remains passive under physiological conditions. Corrosion currents in normal saline are very low: 10^{-8} A/cm^2. Titanium implants remain virtually unchanged in appearance. Ti offers superior corrosion resistance but is not as stiff or strong as steel.

Cobalt-chromium alloys, like titanium, are passive in the human body. They are widely in use in orthopaedic applications. They do not exhibit pittting.

Stainless steels contain enough chromium to confer corrosion resistance by passivity. The passive layer is not as robust as in the case of titanium or the cobalt chrome alloys. Only the most corrosion resistant of the stainless steels are suitable for implants. These are the austenitic types 316, 316L, and 317, which contain molybdenum. Even these types of stainless steel are vulnerable to pitting and to crevice corrosion around screws.

Dental amalgam is an alloy of mercury, silver, and tin. Although the phases are passive at neutral pH, the transpassive potential for the γ_2 phase is easily exceeded, due to interphase galvanic couples or potentials due to differential aeration under dental plaque. Amalgam, therefore, often corrodes and is the most active (corrosion prone) material used in dentistry.

5.6.5. Minimization of Corrosion: Case Studies

Although laboratory investigations are essential in the choice of a metal, clinical evaluation in follow-up is also essential. Corrosion of an implant in the clinical setting can result in symptoms such as local pain and swelling in the region of the implant, with no evidence of infection; cracking or flaking of the implant as seen on x-ray films, and excretion of excess metal ions. At surgery, gray or black discoloration of the surrounding tissue may be seen, and flakes of metal may be found in the tissue. Corrosion also plays a role in the mechanical failures of orthopaedic implants. Most of these failures are due to fatigue, and the presence of a saline environment certainly exacerbates fatigue. The extent to which corrosion influences fatigue in the body is not precisely known. Some specific case histories are as follows.

Case 1: Total hip replacement broke after 1.5 years of service

The prosthesis was x-rayed and found to be broken high in the femoral stem area. The femoral implant was retrieved at surgery and analyzed. It was made of cast stainless steel. Cementation of the implant was not optimal but was adequate. The choice of material was considered poor for this demanding application. Cast stainless steel is mechanically inferior to the wrought steels commonly used in hip nails. Consequently, scratches on the stem during implantation could initiate corrosion fatigue cracks. The stem was found to have failed by fatigue followed by overload fracture. Moreover, in this implant, excessive inclusion content and porosity was found in the area of the failure, indicating poor manufacturing processes.

*Case 2: Patient's arm was x-rayed; a bone plate (shown in Figure 5-23) had been
    left in place for 30 years*

The screws had lost their clear outline due to corrosion and the irritating effect of the corrosion products resulted in osseous proliferation. The plate was found to be vanadium steel, a metal considered suitable in the 1920s but since abandoned for implants.

*Case 3: Radiograph of a mold arthroplasty revealed that the surgeon had used an
    ordinary iron nail (Figure 5-24) to reattach the osteotomized greater trochanter*

The cup was found to be made of Co-Cr-Mo. The nail was removed and found to be grossly corroded. The mold technique is rarely used today, and surgeons would now not even think of using a carpenter's nail.

Case 4: Patient experienced pain and disability in a repaired shoulder fracture (Figure 5-25)

The screws were removed and examined. One was found to be Co-Cr-Mo and the others of stainless steel. Bimetallic corrosion resulted. Such cases can be avoided with better efforts by both manufacturers and surgeons to avoid mixed metals.

Case 5: Broken drill bit found in a radiograph of a repaired broken hip (Figure 5-26)

Removal of the metal revealed severe corrosion. Mixed metals were consequently implanted unintentionally.

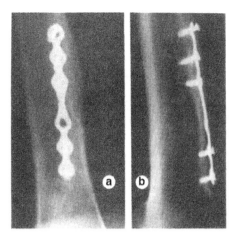

Figure 5-23. Corrosion of an obsolete metal: (**a**) top view, (**b**) side view. Radiograph of a vanadium steel bone plate in place for 30 years. The screws have lost their clear outline as a result of corrosion. Reprinted with permission from Bechtol et al. (1959). Copyright © 1959, Balliere, Tindall, & Cox.

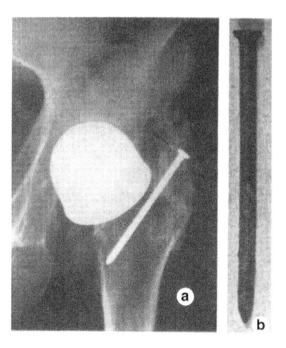

Figure 5-24. Corrosion due to the use of improper metals. (**a**) Radiograph of a CoCrMo mold hip arthroplasty. An ordinary iron nail was used to reattach the osteotomized greater trochanter. (**b**) Retrieved nail is grossly rusty. Reprinted with permission from Bechtol et al. (1959). Copyright © 1959, Balliere, Tindall, & Cox.

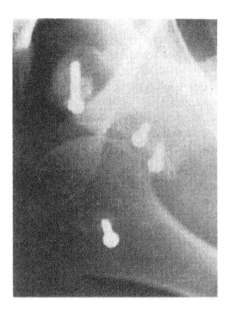

Figure 5-25. Mixed metals. Radiograph of a repaired shoulder. This patient experienced pain and disability. Examination of the screws following removal revealed one to be made of Co-Cr-Mo and the others of stainless steel. Reprinted with permission from Bechtol et al. (1959). Copyright © 1959, Balliere, Tindall, & Cox.

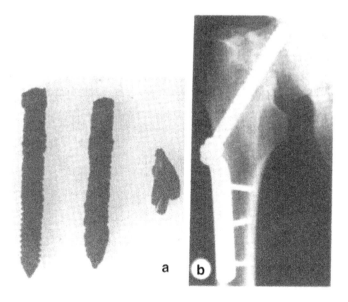

Figure 5-26. Unintentional implantation of mixed metals: (**a**) broken and corroded screws and a broken drill bit removed from a patient; (**b**) Radiograph of fracture fixation device showing screws and drill. Reprinted with permission from Bechtol et al. (1959). Copyright © 1959, Balliere, Tindall, & Cox.

Experience in the orthopaedic setting suggests that corrosion is minimized as follows.

1. Use appropriate metals.
2. Avoid implantation of different types of metal in the same region. In the manufacturing process, provide matched parts from the same batch of the same variant of a given alloy.
3. Design the implant to minimize pits and crevices.
4. In surgery, avoid transfer of metal from tools to the implant or tissue. Avoid contact between metal tools and the implant, unless special care is taken.
5. Recognize that a metal that resists corrosion in one body environment may corrode in another part of the body.

Experience in the dental setting has led to the following suggestions for minimization of corrosion:

1. Avoid using mixed metals to restore teeth in apposition, and if possible, in the same mouth.
2. Use an insulating base when seating a metallic restoration to minimize electrical conduction below the restoration.
3. Avoid conditions that lead to plaque buildup, since regions covered by plaque will experience reduced pH. This could result in corrosion.

PROBLEMS

5-1. A 1-mm diameter eutectoid steel is coated with 1 mm thick aluminum. Using the following data, answer questions a–d.

Material	Young's modulus (GPa)	Yield strength (MPa)	Density (g/cm^3)	Expansion coefficient (/°C)
Eutectic steel	205	300	7.84	10.8×10^{-6}
Aluminum	70	100	2.7	22.5×10^{-6}

a. If the composite bar is loaded in tension, which metal will yield first?

b. How much load can the composite carry in tension without plastic deformation?

c. What is the Young's modulus of the composite?

d. What is the density of the composite?

5-2. Cu has the following characteristics:

Characteristics	Value
Atomic no.	29
Atomic mass (amu)	63.54
Melting point (°C)	1,084.5
Density (g/cm^3)	8.92
Crystal structure	fcc
Atomic radius (nm)	0.1278
Ionic radius (nm)	0.096

a. Calculate the density of Cu.

b. Calculate the packing efficiency (or factor) of Cu.

c. Would the grain size increase, decrease, or remain the same if the Cu was cooled more slowly than the original Cu after remelting?

d. Would the grain size increase, decrease, or remain the same if the metal of (c) was annealed?

5-3.   Silver has the following characteristics:

Characteristics	Value
Atomic no.	47
Atomic mass (amu)	10.7.87
Atomic radius (nm)	0.1444
Ionic radius (nm)	0.126
Crystal structure	fcc
Melting point (°C)	961.9
Density (g/cm^3)	10.5

A bioengineer is trying to construct a binary phase diagram of Cu-Ag and the following additional information was given:

Characteristics	Values
Eutectic temperature	779.4°C
Eutectic composition	28.1 Cu-71.9Ag
Maximum solubility of Cu in Ag	8.8%
Maximum solubility of Ag in Cu	8.0%

a. Construct the binary phase diagram by knowing the information about Cu provided in Problem 5-2.

b. Label all the phases in the phase diagram.

5-5.   In general, pearlite steels are susceptible to corrosion. Explain why this is to be expected and how it may be prevented.

5-6.   Type 316L stainless steel has a maximum carbon content of 0.03%. Welding of finished components is acceptable for this steel but not for type 316. Why? Explain how you would expect the mechanical properties to differ from each other.

5-7.   Calculate the densities of cast and wrought cobalt alloys (CoNiCrMo). The densities for Co, Cr, Mo, and Ni are 8.8, 7.9, 10.22 and 8.91 g/cm^3, respectively. Use the mixture rule.

5-8.   The phase diagram of dental amalgam is shown in Figure 5-10.

a. Calculate the theoretical weight percentage of the silver and tin of the γ phase (Ag$_3$Sn).

b. What is the eutectic temperature?

c. Can you age-harden the amalgam?

5-9.   Suppose a titanium alloy (Ti6Al4V) implant experiences a corrosion current density of 0.1 μA/cm^2.

    a. Determine the depth of metal removed each year.

    b. Suppose no titanium were excreted and that all the metal became lodged in a 1-cm thickness of tissue near the implant. Determine the average concentration of metal in the tissue, in parts per million by weight, during one year.

5-10.   Magnesium is used in airplanes. Is it sensible to use it in an implant? Explain.

5-11.   A patient is to be treated for a fracture from a hunting accident. Lead shotgun pellets remain in the wound. The surgeon decides a bone plate should be used. What is your recommendation?

5-12.   Chromium is considered as a candidate material for a staple to be exposed to stomach contents. Is this suitable? Why?

5-13.   An amalgam filling has been in service for 30 years and has corroded. The filling surface has become black and the patient complains of tooth sensitivity. A gap has formed at the margin of the restoration, allowing leakage. Should the filling be drilled out and replaced by gold?

5-14.   A patient with a stainless steel bone plate complains of pain. The plate has been in service for 30 years and the original fracture has long since healed. Should the plate be removed and replaced by gold?

5-15.   Aluminum and iron are used as electrodes in an electrochemical cell. A voltmeter is connected across the electrodes. What potential difference will be observed?

5-16.   Copper and nickel have face-centered cubic structure and the following data are given:

Element	amu	Atomic radius (nm)	Temperature (°C)
Cu	65	0.1278	1084.5
Ni	59	0.1246	1455

    a. Which one would have a higher bond energy, Cu or Ni?

    b. These metals make a complete solid solution. Give as many reasons for their complete solubility.

    c. The <u>Cu-Ni solid solution</u> would have (a) higher, (b) lower, and (c) the same as the pure metals {Choose one}:

        Hardness _____

        Electrical conductivity _____

        Thermal conductivity _____

        Entropy _____

    d. Calculate the lattice parameter (a) of the unit cell of Cu.

5-17.   From the following list of metallic biomaterials choose the most appropriate one:

        A. Nitinol®   B. Cast Co-based alloy

        C. Grade 1 Ti   D. MP35N

        E. 24K gold   F. 316L stainless steel

        G. HgAgSn alloy   H. Ti6Al4V

a. Cold-worked easily and hot-forged _____

b. Most Ni content _____

c. Lowest eutectic temperature _____

d. SME properties _____

e. CCr_4 forms in the grain boundaries _____

f. Lowest electrical resistivity _____

g. Martensitic phase transformation from austenite _____

h. Impurities determine properties _____

i. Forms γ, γ_1, and γ_2 phases after solidification _____

j. Age harden by forming α–β phase & easily passivated _____

5-18. From the following standard electrochemical potential, answer

Reaction	ΔE_0 (V)
$Ti \Leftrightarrow Ti^{3+}$	2.00
$Al \Leftrightarrow Al^{3+}$	1.70
$Cr \Leftrightarrow Cr^{2+}$	0.56
$H_2 \Leftrightarrow 2H^+$	0.00
$Au \Leftrightarrow Au^+$	−1.68

a. Which metal would be the most anodic?

b. Which metal would corrode the most?

c. Which metal would be the most reactive?

d. What is the maximum voltage one can get from the metals listed in a standard electro-lytic solution?

e. Why could Ti and Au be used for making implants but not Al?

5-19. From the Fe-C phase diagram (Figure 3-24) and data, answer:

Element	Atomic radius (nm)	Structure	Atomic weight (g/mol)
Fe	0.1240	bcc	56
C	0.077	hex	12

a. At what temperature the eutectic reaction takes place?

b. What is the % carbon of the eutectoid composition?

c. Would the austenite (γ) or ferrite (α) steel have a higher density if the atomic radius does not change significantly?

d. What phases exist in (a) and (b)?

For 1080 steel, answer.

a. What are the percentages of carbon and iron?

b. What is the composition of the α phase at room temperative?

c. What is the composition of Fe_3C at room temperature?

    d. Calculate the amount (grams) of α phase at 722°C based on 100 g of 1080 steel.

    e. Calculate the weight of carbon in Fe_3C at 722°C based on 100 g of 1080 steel.

    f. Calculate the density of Fe at room temperature.

Give yes or no for the following questions.

    a. Can you harden the 1080 steel by a <u>precipitation hardening</u> process?

    b. Can you harden the 1080 steel by a <u>work-hardening</u> process?

    c. Can you harden the 1080 steel by a <u>surface compression hardening</u> process?

    d. Can you harden the 1080 steel by a <u>solution hardening</u> process?

    e. Can you harden the 1080 steel by a <u>martensitic phase transformation</u> process?

5-20.   Silver (Ag) is solutionized with Copper (Cu). and the phase diagram (Figure 3-23) was obtained:

Element	Atomic weight (amu)	Crystal structure	Atomic radius (Å)	Density (g/cm³)
Cu	65	fcc	?	8.9
Ag	108	fcc	1.44	10.5

    a. Calculate the radius of Cu atom. Avogadro's No. $= 6.02 \times 10^{23}$.

    b. What is the lowest temperature at which the liquid phase can exist?

    c. What are the eutectic compositions?

    d. What is the maximum amount (%) of Ag that can be solutionized into Cu?

    e. What is the melting temperature of Cu?

For a sterling silver (7.5%Cu) alloy, answer:

    f. Calculate the density by using the simple mixture rule.

    g. What would be the crystal structure of the sterling silver?

    h. Can you harden this alloy by the (cold) work-hardening method?

    i. How can you harden this alloy by the precipitation hardening method?

    j. What phase(s) exist at 900°C?

    k. What is the % Cu in each phase at 900°C?

    l. Calculate the amount (%) of each phase at 900°C.

    m. One is trying to purify the alloy by melting and solidifying. What would be the purity (%Ag) of the solid at 900°C?

    n. If you remelted the solid obtained in the above and cooled, what is the highest-purity Ag one can obtain? Use the detailed phase diagram.

    o. Construct the fraction-versus-temperature diagram for the α phase of the sterling silver. Give the specific temperatures for each change.

SYMBOLS/DEFINITIONS

Greek Letters

γ: Silver–tin intermetallic compound (Ag_3Sn).

γ_1: Silver–mercury intermetallic compound (Ag_2Hg_3).

γ_2: Tin–mercury intermetallic compound (Sn_7Hg).

ω: Bending angle of the three-point mechanical bend test.

Latin Letters

A_s: Temperature for the start of reverse transformation, NiTi(III)–NiTi(II). NiTi(II) and NiTi(III) designate structural variations (phases) produced by diffusionless (martensitic) transformation.

E: Observed electrochemical potential difference.

E_0: Standard electrochemical potential difference, for negligible concentration.

F: Faraday's constant: 96,485 coulombs per mole.

M_d: Maximum temperature for martensite formation by deformation. NiTi(II) structure is thermodynamically stable up to this temperature.

M_f: Temperature where martensitic transformation ceases.

M_s: Martensite transformation starting point in cooling.

n: Number of electron charges removed from ion.

n: Number of moles.

R: Gas constant.

T: Absolute temperature in K.

Words

Amalgam: Alloy obtained by mixing silver tin alloy with mercury.

Anode: Positive electrode in an electrochemical cell.

Cathode: Negative electrode in an electrochemical cell.

Corrosion: Unwanted reaction of metal with environment. In a Pourbaix diagram it is the region in which the metal ions are present at a concentration of more than 10^{-6} molar.

Crevice corrosion: Form of localized corrosion in which concentration gradients around pre-existing crevices in the material drive corrosion processes.

Curie temperature: Temperature above which a ferromagnetic material ceases to be ferromagnetic. For a ferroelectric material used in piezoelectric transducers, the Curie point is the temperature at which the material loses its permanent electric polarization. Control of the temperature of a magnetic metal implant is possible using electromagnetic induction and knowledge of the Curie temperature.

Galvanic corrosion: Dissolution of metal driven by macroscopic differences in electrochemical potential, usually as a result of dissimilar metals in proximity.

Galvanic series: Table of electrochemical potentials associated with ionization of metal atoms. These are called Nernst potentials.

Hyperthermia: Temperature (>42°C) above normal body temperature (37°C). Sufficiently high temperature can denature proteins, kill cells, and cause burns. Prolonged heat may also produce heat shock proteins (hsp), which may resist heat better than normal proteins.

Immunity: Resistance to corrosion by an energetic barrier. In a Pourbaix diagram it is the region in which the metal is in equilibrium with its ions at a concentration of less than 10^{-6} molar.

Nernst potential: Standard electrochemical potential measured with respect to a standard hydrogen electrode.

Noble: Type of metal with a positive standard electrochemical potential. Noble metals resist corrosion by immunity.

Passivation: Production of corrosion resistance by a surface layer of reaction products.

Passivity: Resistance to corrosion by a surface layer of reaction products. In a Pourbaix diagram it is the region in which the metal is in equilibrium with its reaction products at a concentration of less than 10^{-6} molar.

Pitting: Form of localized corrosion in which pits form on the metal surface.

Pourbaix diagram: Plot of electrical potential versus pH for a material in which the regions of corrosion, passivity, and immunity are identified.

Shape memory effect (SME): Thermoelastic behavior of some alloys that can revert to their original shape when the temperature is greater than the phase transformation temperature of the alloy.

BIBLIOGRAPHY

ASM. 1978. *Sourcebook on industrial alloy and engineering data.* Metal Park, Ohio: American Society for Metals.

ASTM. 2000. *Annual book of ASTM standards.* West Conshohocken, PA: ASTM.

Barbucci R. 2002. *Integrated biomaterials science.* New York: Kluwer Academic/Plenum.

Bechtol CO, Ferguson AB, Liang PG. 1959. *Metals and engineering in bone and joint surgery.* London: Balliere, Tindall, & Cox.

Black J, Hastings GW. 1998. *Handbook of biomaterial properties.* London: Chapman & Hall.

Brunette DM, Tengvall P, Textor M, Thomsen P, eds. 2001. *Titanium in medicine: material science, surface science, engineering, biological responses and medical applications (engineering materials).* New York: Springer.

Chang YL, Lew D, Park JB, Keller JC. 1999. Biomechanical and morphometric analysis of hydroxyapatite-coated implants with varying crystallinity. *J Oral Maxillofac Surg* 57(9):1096–1108.

Clark DE, Folz DC, Simmons JH, Hench LL, eds. 2000. *Symposium on surface-active processes in materials.* Ceramic Transactions, Vol. 101. Westerville, OH: American Ceramic Society.

Davidson JA, Poggie RA, Mishra AK. 1994. Abrasive wear of ceramic, metal, and UHMWPE bearing surfaces from third-body bone, PMMA bone cement, and titanium debris. *Biomed Mater Eng* 4(3):213–229.

Davis JR. 2003. *Handbook of materials for medical devices.* Materials Park, OH: ASM International.

Devine TM, Wulff J. 1975. Cast vs. wrought cobalt-chromium surgical implant alloys. *J Biomed Mater Res* 9:151–167.

Duerig TW. 1990. *Engineering aspects of shape memory alloys.* London: Butterworths-Heinemann.

Dumbleton JH, Black J. 1975. *An introduction to orthopedic materials.* Springfield, IL: Thomas.

Eliaz N, Mudali UK, eds. 2003. *Biomaterials corrosion.* Special issue of *Corrosion Reviews* 21(2–3). London: Freund Publishing House.

F67. 1992. Standard specification for unalloyed titanium for surgical implant applications. *Annual book of ASTM standards.* Philadelphia: ASTM. 13.01:3–5.

F136. 1992. Standard specification for wrought titanium 6Al-4Veli alloy for surgical implant applications. *Annual book of ASTM standards*. Philadelphia: ASTM. 13.01:19–21.

F138. 1992. Standard specification for stainless steel bar and wire for surgical implants (special quality). *Annual book of ASTM standards*. Philadelphia: ASTM. 13.01:22–24.

F562. 1992. Standard specification for wrought cobalt-nickel-chromium molybdenum alloy for surgical implant applications. *Annual book of ASTM standards*. Philadelphia: ASTM. 13.01:92–94.

Greener EH, Harcourt JK, Lautenschlager EP. 1972. *Materials science in dentistry*. Baltimore: Williams & Wilkins.

Helsen JA, Breme HJ. 1998. *Metals as biomaterials*. Wiley Series in Biomaterials Science and Engineering. New York: Wiley.

Hille GH. 1966. Titanium for surgical implants. *J Mater* 1:373–383.

Kay JF. 1992. Calcium phosphate coatings for dental implants: current status and future potential. *Dent Clin North Am* **36**:1–18.

Keating FH. 1956. *Chromium-nickel austenitic steels*. London: Butterworths.

Lee JH, Park JB, Andreasen GF, Lakes RS. 1988. Thermomechanical study of Ni–Ti alloys. *J Biomed Mater Res* **22**:573–588.

Levine SN, ed. 1968. *Polymers and tissue adhesives*. New York: New York Academy of Sciences.

Matsuda Y, Yamamuro T. 1994. Metallosis due to abnormal abrasion of the femoral head in bipolar hip prosthesis: implant retrieval and analysis in six cases. *Med Prog Technol* **20**(3–4):185–189.

Mears DC. 1979. *Materials and orthopaedic surgery*. Baltimore: Williams & Wilkins.

Paulus JA. 1994. *Interstitial thermally self-regulating ferromagnetic implants for fractionated hyperthermia treatment of prostate carcinoma*. Iowa City: University of Iowa Press.

Paulus JA, Parida GR, Tucker RD, Park JB. 1997. Corrosion of NiCu and PdCo thermal seed alloys used as interstitial hyperthermia implants. *Biomaterials* **18**:1609–1614.

Perkins J. 1975. *Shape memory effects in alloys*. International Symposium on Shape Memory Effects and Applications, Toronto, Ontario, Canada. New York: Plenum.

Pourbaix M. 1974. *Atlas of electrochemical equilibria in aqueous solutions*. Elmsford, NY: Pergamon Press.

Predel B, Madelung Q, Madelung O. 2003. *Phase equilibria of binary alloys: Ac–Aq ... Zr–An*. New York: Springer.

Puckering FB, ed. 1979. *The metallurgical evolution of stainless steels*. Metals Park, OH: American Society for Metals and the Metals Society.

Schnitman PA, Schulman LB, eds. 1978. *Dental implants: benefits and risk*. NIH Consensus Development Conference Statement. Bethesda, MD: US Department of Health and Human Services.

Semlitsch M. 1980. Properties of wrought CoNiCrMo alloy Protasul-10, a highly corrosion- and fatigue-resistant implant material for joint endoprostheses. *Eng Med* **9**:201–207.

Shreir LL. 1976. *Corrosion*. London: Newnes-Butterworths.

Smith CJE, Hughes AN. 1966. The corrosion fatigue behavior of a titanium-6w/o aluminum-4w/o vanadium alloy. *Eng Med* **7**:158–171.

Smithells CJ, ed. 1976. *Metals reference book*. London: Butterworths.

Van Vlack LH. 1989. *Elements of materials science*. Reading, MA: Addison-Wesley.

Wayman CM, Shimizu K. 1972. The shape memory ("Marmem") effect in alloys. *Metal Sci J* **6**:175–183.

Weinstein A, Horowitz E, Ruff AW, US National Bureau of Standards. 1977. *Retrieval and analysis of orthopaedic implants*. Proceedings of a symposium held at the National Bureau of Standards, Gaithersburg, Maryland, March 5, 1976. Washington, DC: US Department of Commerce, National Bureau of Standards.

Weinstein AM, Spires Jr WP, Klawitter JJ, Clemow AJT, Edmunds JO. 1979. Orthopedic implant retrieval and analysis study. In *Corrosion and degradation of implant materials*, pp. 212–228. Ed BC Syrett, A Acharya. Philadelphia: American Society for Testing and Materials (ASTM STP 684:212–228).

Williams DF, Roaf R. 1973. *Implants in surgery*. Philadelphia: W.B. Saunders.

6

CERAMIC IMPLANT MATERIALS

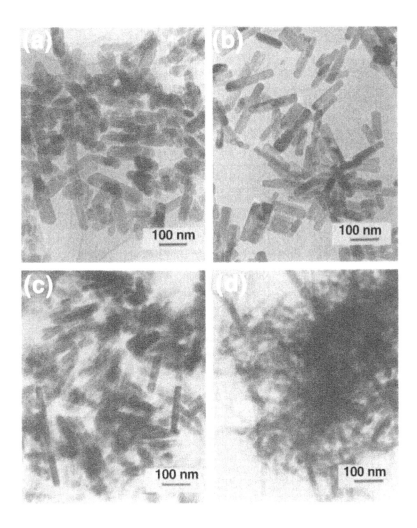

TEM photographs of hydroxyapatite crystals synthesized hydrothermally at 200°C under 2 MPa for 5 hr: (**a**) without additives, (**b**) KOH (10 wt%) added, (**c**) K_4PO_4 (10 wt%) added, and (**d**) EDTA (5 wt%) added. Reprinted with permission from Yoshimura and Suda (1994). Copyright © 1994, Chemical Rubber Co.

Ceramics are refractory, polycrystalline compounds, usually inorganic, including silicates, metallic oxides, carbides, and various refractory hydrides, sulfides, and selenides. Oxides such as Al_2O_3, MgO, SiO_2, etc. contain metallic and nonmetallic elements. Ionic salts (NaCl, CsCl, ZnS, etc.) can form polycrystalline aggregates, but soluble salts are not suitable for structural biomaterials. Diamond and carbonaceous structures like graphite and pyrolized carbons are covalently bonded. The important factors influencing the structure and property relationship of the ceramic materials are radius ratios (§2.2.2) and the relative *electronegativity* between positive and negative ions.

Recently ceramic materials have been given a lot of attention as candidates for implant materials since they possess some highly desirable characteristics for some applications. Ceramics have been used for some time in dentistry for dental crowns by reason of their inertness to body fluids, high compressive strength, and good aesthetic appearance in their resemblance to natural teeth.

Some carbons have also found use as implants, especially for blood interfacing applications such as heart valves. Due to their high specific strength as fibers and their biocompatibility, they are also being used as a reinforcing component for composite implant materials and tensile loading applications such as artificial tendon and ligament replacements. Although the black color can be a drawback in some dental applications, this is not a problem if they are used as implants. They have such desirable qualities as good biocompatibility and ease of fabrication.

6.1. STRUCTURE–PROPERTY RELATIONSHIP OF CERAMICS

6.1.1. Atomic Bonding and Arrangement

As discussed earlier in Chapter 2 (Horwitz et al., 1993), when atoms such as sodium (metal) and chlorine (nonmetal) are ionized, the sodium will lose an electron and the chlorine will gain an electron:

$$Na \Leftrightarrow Na^+ + e^-, \quad e^- + Cl \Leftrightarrow Cl^-. \tag{6-1}$$

Thus, the sodium and chlorine can make an ionic compound by the strong affinity of the positive and negative ions. Again, soluble salts are not suitable for structural biomaterials. The negatively charged ions are much larger than the positively charged ions due to the gain and loss of electrons, as given in Table 6-1. The radius of an ion varies according to the coordination numbers: the higher the coordination number, the larger the radius. For example, the oxygen ion (O^{2-}) has a radius of 0.128, 0.14, and 0.144 nm for coordination numbers 4, 6, and 8, respectively.

Table 6-1. Atomic and Ionic Radii of Some Elements

	Group I			Group II			Group IV			Group VI	
Element	Atomic radius[a]	Ionic radius	Ele.	Atomic radius[a]	Ionic radius	Ele.	Atomic radius[a]	Ionic radius	Ele.	Atomic radius[a]	Ionic radius
Li+	0.152	0.068	Be++	0.111	0.031	O⁻	0.074	1.40	F⁻	0.071	0.130
Na+	0.186	0.095	Mg++	0.160	0.065	S⁻⁻	0.102	1.84	Cl⁻	0.099	0.181
K+	0.227	0.133	Ca++	0.197	0.099	Se⁻⁻	0.116	1.98	Br⁻	0.114	0.195

[a]Covalent. Units are in nm. Reprinted with permission from Starfield and Shrager (1972). Copyright © 1972, McGraw-Hill.

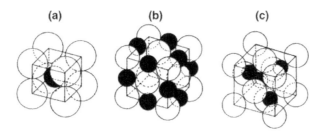

Figure 6-1. AX structures of ceramics. The dark spheres represent positive ions (A^+) and the circled ones represent negative ions (X^-).

Table 6-2. Selected A_mX_n Structures

Prototype compound	Lattice of A (or X)	CN of A (or X) sites	Available sites filled	Minimum r_A/R_x	Other compounds
CsCl	Simple cubic	8	All	0.732	CsI
NaCl	fcc	6	All	0.414	MgP, MnS, LiF
ZnS	fcc	4	1/2	0.225	β-SiC, CdS, AlP
Al_2O_3	hcp	6	2/3	0.414	Cr_2O_3, Fe_2O_3

Ceramics can be classified according to their structural compounds, of which A_mX_n is an example. The A represents a metal and X represents a nonmetal element, and m and n are integers. The simplest case of this system is the AX structure, of which there are three types (see Figure 6-1). The difference between these structures is due to the relative size of the ions (*minimum radius ratio*). If the positive and negative ions are about the same size ($r_A/R_X > 0.732$), the structure becomes a simple cubic (CsCl structure). A face-centered cubic structure arises if the relative sizes of the ions are quite different since the positive ions can be fitted in the tetragonal or octagonal spaces created among larger negative ions. These are summarized in Table 6-2. The aluminum and chromium oxide belong to the A_2X_3 type structure. The O^{2-} ions form hexagonal close packing, while the positive ions (Al^{3+}, Cr^{3+}) fill in 2/3 of the octahedral sites, leaving 1/3 vacant.

6.1.2. Physical Properties

Ceramics are generally hard; in fact, the measurement of hardness is calibrated against ceramic materials. Diamond is the hardest, with a hardness index on the Mohs scale of 10, and talc ($Mg_3Si_4O_{10}COH$) is the softest (Moh's hardness 1), and others such as alumina (Al_2O_3; hardness 9) quartz (SiO_2; hardness 8) and apatite (e.g., fluorapatite, $Ca_5P_3O_{12}F$; hardness 5) are in between. Other characteristics of ceramic materials are their high melting temperatures and low conductivity of electricity and heat. These characteristics come about as a result of the nature of the chemical bonding in ceramics.

Unlike metals and polymers, ceramics are difficult to shear plastically due to the ionic nature of bonding, as shown in Figure 6-2. In order to shear, the planes of atoms should slip past each other. However, for ceramic materials the ions with the same electric charge repel each other; hence moving the plane of atoms is very difficult. This makes the ceramics brittle (non-ductile); moreover, creep at room temperature is almost zero. Some ceramics such as covalently bonded diamond have similar properties as those of ionic ceramics due to the breakages of theses primary bonds when deformed beyond their elastic limit. The ceramics are also very sensitive to notches or microcracks since instead of undergoing plastic deformation (or yield) they will fracture elastically once the crack propagates. This is also the reason why the ceramics have low tensile strength compared to the compressive strength, as discussed in §3.1.2. If the ceramic is made flaw free, then it becomes very strong even in tension. Glass fibers made this way have tensile strengths twice that of a high-strength steel ($\approx$7 GPa).

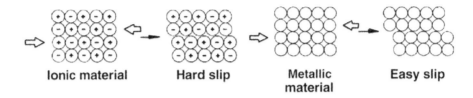

Ionic material   Hard slip   Metallic   Easy slip
                                            material

Figure 6-2. Schematic two-dimensional illustration of slips in ionic and non-ionic bond materials.

Example 6-1

A piece of window glass fails at 70 MPa (10^4 psi). Calculate the largest size elliptic crack which is responsible for the low strength. The stress concentration factor (σ/σ_a) can be expressed as $2\sqrt{c/r}$, where c is the crack depth ($2c$ if away from surface) and r is the crack tip radius.

Answer

Assuming the crack tip radius has the dimension of an oxygen ion (0.14 nm) and the theoretical strength of glass is 7 GPa,

$$c = \frac{r(\sigma/\sigma_a)^2}{4}$$

$$= \frac{(1.4)(7000/70)^2}{4}$$

$$= \underline{3.5 \times 10^2 \text{ nm or } 0.35 \text{ μm}}.$$

So, even a small microcrack significantly weakens the glass.

6.2. ALUMINUM OXIDES (ALUMINA)

Alpha-alumina (α-Al$_2$O$_3$) has a hexagonal close-packed structure (a = 0.4758 nm and c = 1.2991 nm). Natural alumina is known as sapphire or ruby (depending on the types of impurities that give rise to color). The single-crystal form of alumina has been used successfully to make implants. Single-crystal alumina can be made by feeding fine alumina powders onto the surface of a seed crystal, which is slowly withdrawn from an electric arc or oxy-hydrogen flame as the fused powder builds up. Alumina single crystals up to 10 cm in diameter have been grown by this method.

The main source of high-purity alumina (aluminum oxide) is bauxite and native corundum. The commonly available (alpha, α) alumina can be prepared by calcining alumina trihydrate resulting in calcined alumina. The chemical composition and density of commercially available "pure" calcined alumina are given in Table 6-3. The American Society for Testing and Materials (ASTM) specifies 99.5% pure alumina and less than 0.1% of combined SiO_2 and alkali oxides (mostly Na_2O) for implant use.

Table 6-3. Chemical Composition of Calcined Aluminas

Chemicals	Composition (weight %)
Al_2O_3	99.6
SiO_2	0.12
Fe_2O_3	0.03
Na_2O	0.04

Aluminum Company of America. Reprinted with permission from Gitzen (1970). Copyright © 1970, American Ceramic Society.

Table 6-4. Physical Property Requirements of Alumina Implants (ASTM, 2000)

Properties	Values
Flexural strength	> 400 MPa (58,000 psi)
Elastic modulus	380 GPa (55.1 × 10^6 psi)
Density (g/cm^3)	3.8–3.9

The strength of polycrystalline alumina depends on porosity and grain size. Generally the smaller the grains and porosity, the higher the resulting strength. The ASTM standards (F603-78) require a flexural strength of greater than 400 MPa and an elastic modulus of 380 GPa, as given in Table 6-4.

Alumina in general is a quite hard material (Mohs number of 9); the hardness varies from 2,000 kg/mm^2 (19.6 GPa) to 3,000 kg/mm^2 (29.4 GPa). This high hardness permits one to use alumina as an abrasive (emery) and as bearings for watch movements. The high hardness is accompanied by low friction and wear; these are major advantages of using the alumina as joint replacement material in spite of its brittleness.

6.3. ZIRCONIUM OXIDES (ZIRCONIA)

Zirconium oxides or zirconia (ZrO_2) have been tried for application in fabricating implants. Zirconia is called "fake diamond" or "cubic zirconia" since it has a high refractive index (as does diamond) and some zirconia single crystals can be made gem grade. Some mechanical properties are as good or better than alumina ceramics. Zirconia is highly biocompatible, as are other ceramics, and can be made in the form of large implants such as the femoral head and acetabular cup in total hip joint replacement. These materials are strengthened by phase trans-

formation and control of grain sizes. A major drawback is that they may be weakened significantly under stress in the presence of moisture; this weakening occurs at a much faster rate at
elevated temperature such as occurs during steam sterilization (autoclaving).

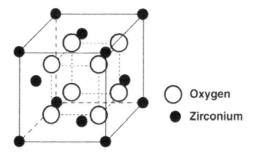

Figure 6-3. Cubic structure of zirconia that belongs to the fluorite structure. Modified with
permission from Kingery et al. (1976). Copyright © 1976, Wiley.

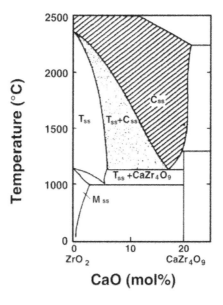

Figure 6-4. Partial phase diagram of ZrO_2–CaO: C_{ss} denotes cubic, T_{ss} tetragonal, and M_{ss}
monoclinic solid solution phase. Reprinted with permission from Drennan and Steele (1986).
Copyright © 1986, Pergamon Press.

6.3.1. Structure of Zirconia

The zirconia is allotropic and the transition from monoclinic ($a \neq b \neq c$, $\alpha = \gamma = 90 \neq \beta$) to
tetragonal ($a = b \neq c$, $\alpha = \gamma = \beta = 90°$) at 1000~1200°C and tetragonal to cubic ($a = b = c$, $\alpha =
\gamma = \beta = 90°$) structure at 2370°C. The monoclinic-to-tetragonal phase transition is a diffu-
sionless transformation accompanying a volume reduction of 7.5%. The cubic structure of the
zirconia belongs to the fluorite (CaF_2) structure, as shown in Figure 6-3. The crystallographic
parameters of the unit cell structures are given in Table 6-5. The partial phase diagram of

Table 6-5. Physical Properties of Zirconia

Property	Values
Polymorphism[a,b]	
Monoclinic tetragonal	1273–1473 (K)
Tetragonal cubic	2643 (K)
Cubic liquid	2953 (K)
Crystallography	
Monoclinic	
a	5.1454 Å
b	5.2075 Å
c	5.3107 Å
	99°14'
Space group	P2$_1$/c
Tetragonal	
a	3.64 Å[c]
c	5.27 Å
Space group	P4$_2$/nmc
Cubic	
a	5.065 Å
Space group	Fm3m
Density (g/cm^3)	
Monoclinic	5.6
Tetragonal	6.10
Cubic	6.29[a]
Thermal expansion coefficient[c] (10^{-6}/K)	
Monoclinic	7
Tetragonal	12
Heat of formationc (kJ/mol)	–1096.7
Boiling point (K)	4549
Thermal conductivity (W /m/K)	
at 100°C	1.675
at 1300°C	2.094
Mohs hardness	6.5
Refractive index	2.15

[a] Calculated value (see Ex. 6-2).

Reprinted with permission from Drennan and Steele (1986). Copyright © 1986, MIT Press.

ZrO_2–CaO is shown in Figure 6-4. The CaO acts as a stabilizing oxide, where Css is the cubic solid solution, called fully stabilized zirconia, which is resistant to most molten metals; thus it is used to make crucibles. Partially stabilized zirconia (PSZ) is resulted in the two-phase region of [Tss + Css]. These materials have enhanced mechanical properties. Another oxide commonly used to stabilize cubic zirconia is yttrium oxide (Y_2O_3), as shown in Figure 6-5. It is critical that the precipitates of tetragonal phase remain small (<0.2 μm) in the cubic zirconia matrix to enhance its mechanical properties. If the tetragonal precipitates become large the phase transforms into monoclinic, causing cracks in the material. To control the phase transformation MgO is used along with Y_2O_3 during sintering and the aging process. Figure 6-6 shows the microstructures of yttrium- and magnesium-stabilized zirconia. The tetragonal precipitates strengthen the structure of the cubic zirconia matrix due to the volume difference during the phase transformation.

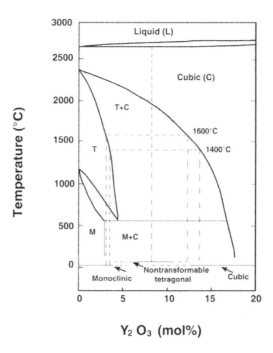

Figure 6-5. Phase diagram of ZrO_2–Y_2O_3. Reprinted with permission from Burger and Willmann (1993). Copyright © 1993, Pergamon Press.

Figure 6-6. Yttria- and magnesium-stabilized zirconia (**A**) and tetragonal precipitates (**B**) in cubic matrix grains. Reprinted with permission from Burger and Willmann (1993). Copyright © 1993, Pergamon Press.

6.3.2. Properties of Zirconia

Properties of various zirconia are summarized in Table 6-6. The strength data for the partially stabilized zirconia with yttrium oxide showed the highest flexural strength and fracture toughness. However, the Weibull modulus was lower than the yttrium magnesium oxide-stabilized

Table 6-6. Properties of Various Zirconia

Properties	CSZ	Y–Mg–PSZ	Y-TZP
Young's modulus (GPa)	210	210	210
Flexural strength (MPa)	200	600	950
Hardness (Vickers, HV0.5)	1250	1250	1250
Fracture toughness (MPa m$^{1/2}$)	–	5.8	10.5
Weibull modulus	8	25	18
Density	6.1	5.85	6

Reprinted with permission from Burger and Willmann (1993). Copyright © 1983, Pergamon.

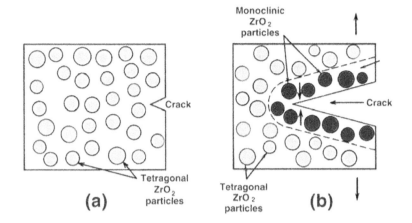

Figure 6-7. Schematic representation of the toughening of zirconia with partially stabilized zirconia: (**a**) crack before phase transformation; (**b**) crack arrestment due to phase transformation of the dispersed PSZ particles. Reprinted with permission from Callister (1994). Copyright © 1994, Wiley.

zirconia. It is also interesting that the increased fracture toughness is due to a phase transformation that operates by arresting the propagation of cracks, as shown in Figure 6-7. Small particles of partially stabilized ZrO_2 are dispersed in the matrix materials, which could be zirconia itself. This partial stabilization enables the retention of a metastable tetragonal structure at ambient temperature. During crack propagation the tetragonal particles in the crack tip region undergo phase transformation, increasing its volume, which sets up a compressive field surrounding the particles and closes the crack opening, resulting in a stronger material. The process is similar to the precipitation of a tetragonal structure in cubic grains.

The yttrium-stabilized zirconia has been used for fabricating the femoral head of total hip joint prostheses and has two advantages over the alumina. One is the finer grain size and a well-controlled microstructure without any residual porosity of the Y-TZP, making it better tribological material than the alumina. The other is higher fracture strength and toughness due to the phase transformation toughening process.

As mentioned, zirconia has many salient features in comparison with alumina. A comparison of the properties is given in Table 6-7. The biocompatibility of zirconia is about the same as alumina ceramic, but its tribological properties are quite different. In one study the friction and wear properties of zirconia, alumina, and 316L stainless steel against ultra-high-molecular-weight polyethylene (UHMWPE) were evaluated by using a uni- and bidirectional wear testing machine in bovine serum, saline, and distilled water. Table 6-8 shows the results of wear of UHMWPE. The wear factor was estimated by the following equation:

$$\text{Wear factor} = \frac{\text{Wear volume (mm}^3)}{\text{Load (N)/Sliding distance (m)}}. \tag{6-2}$$

Table 6-7. Comparison of Properties of Alumina and Zirconia

Property	Alumina	Zirconia
Chemical composition	Al_2O_3 + MgO	ZrO_2 + MgO + Y_2O_3
Purity (%)	99.9	95~97
Density (g/cm^3)	> 3.97	5.74~6.0
Porosity (%)	< 0.1	< 0.1
Bending strength (MPa)	> 500	500~1 000
Compression strength (MPa)	4100	2000
Young's modulus (GPa)	380	210
Poisson's ratio	0.23	0.3
Fracture toughness (MPa m$^{1/2}$)	4	up to 10
Thermal expansion coefficient ($\times 10^{-6}$/K)	8	11
Thermal conductivity (W/m/K)	30	2
Hardness (HV0. 1)	up to 2200	1200
Contact angle (°)	10	50

Reprinted with permission from Willmann (1993). Copyright © 1993, Pergamon.

Table 6-8. Wear of UHMWPE on Two Different Wear Devices

	*Wear factor (mm^3/N-m) $\times 10^{-9}$					
Medium	Bovine serum		Saline		Distilled water	
Counterfaces	Unidirectional	Reciprocate	Unidirectional	Reciprocate	Unidirectional	Reciprocate
Zirconia (3)	10.7 ± 12	0.56 ± 14	7.5 ± 3	0.45 ± 5	8.61 ± 11	0.38 ± 6
Alumina (3)	18.2 ± 6	1.01 ± 8	32.7 ± 7	0.57 ± 2	11.8 ± 4	0.68 ± 4
316L SS (2)	27.7 ± 30	1.81 ± 4	90.5 ± 40	3.89 ± 8	37.1 ± 10	1.12 ± 10

( ) = number of specimens tested. *Average and range.
Reprinted with permission from Kumar et al. (1991). Copyright © 1991, Wiley.

The wear factor for the yttrium oxide-stabilized (Y-PSZ) zirconia showed a smaller value to alumina and 316L stainless steel in all test conditions and modes. Also, the unidirectional wear test showed a greatly higher wear volume than the bidirectional (reciprocating) tests. The actual wear volume versus number of cycles in unidirectional tests is shown in Figure 6-8. The wear factor is the slope of the curve divided by the load (3.45 MPa).

The friction coefficient also showed a lower value for the zirconia (0.028–0.082) than alumina (0.044–0.115) or 316L stainless steel (0.061–0.156). As with the wear factor, the bidirectional reciprocating mode showed somewhat lower friction than the unidirectional arrangement, although it was not as drastic as for wear. Also, the types of lubricant did not influence the friction. One reason for the excellent wear and friction characteristic of the zirconia is attributed to the fact that zirconia has less porosity, as shown in Figure 6-9. Also, the average

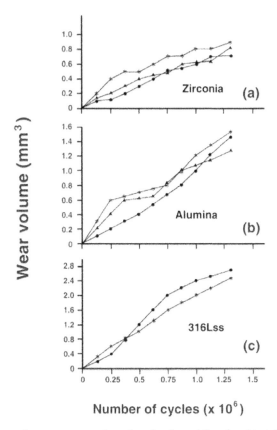

Number of cycles (x 10^6)

Figure 6-8. Wear volume versus number of cycles for unidirectional test (one cycle = 50 mm) for zirconia (**a**), alumina (**b**), and 316L stainless steel (**c**). Reprinted with permission from Kumar et al. (1991). Copyright © 1991, Wiley.

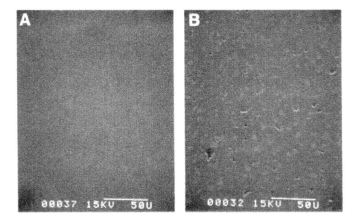

Figure 6-9. Scanning electron microscopic picture of polished surfaces of zirconia (**A**) and alumina (**B**). Note the porosity in the alumina. Reprinted with permission from Kumar et al. (1991). Copyright © 1991, Wiley.

grain size of zirconia (0.3 μm) was about one-tenth that of the alumina (2.5 μm), although the surface roughness was about the same for both (0.005–0.013 μm R_a, the average root mean square value of surface roughness).

Some researchers evaluated the use of zirconia for a hemiarthroplasty femoral head implant and found it suitable due to its low friction with articular cartilage and its excellent biocompatibility. On the other hand, the wear rate of zirconia–zirconia is many times that of the alumina–alumina combination, preempting its use for the femoral head and socket.

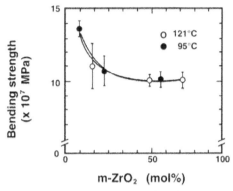

Figure 6-10. Relationship between the bending strength and amount of phase transformation aged in water at 95 and 121°C. Reprinted with permission from Shimizu et al. (1993). Copyright © 1993, Wiley.

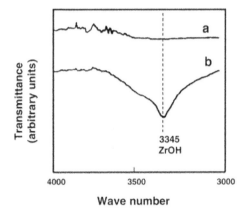

Figure 6-11. Fourier transform IR spectroscopy of zirconia before (**a**) and after (**b**) aging in water at 121°C for 960 hr. Reprinted with permission from Shimizu et al. (1993). Copyright © 1993, Wiley.

There is a direct one-to-one relationship between the amount of phase transformation and bending strength of zirconia, as shown in Figure 6-10, indicating that only the amount of phase transformation influences the mechanical properties. The moisture has an effect on the zirconia by forming Zr–OH bonding that precedes the phase transformation and was detected by infrared (IR) spectroscopy, as shown in Figure 6-11. The yttria-stabilized zirconia is a good candidate to replace the alumina ceramic for orthopedic applications in spite of the effect of aging on the mechanical properties on zirconia. Even after aging, zirconia is a much stronger material than alumina, which has a strength of about 400 MPa.

Example 6-2

Calculate the density of cubic zirconia and compare that with the other forms of zirconia given in Table 6.5.

Answer

$$\text{Density} = \frac{\text{Mass}}{\text{Volume}} = \frac{(4 \times 91 + 8 \times 16)\,\text{gmol}^{-1}}{(5.065 \times 10^{-8}\text{cm})^3\, 6.02 \times 10^{23}\text{mol}^{-1}} = \underline{6.29\ \text{g/cm}^3}.$$

This value seems very reasonable compared to the density of the tetragonal structure, 6.10 g/cm³. However, it is somewhat odd that the density increases with increased temperature since the cubic structure exists at higher temperature than the tetragonal zirconia.

6.3.3. Manufacture of Zirconia

Zircon ($ZrSiO_4$) is a gold-colored silicate of zirconium; zircon is a mineral (baddeleyite) found in igneous and sedimentary rocks and occurring in tetragonal crystals colored yellow, brown, or red, depending on impurities. The zircon is first chlorinated to form $ZrCl_4$ in a fluidized bed reactor in the presence of petroleum coke. A second chlorination is required for high-quality zirconium. Zirconium is precipitated with either hydroxides or sulfates, then calcined to its oxide.

The zirconia is partially stabilized above 1700°C in the cubic phase, which results in large grain sizes (50-70 µm). When it is cooled, a phase transformation takes place and tetragonal precipitates can be formed in the cubic matrix. Combined cubic and tetragonal phase results in enhanced mechanical properties.

Example 6-3

Calculate the wear constant of UHMWPE with zirconia as a mating material for a joint replacement. Use the data in Figure 6-8.

Answer

From Figure 6-8, the average wear volume for zirconia in bovine serum is about 0.6 mm³ after 10^5 cycles, with a sliding distance of 50 mm. Therefore, the wear constant can be calculated, since the wear constant is defined as (ΔV is wear volume, Δl is total sliding distance, P is load, and H is hardness)

$$\text{Wear constant } (K) = \frac{0.6\ \text{mm}^3 \times 3 \times 100\ \text{MPa}}{43.35\ \text{N} \times 50\ \text{mm} \times 10^5} = \underline{8.3 \times 10^{-7}}.$$

Assume the hardness of the UHMWPE is about 100 MPa, and the load applied 43.4 N. This corresponds to a stress of 3.45 MPa.

The wear constant for the UHMWPE with 316L stainless steel would be about 3 times larger according to the wear volume at 100,000 cycles(~2 mm³)

6.4. CALCIUM PHOSPHATE

Calcium phosphate has been used to make artificial bone. Recently, this material has been synthesized and used for manufacturing various forms of implant as well as for solid or porous coatings on other implants. There are mono-, di-, tri-, and tetra-calcium phosphates, in addition to the hydroxyapatite and β-whitlockite, which have ratios of 5/3 and 3/2 for calcium and phosphorus (Ca/P), respectively. The stability in solution generally increases with increasing Ca/P ratios. Hydoxyapatite is the most important among the calcium compounds since it is found in natural hard tissues as mineral phase. Hydroxyapatite acts as a reinforcement in hard tissues and is responsible for the stiffness of bone, dentin, and enamel.

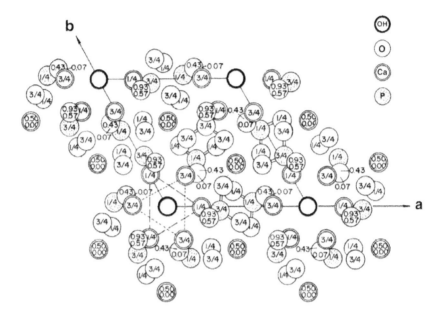

Figure 6-12. Hydroxyapatite structure projected down the c-axis on the basal plane. Reprinted with permission from Posner et al. (1958). Copyright © 1958, Munksgaard International.

6.4.1. Structure of Calcium Phosphate (Hydroxyapatite)

Calcium phosphate can be crystallized into the salts mono-, di- tri-, and tetra-calcium phosphate, hydroxyapatite, and β-whitlockite, depending on the Ca/P ratio, presence of water, impurities, and temperature. The most important is the hydroxyapatite due to its presence in natural bone and teeth. In a wet environment and at lower temperature (<900°C), it is more likely that the (hydroxyl or hydroxy) apatite will form while in a dry atmosphere and at higher temperature the β-whitlockite ($3CaO \cdot P_2O_3$) will be formed. Both forms are very tissue compatible and are used for bone substitute in granular form or as a solid block. We will consider the apatite form of the calcium phosphate since it is considered more closely related to the mineral phase of bone and teeth.

The mineral part of bone and teeth is made of a crystalline form of calcium phosphate similar to hydroxyapatite [$Ca_{10}(PO_4)_6(OH)_2$]. The apatite family of minerals, $A_{10}(BO_4)_6X_2$, crystallizes into hexagonal rhombic prisms and has unit cell dimensions a = 0.9432 nm and c = 0.6881 nm. The atomic structure of hydroxyapatite projected down on the c-axis onto the basal plane is given in Figure 6-12. Note that the hydroxyl ions lie on the corners of the projected basal plane, and they occur at equidistant intervals along half of the cell (0.344 nm), along columns perpendicular to the basal plane and parallel to the c-axis. Six of the ten calcium ions in the unit cell are associated with the hydroxyls in these columns, resulting in strong interactions.

The ideal Ca/P ratio of hydroxyapatite is 10/6 and the calculated density is 3.219 g/cm^3. It is interesting to note that the substitution of OH with F will give greater chemical stability due to the closer coordination of F (symmetric shape) as compared to the hydroxyl (nonsymmetric, two atoms) by the nearest calcium. This is one of the reasons for the better caries resistance of teeth following fluoridation.

6.4.2. Properties of Calcium Phosphates (Hydroxyapatite)

There is a wide variation of the mechanical properties of synthetic calcium phosphates, as given in Table 6-9. The wide variations of properties are due to the variations in the structure of polycrystalline calcium phosphates due to variations in the manufacturing processes. Depending on the final firing conditions, the calcium phosphate can be calcium hydroxyapatite or β–whitlockite. In many instances, however, both types of structure exist in the same final product.

Table 6-9. Physical Properties of Synthetic Calcium Phosphates

Properties	Values
Elastic modulus (GPa)	40–117
Compressive strength (MPa)	294
Bending strength (MPa)	147
Hardness (Vickers, GPa)	3.43
Poisson's ratio	0.27
Density (theoretical, g/cm^3)	3.16

And other sources.

Polycrystalline hydroxyapatite has a high elastic modulus (40–117 GPa). Hard tissues such as bone, dentin, and dental enamel are natural composites that contain hydroxyapatite (or a similar mineral) as well as protein, other organic materials, and water. Enamel is the stiffest hard tissue with an elastic modulus of 74 GPa, and it contains the most mineral. Dentin ($E = 21$ GPa) and compact bone ($E = 12\sim18$ GPa) contain comparatively less mineral. The Poisson's ratio for the mineral or synthetic hydroxyapatite is about 0.27, which is close to that of bone (≈ 0.3).

Among the most interesting properties of hydroxyapatite as a biomaterial is its excellent biocompatibility. Indeed, it appears to form a direct chemical bond with hard tissues. In an experimental trial, new lamellar cancellous bone was formed around implanted hydroxyapatite granules in the marrow cavity of rabbits after 4 weeks, as shown in Figure 6-13.

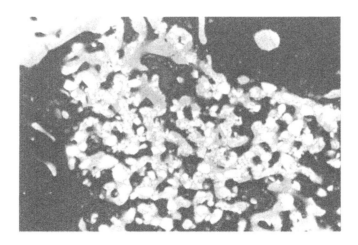

Figure 6-13. X-ray microradiographic picture showing the hydroxyapatite granules and bony tissues surrounding them after 4 weeks in a rabbit marrow cavity (40×). The mottled areas are regions of new bone deposition and the white areas are implants. Reprinted with permission from Niwa et al. (1980). Copyright © 1980, Springer-Verlag.

6.4.3. Manufacture of Calcium Phosphates (Hydroxyapatite)

Many different methods have been developed to make precipitates of hydroxyapatite from an aqueous solution of $Ca(NO_3)_2$ and NaH_2PO_4. One method uses precipitates that are filtered and dried to form a fine particle powder. After calcination for about 3 hours at 900°C to promote crystallization, the powder is pressed into final form and sintered at about 1050–1200°C for 3 hours. Above 1250°C the hydroxyapatite shows a second phase precipitation along the grain boundaries.

Example 6-4

Calculate the theoretical density of hydroxyapatite crystal $[Ca_{10}(PO_4)_6(OH)_2]$.

Answer

From Figure 6-12 one can see that there are 10 Ca atoms in the hexagonal unit cell prism, 4 inside, (2 for 1/2, 2 for 1/4, 3/4 position), 2 for top and bottom (0 and 1 position), and 4 for sides (1/4, 3/4 position). Therefore,

$$\rho = \frac{(10 \times 40 + 6 \times 31 + 26 \times 16 + 2 \times 1)}{9.432 \times \dfrac{\sqrt{3}}{2} \times 9.432 \times 6.881 \times 10^{-8} \times 6.01 \times 10^{23}}$$

$$= \underline{3.16 \text{ g/cm}^3}.$$

(This is very close to the value given in the literature; McConell, 1963).

6.5. GLASS-CERAMICS

Glass-ceramics are polycrystalline ceramics made by controlled crystallization of glasses. They were originally developed by S.D. Stookey of Corning Glass Works in the early 1960s. They were first utilized in photosensitive glasses in which small amounts of copper, silver and gold are precipitated by ultraviolet light irradiation. These metallic precipitates help to nucleate and crystallize the glass into a fine grained ceramic which possess excellent mechanical and thermal properties. Bioglass® and Ceravital® are two glass-ceramics developed for implants.

6.5.1. Formation of Glass-Ceramics

The formation of glass-ceramics is influenced by the nucleation and growth of small (<1-μm diameter) crystals as well as the size distribution of these crystals. It is estimated that about 10^{12} to 10^{15} nuclei per cubic centimeter are required to achieve such small crystals. In addition to the metallic agents mentioned (Cu, Ag, and Au), Pt groups, TiO_2, ZrO_2, and P_2O_5 are widely used for this purpose. The nucleation of glass is carried out at temperatures much lower than the melting temperature. During processing the melt viscosity is kept in the range of 10^{11} and 10^{12} Poise for 1 to 2 hours. In order to obtain a larger fraction of the microcrystalline phase, the material is further heated to an appropriate temperature for maximum crystal growth. Deformation of the product, phase transformation within the crystalline phases, and redissolution of some of the phases are to be avoided. Crystallization is usually more than 90% complete with grain sizes of 0.1 to 1 μm. Grains smaller than one micron are called nanocrystalline. These grains are much smaller than those of the conventional ceramics. Figure 6-14 shows a schematic representation of the temperature–time cycle for a glass-ceramic.

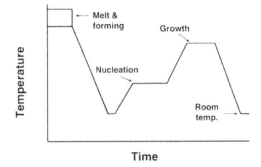

Figure 6–14. Temperature-time cycle for a glass ceramic. Reprinted with permission from Kingery et al. (1976). Copyright © 1976, Wiley.

The glass-ceramics developed for implantation are SiO_2–CaO–Na_2O–P_2O_5 and Li_2O–ZnO–SiO_2 systems. There are two different groups experimenting with the SiO_2–CaO–Na_2O–P_2O_5 glass-ceramic. One group varied the compositions (except for P_2O_5) as given in Table 6-10 in order to obtain the best composition to induce direct bonding with bone. The bonding is related to simultaneous formation of a calcium phosphate and an SiO_2-rich film layer on the surface, as exhibited by 46S5.2 type Bioglass®. If an SiO_2-rich layer forms first and a calcium phosphate film develops later (46–55 mol% SiO_2 samples) or no phosphate film is formed (60 mol% SiO_2), then no direct bonding with bone is observed. The approximate region of the SiO_2–CaO–Na_2O system for the tissue-glass ceramic reaction is shown in Figure 6-15. As can be seen, the best region (Region A) for good tissue bonding is the composition given for 46S5.2 type Bioglass® (Table 6-10).

Table 6-10. Compositions of Bioglass® and Ceravital® Glass-Ceramics

Type	Code	SiO$_2$	CaO	Na$_2$O	P$_2$O$_5$	MgO	K$_2$O
Bioglass							
	42S5.6	42.1	29.0	26.3	2.6	–	–
	(45S5)46S5.2	46.1	26.9	24.4	2.6	–	–
	49S4.9	49.1	25.3	23.0	2.6	–	–
	52S4.6	52.1	23.8	21.5	2.6	–	–
	55S4.3	55.1	22.2	20.1	2.6	–	–
	60S3.8	60.1	19.6	17.7	2.6	–	–
Cervital*							
	Bioactive	40.0–50.	30.0–35.0	5.0–10.0	10.0–15.0	2.5–5.0	0.5–3.0
	**Nonbioactive	30.0–35.0	25.0–30.0	3.5–7.5	7.5–12.0	1.0–2.5	0.5–2.0

*The Ceravital composition is in weight % while the Bioglass compositions are in mol %.

**In addition Al$_2$O$_3$ (5.0–15.0), TiO$_2$(1.0–5.0) and Ta$_2$O$_5$ (5.0–15.0) are added.

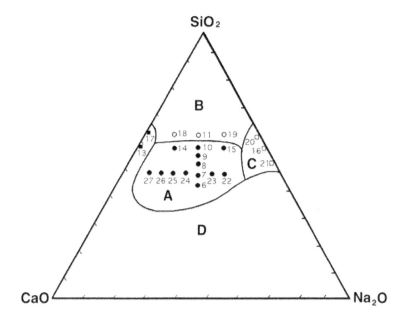

Figure 6-15. The SiO$_2$–CaO–Na$_2$O phase diagram. Region A: bonding in 30 days with bone. Region B: nonbonding—too low reactivity. Region C: nonbonding—too high reactivity. Region D: bonding but does not form glass. Reprinted with permission from Hench and Ethridge (1982). Copyright © 1982, Academic Press.

The composition of Ceravital® is similar to the Bioglass® in terms of SiO$_2$ content but differs somewhat in others, as given in Table 6-10. In addition, Al$_2$O$_3$, TiO$_2$, and Ta$_2$O$_5$ are used for the Ceravital® glass-ceramic in order to control the dissolution rate. The mixtures were melted in a platinum crucible at 1500°C for 3 hours and annealed, and then cooled. The nucleation and crystallization temperatures were 680 and 750°C, respectively, for 24 hours each. When the size of crystallites was about 0.4 nm and the crystals did not exhibit the characteristic needle structure, the process was stopped to obtain a fine grain structure.

6.5.2. Properties of Glass-Ceramics

Glass-ceramics have several desirable properties compared to glasses and ceramics. The thermal coefficient of expansion is very low, typically 10^{-7} to 10^{-5} per degree C, and in some cases it can be made even negative. Due to the controlled grain size and improved resistance to surface damage, the tensile strength of these materials can be increased by at least a factor of two, from about 100 to 200 MPa. The resistance to scratching and abrasion are close to that of sapphire.

In an experimental trial, Bioglass® glass-ceramic was implanted in the femur of rats for 6 weeks. Transmission electron micrographs showed intimate contacts between the mineralized bone and the Bioglass®, as given in Figure 6-16. The mechanical strength of the interfacial bond between bone and Bioglass® ceramic is the same order of magnitude as the strength of the bulk glass-ceramic (850 kg/cm^2 or 83.3 MPa), which is about three-fourths that of the host bone strength.

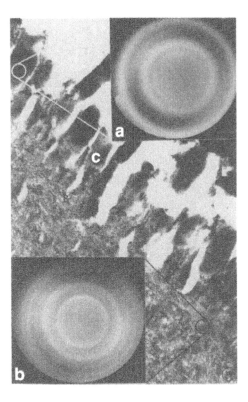

Figure 6-16. Transmission electron micrograph of well-mineralized bone (**b**) juxtaposed to the glass-ceramic (**c**), which was fractured during sectioning (×51,500). Insert (**a**) is the diffraction pattern from the ceramic area and (**b**) is from bone area. Reprinted with permission from Beckham et al. (1971). Copyright © 1971, Springer-Verlag.

The main drawback of the glass-ceramic is its brittleness, as is the case with other glasses and ceramics. Additionally, due to restrictions on the composition for biocompatibility (or osteogenicity), mechanical strength cannot be substantially improved as for other glass-ceramics. Therefore, they cannot be used for making major load-bearing implants such as joint implants. However, they can be used as fillers for bone cement, dental restorative composites, and coating material.

Example 6-5

From the phase diagram of Al_2O_3–SiO_2, answer the following:
a. Determine the exact w/o of Al_2O_3 for mullite, which has a $3Al_2O_3 \cdot 2SiO_2$ composition.
b. Determine the amount of liquid in 50 w/o Al_2O_3–50 w/o SiO_2 at 1588°C.

Answer

a.

$$\frac{3Al_2O_3}{3Al_2O_3 + 2SiO_2} = \frac{6(27) + 9 \times 16}{6 \times 27 \times 9 \times 16 + 2 \times 28 + 4 \times 16}$$

$$= \frac{162 + 144}{306 + 56 + 64}$$

$$= \frac{306}{426}$$

$$= \underline{0.718\ (71.8\%)}.$$

b. Using the lever rule [see §5.1.1]:

$$\%L = \frac{71.8 - 50}{71.8 - 5.5} = \frac{21.8}{66.3} = \underline{0.329\ (32.9\%)}.$$

6.6. OTHER CERAMICS

There are many other ceramic materials studied as well, including titanium oxide (TiO_2), barium titanate ($BaTiO_3$), tricalcium phosphate ($Ca_3(PO_4)_2$), and calcium aluminate ($CaO \cdot Al_2O_3$). Titanium oxide was tried for use in a component of bone cement or as a blood-interfacing material. Porous calcium aluminate was used to induce tissue ingrowth into pores with the aim of achieving better implant fixation. However, this material loses its strength considerably after in vivo and in vitro aging, as shown in Figure 6-17. Tricalcium phosphate together with calcium aluminate were tried as biodegradable implants in the hope of regenerating new bone.

Barium titanate with a textured surface has been used in experimental trials to achieve improved fixation of implants to bone. This material is piezoelectric (following a polarization procedure). Therefore, mechanical loads on the implant will generate electrical signals that are capable of stimulating bone healing and ingrowth. These loads on the implant arise during use of the implanted limb. Alternatively, the implant can be exposed to ultrasound to generate electrical signals.

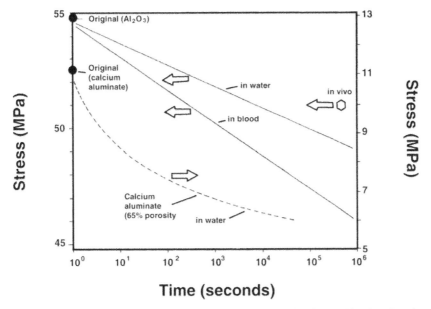

Figure 6-17. Aging effect on the strength of calcium aluminate in vitro and in vivo. Reprinted with permission from Schnittgrund et al. (1973). Copyright © 1973, Wiley.

6.7. CARBONS

Carbons can be made in many allotropic forms: crystalline diamond, graphite, noncrystalline glassy carbon, and partially crystalline (now referred to as icosahedral) pyrolytic carbon. Among these, only pyrolytic carbon is widely utilized for implant fabrication; it is normally used as a surface coating. It is also possible to coat surfaces with diamond-like carbon (DLC). This technique has the potential to improve performance of such medical devices as surgical knives, scissors, and articulating surfaces of joint implants; however, it is not, as of this writing, commercially available. This DLC coating is now used to coat razor blades.

6.7.1. Structure of Carbons

The crystalline structure of carbon as used in implants is similar to the graphite structure shown in Figure 6-18. The planar hexagonal arrays are formed by strong covalent bonds in which one valence electron per atom is free to move, resulting in high but anisotropic electric conductivity. The bonding between layers is stronger than the van der Waals force; therefore, *crosslinks* between them are considered to be present. Indeed, the remarkable lubricating property of graphite cannot be realized unless the crosslinks are eliminated.

The poorly crystalline carbons are thought to contain unassociated or unoriented carbon atoms. The hexagonal layers are not perfectly arranged, as shown in Figure 6-19. The strong bonding within layers and the weaker bonding between layers cause the properties of individual crystallites to be highly anisotropic. However, if the crystallites are randomly dispersed, then the aggregate becomes isotropic.

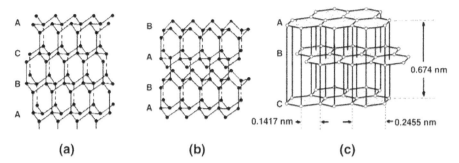

Figure 6-18. Crystal structure of graphite. Reprinted with permission from Shobert (1964). Copyright © 1964, Academic Press.

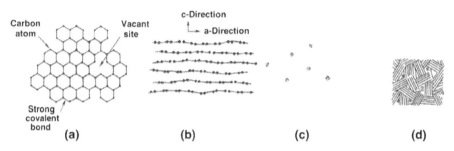

Figure 6-19. Schematic representation of poorly crystalline carbon: (**a**) single-layer plane; (**b**) parallel layers in a crystalline; (**c**) unassociated carbon; (**d**) an aggregate of crystallites, single layers and unassociated carbon. Reprinted with permission from Bokros (1972). Copyright © 1972, Marcel Dekker.

6.7.2. Properties of Carbon

The mechanical properties of carbon, especially pyrolytic carbon, are largely dependent on density, as shown in Figures 6-20 and 6-21. The increased mechanical properties are directly related to the increased density, which indicates the properties depend mainly on the aggregate structure of the material.

Graphite and glassy carbon have much lower mechanical strength than pyrolytic carbon, as given in Table 6-11. However, the average modulus of elasticity is almost the same for all carbons. The strength of pyrolytic carbon is quite high compared to graphite and glassy carbon. This is again due to the lesser amount of flaws and unassociated carbons in the aggregate.

A composite carbon that is reinforced with carbon fiber has been considered for implants. The properties are highly anisotropic, as given in Table 6-12. The density is in the range of 1.4–1.45 g/cm^3, with a porosity of 35–38%.

Carbons exhibit excellent compatibility with tissues. In particular, compatibility with blood has made pyrolytic carbon deposited heart valves and blood vessel walls a widely accepted part of the surgical armamentarium.

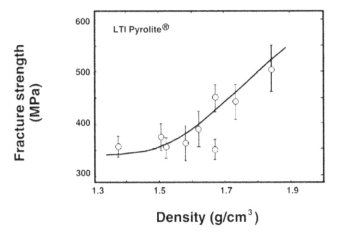

Figure 6-20. Fracture stress versus density for unalloyed LTI pyrolite carbons. Reprinted with permission from Kaae (1971). Copyright © 1971, Elsevier Science.

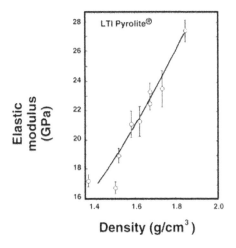

Figure 6-21. Elastic moduli versus density for unalloyed LTI pyrolite carbons. Reprinted with permission from Kaae (1971). Copyright © 1971, Elsevier Science.

6.7.3. Manufacture of Implants

Pyrolytic carbons can be deposited onto finished implants from hydrocarbon gas in a *fluidized bed* at a controlled temperature and pressure, as shown in Figure 6-22. The anisotropy, density, crystallite size, and structure of the deposited carbon can be controlled by temperature, composition of the fluidizing gas, bed geometry, and residence time (velocity) of the gas molecules in the bed. The microstructure of deposited carbon should be particularly controlled since the formation of growth features associated with uneven crystallization can result in a weaker material, as shown in Figure 6-23. It is also possible to introduce various other elements into the fluidizing gas and codeposit them with carbon. Usually silicon (10–20 w/o) is codeposited (or *alloyed*) to increase hardness for applications requiring resistance to abrasion, such as heart valve discs.

Table 6-11. Properties of Various Types of Carbon

	Types of carbon		
Properties	Graphite	Glassy	Pyrolytica
Density (g/cm^3)	1.5-1.9	1.5	1.5-2.0
Elastic modulus (GPa)	24	24	28
Compressive strength (MPa)	138	172	517 (575^a)
Toughness (mN/cm^3)^b	6.3	0.6	4.8

^{a} 1.0 w/o Si-alloyed pyrolytic carbon, Pyrolite® (Carbomedics, Austin, TX)
^{b} 1 m-N/cm^3 = 1.45 × 10^{-3} in-lb/in^3.

Table 6-12. Mechanical Properties of Carbon Fiber-Reinforced Carbon

	Fiber lay-up	
Property	Unidirectional 0–90°	Crossply
Flexural modulus (GPa)		
Longitudinal	140	60
Transverse	7	60
Flexural strength (MPa)		
Longitudinal	1,200	500
Transverse	15	500
Interlaminar shear strength (MPa)	18	18

Reprinted with permission from Adams and Williams (1978). Copyright © 1978, Wiley.

Pyrolytic carbon was deposited onto the surfaces of blood vessel implants made of polymers. This is called ultra-low-temperature isotropic (ULTI) carbon, instead of LTI (low-temperature-isotropic) carbon. The deposited carbon is thin enough not to interfere with the flexibility of grafts yet exhibits excellent blood compatibility.

The vitreous or glassy carbon is made by controlled pyrolysis of polymers such as phenolformaldehyde, rayon (Glasser et al., 1992), and polyacrylonitrile at high temperature in a controlled environment. This process is particularly useful for making carbon fibers and textiles, which can themselves be used or as components of composites.

6.8. DETERIORATION OF CERAMICS

It is of great interest to know whether inert ceramics such as alumina undergo significant static or dynamic fatigue. In one study it was shown that above a critical stress level the fatigue strength of alumina is reduced by the presence of water. This is due to delayed crack growth, which is accelerated by the water molecules. However, another study showed that a reduction in strength occurred if evidence of penetration by water was observed under a scanning electron microscope (SEM). No decrease in strength was observed for samples that showed no watermarks on the fractured surface, as shown in Figure 6-24. It was suggested that the presence of a minor amount of silica in one sample lot may have contributed to permeation of the

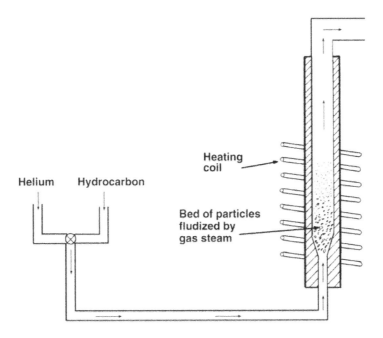

Figure 6-22. Schematic diagram showing particles being coated with carbon in a fluidized bed. Reprinted with permission from Bokros (1972). Copyright © 1972, Marcel Dekker.

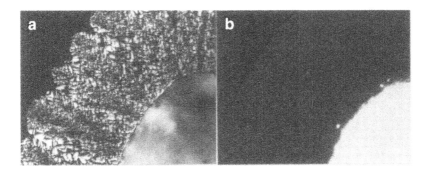

Figure 6-23. Microstructures of carbons deposited in a fluidized bed: (**a**) granular carbon with distinct growth features; (**b**) isotropic carbon without growth features. Both under polarized light, 240×. Reprinted with permission from Bokros (1972). Copyright © 1972, Marcel Dekker.

water molecules, which is detrimental to strength. It is not clear whether the same static fatigue mechanism operates in single-crystal alumina or not. It is, however, reasonable to assume that the same static fatigue will occur if the ceramic contains flaws or impurities, which will act as the source of crack initiation and growth under stress.

A study of the fatigue behavior of vapor-deposited pyrolytic carbon fibers (400–500 nm thick) onto a stainless steel substrate showed that the film did not break unless the substrate underwent plastic deformation at 1.3×10^{-2} strain and up to one million cycles of loading. Therefore, the fatigue is closely related to the substrate, as shown in Figure 6-25. A similar substrate–carbon adherence is the basis for the pyrolytic carbon-deposited polymer arterial grafts, as mentioned earlier.

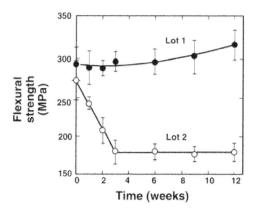

Figure 6-24. Flexural strength of dense alumina rods after aging under stress in Ringer's solution base indicating standard deviation. Lots 1 and 2 are from different batches of production. Reprinted with permission from Krainess and Knapp (1978). Copyright © 1978, Wiley.

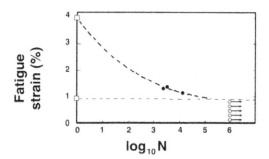

Figure 6-25. Strain versus number of cycles to failure: ○ = absence of fatigue cracks in carbon film; ● = fracture of carbon film due to fatigue failure of substrates; □ = data for substrate determined in single-cycle tensile test. Reprinted with permission from Shim and Haubold (1980). Copyright © 1980, Marcel Dekker.

The fatigue life of ceramics can be predicted by assuming that fatigue fracture is due to the slow growth of preexisting flaws. Generally, the strength distribution (s_i) of ceramics in an inert atmosphere can be correlated with the probability of failure F, by the following equation;

$$\ln \ln \left(\frac{1}{1-F} \right) = m \ln \left(\frac{s_i}{s_0} \right), \tag{6-3}$$

in which m and s_0 are constants. The m is called the Weibull modulus, which indicates the distribution of fracture strength: the higher the value, the narrower the distribution. The metals and polymers have values of 50, while most ceramics and glasses have values less than 20. Figure 6-26 shows a good fit for Bioglass®-coated alumina.

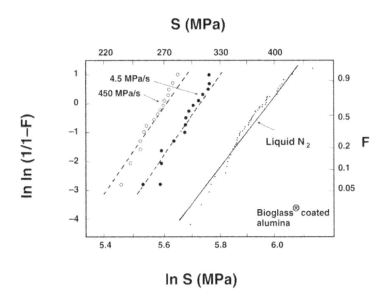

Figure 6-26. Plot of ln $[1/(1 - F)]$ versus ln s for Bioglass®-coated alumina in a tris hydroxyaminomethane buffer and liquid nitrogen. F is the probability of failure and σ is strength. Reprinted with permission from Ritter et al. (1979). Copyright © 1979, Wiley.

A minimum service life (t_{min}) of a specimen can be predicted by means of a proof test wherein it is subjected to stresses greater than those expected in service. Proof tests also eliminate weaker pieces. This minimum life can be predicted from the following equation:

$$t_{min} = B\sigma_p^{N-2}\sigma_a^{-N},\qquad(6\text{-}4)$$

in which σ_p is the proof test stress, σ_a is the applied stress, and B and N are constants. Rearranging Eq. (6-3), we obtain

$$t_{min}\sigma_a^2 = B\left(\frac{\sigma_p}{\sigma_a}\right)^2.\qquad(6\text{-}5)$$

Figure 6-27 shows a plot of Eq. (6-5) for alumina on a logarithmic scale.

Example 6-6
Calculate the proof stress of an alumina sample if it is to last for 20 years at 100 MPa in air and in Ringer's solution.

Answer

$$t_{min}\sigma_a^2 = 20 \text{ yr} \times 3.15 \times 10^7 \text{ s/yr} \times (100 \text{ MPa})^2,$$

$$\log t_{min}\sigma_a^2 = 12.8.$$

This value is comparable to the dotted line for 80 years at 69 MPa in Figure 6-27. Therefore, $\sigma_f/\sigma_a = 2.0$ in air and 2.55 in Ringer's solution. So $\sigma_p = \underline{200 \text{ MPa}}$ in air, and $\sigma_p = \underline{255 \text{ MPa}}$ in Ringer's solution. As one might expect, it is necessary to test the sample more rigorously in Ringer's solution than in air.

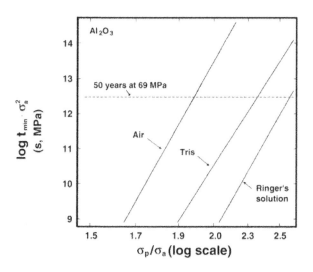

Figure 6-27. Plot of Eq. (6-45) for alumina after proof testing. $N = 43.85$, $m = 13.21$, and $\sigma_0 = 55,728$ psi. Reprinted with permission from Ritter et al. (1979). Copyright © 1979, Wiley.

PROBLEMS

6-1.   A ceramic is used to fabricate a hip joint. Assume a simple ball-and-socket configuration with a surface contact area of 1.0 cm² and continuous static loading in a simulated condition similar to Figure 6-17 (extrapolate the data if necessary). Observe that the actual contact area is smaller than the total surface area.

     a. How long will it last if the loading is a force due to 70 kg (mass), in water and in blood?

     b. Will the implant last a longer or shorter time with dynamic loading? Give reasons.

6-2.   From the phase diagram of Al_2O_3–SiO_2,

     a. Determine the exact mole % of Al_2O_3–SiO_2 for mullite, which has a $3Al_2O_3$–$2SiO_2$ composition.

     b. Determine the amount of mullite in 50 wt % Al_2O_3–50 wt% SiO_2 at 1469°C.

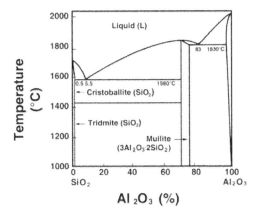

6-3.   Calculate the proof stress of an alumina sample if it is to last for 40 years at 100 MPa in air and in Ringer's solution.

6-4.   One is trying to coat the surface of an orthopedic implant with hydroxyapatite to enhance compatibility with tissues. List the possible problems associated with this technique. Give two methods of applying the coating.

6-5.   Discuss the advantages of coating the femoral stem of a hip prosthesis with hydroxyapatite, alumina, and diamond-like carbon. Give two general advantages to all ceramic coatings and, in addition, one or more specific advantage(s) for each coating. Show them in table form.

6-6.   Zirconia (ZrO_2) is often stabilized with calcium to provide an important refractory. The basic cell is ZrO_2 with 1 Ca^{2+} ion present for every 10 Zr^{4+} ions. Will the vacant sites be anion or cation? What percentage of the total number of all sites will be vacant?

6-7.   Calculate the density of monoclinic and tetragonal zirconia.

6-8.   Calculate the amount of volume change when the zirconia changes from monoclinic to tetragonal structure.

6-9.   Calculate the amount of volume change by adding 3 mole% of Y_2O_3 into cubic zirconia. Make assumptions of ideal mixing.

6-10.   Consider the following mechanical properties.

Properties	Bone	Vitreous carbon
Ultimate tensile strength (MPa)	100	120
Tensile modulus (GPa)	12–18	2.8

This comparison suggests that vitreous carbon would be an excellent material for bone replacement. What is wrong with this idea?

6-11.   Discuss the advantages of coating the femoral stem of a hip prosthesis with hydroxyapatite, alumina, and diamond-like carbon. Give two general advantages to all ceramic coatings and, in addition, one or more specific advantage(s) for each coating. Show them in a table form.

6-12.   CsCl has a simple cubic structure and the following data are given:

Element	Ion radius (nm)	amu
Cl	0.181	35.4
Cs	0.165	132.9

a. Calculate the minimum radius ratio of a simple cubic cell.

b. What is the coordination number of the Cs?

c. What is the density of CsCl?

d. Why is this material brittle?

6-13.   NaCl has a face-centered cubic structure and the following data are given:

Element	Ion radius (Å)	amu
Cl	1.81	35.4
Na	0.97	23

a. Draw the positions of Cl and Na ions in one of the cubic faces.

b. What is the coordination number of the Na?

c. Calculate the lattice parameter (a) of the unit cell.

d. Calculate the density of NaCl.

f. Which direction will have the highest linear density of ions: [100], [110], or [111]?

e. Why is this material brittle?

6-14.   From the list of ceramics choose the most appropriate one:

A. Al_2O_3 B. Hydroxyapatite   C. Tricalcium phosphate
D. Carbon   E. Glass-ceramic

a. Resorbable in vivo.

b. Has best blood compatibility

c. A large single crystal can be made and sometimes is called ruby or sapphire

d. Has a direct bone-bonding ability

e. Bone mineral has the similar structure

6-15.   From the following ceramic materials select the most appropriate one for the questions:

A. Al_2O_3 B. $Ca_3(PO_4)_2$ C. $Ca_{10}(PO_4)_6(OH)_2$ D. 45S5
E. LTI   F. ZrO_2 G. DLC

a. Osteogenic properties.

b. Used to make dental implants.

c. Hardest ceramic among those listed.

d. β-whitlockite forms in dry condition.

e. Readily resorbed in vivo.

f. AmXn structure and 2/3 available lattice sites are filled.

g. Fluidized bed is used to coat the surface of heart valve discs.

h. Glass ceramic and has capacity to bond directly with bone.

    i. Similar to the diamond, coated on a surface.

    j. Single crystal is used for making jewels.

6-16. Collagen and TGF-β (tissue growth factor) are mixed with hydroxyapatite (Ha et al., 1993) for making synthetic bone [BioME®] implants. A 10-mm diameter 20-mm long BioME® rod is made using <u>49–49–2% by weight</u> [2% is TGF-β] of the materials to test its properties. Using the following data, answer the following questions:

Material	Young's modulus (GPa)	Strength (MPa)	Density (g/cc)
Collagen	1	10 (yield)	1.0
HA	100	100 (fracture)	3.2
TGF-β	0	0	1.0

    a. Calculate the density of the BioME®.

    b. Calculate the yield strain of the BioME®.

    c. Calculate the Young's modulus of the BioME®.

    d. Calculate the maximum strength of the BioME®.

    e. Give at least one reason why you would <u>not</u> use TGF-β to make the BioME®.

SYMBOLS/DEFINITIONS

Words

Activation energy of phase transformation: Thermal energy required to overcome an energy barrier. $E_{activation} = RT$ ln(transformation rate), where R is the gas constant and T is temperature (K).

Alumina: Aluminum oxide (Al_2O_3), which is very hard (Mohs hardness of 9) and strong. Single crystals are called sapphire or ruby depending on color. Alumina is used to fabricate hip joint socket components or dental root implants.

Baddeleyite: Mineral containing zircon.

Biolox®: Trade name of alumina ceramic.

Calcium phosphate: A family of calcium phosphate ceramics including hydroxyapatite β-whitlockite, mono-, di-, tri-, and tetra-calcium phosphate, which are used to make substitute or augment artificial bone substitutes.

Cubic zirconia: Partially stabilized zirconia in cubic structure to prevent fracture during cooling.

Electronegativity: Potential of an atom to attract electrons, especially in the context of forming a chemical bond.

Fluorite (CaF_2) structure: AX_2 structure of ceramic, where A is a metal and X a nonmetal.

Glass-ceramics: A glass crystallized by heat treatment. Some of those have the ability of forming chemical bonds with hard and soft tissues. Bioglass® and Ceravital® are well-known examples.

Hydroxyapatite [$Ca_{10}(PO4)_6(OH)_2$]: A calcium phosphate ceramic with a calcium-to-phosphorus ratio of 5/3 and nominal composition. It has good mechanical properties and excellent biocompatibility. Hydroxyapatite is the mineral constituent of bone.

LTI carbon: A silicon alloyed pyrolytic carbon deposited onto a substrate at low temperature with isotropic crystal morphology. Highly blood compatible and used for cardiovascular implant fabrication, such as artificial heart valves.

Maximum radius ratio: The ratio of atomic radii computed by assuming the largest atom or ion which can be placed in a crystal's unit cell structure without deforming the structure.

Mohs scale: A hardness scale in which 10 (diamond) is the highest and 1(talc) is the softest.

Monoclinic structure: One crystal system having $a \neq b \neq c$, $\alpha = \gamma = 90°$.

Partially stabilized zirconia: See cubic zirconia.

Phase transformation toughening process: By transforming phases under stress the material becomes stronger due to the volume expansion of the transformed phase.

Tetragonal structure: One crystal system having $a = b \neq c$, $\alpha = \beta = \gamma = 90°$.

Wear factor: Similar to the wear constant; wear volume generated by given load and sliding distance .

Weibull modulus: Slope of the Weibull plot. Larger values indicate predictable failure stress. Ceramics are 10~20, metals are >50.

Weibull plot: Plot of the logarithm of probability of failure (F) [actually, ln ln $1/(1 - F)$] and test stress/strength fraction.

β-whitlockite ($3CaO \cdot P_2O_3$): One of the calcium phosphate compounds, similar to tricalcium phosphate, $3Ca \cdot PO_4$.

Yttrium oxide (Y_2O_3): Oxide used to partially stabilize zirconia.

Zircon ($ZrSiO_4$): Crystalline mineral, a silicate of zirconium with tetragonal structure.

Zirconia (ZrO_2): Zirconium oxide, which is hard and strong.

BIBLIOGRAPHY

Adams D, Williams DF. 1978. Carbon fiber-reinforced carbon as a potential implant material. *J Biomed Mater Res* **12**:35–42.

ASTM. 2000. *Annual book of ASTM standards*. West Conshohocken, PA: ASTM.

Beckham CA, Greenlee Jr TK, Crebo AR. 1971. Bone formation at a ceramic implant interfacing. *Calcif Tissue Res* **8**:165–171.

Blencke BA, Bromer H, Deutscher KK. 1978. Compatibility and long-term stability of glass-ceramic implants. *J Biomed Mater Res* **12**(3):307–316.

Bokros JC, LaGrange LD, Schoen GJ. 1972. Control of structure of carbon for use in bioengineering. In *Chemistry and physics of carbon*, pp. 103–171. Ed PL Walker. New York: Dekker.

Bokros JC, Atkins RJ, Shim HS, Haubold AD, Agarwal MK. 1976. Carbon in prosthetic devices. In *Petroleum derived carbons*, pp. 237–265. Ed ML Deviney, TM O'Grady. Washington, DC: American Chemical Society.

Burger W, Willmann G. 1993. Advantages and risks of zirconia ceramics in biomedical applications. In *Bioceramics*, Vol. 6, pp. 299-304. Ed P Ducheyne, D Christiansen. Oxford: Pergamon.

Callister JWD. 1994. *Materials science and engineering: an introduction*, 3rd ed. New York: Wiley.

Drennan J, Steele BCH. 1986. Zirconia and hafnia. In *Encyclopedia of materials science and engineering*, pp. 5542–5545. Ed MB Beaver. Oxford: Pergamon/MIT Press.

Gill RM. 1972. *Carbon fibres in composite materials*. London: Butterworths.

Gilman JJ. 1967. The nature of ceramics. *Sci Am* **218**(3):113.

Gitzen WH, ed. 1970. *Alumina as a ceramic material*. Westerville, OH: American Ceramic Society.

Glasser WG, Hatakeyama H, eds. 1992. *Viscoelasticity of biomaterials*. Washington, DC: American Chemical Society.

Ha S-W, Mayer J, Wintermantel E. 1993. Micro-mechanical testing of hydroxyapatite coatings on carbon fiber reinforced thermoplastics. In *Bioceramics*, Vol. 6, 489–493. Ed P Ducheyne, D Christiansen. Oxford: Pergamon.

Hastings GW, Williams, DF, eds. 1980. *Mechanical properties of biomaterials*, Part 3. Chichester: Wiley.

Hench LL, Ethridge EC. 1982. *Biomaterials: an interfacial approach*. New York: Academic Press.

Horwitz BR, Rockowitz NL, Goll SR, Booth Jr RE, Balderston RA, Rothman RH, Cohn JC. 1993. A prospective randomized comparison of two surgical approaches to total hip arthroplasty. *Clin Orthop Relat Res* **291**:154–163.

Hulbert SF, Young FA, eds. 1978. *Use of ceramics in surgical implants*. New York: Gordon & Breach.

Hulbert SF, King FW Jr, Klawitter JJ. 1972. Initial surface interaction of blood and blood components with Al_2O_3 and TiO_2. *J Biomed Mater Res Symp* **2**:69–89

Kaae JL. 1971. Structure and mechanical properties of isotropic pyrolytic carbon deposited below 1600 degrees. *J Nucl Mater* **38**:42–50.

Kawahara H, Hirabayashi M, Shikita T. 1980. Single crystal alumina for dental implants and bone screws. *J Biomed Mater Res* **14**:597–606.

Kingery WD, Bowen HK, Uhlmann DR. 1976. *Introduction to ceramics*, 2nd ed. New York: Wiley.

Krainess FE, Knapp WJ. 1978. Strength of a dense alumina ceramic after aging in vitro. *J Biomed Mater Res* **12**:241–246.

Kumar P, Shimizu K, Oka M. 1991. Low wear rate of UHMWPE against zirconia ceramic (Y-PSZ) in comparison to alumina ceramic and SUS 316L alloy. *J Biomed Mater Res* **25**:813–828.

McConell D. 1973. *Apatite: its crystal chemistry, mineralogy, utilization, and biologic occurrence*. Berlin: Springer-Verlag.

McMillan PW. 1979. *Glass-ceramics*. New York: Academic Press.

Niwa S, Sawai K, Takahashie S, Tagai H, Ono M, Fukuda Y. 1980. Experimental studies on the implantation of hydroxyapatite in the medullary canal of rabbits. In *Proceedings of the first world biomaterials congress* (Baden, Austria). Berlin: Springer-Verlag.

Norton F. 1974. *Elements of ceramics*. Reading, MA: Addison-Wesley.

Ogino M, Ohuchi F, Hench LL. 1980. Compositional dependence of the formation of calcium phosphate film on Bioglass. *J Biomed Mater Res* **14**:55–64.

Posner AS, Perloff A, Diorio AD. 1958. Refinement of the hydroxyapatite structure. *Acta Crystallogr* **11**:308–309.

Ritter Jr JE, Greenspan DC, Palmer RA, Hench LL. 1979. Use of fracture of an alumina and Bioglass-coated alumina. *J Biomed Mater Res* **13**:251–263.

Schnittgrund GS, Kenner GH, Brown SD. 1973. In vivo and in vitro changes in strength of orthopedic calcium aluminate. *J Biomed Mater Res Symp* **4**:435–452.

Shim HS, Haubold AD. 1980. The fatigue behavior of vapor deposited carbon films. *Biomater Med Devices Artif Organs* **8**:333–344.

Shimizu K, Oka M, Kumar P, Kotoura Y, Yamamuro T, Makinouchi K, Nakamura T. 1993. Time-dependent changes in the mechanical properties of zirconia ceramic. *J Biomed Mater Res* **27**:729–734.

Shobert EI. 1964. *Carbon and graphite*. New York: Academic Press.

Starfield MJ, Shrager AM. 1972. *Introductory to materials science*. New York: McGraw-Hill.

Willmann G. 1993. Zirconia: a medical-grade material? In *Bioceramics*, Vol. 6. pp. 271–276. Ed P Ducheyne, D Christiansen. Oxford: Pergamon.

Yoshimura M, Suda H, eds. 1994. *Hydrothermal processing of hydroxyapatite: past, present, and future. Hydroxyapatite and related materials*. Boca Raton, FL: CRC Press.

7

POLYMERIC IMPLANT MATERIALS

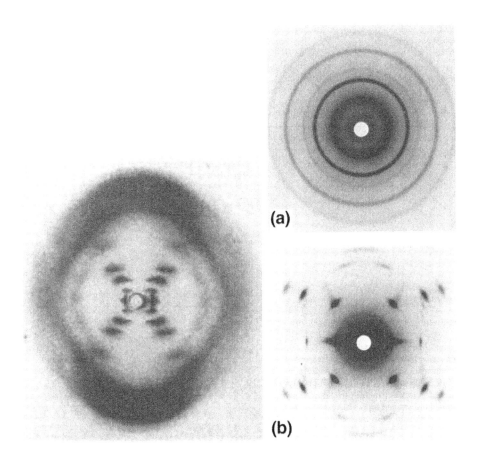

(a)

(b)

Left: X-ray diffraction of wet DNA showing B form double helix. Photo taken by R. Franklin and R. Gosling on May 2, 1952 (http://www.ba-education.demon.co.uk/for/science/dnamain.html). **Right**: X-ray diffraction pattern of POM (polyoxymethylen) (**a**) before and (**b**) after orientation. Reprinted with permission from Billmeyer (1962). Copyright © 1962, Wiley.

Polymers (poly = many, mer = unit) are made by linking small molecules (mers) through *primary covalent* bonding in the main molecular chain backbone with C, N, O, Si, etc. One example is polyethylene, which is made from ethylene ($CH_2=CH_2$), where the carbon atoms share electrons with two other hydrogen and carbon atoms: $-CH_2-(CH_2-CH_2)_n-CH_2-$, in which n indicates the number of repeating units. Also note (in view of the monomer structure) that the repeating unit is $-CH_2CH_2-$, not $-CH_2-$.

In order to make a strong solid, the *repeating unit* (n) should be well over 1,000. For example, the molecular weight (m.w.) of the polyethylene is over 28,000 grams per mole. This is why the polymers are made of *giant molecules*. At low m.w. the material behaves as a wax (paraffin wax used for household candles) and at still lower m.w. as an oil or gas.

The main backbone chain can be made up entirely of different atoms; for example, the polydimethyl siloxane (silicone rubber) $-Si(CH_3)_2[O-Si(CH_3)_2]_nO-$. The side group atoms can be changed; thus, if we substitute the hydrogen atoms in polyethylene with fluorine (F), the resulting material is well known as Teflon® (polytetrafluoroethylene).

Absorbable polymers in the body (bioabsorbable or biodegradable) will be studied in the tissue engineering chapter (Chapter 16) since these are widely used for that purpose.

The elastomeric polymers were first developed for making synthetic rubbers for military purpose. A good understanding and synthesis of various polymers were accelerated since World War II as, shown in Table 7-1.

Table 7-1. History of Some Commercially Important Polymers

Date	Polymer	Date	Polymer
1930	Styrene-butadiene rubber	1944	Polyethylene terephthalate
1936	Polyvinyl chloride	1947	Epoxies
1936	Polychloroprene (Neoprene)	1955	Polyethylene, linear
1936	Polymethylmethacrylate	1956	Polyoxymethylene
1937	Polystyrene	1957	Polypropylene
1939	Nylon 66	1957	Polycarbonates
1941	Polytetrafluoroethylene	1964	Ionomer resins
1942	Unsaturated polyesters	1965	Polyimides
1943	Polyethylene, branched	1970	Thermoplastic elastomers
1943	Nylon 6	1974	Aromatic polyamides
1943	Silicones	1980s	Ultrahigh-molecular-weight polyethylene

Modified with permission from Billmeyer (1984). Copyright © 1984, Wiley.

7.1. POLYMERIZATION AND PROPERTIES

7.1.1. Polymerization

In order to link small molecules, one has to force them to lose their electrons by the chemical processes of condensation and addition. By controlling the reaction temperature, pressure, and time in the presence of catalyst(s), the degree to which repeating units are put together into chains can be manipulated.

7.1.1.a. Condensation or Step Reaction Polymerization

During condensation polymerization, a small molecule such as water will be condensed out by the following chemical reaction:

$$R-NH_2 \quad + \quad R'COOH \quad \Leftrightarrow \quad R'CONHR \quad + \quad H_2O \qquad (7\text{-}1)$$

| (amine) | (carboxylic acid) | (amide) | (condensation molecule) |

This particular process is used to make polyamides (nylons). Nylon was the first commercial polymer, made in the 1930s.

Table 7-2. Typical Condensation Polymers

Type	Interunit linkage
Polyester	$\begin{array}{c} O \\ \parallel \\ -C-O- \end{array}$
Polyamide	$\begin{array}{cc} O & H \\ \parallel & \vert \\ -C- & N- \end{array}$
Polyurea	$\begin{array}{ccc} H & O & H \\ \vert & \parallel & \vert \\ -N- & C- & N- \end{array}$
Polyurethane	$\begin{array}{cc} O & H \\ \parallel & \vert \\ -O-C- & N- \end{array}$
Polysiloxane	$\begin{array}{c} R \\ \vert \\ -Si-O- \\ \vert \\ R \end{array}$
Protein	$\begin{array}{cc} O & H \\ \parallel & \vert \\ -C- & N- \end{array}$
Cellulose	$-C-O-C-$

Some typical condensation polymers and their interunit linkages are given in Table 7-2. One major drawback of condensation polymerization is the tendency for the reaction to cease before the chains grow to a sufficient length. This is due to the decreased mobility of the chains and reactant chemical species as polymerization progresses. This results in short chains. However, in the case of nylon the chains are polymerized to a sufficiently large extent before this occurs and the physical properties of the polymer are preserved.

Natural polymers, such as polysaccharides and proteins are also made by condensation polymerization. The condensing molecule is always water (H_2O).

7.1.1.b. Addition or Free Radical Polymerization

Additional polymerization can be achieved by rearranging the bonds within each monomer. Since each "mer" has to share at least two covalent electrons with other mers, the monomer has to have at least one double bond. For example, in the case of ethylene:

$$n \begin{array}{c} H \quad H \\ \vert \quad \vert \\ C = C \\ \vert \quad \vert \\ H \quad H \end{array} \Rightarrow \begin{array}{c} H \quad H \quad H \quad H \\ \vert \quad \vert \quad \vert \quad \vert \\ -C-(C-C)_n-C- \\ \vert \quad \vert \quad \vert \quad \vert \\ H \quad H \quad H \quad H \end{array} \qquad (7\text{-}2)$$

The breaking of a double bond can be made with an *initiator*. This is usually a free radical such as benzoyl peroxide:

$$C_6H_5COO-OOCC_6H_5 \Rightarrow 2C_6H_5COO\cdot \Rightarrow 2\ C_6H_5\cdot\ + 2CO_2 \qquad (7\text{-}3)$$
$$(R\cdot)$$

The initiation can be activated by heat, ultraviolet light, and other chemicals.

The free radicals (initiators) can react with monomers:

$$
\begin{array}{c}
\text{H} \\
| \\
R\cdot + CH_2{=}CHX \Rightarrow RCH_2{-}C\cdot \\
| \\
\text{X}
\end{array}
\qquad (7\text{-}4)
$$

and this free radical can react with another monomer:

$$
\begin{array}{cc}
\text{H} & \text{H} \\
| & | \\
RCH_2{-}C\cdot + CH_2{=}CHX \Rightarrow RCH_2{-}CHX{-}CH_2{-}C\cdot \\
| & | \\
\text{X} & \text{X}
\end{array}
\qquad (7\text{-}5)
$$

and the process can continue on.

This process is called *propagation* and can be written in short form as

$$R\cdot + M \Rightarrow RM\cdot$$
$$RM\cdot + M \Rightarrow RMM\cdot \qquad (7\text{-}6)$$

where M is a monomer.

Table 7-3. Monomers for Addition Polymerization and Suitable Processes

Monomer name	Chemical formula	Polymerization mechanism			
		Radical	Cationic	Anionic	Coordin.
Acrylonitrile	$CH_2{=}CH{-}CN$	+	−	+	+
Ethylene	$CH_2{=}CH_2$	+	+	−	−
Methyacrylate	$CH_2{=}CH{-}COOCH_3$	+	−	+	+
Methylmethacrylate	$CH_2{=}CCH_3COOCH_3$	+	−	+	+
Propylene	$CH_2{=}CHCH_3$	−	−	−	+
Styrene	$CH_2{=}CH{-}C_6H_5$	+	+	+	+
Vinyl chloride	$CH_2{=}CHCl$	+	−	−	+
Vinylidene chloride	$CH_2{=}CCl_2$	+	−	+	−

+ = high polymer formed; − = no reaction or oligomers only.

Reprinted with permission from Billmeyer (1984). Copyright © 1984, Wiley.

The propagation process can be *terminated* by combining two free radicals, by transfer or by disproportionate processes, respectively, in the following reactions:

$$RM_nM\cdot + R\cdot \text{ (or } RM\cdot) \Rightarrow RM_{n+1}R \text{ (or } RM_{n+2}R) \qquad (7\text{-}7)$$

$$RM_nM\cdot + RH \Rightarrow RM_{n+1}H + R\cdot \qquad (7\text{-}8)$$

$$RM_nM\cdot + \cdot MM_nR \Rightarrow RM_{n+1} + M_{n+1}R \qquad (7\text{-}9)$$

An example of the disproportionate termination is given below:

$$
\begin{array}{cccc}
\text{H} & \text{H} & \text{H} & \text{H} \\
| & | & | & | \\
-\text{CH}_2\text{C}\cdot + \cdot\text{C}-\text{CH}_2- & \Rightarrow & -\text{CH}_2\text{CH} + \text{C}=\text{CH}- \\
| & | & | & | \\
\text{X} & \text{X} & \text{X} & \text{X}
\end{array}
$$

Some of the commercially important monomers for addition polymers are given in Table 7-3.

There are three more types of initiating species for additional polymerization aside from free radicals; cations, anions, and coordination (stereospecific) catalysts. Some monomers can employ two or more of the initiation processes, but others can use only one process, as given in Table 7-3.

Example 7-1

A polytetrafluoroethylene (PTFE) crystal has a rhombohedral unit cell structure as with the polyethylene which has a repeating distance of 2.54 Å. By contrast, PTFE has values of 16.9 to 19.5 Å depending on the temperature (below 19°C, 13 repeating units $(-\text{CF}_2)-$, between 19 and 30°C, 15 repeating units). A 15-repeating-unit PTFE crystal has a hexagonal structure with a = 5.65 Å and c = 19.5 Å (Geil, 1963).

Calculate its theoretical density.

Answer

$$\text{Mass } M = 15(12+19*2)/6.02*1023 = 1.25*10\text{-}21 \text{ g/mol},$$

$$\text{Volume } V = (\sqrt{3/2})*5.65\text{C}^2*19.5\text{C} = 5.39*10^{-22} \text{ cm}^3/\text{mol}.$$

Thus, $\rho = M/V = \underline{2.32 \text{ g/cm}^3}$ (theoretical value for single-crystal PTFE, i.e., 100% crystallinity). The following illustrations shows the repeating units for polyethylene and PTFE at low and high temperatures:

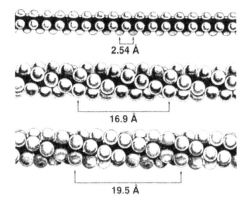

7.2. EFFECT OF STRUCTURAL MODIFICATION AND TEMPERATURE ON PROPERTIES

The physical properties of polymers can be affected in many ways. In particular, the chemical composition and arrangement of chains will have a great effect on the final properties. By such means we can tailor polymers to meet a specific end use.

7.2.1. Effect of Molecular Weight and Composition

The molecular weight and its distribution have a great effect on the properties of a polymer since its rigidity is primarily due to the immobilization or *entanglement* of the chains. This is because the chains are arranged like cooked spaghetti strands in a bowl. By increasing the molecular weight, the polymer chains become longer and less mobile, and a more rigid material results, as shown in Figure 7-1. Equally important is that all chains should be equal in length, since if there are short chains they will act as *plasticizers*. Plasticizers can lower the melting and glass transition temperatures and rigidity (i.e. modulus of elasticity, density, etc.), since the small molecules will facilitate the movement of chains and interfere with (efficient) packing of long-chain molecules.

Figure 7-1. Approximate relations among molecular weight, T_g, T_m, and polymer properties.

Another obvious way of changing properties is to change the *chemical composition* of the backbone (the main molecular chain) or of side chains. Substituting the backbone carbon of a polyethylene with divalent oxygen or sulfur will decrease the melting and glass transition temperatures since the chain becomes more flexible due to the increased rotational freedom.

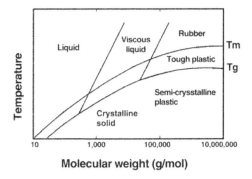

(7-10)

On the other hand, if the backbone chains can be made more rigid, then a stiffer polymer will result, as with the case of polyester. The chemical structure of polyethylene terephthalate (polyester, Dacron®) is as follows:

(7-11)

7.2.2. Effect of Side-Chain Substitution, Crosslinking, and Branching

Increasing the size of side groups in linear polymers such as polyethylene will decrease the melting temperature due to the lesser perfection of molecular packing, i.e., decreased crystallinity. This effect is seen until the side group itself becomes large enough to hinder the movement of the main chain, as shown in Table 7-4. Very long side groups can be thought of as branches.

Table 7-4. Effect of Side Chain Substitution on Melting Temperature in Polyethylene

Side chain	T_m (°C)
–H	140
–CH$_3$	165
–CH$_2$CH$_3$	124
–CH$_2$CH2CH3	75
–CH$_2$CH$_2$CH$_2$CH$_3$	–55
–CH$_2$CH CH$_2$CH$_3$ $\quad\quad$│ $\quad\quad$CH$_3$	196
$\quad\quad$CH$_3$ $\quad\quad$│ –CH$_2$–C–CH$_2$CH$_3$ $\quad\quad$│ $\quad\quad$CH$_3$	350

Crosslinking (see also §7.3.6 on rubbers and §7.2.3) of the main chains (Figure 2-12) is in effect similar to side-chain substitution with a small molecule, i.e., it lowers the melting temperature. This is due to the interference of crosslinking, which causes lesser mobility of the chains, resulting in further retardation of the crystallization rate. In fact, a large degree of crosslinking can prevent crystallization completely. However, when the crosslinking density increases for a rubber, the material becomes harder and the glass transition temperature also increases.

7.2.3. Effect of Temperature on Properties

Amorphous polymers undergo a substantial change in their properties as a function of temperature. The *glass transition temperature* (T_g) is a demarcation between the glassy region of behavior, in which the polymer is relatively stiff, and the rubbery region, in which it is very compliant. T_g can also be defined as the temperature at which the slope of volume change versus temperature has a discontinuity in slope, as discussed in Chapter 2. Since the polymers are viscoelastic, the value obtained in this measurement depends on how fast it is taken. An alternative definition of the glass transition temperature is the temperature of the peak in the viscoelastic loss tangent, as shown in Figure 7-2. Materials considered as hard polymers, such as PMMA, have glass transition temperatures well above room temperature. Rubbery materials have glass transition temperatures below room temperature; they become rigid at sufficiently low temperatures but are flexible near room temperature.

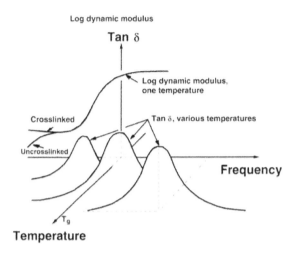

Figure 7-2. Glass transition temperature obtained from the peak in the loss tangent curve at a given frequency. The stiffness undergoes an abrupt change at the glass transition temperature. An uncrosslinked polymer exhibits continued mobility of chains, resulting in a decrease of stiffness at low frequency corresponding to flow or creep at long times.

As for crosslinks, as discussed in §7.2.2, they serve the function of maintaining the solidity of the polymer when it is subjected to prolonged load. An un-crosslinked polymer exhibits continued mobility of chains, resulting in a decrease of stiffness at low frequency, corresponding to flow or creep at long times, as shown in Figure 7-2. If the characteristic retardation times of this creep are less than the expected time of service of the implant, undesired deformation will occur. Vulcanization, or heat treatment with sulfur, is a crosslinking process used in industry to prevent this progressive creep in rubbery polymers.

7.3. POLYMERIC IMPLANT MATERIALS

Polymeric materials have a wide variety of applications for implantation since they can be easily fabricated into many forms: fibers, textiles, films, rods, and viscous liquids. Polymers bear a close resemblance to natural polymeric tissue components such as collagen. In some cases it is possible to achieve a bond between synthetic polymers and natural tissue polymers. Examples include the bonding of heparin protein on the surface of polymers (silicone, urethane rubbers, etc.) for the prevention of blood clotting, and the use of cyanoacrylates as tissue adhesives. Adhesive polymers can be used to close wounds or lute orthopedic implants in place.

Whenever possible, ASTM standards for the specifications and tests should be utilized to get more uniform and better results when the material is implanted. Only a few plastics and rubbers are listed by the ASTM, and these are indicated in the text whenever possible.

7.3.1. Polyamides (Nylons)

Polyamides are known as nylons and are designated by the number of carbon atoms in the repeating units. Nylons can be polymerized by step-reaction (or condensation) and ring-scission polymerization. They have excellent fiber-forming ability due to interchain hydrogen bonding and a high degree of crystallinity, which increases strength in the fiber direction.

The basic chemical structure of the repeating unit of polyamides can be written in two ways,

$$-[NH(CH_2)_x NHCO(CH_2)_y CO]_2- \tag{7-12}$$

and

$$-[NH(CH_2)_x CO]_n- \tag{7-13}$$

Structure (7-12) represents polymers made from diamine and diacids such as type 66 ($x = 6$, $y = 4$) and 610 ($x = 6$, $y = 8$). Polyamides made from ω-acids (caprolactums, Eq. (7-13)) are designated as nylon 6 ($x = 5$), 11 ($x = 10$), and 12 ($x = 11$). These types of polyamides can be produced by ring-scission polymerization, for example:

$$\text{(7-14)}$$

Caprolactam   Nylon 6

The presence of –CONH– groups in polyamides attracts the chains strongly toward one another by hydrogen bonding. as shown in Figure 7-3. Since the hydrogen bond plays a major role in determining properties, the number and distribution of –CONH– groups are important factors. For example, the glass transition temperature (T_g) can be decreased by decreasing the number of CONH groups, as given in Table 7-5. On the other hand, an increase in the number of –CONH– groups improves such physical properties as strength: Nylon 66 is stronger than Nylon 610, and 6 is stronger than Nylon 11.

In addition to the higher nylons (610 and 11), there are aromatic polyamides named aramids. One of them is poly (p-phenylene terephthalate) commonly known as Kevlar®, made by DuPont:

$$\text{(7-15)}$$

This material is readily made into fibers. The specific strength (strength per unit weight) of such fibers is five times that of steel, and it is consequently most suitable to make composites.

The nylons are hygroscopic and lose their strength in vivo when implanted. The water molecules serve as *plasticizers*, which attack the amorphous region. Proteolytic enzymes also aid to hydrolyze by attacking the amide group. This is probably due to the fact that the proteins also contain the amide group along their molecular chains, which the proteolytic enzymes can attack.

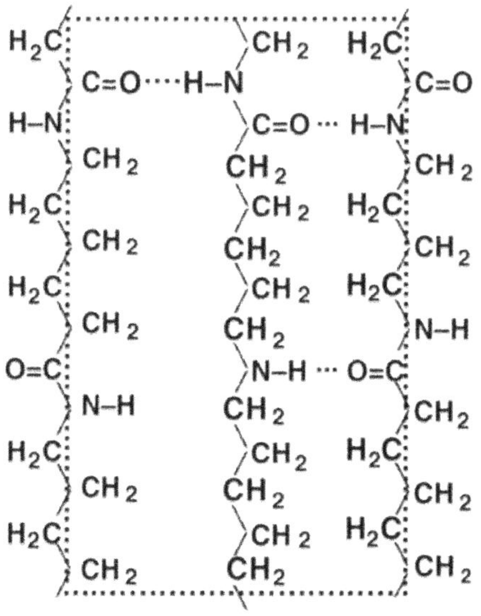

Figure 7-3. Hydrogen bonding in polyamide chains (nylon 6).

Table 7-5. Properties of Polyamides

Properties	66	610	6	11	Aramid[a]	Kevlar[b]
Density (g/cm³)	1.14	1.09	1.13	1.05	1.30	1.45
Tensile strength (MPa)	76	55	83	59	120	2700
Elongation (%)	90	100	300	120	<80	2.8
Modulus of elasticity (GPa)	2.8	1.8	2.1	1.2	>2.8	130
Softening temperature (°C)	265	220	215	185	275	–

[a] Molded parts, unfilled.

[b] Kevlar® 49 (duPont) fibers.

7.3.2. Polyethylene

Polyethylene and polypropylene and their copolymers are called polyolefins. These are linear thermoplastics. Polyethylene is available commercially in three major grades: low and high density and ultra high molecular weight (UHMWPE). Polyethylene has the repeating unit structure

$$-(-\underset{\underset{H}{|}}{\overset{\overset{H}{|}}{C}}-\underset{\underset{H}{|}}{\overset{\overset{H}{|}}{C}}-)_n- \tag{7-16}$$

which can be readily crystallized. In fact, it is almost impossible to produce noncrystalline polyethylene because of the small hydrogen side groups, which causes high mobility of chains.

The first polyethylene was synthesized by reacting ethylene gas at high pressure (100–300 MPa) in the presence of a catalyst (peroxide) to initiate polymerization. The process yields the *low*-density polyethylene. By using a *Ziegler catalyst* (stereospecific), high-density polyethylene can be produced at low pressure (10 MPa). Unlike the low-density variety, high-density polyethylene does not contain branches. The result is better packing of chains, which increases density and crystallinity. The crystallinity is usually 50 to 70% and 70 to 80% for the low- and high-density polyethylenes, respectively. Some important physical properties of polyethylenes are given in Table 7-6.

Table 7-6. Properties of Polyethylene

Properties	Low density	High density	UHMWPE[a]	Enhanced UHMWPE[b]
Molecular weight (g/mol)	$3\sim4 \times 10^3$	5×10^5	2×10^6	same
Density (g/cm^3)	0.90–0.92	0.92–0.96	0.93–0.94	same
Tensile strength (MPa)	7.6	23–40	27 min.	higher
Elongation (%)	150	400–500	200–250	same
Modulus of elasticity (MPa)	96–260	410–1,240	c	2,200
Crystallinity (%)	50–70	70–80	d	e

[a] Data from ASTM F648; also. 2% deformation after 90 minutes recovery subjected to 7 MPa for 24 hours (D621).

[b] Same as the conventional UHMWPE (ASTM, F648). Data from *A new enhanced UHMWPE for orthopaedic applications: a technical brief*, Warsaw, IN: DePuy, 1989.

[c] Close to 2,200 MPa.

[d] Higher than high-density polyethylene

[e] Equal or slightly higher than d.

The *ultrahigh-molecular-weight* polyethylene (m.w. $> 2 \times 10^6$ g/mol) has been used extensively for orthopedic implant fabrications, especially for such load-bearing surfaces as total hip and knee joints. A modified UHMWPE introduced by duPont with DePuy (Warsaw, Indiana) has been claimed to have a longer chain fold length than the conventional UHMWPE, thus increasing crystallinity. Since the folded chains are crystalline, the amount of amorphous region is reduced, thus reducing the possibility of an environmental attack (usually oxidation). A more crystalline polyethylene also exhibits enhanced mechanical properties (higher hardness, modulus of elasticity, tensile yield strength). Moreover, creep properties are superior (i.s., less creep), which is an important factor for designing joint implants. Other properties such as coefficient of friction and wear are improved marginally. This material has no known effective solvent at room temperature; therefore, only high temperature and pressure sintering may be used to produce desired products. Conventional extrusion or molding processes are difficult to employ.

Recently, crosslinked ultrahigh-molecular-weight polyethylene has been developed for the use of articulating joint materials such as the acetabular cup of a hip joint prosthesis. A particular advantage of crosslinked UHMWPE in this application is that it has a low wear rate compared to the conventional uncrosslinked material. Although the crosslinking is similar to that performed for elastomeric (rubbery) polymers used in tires, the polyethylene is much stiffer than rubber. During crosslinking the material also becomes weaker and sometimes changes color from white to brown. Crosslinking of polyethylene can be achieved by ionizing radiation

(γ rays from ^{60}Co or electron beam irradiation) or by chemical reactions. Crystallinity decreases after crosslinking.

Polyethylene has been used in solid or porous form. Biocompatibility tests for nonporous (F981) and porous polyethylene (F639 and 755) are given by ASTM standards.

7.3.3. Polypropylene

Polypropylene can be synthesized by using a Ziegler type stereospecific catalyst, which controls the position of each side group as it is being polymerized to allow the formation of a regular chain structure from the asymmetric repeating unit:

$$\begin{array}{cc} H & CH_3 \\ | & | \\ -(C\!-\!C)_n\!- \\ | & | \\ H & H \end{array} \tag{7-17}$$

Three types of structure can exist, depending on the position of the methyl (CH_3) group along the polymer chain. The random distribution of methyl groups in the *atactic* polymer prevents close packing of chains and results in amorphous polypropylene. In comparison, the *isotactic* and *syndiotactic* structures have a regular position of the methyl side groups in the same side and alternate side, respectively. They usually crystallize. However, the presence of the methyl side groups restricts movement of the polymer chains, and crystallization rarely exceeds 50–70% for material with isotacticity over 95%. Table 7-7 lists typical properties of commercial polypropylenes, which are largely an atactic type of structure.

Table 7-7. Properties of Polypropylene

Properties	Values
Density (g/cm^3)	0.90–0.91
Tensile strength (MPa)	28–36
Elongation (%)	400–900
Modulus of elasticity (GPa)	1.1–1.55
Softening temperature (°C)	150

Polypropylene has an exceptionally high flexural fatigue life; hence it has been used to make integrally molded hinges for finger joint prostheses. It also has excellent environmental stress-cracking resistance. The permeability of polypropylene to gases and water vapor is between that of low- and high-density polyethylene.

Example 7-2

Calculate the percent crystallinity of UHMWPE assuming that the noncrystalline and 100% crystalline polyethylene have densities of 0.85 and 1.01 g/cm^3, respectively.

Answer

From Table 7-6, the density of UHMWPE is 0.93~0.94; therefore,

$$\% \text{ crystallinity} = \frac{0.95 - 0.85}{1.01 - 0.85} = \underline{0.56, \text{ or } 56\%}.$$

This value is lower than that of the low- and high-density polyethylene, as given in Table 7-6. The tabulated crystallinities were based on an x-ray diffraction measurement technique instead of the density method.

7.3.4. Polyacrylates

These polymers are used extensively in medical applications as (hard) contact lenses, implantable ocular lenses, and as bone cement for joint prosthesis fixation. Dentures and maxillofacial prostheses are also made from acrylics because they have excellent physical and coloring properties and are easy to fabricate.

7.3.4.a. Structure and Properties of Acrylics and Hydrogels

The basic chemical structure of repeating units of acrylics can be represented by

$$-(CH_2-\underset{\underset{COOR_2}{|}}{\overset{\overset{R_1}{|}}{C}})_n- \qquad (7-18)$$

The only difference between polymethyl acrylate (PMA) and polymethyl methacrylate (PMMA) is the R groups of structure (7-18). The R_1 and R_2 groups for PMA are H and CH_3, and for PMMA they are both CH_3. These polymers are additionally (or free radical) polymerized. These polymers can be obtained in liquid monomer or fully polymerized beads, sheets, rods, etc.

The bulky side groups inhibit crystallization; therefore, these polymers are usually amorphous, and lacking heterogeneities to scatter light, are transparent. PMMA has bulkier side groups than the PMA, and the PMMA thus has a higher tensile strength (60 MPa) and softening temperature (125°C) than PMA (7 MPa and 33°C) if the molecular weights are similar. PMMA has an excellent light transparency (92% transmission), a relatively high index of refraction (1.49), and excellent weathering properties. This material can be cast, molded, or machined with conventional tools. It has an excellent chemical resistivity and is highly biocompatible in pure form. The material is very hard and brittle in comparison with other polymers.

The first *hydrogel* polymer developed is the poly(hydroxyethyl methacrylate) or poly-HEMA, which can absorb water more than 30% of its weight. This property makes it useful for soft lens applications. The chemical formula is similar to that of (7-18),

$$-(CH_2-\underset{\underset{COOCH_2OH}{|}}{\overset{\overset{CH_3}{|}}{C}})_n- \qquad (7-19)$$

where the OH group is the hydrophilic group responsible for hydration of the polymer. Generally, hydrogels for contact lenses are made by polymerization of certain hydrophilic monomers with small amounts of crosslinking agent such as ethylene glycol dimethacrylate (EGDM),

$$\begin{array}{c} CH_3 \\ | \\ CH_2=C-COOCH_2 \\ | \\ CH_2=C-COOCH_2 \\ | \\ CH_3 \end{array} \qquad (7-20)$$

Another type of hydrogel was developed in the United States at about the same time as the polyHEMA hydrogels were developed. The monomer of polyacrylamide hydrogel has the following chemical formula:

$$
\begin{array}{c}
H \\
| \\
CH_2{=}CCONH_2
\end{array}
\qquad (7\text{-}21)
$$

The water content of the copolymer can be increased to over 60%, while the normal water content for polyHEMA is 40%.

The hydrogels have a relatively low oxygen permeability in comparison with silicone rubber (see Table 7-8). However, the permeability can be increased with increased hydration (water content) or decreased (lens) thickness. Silicone rubber is not a hydrophilic material, but its high oxygen permeability and transparency make it an attractive lens material. It is usually used after coating with hydrophilic hydrogels by grafting.

Table 7-8. Oxygen Permeability Coefficient for Contact Lens Materials

Polymers	$P_g \times 10^4$ (μl cm/cm^{2}h kPa)	Comments
Poly(methylmethacrylate)	0.27	Hard contact lens
Poly(dimethylsiloxane)	1750	Flexible
Poly(hydroxyethylmethacrylate)	24	39% H$_2$O, soft contact lens

At STP to convert μl cm/cm^{2}h kPa to μl cm/cm^{2}h mm Hg, divide by 7.5.

Reprinted with permission from Refojo (1979). Copyright © 1979, Wiley.

Table 7-9. Composition of Bone Cement*

Liquid component (20 ml)	
Methyl methacrylate (monomer)	97.4 v/o (volume %)
N,N,-dimethyl-p-toluidine	2.6 v/o
Hydroquinone	75 ± 15 ppm
Solid powder component (40 g)	
Polymethyl methacrylate	15.0 w/o (weight %)
Methyl methacrylate-styrene-copolymer	73.5 w/o
Barium sulfate (BaSO$_4$), USP	10.0 w/o
Dibenzoyl peroxide	1.5 w/o

*Surgical Simplex® P Radiopaque Bone Cement (Howmedica Inc. Rutherford, NJ) (1977).

7.3.4.b. Bone Cement (PMMA)

Bone cement has been used for clinical applications to secure firm fixation of joint prostheses for hip and knee joints. Bone cement is primarily made of poly(methylmethacrylate) powder and monomer methylmethacrylate liquid, as given in Table 7-9. Hydroquinone is added to prevent premature polymerization, which may occur under certain conditions; e.g., exposure to light, elevated temperatures, etc. N,N-dimethyl-p-toluidine is added to promote or accelerate (cold) curing of the finished compound. The name cold curing is used here to distinguish it from the high temperature and pressure (hot) molding technique used to make articles in dental laboratories. The liquid component is sterilized by membrane filtration. The solid component is a finely ground white powder (mixture of polymethyl methacrylate, methyl methacrylate-styrene-copolymer, barium sulfate, and benzoyl peroxide).

When the powder and liquid are mixed together, the monomer liquid is polymerized by the free radical (addition) polymerization process. An activator, dibenzoyl peroxide (see Eq. (7-3)), which is mixed with the powder, will react with a monomer to form a monomer radical that will then attack another monomer to form a dimer radical. The process will continue until long-chain molecules are produced. The monomer liquid will wet the polymer powder particle surfaces and link them together after polymerization, as shown in Figure 7-4. ASTM standard F451 specifies the characteristics of the powder–liquid mixture and cured polymer after setting, as given in Tables 7-10 and 11.

Table 7-10. Requirements for Powder Liquid Mixture

Maximum dough time (min)	Setting time range (min)	Maximum exotherm (°C)	Minimum intrusion (mm)
5.0	5-15	90	2.0

From ASTM F451.

Table 7-11. Requirements for Cured Polymer after Setting

Minimum compressive strength (MPa)	Maximum indentation (mm)	Minimum recovery (%)	Maximum water sorption (mg/cm^3)	Maximum water solubility (mg/cm^3)
70	0.14	60	0.7	0.05

From ASTM F451.

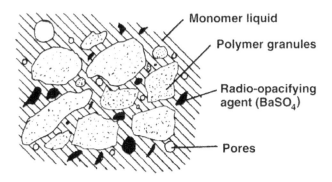

Figure 7-4. Two-dimensional representation of bone cement structure after curing. The monomer liquid will be polymerized and become solid.

Polymerization during curing obviously increases the degree of polymerization, that is, an increase in molecular weight as given in Table 7-12. However, the molecular weight distribution does not change significantly after curing, as shown in Figure 7-5. The properties of cured bone cement are compared with that of commercial acrylic resins in Table 7-13. These studies show that bone cement properties can be affected by the intrinsic and extrinsic factors listed in Table 7-14. The most important factor controlling the acrylic bone cement properties is the porosity developed during curing. Large pores (pores on the order of a millimeter in diameter have been observed in clinical settings) are detrimental to the mechanical properties. Monomer vapors and air trapped during mixing are two reasons for the porosity. Obviously, one can reduce porosity by exposure to a vacuum and by centrifugation during mixing of monomer and powder. However, both techniques have some disadvantages — such as depletion of monomer, difficulty of mixing while under vacuum, segregation of constituents by centrifuging, etc. — in addition to the extra equipment needed.

Table 7-12. Molecular Weight of Bone Cement

Types of m.w. (g/mol)	Monomer	Powder	Cured
M_n (number average)	100	44,000	51,000
M_w (weight average)	100	198,000	242,000

Reprinted with permission from Haas et al. (1975). Copyright © 1975, Charles C. Thomas.

Table 7-13. Physical Properties of Bone Cement: Commercial Acrylic Resins

Properties	Radiopaque Bone cement[a]	Commercial acrylic resins[b]
Tensile strength (MPa)	28.9 ± 1.6	55-76
Compressive strength (MPa)	91.7 ± 2.5	76–131
Young's modulus (compressive loading, MPa)	2,200 ± 60	2,960–3,280
Endurance limit[c]	0.3 uts[d]	0.3 uts
Density (g/cm³)	1.10–1.23	1.18
Water sorption (%)	0.5	0.3–0.4
Shrinkage after setting (%)	2.75 – 5	–

[a] Haas, Brauer, and Dickson (Haas et al., 1975); [b] *Modern plastics encyclopedia* (1980), [c] Kusy (Kusy, 1978); [d] ultimate tensile strength.

Table 7-14. Factors Affecting Bone Cement Properties

Intrinsic Factors
— Composition of monomer and powder
— Powder particle size, shape, and distribution: degree of polymerization
— Liquid/powder ratio

Extrinsic Factors
— Mixing environment: temperature, humidity, type of container
— Mixing technique: rate and number of beating with spatula
— Curing environment; temperature, humidity, pressure, contacting surface (tissue, air, water, etc.)

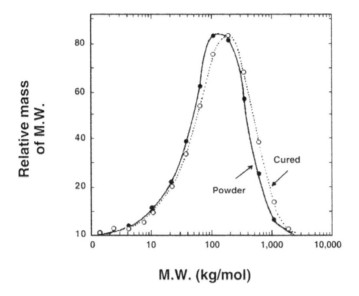

Figure 7-5. Molecular weight distribution of bone cement powder and after curing. Reprinted with permission from Haas et al. (1975). Copyright © 1975, Charles C. Thomas.

7.3.5. Fluorocarbon Polymers

The best-known fluorocarbon polymer is polytetrafluoroethylene (PTFE), commonly known as Teflon® (DuPont). Other polymers containing fluorine are polytrifluorochloroethylene (PTFCE), polyvinylfluoride (PVF), and fluorinated ethylenepropylene (FEP). Only PTFE will be discussed here since the others have rather inferior chemical and physical properties and are rarely used for implant fabrication.

PTFE is made from tetrafluoroethylene under pressure with a peroxide catalyst in the presence of excess water for removal of heat. The repeating unit is similar to that of polyethylene, except that the hydrogen atoms are replaced by fluorine atoms:

$$-(\overset{\displaystyle \overset{F}{|}\,\,\overset{F}{|}}{\underset{\displaystyle \underset{F}{|}\,\,\underset{F}{|}}{C\text{--}C}})_n- \tag{7-22}$$

The polymer is highly crystalline (over 94% crystallinity), with an average molecular weight of $0.5 \sim 5 \times 10^6$ g/mol. This polymer has a very high density ($2.15 \sim 2.2$ g/cm^3), and a low modulus of elasticity (0.5 GPa) and tensile strength (14 MPa). It also has a very low surface tension (18.5 erg/cm^2) and friction coefficient (0.1).

Standard specifications for implantable PTFE are given by ASTM F754. PTFE also has the unusual property of being able to expand on a microscopic scale into a microporous material, which is an excellent thermal insulator, as depicted in Figure 7-6.

PTFE cannot be injection molded or melt extruded because of its very high melt viscosity, and it cannot be plasticized. Usually the powders are sintered to above 327°C under pressure to produce implants. Fibrous expanded PTFE is used to knit arterial grafts that do not leak blood during and after replacement. The high density is a drawback compared to a polyester (poly-ethyleneterephthalate) graft. However, the polyester graft leaks blood, which has to be prevented by pre-clotting with a patient's own blood.

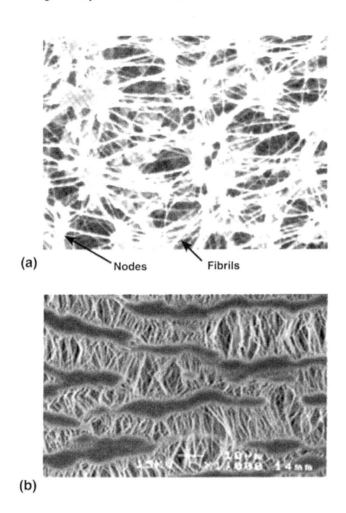

Figure 7-6. Microstructure of PTFE after being expanded. Note the nodes and interconnecting fibrils (**a**) (6,700×). Courtesy of Gore & Associates. (**b**) Similar ePTFE but at lower magnification. Courtesy of Dr. Insup Noh, Department of Chemical Engineering, Seoul National University of Technology.

7.3.6. Rubbers

Silicone, natural, and synthetic rubbers have been used for fabrication of implants. Rubbers or elastomers are defined by ASTM as "a material that at room temperature can be stretched repeatedly to at least twice its original length and upon release of the stress, returns immediately with force to its approximate original length." Rubbers are stretchable because of the kinks of the individual chains, such as that seen in *cis*-1,4 polyisoprene (discussed earlier in §2.4).

The *repeated* stretchability and retractability is due in part to the crosslinks between chains that hold the chains together. The amount of crosslinking for natural rubber controls the flexibility of the rubber: the addition of 2–3% sulfur results in a flexible rubber, while adding as much as 30% sulfur makes it a hard rubber.

Rubbers contain antioxidants to protect them against decomposition by oxidation, hence improving aging properties. Fillers such as carbon black or silica powders are also used to improve their physical properties.

Natural rubber is made mostly from the latex of the Hevea brasiliensis tree, and the chemical formula is the same as that of *cis*-1,4-polyisoprene. Natural rubber in its pure form was found to be compatible with blood. Also, crosslinking by x-ray and organic peroxides produces rubber with superior blood compatibility compared with rubbers made by the conventional sulfur vulcanization.

Synthetic rubbers were developed to substitute for natural rubber. The Natta and Ziegler types of stereospecific polymerization techniques have made this variety possible. The synthetic rubbers have been used rarely to make implants. One of these rubbers, neoprene (polychloroprene, $-CH_2-C(Cl)=CH-CH_2-$) is listed in Table 7-15 for comparison. The physical properties vary widely due to the wide variations in the preparation recipes of these rubbers.

Table 7-15. Properties of Rubbers

Properties	Natural	Neoprene	Silicone	Urethane
Tensile strength (MPa)	7–30	20	6-7	35
Elongation (%)	100–700	–	350–600	650
Hardness (Shore A durometer)	30–90	40–95	–	65
Density (g/cm^3)	0.92	1.23	1.12–1.23	1.1–1.23

See also ASTM standards F604, F881 (silicone rubber) and F624 (urethane).

Silicone rubber, developed by Dow Corning Company, is one of the few polymers developed for medical use. The repeating unit is dimethyl siloxane, which is polymerized by a condensation polymerization. The reaction product is unstable and condenses, resulting in polymers:

$$
\begin{array}{ccc}
CH_3 & & CH_3 \\
| & & | \\
n\ HO-Si-OH & \Rightarrow & -(Si-O)_n- + n\ H_2O \\
| & & | \\
CH_3 & & CH_3
\end{array}
\qquad (7\text{-}23)
$$

Low-molecular-weight polymers have low viscosity and can be crosslinked to make a higher-molecular-weight rubber-like material. Medical grade silicone rubbers contain stannous octate as a catalyst and can be mixed with base polymer at the time of implant fabrication.

Silicone rubbers may contain silica (SiO_2) powder as fillers to improve their mechanical properties. The density and stiffness of filled rubber increase with the volume fraction of filler. Filled rubber is actually a particle reinforced composite.

Polyurethanes are usually thermosetting polymers; they are widely used to coat implants. Polyurethane rubbers are produced by reacting a prepared prepolymer chain A with an aromatic diisocyanate to make very long chains possessing active isocyanate groups for crosslinking. The polyurethane rubber is quite strong and has good resistance to oil and chemicals. Some physical properties of rubbers are summarized in Table 7-15. Polyurethane rubber can be (block) copolymerized with urethanes, carbonates, etc. in order to improve its properties. Depending on the applications, the copolymers can be tailor made to achieve a particular application such as contact lenses, as shown in Figure 7-7. Oxygen permeability is one of the most important parameters for contact lens applications.

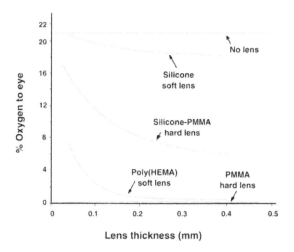

Figure 7-7. Oxygen permeability of contact lenses made of various materials versus thickness. Reprinted with permission from Arkles and Redinger (1983). Copyright © 1983, Technomics Publishing.

7.4. HIGH-STRENGTH THERMOPLASTICS

Strong polymeric materials have been developed to match the strength (but not the stiffness) properties of light metals. These polymers have excellent mechanical, thermal, and chemical properties due to their stiffened main backbone chains. Polyacetals and polysulfones can be used as implant materials, while polycarbonates have found their applications in heart/lung assist devices, food packaging, etc. Polysulfones and polycarbonates are nearly transparent due to their large side groups, as in polymethylmethacrylate, while the polyacetals are not transparent due to their high degree of crystallization (75%).

Polyacetals are produced by reacting formaldehyde as follows

$$n \begin{array}{c} H \\ | \\ C=O \\ | \\ H \end{array} \implies \cdots O^{CH_2} O^{CH_2} O \left({}^{CH_2} O \right)_n \qquad (7\text{-}24)$$

The polyformaldehyde is also sometimes called polyoxymethylene (POM) and is known widely as Delrin® (DuPont). These polymers have a reasonably high molecular weight (> 20,000 g/mol) and have excellent mechanical properties. More importantly, they display an excellent resistance to most chemicals and to water over wide temperature ranges. *Polysulfones* were developed by Union Carbide in the 1960s. The chemical formula for a polysulfone is given as

$$\tag{7-25}$$

These polymers have a high thermal stability due to the bulky side groups (therefore, they are amorphous) and rigid main backbone chains. They are also highly stable to most chemicals but are not so stable in the presence of polar organic solvents such as ketones and chlorinated hydrocarbons.

Polycarbonates are tough, amorphous, and transparent polymers made by reacting bisphenol A and diphenyl carbonate:

$$\tag{7-26}$$

The best-known commercial polycarbonate is Lexan® (General Electric). It has excellent mechanical and thermal properties. Some physical properties of the high-strength thermoplastics are summarized in Table 7-16 for comparison.

Table 7-16. Properties of Polyacetal, Polysulfones, and Polycarbonates

Properties	Polyacetal (Delrin®)	Polysulfone (UDEL®)	Polycarbonate (Lexan®)
Density (g/cm^3)	1.425	1.24	1.20
Tensile strength (MPa)	70	70	63
Elongation (%)	15–75	50–100	60–100
Tensile modulus (GPa)	3.65	2.52	2.45
Water adsorption (%, 24 h)	0.25	0.3	0.3

Example 7-3

Silica flour (finely ground SiO_2, $\rho = 2.65$ g/cm^3) is used as a filler for polydimethyl siloxane (silastic rubber).

a. What volume fraction of SiO_2 is required to make a silastic rubber with a density of 1.25 g/cm^3?

b. What is the weight percent of SiO2?

Answer

a.

$$\rho = \rho_1 V_1 + \rho_2 V_2 + ...,$$

$$\rho_{rubber} = 1.25 = 2.65 \times V_1 + 0.90 \times V_2,$$

$$V_1 + V_2 = 1,$$

$$1.25 = 2.65V_1 + 0.90 (1 - V_1) = 1.75V_1 + 0.90.$$

$$V_1 = \underline{0.20 \ (20 \ v/o)}.$$

b. Since the weight of 1 cm^3 of rubber is 1.25 g and the volume fraction is 0.2 (or 0.2 cm^3/1 cm^3), therefore,

$$W_1 = \frac{0.2 \times 2.65}{1.25} = \underline{0.42 \ (42 \ w/o)}.$$

7.5. DETERIORATION OF POLYMERS

Polymers deteriorate due to chemical, thermal, and physical factors. These factors may act synergistically, hence accelerating the deterioration process. The deterioration affects the main backbone chain, side groups, crosslinks, and their original molecular arrangement.

7.5.1. Chemical Effects

If a linear polymer is undergoing deterioration, the main chain will usually be randomly scissioned (cut). Sometimes depolymerization occurs, which differs from random chain scission. This process is the inverse of the chain termination of addition polymerization (cf. Eqs. (7-8) and (7-9)).

Crosslinking of a linear polymer may result in deterioration. An example of this is low-density polyethylene, in which the crosslinking interferes with regular orderly arrangement of chains, resulting in lower crystallinity, which decreases mechanical properties. On the other hand, if crosslinking is broken by oxygen or ozone attack on (poly-)isoprene rubber, the rubber becomes brittle.

It is also undesirable to change the nature of bonds, as in the case of polyvinylchloride:

$$
\begin{array}{c}
\text{H H H H H} \\
\text{| | | | |} \\
\text{–C–C–C–C–C–} \\
\text{| | | | |} \\
\text{H Cl H Cl H}
\end{array}
\Rightarrow
\begin{array}{c}
\text{H H H H H} \\
\text{| | | | |} \\
\text{–C–C–C=C–C–} \\
\text{| | \quad |} \\
\text{H Cl \quad H}
\end{array}
+ \ \text{HCl}
\tag{7-27}
$$

The byproducts of degradation (HCl) can certainly be irritable to tissues since they are, in this case, acids.

7.5.2. Sterilization Effects

Sterilization is essential for all implanted materials. Some methods of sterilization may result in polymer deterioration. In *dry heat sterilization* the temperature varies between 160 to 190ºC. This is above the melting and softening temperature of many linear polymers like polyethylene and polymethylmethacrylate, so that sterilization of these polymers by heat is inappropriate. In the case of polyamide (nylon), oxidation will occur at the dry sterilization temperature although it is below its melting temperature. The only polymers that can be safely dry sterilized are polytetrafluoroethylene (Teflon®) and silicone rubber.

Steam sterilization (autoclaving) is performed under high steam pressure at relatively low temperature (120–135ºC). However, if the polymer is subjected to attack by water vapor, this method cannot be employed. Polyvinylchloride, polyacetals, polyethylenes (low-density variety), and polyamides (nylons) belong to this category.

Chemical agents such as ethylene and propylene oxide gases and phenolic and hypochloride solutions are widely used for sterilizing polymers since they can be used at low temperatures. Chemical sterilization takes a longer time than the above heating methods and is more costly. Sometimes chemical agents cause polymer deterioration even when sterilization takes place at room temperature. However, the time of exposure is relatively short (a few hours to overnight), and most polymeric implants can be sterilized with this method.

Radiation sterilization using the isotope cobalt 60 can also deteriorate polymers since at high dosage the polymer chains can be broken and recombined. In the case of polyethylene, at high dosage (above 10^6 Gy) it becomes a brittle, hard material. This is due to a combination of random chain scission and crosslinking.

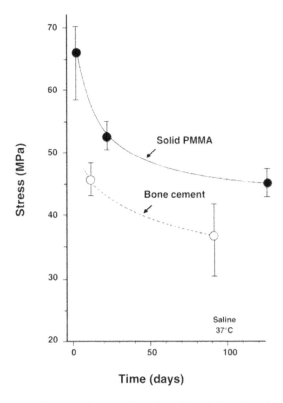

Figure 7-8. Ultimate tensile strength versus time for polymethylmethacrylate in saline solution at 37°C. Note the large decrease in tensile strengths for both solid and porous bone cement. Unpublished data from T. Parchinski, G. Cipoletti, and F.W. Cooke, Clemson University, 1977.

7.5.3. Mechanochemical Effects

It is well known that cyclic or constant loading deteriorates polymers. This effect can be accelerated if the polymer is simultaneously subjected to chemical processes under mechanical loading. Thus, if the polymer is stored in water or in saline solution, its strength will decrease, as shown in Figure 7-8. Another reason for the decreased strength is the plasticizing effect of the water molecules at higher temperature. However, the plasticizing effect compensates for the deleterious effect of the saline solution under cyclic loading, so that there is no difference between samples stored in saline or in air, as shown in Figure 7-9.

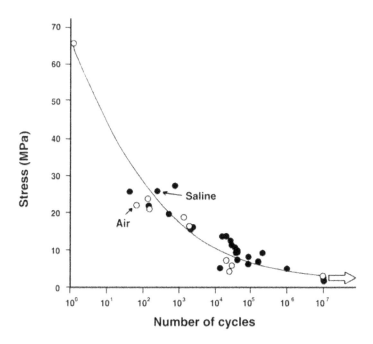

Figure 7-9. Fatigue test of solid polymethylmethacrylate. Note the S–N curve is the same for both samples, non-treated and soaked, in saline solution at 37°C. Unpublished data from T. Parchinski and F.W. Cooke, Clemson University, 1977.

7.5.4. In-Vivo Environmental Effects

Even though materials implanted inside the body are not subjected to light (except in the eye and teeth), radiation, gaseous oxygen, ozone, and extreme temperature variations (except for the teeth), the body environment is very hostile, and all polymers begin to deteriorate as soon as they are implanted. The most probable cause of polymer deterioration is ionic attack (especially hydroxyl ion, OH^-) and dissolved oxygen. Enzymatic degradation may also play a significant role if the implant is made from natural polymeric materials like reconstituted collagen.

It is safe to predict that if a polymer deteriorates in physiological solution in vitro the same will be true in vivo. Most hydrophilic polymers such as polyamides and polyvinyl alcohol will react with body water and undergo rapid deterioration. The hydrophobic polymers like polytetrafluoroethylene (Teflon®) and polypropylene are less prone to deteriorate in vivo.

The deterioration products may induce tissue reactions. In the case of in-vivo deterioration, the original physical properties will be changed if the implant deteriorates. For example, polyolefins (polyethylene and polypropylene) will lose their flexibility and become brittle. For polyamides, the amorphous region is selectively attacked by water molecules, which act as plasticizers, making polyamides more flexible. Table 7-17 shows the effects of implantation on several polymeric materials. More detail on these topics is given in subsequent chapters, which deal in more depth with applications.

Table 7-17. Effect of Implantation on Polymers

Polymers	Effects of implantation
Polyethylene	Low-density ones absorb some lipids and lose tensile strength. High-density ones are inert and no deterioration occurs.
Polypropylene	Generally no deterioration.
Polyvinylchloride (rigid)	Tissue reaction, plasticizers may leach out and material becomes brittle.
Polyethyleneterephthalate	Susceptible to hydrolysis and loss of tensile strength (in polyester).
Polyamides (nylon)	Absorb water and irritate tissue, lose tensile strength rapidly.
Silicone rubber	No tissue reaction, very little deterioration.
Polytetrafluoroethylene	Solid specimens are inert. If is fragmented into pieces, irritation will occur.
Polymethylmethacrylate	Rigid form: crazing, abrasion, and loss of strength by heat sterilization. Cement form: high heat generation, unreacted monomers during and after polymerization may damage tissues.

Reprinted with permission from Bloch and Hastings (1972). Copyright © 1972, Charles C. Thomas.

Example 7-4

Porous polyethylene (with interconnecting pores, 50–100 μm in diameter) is tested in the form of rods with a diameter of 3.4 mm following 2 months of exposure in vitro and in vivo. Typical force-elongation curves are obtained with a 1-cm gauge length and as shown in the accompanying figure:

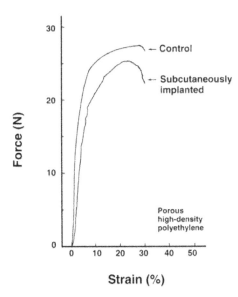

a. Calculate the modulus of elasticity and tensile strength form both curves.

b. Explain why the curve for the sample tested in vivo is not as smooth as the control sample, and why the strength is lower in vivo.

Load-versus-strain curves obtained after 2 months implanted and control (not implanted).

Answer

a.   —For control:

$$\text{Tensile strength } (\sigma) = F/A = 27 \text{ N}/(\pi * 0.17 * 10^{-2}\text{m})^2 = \underline{2.97 \text{ MPa}},$$

$$\text{Modulus of Elasticity } (E) = \sigma/\varepsilon = 2.2 \text{ MPa}/0.04 = \underline{55 \text{ MPa}}.$$

— For implanted material:

$$\text{Tensile strength } (\sigma) = F/A = 25 \text{ N}/(\pi * 0.17 * 10^{-2}\text{m})^2 = \underline{2.75 \text{ MPa}}.$$

$$\text{Modulus of Elasticity } (E) = \sigma/\varepsilon = 2.2 \text{ MPa}/0.09 = \underline{24.4 \text{ MPa}}.$$

b. The implanted material undergoes bone/tissue ingrowth in the porous region. Thus, when tested, it shows an uneven curve due to uneven distribution of strength between the implanted material and ingrowth portion.

PROBLEMS

7-1.   A sample of methacrylate ($CH_2=CHCOOCH_3$) is polymerized. The resulting polymer has a DP (degree of polymerization, n) of 1000. Draw the structure for the repeating unit of the polymer and calculate the polymer molecular weight.

7-2.   An applied strain of 0.3 produces an immediate stress of 10 MPa in a piece of rubber, but after 42 days the stress is only 5 MPa.

a.   What is relaxation time assuming a single relaxation time?

b.   What is the stress after 90 days?

7-3.   Polytetrafluoroethylene crystal has a rhombohedral unit cell structure as shown. Calculate its theoretical density.

7-4.   Polymers containing chemical bonds similar to those found in the body, e.g., amide and ester groups are less biodegradable than polymers containing C–C and C–F bonds. Is this statement true?

7-5.   The average bond energy of C–Cl for polyvinyl chloride is 340,000 J/mol. Can visible light ($\lambda = 4000\sim7000$ Å) have enough energy to break the bond? Recall $E = h\nu$, where h is Planck's constant and ν is vibration frequency.

7-6.   The average and end-to-end distance (L) of a chain of an amorphous polymer can be expressed as $L = l\sqrt{m}$, where l is the interatomic distance (1.54 Å for C–C) and m is the number of bonds. If the average molecular weight of a polystyrene is 20,800 g/mol, what is the average end-to-end distance (L) of a chain?

7-7.   Name the following polymers with the repeating unit given.

a. $-CH_2CH_2-$

b. $-CH_2-O-$

c. $-CF_2-CF_2-$

d.   CH₃ H
        \   /
         C=C
        /   \
   –H₂C   CH₂–

e.   CH₃
      |
     –Si–O–
      |
     CH₃

f.   CH₃
              |
     –CH₂–CH–

g.   Cl
              |
     –CH₂–CH–

h.   H   O
      | ‖
     –N(CH₂)₅–C–

i.   OH
           |
     –CH₂–CH–

j.   CH₃
             |
     –CH₂–C–
             |
           COOCH₃

7-8   From the following data obtained from polypropylene, answer.

Mean m.w. (g/mol)	Weight (g)
10,000	3
20,000	5
30,000	2

a. Calculate the weight average m.w.

b. Calculate the number average m.w.

c. What is its polydispersity?

d. What is the degree of polymerization? [C = 3, H = 6]/mer; C = 12, H =1 amu.

e. Give a schematic diagram for the syndiotactic structure of the polymer.

f. What type of polymerization would you use for this polymer?

g. Is this polymer thermoplastic or thermoset?

h. Is this polymer crystalline, semicrystalline, or noncrystalline?

7-9   A polyethylene is made of the following fractional distribution.

W_i	0.10	0.20	0.30	0.30	0.10
M_i (kg/mol)	10	20	30	40	50

   a. Calculate M_w.

   b. Calculate M_n.

   c. Can you make a suture from this polymer?

   d. What is the best way of making this polymer: free radical, step reaction, or ionic polymerization?

   e. Would the polymer become crystalline, noncrystalline, or semicrystalline when cooled from melt?

   f. What effect will the branching of chains have on the density of the polymer? Would it increase, decrease, or remain the same?

7-10.   List four methods of sterilization and from the following list of materials, which one would be suitable for a particular method of sterilization? Give one reason for each answer.
   a. Polyethylene
   b. Polytetrafluoroethylene
   c. Silicone rubber
   d. Hydroxyapatite powder

7-11.   The plastic was stretched from 100.00 to 120.00 cm and released at room temperature. The length changes with time were obtained as follows:

Time (hour)	Length (cm)
0	120.0
1	110.0
3	103.0
6	100.4
20	100.1

Assuming the Voigt model can be applied for this mechanical behavior, answer.

$$\varepsilon_{recovery} = \varepsilon_0 \exp[-(E/\eta)t],$$

where E = modulus, η = viscosity, t = time.

   a. What is the retardation time (hours)?

   b. What is the % strain at 15 minutes based on the original length of the sample?

   c. What is the amount of strain (%) recovered after 15 minutes?

   d. If the test was made at body temperature, would the recovery be faster, slower, or the same compared to the room temperature test?

7-12.   PMMA is polymerized from MMA.

```
    H   CH₃
    |   |
    C = C
    |   |
    H   COOCH₃
```

The molecular weight distribution is given as follows:

Mean m.w. (amu)	Weight (g)
10000	50
20000	50

a. What type of polymerization does it undergo?

b. Calculate the molecular weight of the repeating unit.

c. Calculate the weight average molecular weight.

d. Calculate the degree of polymerization: C = 12, O = 16, H = 1 amu.

e. What type of polymer is this?

Circle the correct one.

In terms of its morphology: A. amorphous, B. crystalline, C. semicrystalline

In terms of its molecular structure: A. linear, B. network, C. crosslinked

In terms of its thermal behaviour: A. thermoplastic, B. thermosetting

7-13. A piece of suture was tested for its stress relaxation properties. The initial force recorded after stretching 0.1 cm between grips was 10 Newtons. Assume the material behaves as if it has one relaxation time. The original length was 10 cm and diameter was 1.0 mm.

a. Calculate the initial stress.

b. Calculate the initial strain.

c. Calculate the modulus of elasticity of the suture if the initial stretching can be considered as linear and elastic.

d. Calculate the relaxation time if the force recorded after 100 hours was 5 Newtons:

$$\sigma = \sigma_0 \exp[-t/\tau].$$

e. Calculate the viscosity.

7-14. A polyethylene is made of the following fractional distribution.

Weight (g)	10	20	30	30	10
M_i (kg/mol)	10	20	30	40	50

a. Calculate the molecular weight of the *repeating unit* of the polyethylene: $-[CH_2CH_2]_n-$, C = 12 amu, H = 1 amu.

b. Calculate the weight average molecular weight (m.w.) of the polymer.

c. Calculate the number average m.w. of the polymer.

d. Calculate the degree of polymerization.

e. Calculate the polydispersity.

f. What would be the volume of the sample (100 g) if the density of the polymer is 0.95 g/cc?

g. The solid polyethylene can be classified as (choose one):
   a. crystalline   b. noncrystalline   c. semicrystalline

7-15. A piece of polypropylene suture (10 cm long, 2 mm diameter) is tested for its creep recovery properties at room temperature. The initial force recorded after stretching 0.1 cm between grips was 3.14 Newtons. Assume the material behaves as if it has one retardation time.

    a. Calculate the initial strain (%).

    b. Calculate the retardation time if the specimen recovered 50% of its original length strain after 1 hour.

    c. Calculate its *length* (cm) after 10 hours.

    d. Calculate its stress at 0.1-cm stretch.

    e. Calculate its viscosity.

7-16. Polymethylmethacylate $-[CH_2(CH_3)(COOCH_3)C]_n-$ is used to fabricate implants. The following molecular weight distributions (number fraction) were obtained from three different companies. C = 12 amu, H = 1 amu, O = 16 amu. Avog. no. $= 6.02 \times 10^{23}$.

Average m.w. (g/mol)	A Company	B Company	C Company
20,000	0.5	0.1	0.9
40,000	0.5	0.9	0.1

    a. Calculate the weight average m.w. (M_w) of the polymer from company A.

    b. Calculate the number average m.w. (M_w) of the polymer from company A.

    c. Calculate the m.w. of the repeating unit.

    d. Calculate the degree of polymerization of the polymer from company A.

    e. Calculate the polydispersity of the polymer from company A.

    f. Which company's polymer would have the highest molecular weight?

    g. Is this polymer thermoplastic or thermosetting?

    h. Is this polymer crystalline, semicrystalline, or noncrystalline?

    i. Is this polymer addition, step-reaction, or ring-scission polymerized?

7-17. The PMMA of Company A is used to make bone cement.

    A. powdered PMMA   B. benzoyl peroxide   C. TiO$_2$
    D. N,N-dimethyl-p-toluidine   E. hydroquinone,
    F. monomer liquid

    a. Which material is used for the initiator of polymerization?

    b. Which material is used for the accelerator of polymerization?

    c. Which material is used for the inhibitor of polymerization?

    d. Which material is used for the radio-opacifier of bone cement?

    e. What chemicals (two) can be put together with liquid monomer?

    f. What is the powder: liquid ratio for ordinary bone cement?

    g. Give two reasons for the formation of pores in the bone cement.

7-18.   A piece of polyethylene suture (10 cm long, 2 mm × 2 mm cross-section) is tested for its viscoelastic properties at room temperature. The initial force recorded after stretching 1 mm between grips was 16 N force.

a.   Calculate the initial strain (%).

b.   Calculate the initial stress (MPa).

c.   Calculate Young's modulus if the PE deformed linearly for the 1-mm stretch.

d.   Calculate the relaxation time (τ) if the specimen was held at constant length between grips but its force decreased to 8 Newtons after 10 hours.

e.   Calculate the viscosity of the PE.

When the stress was released the length became 10.5 mm after 10 hour.

f.   Calculate its retardation time (τ).

g.   Calculate its length after 24 hours.

7-19.   Polyethylenes (PE), [$-CH_2CH_2-$], A and B are mixed thoroughly as follows:

	A	B
Weight fraction	0.5	0.5
(mega gram/mol)	?	2

a.   Calculate the weight average molecular weight (M_w) of the A PE if the weight average molecular weight of the mixed PE is 1.8 Mg/mol.

b.   Calculate the molecular weight of the repeating unit of the PE: C = 12, H = 1 g/mol.

c.   Calculate the degree of polymerization (DP) of the mixed PE.

d.   Calculate the number of chains (molecules) in 1 gram of the mixed PE.

e.   Calculate the number average molecular weight (M_n) of the mixed PE.

7-20.   MMA monomer liquid is mixed with the powder PMMA to make a bone cement. Choose one:

   A. Methyl methacrylate (monomer)   B. N,N,-dimethyl-p-toluidine
   C. Hydroquinone   D. Barium sulfate ($BaSO_4$), USP   E. Dibenzoyl peroxide

a.   Which is the initiator?

b.   Which is the accelerator?

c.   Which is the inhibitor?

d.   Which is the radioopacifying agent?

e.   What is the liquid: powder ratio?

What type of polymer is this? Check the correct one

f.   In terms of its crystallinity: A. amorphous, B. crystalline, C. semicrystalline

g.   In terms of its molecular structure: A. linear, B. network, C. crosslinked

h.   In terms of its thermal behavior: A. thermoplastic, B. thermosetting, C. neither

DEFINITIONS

Addition (or free radical) polymerization: Polymerization in which monomers are added to the growing chains, initiated by free radical agents.

Backbone: The main molecular chain of a polymer.

Barium sulfate: Inert ceramic mixed in the bone cement as powder to make it radio opaque to x-rays for better visualization of the implant on x-ray film.

Bone cement: Mixture of polymethylmethacrylate powder and methylmethacrylate monomer liquid to be used as a grouting material for the fixation of orthopedic implants.

Branching: Chains grown from the sides of the main backbone chains.

Condensation (step reaction) polymerization: Polymerization in which two or more chemicals are reacted to form a polymer by condensing out small molecules such as water and alcohol.

Delrin®: Polyacetal made by Union Carbide.

Filler: Materials added as a powder to a rubber to improve its mechanical properties.

Hydrogel: Polymer that can absorb water 30% or more of its weight.

Hydroquinone: Chemical inhibitor added to the bone cement liquid monomer to prevent accidental polymerization during storage.

Initiator: Chemical used to initiate the addition polymerization by becoming a free radical which in turn reacts with a monomer.

Kevlar®: Aromatic polyamides trademarked by DuPont.

Lexan®: Polycarbonate trademarked by General Electric.

N,N-dimethyl-p-toluidine: Chemical added to the powder portion of the bone cement for the acceleration of its polymerization.

Repeating unit: Basic molecular unit that can represent a polymer backbone chain. The average number of repeating unit is called the degree of polymerization.

Relaxation time: A time constant for stress relaxation assuming the polymer exhibits an exponential behavior. Real polymers exhibit a distribution of relaxation times. This term is sometimes taken to refer to a time constant for other time-dependent processes.

Retardation time: A time constant for creep assuming the polymer exhibits an exponential creep behavior. Real polymers exhibit a distribution of retardation times.

Side group: Chemical group attached to the main backbone chain. It is usually shorter than the branches and exists before polymerization.

Simplex®: Acrylic bone cement made by Howmedica Inc.

Tacticity: Arrangement of side groups of a linear polymer chain that can be atactic (random), isotactic (same side), and syndiotactic (alternating side).

Teflon®: Polytetrafluoroethylene made by DuPont.

Udel®: Polysulfone made by General Electric.

Vulcanization: Crosslinking of a (natural) rubber by adding sulfur.

Ziegler catalyst: Organometallic compounds which have the remarkable capacity of polymerizing a wide variety of monomers to linear and stereoregular polymers.

BIBLIOGRAPHY

Arkles B, Redinger P. 1983. Silicones in biomedical applications. In *Biocompatible Polymers, Metals and Composites*, chapter. 32. Ed M Szycher. Westport, CT: Technomics Publishing.

Barbucci R. 2002. *Integrated biomaterials science*. New York: Kluwer Academic/Plenum.

Billmeyer Jr FW. 1984. *Textbook of polymer science*. New York: Wiley.

Bloch B, Hastings GW. 1972. *Plastics materials in surgery*. Springfield, IL: Thomas.

Bruck SD. 1974. *Blood compatible synthetic polymers: an introduction*. Springfield, IL: Thomas.

Davis JR. 2003. *Handbook of materials for medical devices*. Materials Park, OH: ASM International.

Dawids S, ed. 1989. *Polymers: their properties and blood compatibility*. Dordrecht: Kluwer Academic.

Geil P. 1963. *Polymer single crystals*. New York: Wiley.

Haas SS, Brauer GM, Dickson G. 1975. A characterization of PMMA bone cement. *J Bone Joint Surg Am* **57A**:380–391.

Heijkants RG, van Calck RV, De Groot JH, Pennings AJ, Schouten AJ, van Tienen TG, Ramrattan N, Buma P, Veth RP. 2004. Design, synthesis and properties of a degradable polyurethane scaffold for meniscus regeneration. *J Mater Sci Mater Med* **15**(4):423–427.

Hoffman AS. 1981. A review of the use of radiation plus chemical and biochemical processing treatments to prepare novel biomaterials. *Radiat Phys Chem* **18**:323–342.

Kronenthal RL, Oser Z, eds. 1975. *Polymers in medicine and surgery*. New York: Plenum.

Kusy RP. 1978. Characterization of self-curing acrylic bone cements. *J Biomed Mater Res* **12**:271–305.

Lee H, Neville K. 1971. *Handbook of biomedical plastics*. Pasadena, CA: Pasadena Technology Press (chapters 3–5 and 13).

Leinninger RI. 1972. Polymers as surgical implants. *CRC Crit Rev Bioeng* **2**:333–360.

Levine SN, ed. 1968. *Polymers and tissue adhesives*. New York: New York Academy of Sciences.

Modern Plastics Encyclopedia. 1980. New York, McGraw-Hill.

NIH. 1980. *Guidelines for physicochemical characterization of biomaterials*. Bethesda, MD: National Heart, Lung, and Blood Institute Working Group, Devices and Technology Branch.

Park JB. 1983. Acrylic bone cement: In vitro and in vivo property–structure relationship: a selective review. *Ann Biomed Eng* **11**:297–312.

Park KD, Lewis G, Park JB. 2004. Ultra-high molecular-weight polyethylene (UHMWPE). In *encyclopedia of biomaterials and biomedical engineering (EBBE)*, pp. 1690–1696. New York: Marcel Dekker.

Refojo MF. 1979. Contact lenses. In *Encyclopedia of chemical technology*, Vol. 16, pp. 720–742. New York: Wiley.

Shalaby SW, Burg KJL, eds. 2004. *Absorbable and biodegradable polymers: advances in polymeric biomaterials*. Boca Raton, FL: CRC Press.

Shtil'man MI. 2003. *Polymeric biomaterials*. Utrecht: VSP.

Surgical Simplex P bone cement technical monograph. 1977. Rutherford, NJ: Howmedica.

Szycher M, Robinson WJ, eds. 1993. *Synthetic biomedical polymers, concepts and applications*. Lancaster, PA: Technomics Publishing.

Ward IM, Hadley DW. 1993. *Mechanical properties of solid polymers*. New York: Wiley.

8

COMPOSITES AS BIOMATERIALS

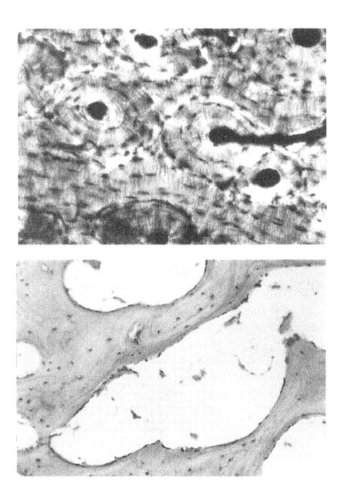

Bone is a perfect example of a composite material designed by nature. Minerals are embedded as reinforcing elements while the collagen serves as matrix. (See Chapter 9 for more studies on bone.) **Top**: compact bone; **bottom**: sponge bone. (http://cellbio.utmb.edu/microanatomy/bone/compact_bone_histology.htm and http://cellbio.utmb.edu/microanatomy/bone/spongy_bone_histology.htm)

Composite materials are those that contain two or more distinct constituent materials or phases, on a microscopic or macroscopic size scale. The term "composite" is usually reserved for those materials in which the distinct phases are separated on a scale larger than the atomic, and in which properties such as the elastic modulus are significantly altered in comparison with those of a homogeneous material. Accordingly, fiberglass and other reinforced plastics as well as bone are viewed as composite materials, but alloys such as brass or metals such as steel with carbide particles are not. Natural biological materials tend to be composites; these are discussed in Chapter 9. Natural composites include bone, wood, dentin, cartilage, and skin. Natural foams include lung, cancellous bone, and wood. Natural composites often exhibit hierarchical structures in which particulate, porous, and fibrous structural features are seen on different length scales. In this chapter, composite material fundamentals and applications in biomaterials are explored.

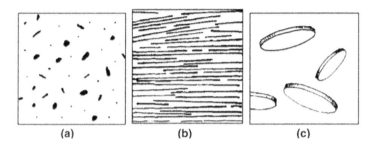

Figure 8-1. Morphology of basic composite inclusions: (**a**) particle; (**b**) fiber; (**c**) platelet.

8.1. STRUCTURE

The properties of composite materials depend very much upon *structure* (see Chapter 2), as they do in homogeneous materials. Composites differ in that considerable control can be exerted over the larger scale structure, and hence over the desired properties. (Christensen, 1979; Agarwal and Broutman, 1980). In particular, the properties of a composite material depend upon the *shape* of the heterogeneities, upon the *volume fraction* occupied by them, and upon the *interface* among the constituents. The shape of the heterogeneities in a composite material is classified as follows. The principal inclusion shape categories are the particle, with no long dimension; the fiber, with one long dimension, and the platelet or lamina, with two long dimensions, as shown in Figure 8-1. The inclusions may vary in size and shape within a category. For example, particulate inclusions may be spherical, ellipsoidal, polyhedral, or irregular. Cellular solids (Gibson and Ashby, 1997) are those in which the "inclusions" are voids, filled with air or liquid. In the context of biomaterials, it is necessary to distinguish the above cells, which are structural, from biological cells, which occur only in living organisms. We moreover make the distinction, in each composite structure, between random orientation and preferred orientation.

Several examples of composite material structures were presented in Chapter 2. The dental composite filling material shown in Figure 2-21 has a particulate structure. Figure 2-22 shows a fibrous material fracture surface, with fibers that have been pulled out. Figure 2-23 shows a section of a cross-ply laminate. Figure 2-24 shows several types of cellular solid.

The degree of *adhesion* of the reinforcing materials with the matrix is another important factor in the performance of composites.

8.2. MECHANICS OF COMPOSITES

In the context of mechanical properties, we may classify two-phase composite materials according to their microstructure. The inclusions within the matrix may be particles, fibers, or platelets. If either the inclusions or the matrix consists of air or liquid, the material is a cellular solid [foam]. In each of the above types of structure, we may moreover make the distinction between random orientation and preferred orientation. The use of composite materials is motivated by the fact that they can provide more desirable material properties than those of homogeneous materials.

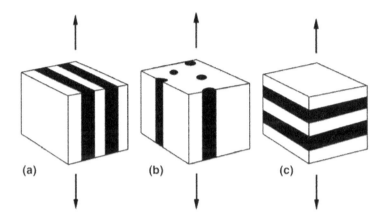

(a)   (b)   (c)

Figure 8-2. Tension force indicated by arrows, applied to Voigt (**a**, laminar; **b**, fibrous) and Reuss (**c**) composite models.

Mechanical properties in many composite materials depend on structure in a complex way; however, for some structures the prediction of properties is relatively simple. The simplest composite structures are the idealized Voigt and Reuss models, shown in Figure 8-2. The dark and light areas in these diagrams represent different constituent materials in the composite. In contrast to most composite structures, it is easy to calculate the stiffness of materials with the Voigt and Reuss structures. Young's modulus E of the Voigt composite is (neglecting restraint due to Poisson's ratio)

$$E = E_i V_i + E_m [1 - V_i]. \tag{8-1}$$

Here E_i is Young's modulus of the inclusions, V_i is the volume fraction of inclusions, and E_m is Young's modulus of the matrix. The Voigt relation for stiffness is also called the rule of mixtures.

Young's modulus for the Reuss model is

$$E = \left[(V_i / E_i) + (1 - V_i) / E_m) \right]^{-1}. \tag{8-2}$$

This is less than that of the Voigt model. The Voigt and Reuss formulae constitute upper and lower bounds, respectively, upon the stiffness of a composite of arbitrary phase geometry.

Example 8-1

Determine Young's modulus of materials with the Voigt structure, assuming that Young's modulus is known for each constituent, and the volume fraction of each.

Answer

For the Voigt model, if tension is applied as in Figure 8-2, the inclusions (dark) and the matrix (light) deform together, with equal strain, so $\varepsilon_c = \varepsilon_i = \varepsilon_m$, in which c refers to the composite, i refers to the inclusions, and m to the matrix. Assume linearly elastic behavior so that $\sigma_c = E_c \varepsilon_c$ and similarly for the inclusions and matrix. The force F_c on the composite block is the sum of the forces on the inclusions and the matrix. Therefore, $F_c = \sigma_c A_c = \sigma_i A_i + \sigma_{mAm} = E_i \varepsilon_i A_i + E_m \varepsilon_m A_m$. Now divide both sides by the total cross-sectional area of the composite block, A_c, to get the stress in the composite. Then divide by the strain, which is equal in both constituents, and observe that for this geometry the volume fraction V_i is the same as the cross-sectional area fraction A_i/A_c. Therefore, $E_c = E_f V_f + E_m V_m = E_f V_f + E_m(1 - V_f)$. This is the "rule of mixtures."

The stiffness of the Reuss model can be obtained in a similar manner (see Problem 8.1). This stiffness is quite different from that of the Voigt model. However, the Reuss laminate is identical to the Voigt laminate, except for a rotation with respect to the direction of load. Therefore, the stiffness of the laminate is *anisotropic*, that is, dependent on direction. Anisotropy is characteristic of composite materials. The relationship between stress σ_{ij} and strain ε_{kl} in anisotropic materials is given by the tensorial form of Hooke's law:

$$\sigma_{ij} = \sum_{k=1}^{3} \sum_{l=1}^{3} C_{ijkl}\varepsilon_{kl}. \tag{8-3}$$

Here C_{ijkl} is the elastic modulus tensor. It has $3^4 = 81$ elements; however, since the stress and strain are represented by symmetric matrices with six independent elements each, the number of independent modulus tensor elements is reduced to 36. An additional reduction to 21 is achieved by considering elastic materials for which a strain energy function exists. The physical meaning of the tensor elements is as follows. For example, C_{2323} represents a shear modulus since it couples a shear stress with a shear strain. The modulus element C_{1111} couples axial stress and strain in the 1- or x-direction. It is not the same as Young's modulus. The reason is that Young's modulus is measured using a slender specimen, in which the lateral strains are free to occur by the Poisson effect. By contrast, C_{1111} is the ratio of axial stress to strain when there is only one nonzero strain value; there is no lateral strain. Ultrasonic longitudinal wave measurements give C_{1111}; cartilage has the same constrained modulus.

A *triclinic* crystal, which is the least symmetric crystal form, would be described by such a modulus tensor with 21 elements. The unit cell has three different oblique angles and three different side lengths. Triclinic modulus elements such as C_{1123} couple shearing deformations with normal stresses; this is undesirable in many applications. An *orthorhombic* crystal or an *orthotropic* composite has a unit cell with orthogonal angles. There are nine elastic moduli. The associated engineering constants are three Young's moduli, three Poisson's ratios, and three shear moduli; there are no cross-coupling constants. An example of such a composite is a unidirectional fibrous material with a rectangular pattern of fibers in the cross-section. Bovine bone, which has a laminated structure, exhibits orthotropic symmetry, as does wood. In *hexagonal* symmetry, there are five independent elastic constants out of the nine remaining C

elements. For directions in the transverse plane the elastic constants are the same; hence the alternate name transverse isotropy. A unidirectional fiber composite with a hexagonal or random fiber pattern has this symmetry, as does human Haversian bone. In *cubic* symmetry, there are three independent elastic constants, Young's modulus E, shear modulus G, and an independent Poisson's ratio v. Cross-weave fabrics have cubic symmetry. Finally, an *isotropic* material has the same material properties in any direction. There are only two independent elastic constants. The others are related by equations such as

$$E = 2G(1+v).\tag{8-4}$$

Random fibrous and random particulate composite materials are isotropic.

The properties of several composite material structures are shown in Table 8-1. V_i is the volume fraction [between zero and one] of inclusions, V_s is the volume fraction of a solid in the case of foams, E is Young's modulus, and m refers to the matrix. As for strength, relationships are given only when they are relatively simple. The strength of composites depends not only on the strength of the constituents, but also on the stiffness and degree of ductility of the constituents. The Voigt relation for the stiffness is referred to as the rule of mixtures; related rules of mixtures are discussed elsewhere in this book. The Voigt and Reuss models provide upper and lower bounds, respectively, upon the stiffness of a composite of arbitrary phase geometry. For composite materials that are isotropic, the more complex Hashin–Shtrikman relations provide tighter bounds upon the moduli; both Young's and shear moduli must be known for each constituent. The relations given in Table 8-1 for inclusions are valid for small volume fractions; the relations become much more complex in the case of large volume fraction. As for particles, they are assumed to be spherical and the matrix to have a Poisson's ratio of 0.5.

Table 8-1. Theoretical Properties of Composites (Gibson and Ashby, 1988)

Structure	Stiffness	Strength
Voigt model	$E = E_i V_i + E_m[1 - V_i]$	
Reuss model	$E = [V_i/E_i + (1 - V_i)/E_m]^{-1}$	
Isotropic: 3D random orientation		
Particulate, dilute	$E = [5(E_i - E_m)V_i]/[3 + 2E_i/E_m] + E_m$	
Fibrous, dilute	$E = E_i V_i/6 + E_m$	
Platelet, dilute	$E = E_i V_i/2 + E_m$	
Foam, open cell	$E = E_s[V_s]^2$	
Crushing strength		$\sigma_{crush} = \sigma_{I,s} 0.65 \, [V_s]^{3/2}$
Elastic collapse		$\sigma_{coll} = 0.05 \, E_s[V_s]^2$
Anisotropic, oriented		
Unidirectional, fibrous	$E_{long} = E_i V_i + E_m[1 - V_i]$	$\sigma_{long} = \sigma_i V_i + \sigma_m[1 - V_i]$
	$E_{transv} = E_m[1 + 2nV_i/(1 - nV_i)]$	
	where $n = (E_i/E_m - 1)/(E_i/E_m + 2)$	

Reprinted with permission from Agarwal and Broutman (1980). Copyright © 1980, Wiley.

Observe that in isotropic systems stiff platelet inclusions are the most effective in creating a stiff composite, followed by fibers, and the least effective geometry for stiff inclusions is the spherical particle. Even if the particles are perfectly rigid, their stiffening effect at low concentrations is modest:

$$E = E_m[1 + 5V_i/2].\tag{8-5}$$

Although particle inclusions do not increase the composite stiffness as much as inclusions of other shapes, they are often used for reasons of simplicity of preparation or availability of inclusions of that shape. Conversely, when the inclusions are more compliant than the matrix, spherical ones are the least harmful and platelet ones are the most harmful. Indeed, platelets in this case are suggestive of crack-like defects. Soft platelets therefore result not only in a compliant composite, but also in a weak one. Soft spherical inclusions are used intentionally as crack stoppers to enhance the toughness of polymers such as polystyrene (high-impact polystyrene), with a small sacrifice in stiffness.

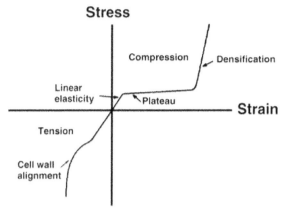

Figure 8-3. Representative stress–strain curve for a cellular solid. The plateau region for compression in the case of elastomeric foam (a rubbery polymer) represents elastic buckling; for an elastic-plastic foam (such as metallic foam) it represents plastic yield, and for an elastic-brittle foam (such as ceramic) it represents crushing. On the tension side, point A represents the transition between cell wall bending and cell wall alignment. In elastomeric foam the alignment occurs elastically; in elastic plastic foam it occurs plastically, and an elastic-brittle foam fractures at A.

As for cellular solids, representative cellular solid structures are shown in Figure 2-24; measurement of density and porosity is described in Chapter 4 (§4). The stiffness relationships in Table 8-1 for cellular solids are valid for all solid volume fractions; the strength relationships, only for relatively small density. The derivation of these relations is based on the concept of *bending* of the cell ribs and is presented in (Gibson and Ashby, 1988). Most man-made closed cell foams tend to have a concentration of material at the cell edges, so that they behave mechanically as open cell foams. The salient point in the relations for the mechanical properties of cellular solids is that the *relative density* dramatically influences stiffness and strength. As for the relationship between stress and strain, a representative stress–strain curve is shown in Figure 8-3. Observe that the physical mechanism for the deformation mode beyond the elastic limit depends on the material from which the foam is made. Trabecular bone, for example, is a natural cellular solid, which tends to fail in compression by crushing. It is of interest to compare the predicted strength of an open cell foam from Table 8-1 with the observed dependence of strength upon density for trabecular bone. Although many kinds of trabecular bone appear to behave as a normal open cell foam, there are different structures of trabecular bone that may behave differently.

Anisotropic composites offer superior strength and stiffness in comparison to isotropic ones. Material properties in one direction are gained at the expense of properties in other directions. It is sensible, therefore, to use anisotropic composite materials only if the direction of application of the stress is known in advance.

The strength of composites depends on such particulars as the brittleness or ductility of the inclusions and the matrix. In fibrous composites failure may occur by fiber breakage, buckling, or pullout, matrix cracking, or debonding of fiber from matrix. While unidirectional fiber composites can be made very strong in the longitudinal direction, they are weaker than the matrix alone when loaded transversely, as a result of stress concentration around the fibers. In many applications, short-fiber composites are used. While they are not as strong as those with continuous fibers, they can be formed economically by injection molding or by in situ polymerization. Choice of an optimal fiber length can result in improved toughness, due to the predominance of fiber pull-out (Figure 2-20) as a fracture mechanism.

Example 8-2

Determine Young's modulus of trabecular bone of density 0.2 g/cm^3. Assume that the tissue behaves as an isotropic open cell foam. Observe that Young's modulus for human compact tibial bone is about 18 GPa and its tensile strength is about 140 MPa.

Answer

From Table 8-1, $E/E_s = [\rho/\rho_s]^2$ for open cell foams. So E = 18 GPa $[0.2/2]^2$ = <u>180 MPa</u>. We have ignored the effect of tissue fluid or bone marrow in the pores. At low strain rates, it has been found that their effect on mechanical properties is negligible. However, at higher strain rates, the marrow and tissue fluid can contribute to viscoelastic behavior in trabecular bone and to its energy absorbing capacity. Moreover, trabecular bone varies in structure throughout the skeleton. If the trabeculae are highly oriented, as in the interior of vertebrae, the modulus can be proportional to the first power of the density, not the second power.

8.3. APPLICATIONS OF COMPOSITE BIOMATERIALS

Composite materials offer a variety of advantages in comparison with homogeneous materials. However, in the context of biomaterials, it is important that each constituent of the composite be biocompatible, and that the interface between constituents not be degraded by the body environment. Composites currently used in biomaterial applications include the following: dental filling composites; bone particle or carbon fiber reinforced methyl methacrylate bone cement and ultrahigh-molecular-weight polyethylene; and porous surface orthopedic implants. Moreover, rubber used in catheters, rubber gloves, etc. is usually filled with very fine particles of silica to make the rubber stronger and tougher.

8.3.1. Dental Filling Composites and Cements

While metals such as silver amalgam and gold are commonly used in the restoration of posterior teeth, they are not considered desirable in anterior teeth for cosmetic reasons. Acrylic resins and silicate cements had been used for anterior teeth, but their poor material properties led to short service life and clinical failures. Dental composite resins have virtually replaced these materials for restorations in anterior teeth and are very commonly used to restore posterior teeth as well as anterior teeth.

The composite resins consist of a polymer matrix and stiff inorganic inclusions. Representative structures are shown in Figures 2-18 and 8-4. Observe that the particles are very angular in shape. The inorganic inclusions confer a relatively high stiffness and high wear resistance to the material. Moreover, by virtue of their translucence and index of refraction similar to that of dental enamel, they are cosmetically acceptable. The inorganic inclusions are typically barium glass or silica [quartz, SiO_2]. Inclusions, also called fillers, have a particle size from 0.04 to 13 μm, and concentrations from 33 to 78% by weight. The matrix consists of BIS-GMA, an addition reaction product of bis(4-hydroxyphenol), dimethylmethane, and glycidyl methacrylate. Since the material is mixed, then placed in the prepared cavity to polymerize, the viscosity must be sufficiently low and the polymerization controllable. Low-viscosity liquids such as triethylene glycol dimethacrylate (TEGDMA) are used to lower the viscosity, and inhibitors such as BHT (butylated trioxytoluene, or 2,4,6-tri-tert-butylphenol) are used to prevent premature polymerization. To fill a cavity, the dentist mixes several constituents, then places them in the prepared cavity to polymerize. Polymerization can be initiated by a thermochemical initiator such as benzoyl peroxide, or by a photochemical initiator (benzoin alkyl ether), which generates free radicals when subjected to ultraviolet light from a lamp used by the dentist.

10 μm

Figure 8-4. Microstructure of a dental composite. Miradapt® (Johnson & Johnson) 50% by volume filler: barium glass and colloidal silica.

The compositions and stiffnesses of several representative commercial dental composite resins are given in Table 8-2. In view of the greater density of the inorganic filler phase, a 77 weight percent of filler corresponds to a volume percent of about 55. Typical mechanical and physical properties of dental composite resins of about 50% filler by volume are shown in Table 8-3. Dental composites are considerably less stiff than natural enamel, which contains about 99% mineral. One cannot easily obtain such high concentrations of mineral particles in synthetic composites. The particles do not pack densely. Moreover, the viscosity of the unpolymerized paste increases with particle concentration. Too high a viscosity would prevent the dentist from adequately packing the paste into the prepared cavity.

The thermal expansion of dental composites exceeds that of tooth structure. The same is true of other dental materials. There is also a contraction up to 1.6% during polymerization. The contraction is thought to contribute to leakage of saliva, bacteria, etc., at the interface margins. Such leakage in some cases can cause further decay of the tooth. For some materials the contraction is counteracted by swelling due to absorption of water in the mouth. Use of colloidal silica in the so-called "microfilled" composites allows these resins to be polished, so that less wear occurs and less plaque accumulates. It is more difficult, however, to make these with a high fraction of filler, since the tendency for high viscosity of the un-polymerized paste must be counteracted. An excessively high viscosity is problematical since it prevents the dentist from adequately packing the paste into the prepared cavity; the material will then fill in crevices less effectively. All the dental composites exhibit creep (Papadogianis et al., 1984, 1985). The stiffness changes by a factor of from 2.5 to 4 (depending on the particular material) over a time period from 10 seconds to 3 hours under steady load. This creep may result in indentation of the restoration, but wear seems to be a greater problem.

Table 8-2. Composition and Shear Modulus of Dental Composites

Name	Fillers	Filler amount (w/o)	Particle size (μm)	G (GPa), 37°C
Adaptic	quartz	78	13	5.3
Concise	quartz	77	11	4.8
Nuva-fil	barium glass	79	7	–
Isocap	colloidal silica	33	0.05	–
Silar	colloidal silica	50	0.04	2.3

Reprinted with permission from Papadogianis et al. (1984). Copyright © 1984, Wiley.

Table 8-3. Typical Properties of Dental Composites

Property	Values
Young's modulus E (GPa)	10–16
Poisson's ratio ν	0.24–0.30
Compressive strength (MPa)	170–260
Shear strength (MPa)	30–100
Porosity (vol%)	1.8–4.8
Polymerization contraction (%)	1.2–1.6
Thermal expansion α (10^{-6}/°C)	26–40
Thermal conductivity k (10^{-4} cal/sec/cm^2 (°C/cm)	25–33
Water sorption coeff. (mg/cm^2, 24 hr, rm. temp.	0.6–0.8

Reprinted with permission from Cannon (1988). Copyright © 1988, Wiley.

Dental composites tend to be brittle and relatively weak in tension (Ban and Anusavice 1990). Moreover, they are subject to mechanical fatigue, so they can break or become loose at stress levels below the static fracture strength (Braem et al., 1995). Therefore, their use is restricted to certain types of dental restorations.

More recently, "packable" or condensable dental composites have been introduced as better alternatives to amalgam in restorations of posterior teeth (Leinfelder et al., 1999). These materials are designed to be less sticky and more viscous than prior composites, so that they can be packed more easily into the prepared cavity, and to produce tighter contacts between restored teeth. The elastic moduli of these composites range from about 9.5 to 21 GPa.

Dental cements (Rosenstiel et al., 1998) are used to attach dental crowns to the remaining tooth structure. A variety of filled resin-based cements, with 65 to 74% by weight filler, is available for this purpose. A high elastic modulus is considered beneficial in the ability of the cement to prevent loss of a crown.

Dental composite resins have become established as restorative materials for both anterior and posterior teeth and as cements. The use of these materials is likely to increase as improved compositions are developed and in response to concern over the long-term toxicity of silver–mercury amalgam fillings.

Example 8-3

Consider an isotropic composite in which the spherical particles are silica with a Young's modulus of 72 GPa and a polymer matrix with a Young's modulus of 1 GPa. Determine the modulus of the composite for an inclusion fraction of 33% by volume. Compare with the Reuss model.

Answer

Using the relation given in Table 8-1,

$$E = [5(72 - 1) \, 0.33] \, [3 + 2(72) \, / \, 1]^{-1} + 1 = \underline{1.8 \text{ GPa}}.$$

The calculation is approximate in that 33% particle concentration is not really dilute. For comparison, the Reuss model (see Problem 8.1) gives

$$E = [0.33/72 + 0.67/1]^{-1} = \underline{1.5 \text{ GPa}}.$$

The stiffness of the particulate composite is not much greater than the Reuss lower bound; this is representative of spherical inclusions.

8.3.2. Porous Implants

Porous implants allow tissue ingrowth. The ingrowth is considered desirable in many contexts, since it allows a relatively permanent anchorage of the implant to the surrounding tissues; see Chapter 14 for more details. There are actually two composites to be considered in porous implants: the implant prior to ingrowth, in which the pores are filled with tissue fluid which is ordinarily of no mechanical consequence; and the implant filled with tissue. In the case of the implant prior to ingrowth, it must be recognized that the stiffness and strength of the porous solid are much less than in the case of the solid from which it is derived, as described by the relationships in Table 8-1.

Porous layers are used on bone-compatible implants to encourage bony ingrowth. The pore size of a cellular solid has no influence on its stiffness or strength (though it does influence toughness); however, pore size can be of considerable biological importance. Specifically, in orthopedic implants with pores larger than about 150 μm, bony ingrowth into the pores occurs and this is useful to anchor the implant. This minimum pore size is on the order of the diameter of osteons in normal Haversian bone. It was found experimentally that smaller pores less than 75 μm in size did not permit the ingrowth of bone tissue. Moreover, it was difficult to maintain fully viable osteons within pores in the 75–150 μm size range. The representative structure of such a porous surface layer is shown in Figure 8-5. Porous coatings are also under study for application in anchoring the artificial roots of dental implants to the underlying jawbone. When a porous material is implanted in bone, the pores become filled first with blood, which clots, then with *osteoprogenitor mesenchymal cells*, and then, after about 4 weeks, bony trabeculae. The ingrown bone then become remodelled in response to mechanical stress. The bony ingrowth process depends on a degree of mechanical stability in the early stages. If too much motion occurs, the ingrown tissue will be collagenous scar tissue, not bone.

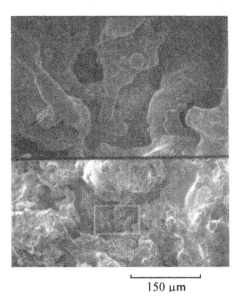

150 µm

Figure 8-5. Structure of porous coating for bony ingrowth. The scanning electron microscopic picture is 5× magnification of the rectangular region of the bottom picture (200×, Ti6Al4V alloy). Note the irregular pore structure. Unpublished data from S.H. Park and J.B. Park, University of Iowa, 1986.

Porous materials used in soft tissue applications include polyurethane, polyimide, and polyester velours used in percutaneous devices. Porous reconstituted collagen has been used in artificial skin, and braided polypropylene has been used in artificial ligaments. As in the case of bone implants, porosity encourages tissue ingrowth, which anchors the device. The healing and tissue response to a porous implant generally follows the sequence described elsewhere in this book. It must, however, be borne in mind that porous materials have a high ratio of surface area to volume. Consequently, the demands upon inertness and biocompatibility are likely to be greater for a porous material than a homogeneous one.

Ingrowth of tissue into implant pores is not always desirable: sponge (polyvinyl alcohol) implants used in early mammary augmentation surgery underwent ingrowth of fibrous tissue, and contracture and calcification of that tissue, resulting in hardened, calcified breasts. Current mammary implants make use of a balloon-like nonporous silicone rubber layer enclosing saline solution (silicone oil or gel was prohibited for such use by the American Food and Drug administration (FDA) due to litigation in the 1990s). A porous layer of polyester felt or velour attached to the balloon is provided at the back surface of the implant so that limited tissue ingrowth will anchor it to the chest wall and prevent it from migrating.

Porous blood vessel replacements encourage soft tissue to grow in, eventually forming a new lining, or *neointima* or *pseudointima*. This is another example of the biological role of porous materials as contrasted with the mechanical role. As discussed in Chapter 13, a material in contact with the blood should be nonthrombogenic. The role of the neointima encouraged to grow into a replacement blood vessel is to act as a natural nonthrombogenic surface resembling the lining of the original blood vessel.

Porous materials are produced in a variety of ways such as by sintering of beads or wires in the case of bone compatible surfaces. Vascular and soft tissue implants are produced by weaving or braiding fibers as well as by nonwoven "felting" methods. Protective foams for external use are usually produced by use of a "blowing agent," which is a chemical that evolves gas during polymerization of the foam. An interesting approach to producing micro-porous materials is the replication of structures found in biological materials: the *replamine-form* process. The rationale is that the unique structure of communicating pores is thought to offer advantages in the induction of tissue ingrowth. The skeletal structure of coral or echino-derms (such as sea urchins) is replicated by a casting process in metals and polymers; these have been tried in vascular and tracheal prostheses as well as in bone substitutes.

8.3.3. Fibrous and Particulate Composites in Orthopedic Implants

The rationale for incorporating stiff inclusions in a polymer matrix is to increase stiffness, strength, fatigue life, and other properties. For that reason, carbon fibers have been incorporated in the high-density polyethylene used in total knee replacements. The reason for wishing to modify the standard ultrahigh-molecular-weight polyethylene (UHMWPE) used in these implants is that it should provide adequate wear resistance over ten years' use. While this is sufficient for implantation in older patients, a longer wear-free lifetime is desirable in implants to be used in younger patients. Improvement in the resistance to creep of the polymeric com-ponent is also considered desirable, since excessive creep results in an indentation of the poly-meric component after long-term use. Representative properties of carbon-reinforced ultra-high-molecular-weight polyethylene are shown in Table 8-4. Enhancements of various properties by a factor of two are feasible.

Table 8-4. Properties of Carbon-Reinforced UHMWPE

Fiber amount (%)	Density (g/cm^3)	Young's modulus (GPa)	Flexural strength (MPa)
0	0.94	0.71	14
10	0.99	1.01	20
15	1.00	1.4	23
20	1.03	1.5	25

Reprinted with permission from Sclippa and Piekarski (1973). Copyright © 1973, Wiley.

Fibers have also been incorporated into polymethyl methacrylate (PMMA) bone cement on an experimental basis. Significant improvements in mechanical properties can be achieved. However, this approach has not found much acceptance since the fibers also increase the vis-cosity of unpolymerized material. It is consequently difficult to form and shape the polymeriz-ing cement during the surgical procedure. Metal wires have been used as macroscopic "fibers" to reinforce PMMA cement used in spinal stabilization surgery, but such wires are not useful in joint replacements owing to the limited space available.

Particle reinforcement has also been used in experiments to improve the properties of PMMA bone cement. For example, inclusion of bone particles in PMMA cement somewhat improves stiffness and fatigue life considerably, as shown in Figure 3-8. Moreover, the bone particles at the interface with the patient's bone are ultimately resorbed and replaced by in-grown new bone tissue. Short, slender titanium fibers have been embedded in PMMA cement

(Topoleski et al., 1992). A 5% volumetric fiber content gave rise to a toughness increase of 51%. Reinforcement of PMMA cement has not found much acceptance since any inclusions also increase the viscosity of the unpolymerized material. It is consequently difficult for the surgeon to form and shape the polymerizing cement prior to insertion.

Composites have been considered for bone plates and in the femoral component of total hip replacements. Metals are currently used in these applications, as discussed in previous chapters. Currently used implant metals are much stiffer than bone. They therefore shield the nearby bone from mechanical stress. Such stress-shielding results in a kind of disuse atrophy: the bone resorbs (Engh and Bobyn, 1988). Composite materials can be made more compliant than metal, and deform elastically to a higher strain (to about 0.01 compared with 0.001 for a mild steel): a potential advantage in this context (Bradley et al., 1980; Skinner, 1988). Flexible composite bone plates are effective in promoting healing (Jockisch, 1992). Hip replacement prostheses have been made with composites containing carbon fibers in a matrix of polysulfone and polyetherether ketone (Guyer et al., 1988). A polysulfone–carbon composite femoral stem is shown in Figure 8-6. In polymer matrix composites, creep behavior due to the polymer component is a matter of concern. Prototype composite femoral components exhibited creep of small magnitude limited by the fibers, which do not creep much. Creep is not expected to limit the life of the implant (Maharaj and Jamison, 1993).

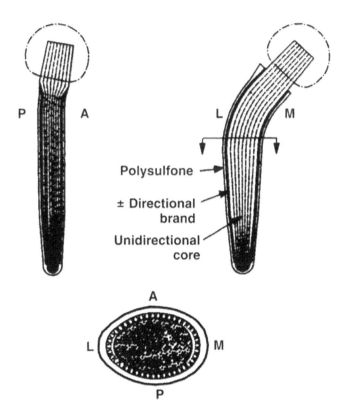

Figure 8-6. Details of carbon-polysulfone composite femoral stem construction. Reprinted with permission from Magee et al. (1988). Copyright © 1988, Lippincott, Williams & Wilkins.

Figure 8-7. Knee prostheses with black carbon fiber-reinforced polyethylene tibial components.

Example 8-4

How long should the fibers be in a short-fiber composite of graphite fibers in a polyethylene matrix, to be used for a knee replacement prosthesis (see Figure 8-7)?

Answer

To deal with the mechanical aspects of this question, we develop the "shear lag" model due to Cox. Consider the equilibrium conditions of a *segment* of a circular cross-section fiber of radius r and length dz acted upon by normal stress $\sigma_f l_{end}$ on the end, a normal stress distribution σ_f in the fiber to be found, and shear stress τ on the lateral surfaces. The equilibrium equation is

$$\pi r^2 \sigma_f + \tau 2\pi r dz = \pi r^2 (\sigma_f + d\sigma_f),$$

so

$$2\tau dz = d\sigma f, \quad \text{so} \quad d\sigma_f / dz = 2\tau / r,$$

so that

$$\sigma_f(z) = \sigma_f l_{end} + \frac{2}{4} \int_0^z \tau(z) dz.$$

Further progress is facilitated by assuming that the matrix is rigid-perfectly plastic, so that the stress upon the lateral surfaces of the fiber is equal to the matrix yield strength in shear τ_y. Such an assumption would approximate reality in the case of a ductile matrix subjected to a large load. Moreover, suppose that the fiber is sufficiently long that the stress in it is mostly due to shear from the lateral surfaces rather than tension at the ends. Then the above integral becomes simplified to $\sigma_f(z) = 2\tau_y z/r$, so that the fiber stress increases linearly with position on the fiber, up to half the fiber length L. The maximum fiber stress occurs at the fiber midpoint, $z = L/2$, so that

$$\frac{L}{2r} = \frac{\sigma_f^{max}}{2\tau_y}.$$

If we set the maximum fiber stress equal to its ultimate strength σ_f^{ult}, we obtain the *critical fiber length* L_c:

$$L_c = 2r\frac{\sigma_f^{ult}}{2\tau_y}.$$

For fibers longer than this length, load is efficiently transferred to them, to achieve the maximum strength of the composite. Fibers about as long as the critical fiber length can either break or pull out of the matrix, leading to enhanced toughness. In applications such as the knee replacement, considerations such as the manufacture of the composite also are important in determining the fiber length.

8.4. BIOCOMPATIBILITY OF COMPOSITE BIOMATERIALS

Each constituent of the composite must be biocompatible, and the interface between constituents must not be degraded by the body environment. As for inclusion materials, carbon itself has good compatibility and is used successfully. Carbon fibers used in composites are known to be inert in aqueous and even seawater environments; however, they do not have a long track record as biomaterials. Substantial electrochemical activity occurs in carbon fiber composites in an aqueous environment (Kovacs, 1993). There is thus concern that composites, if placed near a metallic implant, may cause galvanic corrosion. Inclusions in dental composites are minerals and ceramics with a good record of compatibility. As for the matrix material, polymers tend to absorb water when placed in a hydrated environment. Water acts as a plasticizer of the matrix and shifts the glass transition temperature toward lower values. This causes a reduction in stiffness and an increase in mechanical damping. Water absorption also causes swelling in polymers; this can be beneficial in dental composites since it neutralizes some of the shrinkage due to polymerization.

PROBLEMS

8-1.   For the Reuss model, show that the Young's modulus of the composite is given by $E_c = [E_f/V_f + E_m/V_m]^{-1}$. Hint: the stress is the same in each constituent and the elongations are additive [explain why].

8-2.   Derive Eq. (8-3). Assume that the particulate inclusions are perfectly rigid.

8-3.   Consider a bone plate made of a unidirectional fibrous composite. What fiber and matrix materials are suitable in view of the need for biocompatibility? Assume 50% fibers by volume and determine Young's modulus in the longitudinal direction. Compare with a metal implant.

8-4.   Consider an isotropic composite in which the spherical particles are silica with a Young's modulus of 72 GPa and a polymer matrix with a Young's modulus of 1 GPa. Determine the modulus of the composite for an inclusion fraction of 20% by volume. Compare with the Reuss model. Compare with the Voigt model and with Example 8.3.

8-5.   Calculate and plot the Voigt and Reuss bounds for a collagen–hydroxyapatite composite versus volume fraction hydroxyapatite. Plot on the same graph the Young's moduli for compact bone, dentin, and enamel. Use mechanical properties given elsewhere in the book.

8-6.   Calculate the shear modulus of dental composites given in Table 8-3 using the relation $E = 2G(1 + v)$. Discuss the validity of this equation in the context of this problem.

8-7.   Consider a bone plate made of a unidirectional fibrous composite. What fiber and matrix materials are suitable in view of the need for biocompatibility? Assume 50% fibers by volume and determine the Young's modulus in the longitudinal direction. Compare with a metal implant.

Material	Young's modulus (GPa)
Carbon fiber	250~500
PMMA	2
Bulk carbon	20

8-8.   A 2-mm-diameter 316L stainless steel wire is coated with 1-mm thick titanium. Using the following data, answer.

Material	Young's modulus (GPa)	Yield strength (MPa)	Density (g/cm³)
Stainless steel	200	400	7.84
Titanium	100	200	2.7

a. What is the Young's modulus of the composite?

b. If the composite wire is loaded in the longitudinal direction, what will the yield strength be?

c. How much load can the composite carry in tension without plastic deformation?

d. What is the density of the composite?

SYMBOLS/DEFINITIONS

Greek Letters

ε Strain

ρ Density

σ Stress

$\sigma_{f,s}$ Fracture strength of a solid

Latin letters

C_{ijkl} Elastic modulus tensor
E Young's modulus
V Volume fraction of a constituent

Words

Anisotropic: Dependent upon direction, referring to the material properties of composites.

Closed cell: A type of cellular solid in which a cell wall isolates the adjacent pores.

Composite: Composite materials are those that contain two or more distinct constituent materials or phases, on a microscopic or macroscopic size scale.

Cubic: A type of anisotropic symmetry in which the unit cells are cube shaped. There are three independent elastic constants.

Hexagonal: A type of anisotropic symmetry in which the unit cells are hexagonally shaped. There are five independent elastic constants. Transverse isotropy is mechanically equivalent to hexagonal although the structure may be random in the transverse direction.

Inclusion: Embedded phase of a composite.

Isotropic: Independent of direction, referring to material properties.

Matrix: The portion of a composite in which inclusions are embedded. The matrix is usually less stiff than the inclusions.

Neointima: New lining of a blood vessel. It is stimulated to form by fabric-type blood vessel replacements.

Open cell: A type of cellular solid in which there is no barrier between adjacent pores.

Orthotropic: A type of anisotropic symmetry in which the unit cells are shaped like rectangular parallelepipeds. In crystallography, this is called orthorhombic. There are nine independent elastic constants.

Porous ingrowth: Growth of tissue into the pores of an implanted porous biomaterial. Such ingrowth may or may not be desirable.

Replamineform: Cellular solid made using a biological material as a mold.

Transverse isotropy: See hexagonal.

Triclinic: A type of anisotropic symmetry in which the unit cells are oblique parallelepipeds with unequal sides and angles. There are 21 independent elastic constants.

BIBLIOGRAPHY

Agarwal BD, Broutman LJ. 1980. *Analysis and performance of fiber composites*. New York: Wiley.

Ashby MF. 1983. The mechanical properties of cellular solids. *Metallurg Trans* **14A**:1755–1768.

Ban S, Anusavice KJ. 1990. Influence of test method on failure stress of brittle dental materials. *J Dent Res* **69**:1791–1799.

Bradley JS, Hastings GW, Johnson-Hurse C. 1980. Carbon fibre reinforced epoxy as a high strength, low modulus material for internal fixation plates. *Biomaterials* **1**:38-40.

Braem M, Davidson CL, Lambrechts P, Vanherle G. 1994. In vitro flexural fatigue limits of dental composites. *J Biomed Mater Res* **28**:1397–1402.

Cannon ML. 1988. Composite resins. In *Encyclopedia of medical devices and instrumentation*, Ed JG Webster. New York: Wiley.

Christensen RM. 1979. *Mechanics of composite materials*. New York: Wiley.

Craig R. 1981. Chemistry, composition, and properties of composite resins. *Dent Clin North Am* **25**(2):219–239.

Engh CA, Bobyn JD. 1988. Results of porous coated hip replacement using the AML prosthesis. In *Noncemented total hip arthroplasty*, pp. 393–245. Ed RH Fitzgerald. New York: Raven Press.

Gibson LJ, Ashby MF. 1982. The mechanics of three dimensional cellular materials. *Proc Royal Soc London* **A382**:43–59.

Gibson LJ, Ashby MF. 1988. *Cellular solids*, 2nd ed. Oxford: Pergamon.

Guyer DW, Wiltse LL, Peek RD. 1988. The Wiltse pedicle screw fixation system. *Orthopedics* **11**(10): 455–1460.

Jockisch KA, Brown SA, Bauer TW, Merritt K. 1992. Biological response to chopped-carbon-fiber-reinforced PEEK. *J Biomed Mater Res* **26**(2):133–146.

Kovacs P. 1993. In vitro studies of the electrochemical behavior of carbon-fiber composites. In *Composite materials for implant applications in the human body: characterization and testing*, pp. 41–52. Ed RD Jamison, LN Gilbertson. Philadelphia: ASTM (STP 1178:41–52).

Leinfelder KF, Bayne SC, Swift EJ. 1999. Packable composites: overview and technical considerations. *J Esthet Dent* **11**:234–249.

Magee FP, Weinstein AM, Longo JA, Koeneman JB. 1988. A canine composite femoral stem: an *in vivo* study. *Clin Orthop Relat Res* **235**:237–252.

Maharaj GR, Jamison RD. 1993. Creep testing of a composite material human hip prosthesis. In *Composite materials for implant applications in the human body: characterization and testing*, pp. 86–97. Ed RD Jamison, LN Gilbertson. Philadelphia, PA: ASTM (STP 1178).

Papadogianis Y, Boyer DB, Lakes RS. 1984. Creep of conventional and microfilled dental composites. *J Biomed Mater Res* **18**:15–24.

Papadogianis Y, Boyer DB, Lakes RS. 1985. Creep of posterior dental composites. *J Biomed Mater Res* **19**:85–95.

Sclippa E, Piekarski K. 1973. Carbon fiber reinforced polyethylene for possible orthopaedic usage. *J Biomed Mater Res* **7**:59–70.

Spector M, Miller M, Beals N. 1988. Porous materials. In *Encyclopedia of medical devices and instrumentation*, Ed JG Webster. New York: Wiley.

Topoleski LDT, Ducheyne P, Cackler JM. 1992. The fracture toughness of titanium fiber reinforced bone cement. *J Biomed Mater Res* **26**:1599–1617.

9

STRUCTURE–PROPERTY RELATIONSHIPS OF BIOLOGICAL MATERIALS

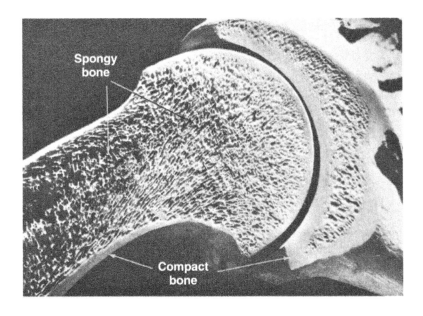

Sagittal section of the proximal end of the humerus in relation to the glenoid fossa of the scapula at the shoulder joint. These are dry bones, and the cartilaginous articular surfaces of the joint are not present. After A. Feininger, from *Anatomy of Nature*, Crown Publishers, with permission of Time Inc. Reprinted with permission from Bloom and Fawcett (1968). Copyright © 1968, W.B. Saunders.

The major difference between biological materials and biomaterials (implants) is *viability*. There are other equally important differences that distinguish living materials from artificial replacements. First, most biological materials are continuously bathed with body fluids. Exceptions are the specialized surface layers of skin, hair, nails, hooves, and the enamel of teeth. Second, most biological materials can be considered as *composites*.

Structurally, biological tissues consist of a vast network of intertwining fibers with polysaccharide ground substances immersed in a pool of ionic fluid. Attached to the fibers are cells that comprise the living tissues. Physically, ground substances function as a glue, lubricant, and shock absorber in various tissues.

The structure and properties of a given biological material are dependent upon the chemical and physical nature of the components present and their relative amounts. For example, neural tissues consist almost entirely of cells, while bone is composed of organic materials and calcium phosphate minerals with minute quantities of cells and ground substances as a glue.

An understanding of the exact role played by a tissue and its interrelationship with the function of the entire living organism are essential if biomaterials are to be used intelligently. Thus, a person who wants to design an artificial blood vessel prosthesis has to understand not only the property–structure relationship of the blood vessel wall but also its systemic function. This is because the natural artery is not only a blood conduit, but a component of a larger system, including a pump (heart), an oxygenator (lung), as well as neural systems that control stress in its walls, and a complex feedback system that governs cellular remodeling of the vessel structure.

9.1. PROTEINS

Proteins are polyamides formed by step-reaction polymerization between amino and carboxyl groups of amino acids:

$$-(-C-N-C-)_n-$$

where the structure is

$$\begin{array}{c} O\ H\ H \\ \|\ \ |\ \ \ | \\ -(-C-N-C-)_n- \\ | \\ R \end{array} \qquad (9\text{-}1)$$

where R is a side group. Depending on the side group, the molecular structure changes drastically. The simplest side group is hydrogen (H), which will form *glycine* (Gly). The geometry of the glycine is shown in Figure 9-1a, where the hypothetical flat sheet structure is shown. The structure has a repeating distance of 0.72 nm, and the side groups (R) are crowded, except for polyglycine, which has the smallest atom for the side group, H. If the side groups are larger, then the resulting structure is an α-*helix*, where the hydrogen bonds occur between different parts of the same chain and hold the helix together, as shown in Figure 9-1b.

9.1.1. Collagen

Collagen is a structural protein found in bone, cartilage, tendon, ligament, skin, and in the structural fibers of various organs. One of the basic constituents of protein is collagen, which has the general amino acid sequence–X–Gly–Pro–Hypro–Gly–X– (X can be any other amino acid) arranged in a triple α-helix. It has a high proportion of proline (Pro) and hydroxyproline (Hypro), as given in Table 9-1. Since the presence of hydroxyproline is unique in collagen (elastin contains a minute amount), the determination of collagen content in a given tissue is readily made by assaying the hydroxyproline.

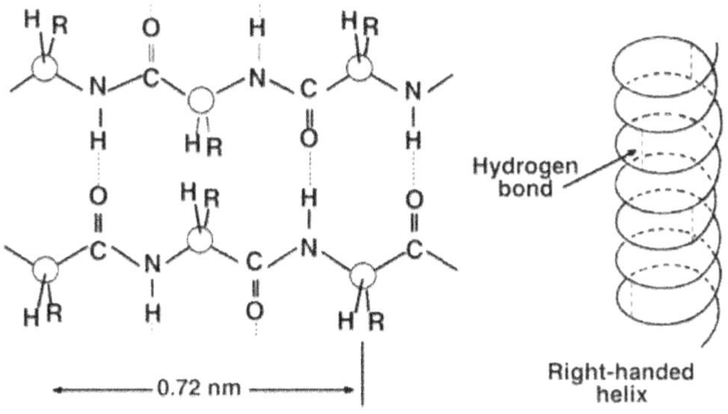

Figure 9-1. (a) Hypothetical flat sheet structure of a protein. (b) Helical arrangement of a protein chain.

Table 9-1. Amino Acid Content of Collagen

A.A. and component	Content (mol/100 mol amino acids)
Gly	31.4–338
Pro	11.7–13.8
Hypro	9.4–10.2
Acid polar a.a.s. (aspt. glut. asparagine)	11.5–12.5
Basic polar a.a.s. (lys. agr. his.)	8.5–8.9
Other a.a.s.	Residue

Reprinted with permission from Chvapil (1967). Copyright © 1967, Butterworths.

Three left-handed-helical peptide chains are coiled together to give a right-handed coiled superhelix with periodicity of 2.86 nm. This triple super helix is the molecular basis of *tropocollagen*, the precursor of collagen (see Figure 9-2). The three chains are held strongly to each other by H bonds between glycine residues and between hydroxyl (OH) groups of hydroxyproline. In addition, there are crosslinks via lysine among the (three) helices.

The primary factors stabilizing the collagen molecules are invariably related to the interactions among the α-helices. These factors are H bonding between the C=O and NH groups, ionic bonding between the side groups of polar amino acids, and the interchain crosslinks between helices.

The collagen fibrils (20–40 nm in diameter) form fiber bundles with a 0.2–1.2 μm diameter. Figure 9-3 shows scanning and transmission electron microscopic pictures of collagen fibrils in bone, tendon, and skin. Note the straightness of tendon collagen fibrils compared to the more wavy skin fibers.

The side groups of some amino acids are highly non-polar in character and hence *hydrophobic*; therefore, chains with these amino acids avoid contact with water molecules and seek the greatest number of contacts with the non-polar chains of amino acids. If we destroy the hydrophobic nature by an organic solvent solution (e.g., urea), the characteristic structure is lost, resulting in microscopic changes such as shrinkage of collagen fibers. The same effect can be achieved by simply warming the collagen fibers. Another factor affecting the stability of the collagen is the incorporation of water molecules into the intra- and inter-chain structure. If the water content is lowered, the structural stability decreases. If the collagen is dehydrated completely (lyophilized), then the solubility also decreases (so-called in vitro aging of collagen).

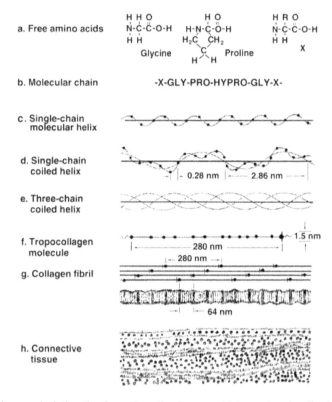

Figure 9-2. Diagram depicting the formation of collagen, which can be visualized as taking place in seven steps. The starting materials (**a**) are amino acids, of which two are shown, and the side chain of any others is indicated by R in amino acid X. (**b**) The amino acids are linked together to form a molecular chain. (**c**) This then coils into a left-handed helix (**d** and **e**). Three such chains then intertwine in a triple stranded helix, which constitutes the tropocollagen molecule (**f**). Many tropocollagen molecules become aligned in staggered fashion, overlapping by a quarter of their length to form a cross-striated collagen fibril (**g**). Reprinted with permission from Gross (1961). Copyright © 1961, Scientific American.

It is known that the acid mucopolysaccharides also affect the stability of collagen fibers by mutual interactions by forming mucopolysaccharide–protein complexes. It is believed that the water molecules affect the polar region of the chains, making the dried collagen more disoriented than it is in the wet state.

9.1.2. Elastin

Elastin is another structural protein found in a relatively large amount in elastic tissues such as *ligamentum nuchae* (major supporting tissue in the head and neck of grazing animals), aortic wall, skin, etc. The chemical composition of elastin is somewhat different from that of collagen.

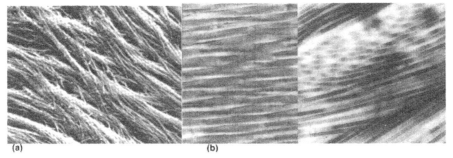

Figure 9-3. (a) Scanning electron micrograph of the surface of adult rabbit bone matrix, showing how the collagen fibrils branch and interconnect in an intricate, woven pattern (×4800). Reprinted with permission from Tiffit (1980). Copyright © 1980, J.B. Lippincott. (b) Transmission electron micrographs of (**left**) parallel collagen fibrils in a tendon and (**right**) mesh work of fibrils in skin (×24,000). Reprinted with permission from Fung (1981). Copyright © 1981, Springer-Verlag.

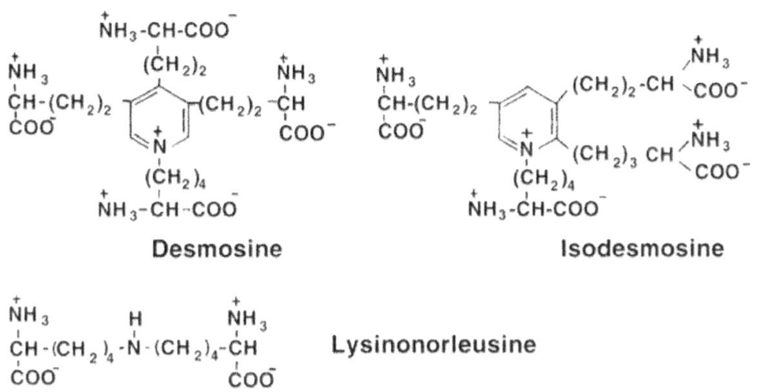

Desmosine

Isodesmosine

Lysinonorleusine

Figure 9-4. Structure of desmosine, isodesmosine, and lysinonorleucine.

The high elastic compliance and extensibility of elastin is due to the crosslinking of lysine residues via *desmosine, isodesmosine*, and *lysinonorleusine* (shown in Figure 9-4). The formation of desmosine and isodesmosine is only possible by the presence of copper and lysyl oxidase enzyme; hence, deficiency of copper in the diet may result in non-crosslinked elastin. This, in turn, will result in tissue that is viscous rather than normal rubber-like elastic tissue, and abnormality that can lead to rupture of the aortic walls.

Elastin is very stable at high temperature in the presence of various chemicals due to the very low content of polar side groups (hydroxyl and ionizable groups). The specific staining of elastin in tissue (prepared for microscopic study), by lipophilic stains such as Weigert's resorcin–fuchsin is due to the same reason. Elastin contains a high percentage of amino acids with aliphatic side chains such as *valine* (6 times that of collagen). It also lacks all the basic and acidic amino acids, so that it has very few ionizable groups. The most abundant of these, glutamic acid, occurs only a sixth as often as in collagen. *Aspartic acid, lysine*, and *histidine* are all below 2 residues per 1,000 in mature elastin. The composition of elastin is given in Table 9-2.

Table 9-2. Amino Acid Content of Elastin

Content	Amount (residues/1000)
Gly	324
Hypro	26
Cationic residues (Asp, Glu)	21
Anionic residues (His, Lys, Arg)	13
Nonpolar residues	595
(Pro, Ala, Val, Met, Leu, Ile, Phe, Tyr)	
Half-cystine	4

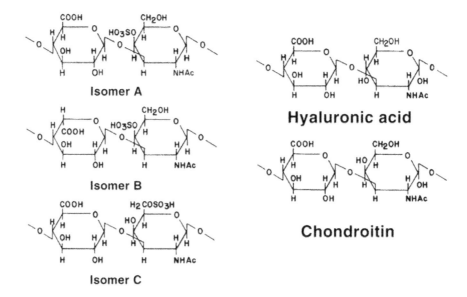

Figure 9-5. Structure of hyaluronic acid, chondroitin, and chondroitin sulfates.

9.2. POLYSACCHARIDES

Polysaccharides are polymers of simple sugars. They exist in tissues as a highly viscous material that interacts readily with proteins, including collagen, resulting in *glycosamino–glycans*
(also known as *mucopolysaccharides)* or *proteoglycans*. These molecules readily bind
both water and cations due to the large content of anionic side chains. They also exist at
physiological concentrations not as viscous solids but as viscoelastic gels. All of these polysaccharides consist of disaccharide units polymerized into unbranched macromolecules, as
shown in Figure 9-5.

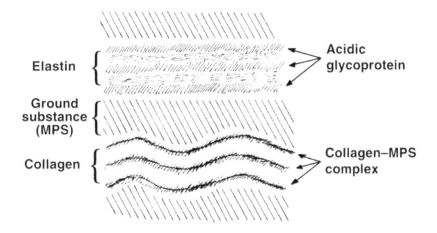

Figure 9-6. Schematic representation of mucopolysaccharides—protein molecules in connective tissues. Note the wavy nature of collagen fibers and the straighter form of elastin.

9.2.1. Hyaluronic Acid and Chondroitin

Hyaluronic acid is found in the vitreous humor of the eye, synovial fluid, skin, umbilical cord, and aortic walls. Hyaluronic acid is made of residues of N-acetylglucosamine and D-glucuronic acid, but it lacks the sulfate residues. The animal hyaluronic acid contains a protein component (0.33 w/o or more) and is believed to be chemically bound to at least one protein or peptide that cannot be removed. This, in turn, will result in proteoglycan molecules that may behave differently from the pure polysaccharides. Chonodroitin is similar to hyaluronic acid in its structure and properties and is found in the cornea of the eyes.

9.2.2. Chondroitin Sulfate

This is the sulfated mucopolysaccharide that resists the hyaluronidase enzyme. It has three isomers, as shown in Figure 9-5. Isomer A (chondroitin 4-sulfate) is found in cartilage, bones, and the cornea, while isomer C (chondroitin 6-sulfate) can be isolated from cartilage, umbilical cord, and tendon. Isomer B (dermatan sulfate) is found in skin and the lungs and is resistant to testicular hyaluronidase enzyme.

The chondroitin sulfate chains in connective tissues are bound covalently to a polypeptide backbone through their reducing ends. Figure 9-6 shows a proposed macromolecular structure of protein–polysaccharides from which one can imagine the nature of the viscoelastic properties of the ground substance. These complexes of protein and mucopolysaccharides (ground substance) play an important role in the physical behavior of connective tissues either as lubricating agents between tissues (e.g., joints) or between elastin and collagen microfibrils.

Example 9-1

Calculate the degree of polymerization of a chondroitin compound that has an average molecular weight of 100,000 g/mole.

Answer

From Figure 9-5 (also Ac is $COOCH_3$, acetyl) it can be seen that there are 14 carbon atoms, 13 oxygen atoms, 21 hydrogen atoms, and 1 nitrogen atom in a repeating unit; therefore, $12 \times 14 + 14 \times 16 + 21 \times 1 + 1 \times 14 = 427$, and

$$\text{D.P.} = 100,000/427 = \underline{234}$$

9.3. STRUCTURE–PROPERTY RELATIONSHIP OF TISSUES

Understanding the structure–property relationship of various tissues is important since one has to know what is being replaced by the artificial materials (biomaterials). Also, one may want to use natural tissues as biomaterials (for example, porcine heart valves). The property measurements of any tissues are confronted with many of the following limitations and variations:

1. Limited sample size,
2. Original structure can undergo change during sample collection or preparation,
3. Inhomogeneity,
4. Complex nature of the tissues makes it difficult to obtain fundamental physical parameters,
5. Tissue cannot be frozen or homogenized without altering its structure or properties.
6. The in-vitro and in-vivo property measurements are sometimes difficult, if not impossible, to correlate.

The main objective of studying the property–structure relation of tissues is to design better performing implants in our body. Therefore, one should always ask, "What kind of physiological functions are being performed by the tissues or organs under study in vivo and how can one best assume their lost function?" Keeping this in mind, let us study the tissue structure–property relationships.

9.3.1. Mineralized Tissue (Bone and Teeth)

9.3.1.a. Composition and Structure

Bone and teeth are mineralized tissues whose primary function is "load carrying." Teeth are in more extraordinary physiological circumstances since their function is carried out in direct contact with ex-vivo substances, while the functions of bone are carried out inside the body in conjunction with muscles, ligaments, and tendons. A schematic anatomical view of a long bone is shown in Figure 9-7.

Wet cortical bone is composed of 22 w/o organic matrix, of which 90–96 w/o is collagen, mineral (69 w/o), and water (9 w/o), as given in Table 9-3. The major subphase of the mineral consists of submicroscopic (nanoscale) crystals of an apatite of calcium and phosphate, resembling hydroxyapatite in its crystal structure — $(Ca_{10}(PO_4)_6(OH)_2)$. There are other mineral ions such as citrate $(C_6H_5O_7^{-4})$, carbonate (CO_3^{-2}), fluoride (F^-), and hydroxyl ions (OH^-), which may yield some other subtle differences in microstructural features of the bone. The apatite crystals are formed as slender needles, 20–40 nm in length by 1.5–3 nm in thickness, in the collagen fiber matrix. Plate-like crystals are also found in bone. Collagen fibrils, which contain mineral, are arranged into lamellar sheets (3-7 μm) that run helically with respect to the long axis of the cylindrical *osteons* (or sometimes called *Haversian systems*). The osteon is made up of 4 to 20

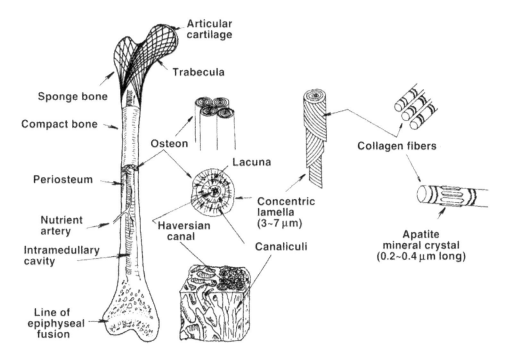

Figure 9-7. Organization of a typical bone.

Table 9-3. Composition of Bone

Component	Amount (w/o)
Mineral (apatite)	69
Organic matrix	22
collagen	(90–96% of organic matrix)
others	(4–10% of organic matrix)
Water	9

Reprinted with permission from Tiffit (1980). Copyright © 1980, J.B. Lippincott.

lamellae that are arranged in concentric rings around the Haversian canal. Osteons are typically from 150 to 250 μm in diameter. Between these osteons the interstitial systems are sharply divided by the *cementing line*, or *cement line*. The metabolic substances can be transported by the intercommunicating pore systems, known as *canaliculi, lacunae,* and *Volksmann's canals,* which are connected with the marrow cavity. These various interconnecting systems are filled with body fluids, and their volume can be as high as 18.9 ± 0.45% according to one estimate for compact beef bone. The external and internal surfaces of the bone are called the periosteum and endosteum, respectively, and both have osteogenic properties.

It is interesting to note that the mineral phase is not a completely discrete aggregation of the calcium phosphate mineral crystals. Rather, it is made of a somewhat contiguous phase, as evidenced in Figure 9-8, and by the fact that complete removal of the organic phase of the bone still gives a material with good strength.

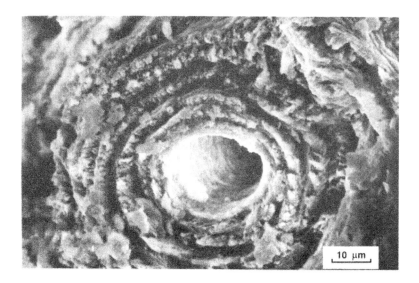

Figure 9-8. Scanning electron microscope photomicrograph showing the mineral portion of osteon lamellae. The organic phase has been removed by ethylenediamine in a Soxhlet apparatus.
Reprinted with permission from Pierkarski (1978). Copyright © 1978, Academic Press.

Long bones such as the femur contain cancellous (or spongy) and compact bone. The
spongy bone consists of three-dimensional branches or bony trabeculae interspersed by the
bone marrow. More spongy bone is present in the epiphyses of long bones and within vertebrae, while compact bone is the major form present in the diaphysis of a long bone, as shown
in Figure 9-7.

There are two types of teeth — deciduous or primary, and permanent — of which the latter is more important for us from the biomaterials point of view. All teeth are made of two portions — the crown and the root — usually demarcated by the gingiva (gum). The root is placed
in a socket called the alveolus in the maxillary (upper) or mandibular (lower) bones. A sagittal
cross-section of a permanent tooth is shown in Figure 9-9 to illustrate various structural features. The enamel is the hardest substance found in the body and consists almost entirely of
calcium phosphate salts (97%) in the form of large apatite crystals.

The dentin is another mineralized tissue whose distribution of organic matrix and mineral
is similar to that of compact bone. Consequently, its physical properties are also similar. The
collagen matrix of the dentin might have a somewhat different molecular structure than the
normal bone: it is more crosslinked than that found in other tissues. Dentinal tubules (3–5 μm
wide) radiate from the pulp cavity toward the periphery and penetrate every part of the dentin.
The dentinal tubules contain collagen fibrils (2-4 μm thick) in the longitudinal direction, and
the interface is cemented by a protein–polysaccharide complex substance, as well as the processes of the odontoblasts, which are cells lining the pulp cavity.

Cementum covers most of the root of the tooth with a coarsely fibrillated bone-like substance devoid of canaliculi, Haversian systems, and blood vessels. The pulp occupies the cavity and contains thin collagenous fibers running in all directions and not aggregated into
bundles. Ground substance, nerve cells, blood vessels, etc. are also contained in the pulp. The
periodontal membrane anchors the root firmly into the alveolar bone and is made of mostly
collagenous fibers plus glycoproteins (protein–polysaccharides complex).

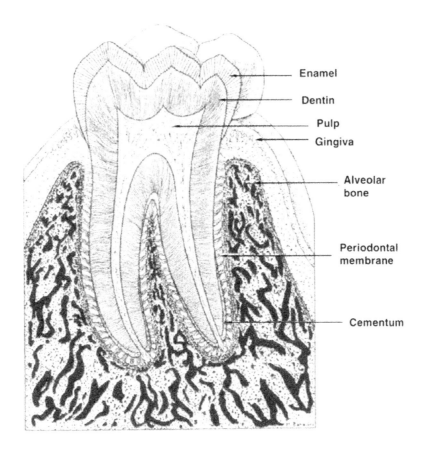

Figure 9-9. Sagittal section of a molar tooth.

Some of the physical properties of teeth are given in Table 9-4. As can be expected the strength is the highest for enamel and dentin is between bone and enamel. The thermal expansion and conductivity are higher for enamel than for dentin as given in Table 9-4.

Example 9-2

Calculate the volume percentage of each major component of a wet bone based on the weight percentage, i.e., 9, 69, and 22 w/o for water, mineral, and organic phase, respectively. Assume the densities of mineral and organic phase to be 3.16 and 1.03 g/cm^3, respectively.

Answer

Based on 100 g of bone, the volume of each component can be calculated dividing the weight by its density:

Component	w/o	Wt (g)	V (cm^3)	v/o
Mineral	69	69	21.8	41.8
Organic	22	22	21.4	41.0
Water	9	9	9.0	17.2
Total	100	100	52.2	100

Table 9-4. Physical Properties of Teeth

Tissue	Density (g/cm³)	Modulus of elasticity elasticity (GPa)	Compressive strength strength (MPa)	Coefficient of thermal expansion (/°C)	Thermal conductivity (W/m·K)
Enamel	2.2	48	241	11.4×10^{-6}	0.82
Dentin	1.9	13.8	138	8.3×10^{-6}	0.59

Table 9-5. Properties of Bone

Tissue	Direction of test	Modulus of elasticity (GPa)	Tensile strength (MPa)	Compressive strength (MPa)
Leg bone				
Femur	Longitudinal	17.2	121	167
Tibia	"	18.1	140	159
Fibula	"	18.6	146	123
Arm Bone				
Humerus	Longitudinal	17.2	130	132
Radius	"	18.6	149	114
Ulna	"	18.0	148	117
Vertebra				
Cervical	Longitudinal	0.23	3.1	10
Lumbar	"	0.16	3.7	5
Spongy bone	"	0.09	1.2	1.9
Skull	Tangential	–	25	–
	Radial	–	–	97

Reprinted with permissioni from Yamada (1970). Copyright © 1970, Williams & Wilkins.

9.3.1.b. Mechanical Properties of Bone and Teeth

As with most other biological materials, the properties of bone and teeth depend substantially on the humidity, sign of load (compressive or tensile), rate of loading, and the direction of the applied load with respect to the orientation of the microstructure. Therefore, one usually studies the effect of the above-mentioned factors and correlates the results with structural features. We will follow the same practice. Table 9-5 gives some general idea of the mechanical properties of various bones.

The *effect of drying* of the bone can be seen easily from Figure 9-10, where the dry sample shows a slightly higher modulus of elasticity but lower toughness, fracture strength, and strain to failure. Thus, the wet bone in the laboratory that behaves similarly to in-vivo bone can absorb more energy and elongate more before fracture.

The *effect of anisotropy* is expected since the osteons are longitudinally arranged along the long axis of the bone, and the load is mostly borne in that direction. The Young's modulus and the tensile and compressive strengths in the longitudinal direction are approximately 2 and 1.5 times higher than those in the radial or tangential directions, respectively. The tangential and radial directions differ little in their mechanical properties in human Haversian bone, but they do differ in bovine bone, which has a different structure.

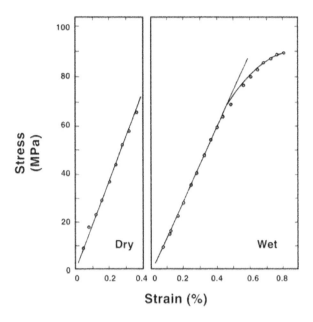

Figure 9-10. Effect of drying on the behavior of human compact bone. Reprinted with permission from Evans and Lebow (1952). Copyright © 1952, Excerpta Medica.

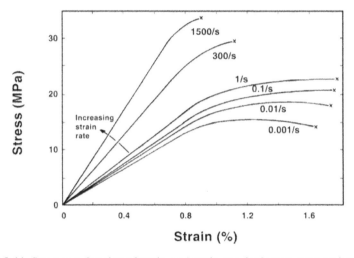

Figure 9-11. Stress as a function of strain, and strain rate for human compact bone. Reprinted with permission from McElhaney (1966). Copyright © 1966, American Institute of Physics.

The *effect of the rate of loading* on the bone is shown in Figure 9-11. As can be seen, the Young's modulus, ultimate compressive, and yield strength increase with increased rate of loading. However, the failure strain and the fracture toughness of the bone reach a maximum and then decrease. This implies that there is a critical rate of loading.

The *effect of mineral content* on the mechanical properties is given in Table 9-6. More highly mineralized bone has a higher modulus of elasticity and bending strength but lower toughness. This illustrates the importance of the organic phase in providing toughness and energy absorption capability in bone. This table also illustrates the adaptation of bony tissue to different roles in biological systems. Antlers optimize toughness, while whale ear bones maximize stiffness; limb bones are intermediate.

Table 9-6. Properties of Three Different Bones with Varying Mineral Contents

Type of bone	Work of fracture (J/m^2)	Bending strength (MPa)	Young's modulus (GPa)	Mineral content (w/o)	Density (g/cm^3)
Deer antler	6190	179	7.4	59.3	1.86
Cow femur	1710	247	13.5	66.7	2.06
Whale tympanic bulla	200	33	31.3	86.4	2.47

Reprinted with permission from Currey (1981). Copyright © 1981, American Society of Mechanical Engineers.

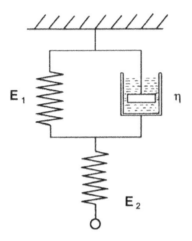

Figure 9-12. Three-element viscoelastic model of bone.

The *effect of viscoelasticity* includes the loading rate effect described above. In viscoelastic materials, the stiffness depends on time under constant load, upon rate of loading, and upon frequency of (sinusoidal) loading. The viscoelastic properties of bone can be viewed conceptually in terms of mechanical models such as the one shown in Figure 9-12. The differential equation for the three element model can be derived and be applied for various testing conditions (see Problem 9-5). In the case of bone, this is not particularly realistic unless only a small portion, less than a factor ten, of the time/frequency range is considered. The measured effect of frequency of loading upon the stiffness and dynamic mechanical damping of compact bone is shown in Figure 9-13. A simple spring-dashpot model such as that shown in Figure 9-12 would give substantial loss and modulus variation only over about a factor ten in time or frequency: a small part of the range shown in Figure 9-13. The stiffness increases with frequency (and hence also with rate of loading), while the damping (loss tangent) has a minimum at frequencies associated with walking, running, and other activities. Impact as well as ultrasonic waves used in medical diagnosis are absorbed in bone.

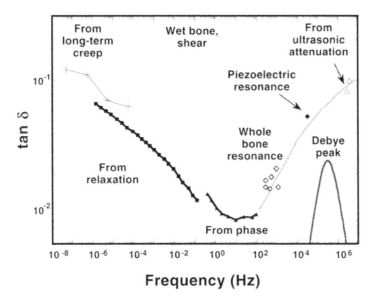

Figure 9-13. Loss tangent of compact bone over a wide range of frequency. Also shown is a Debye peak that is the loss tangent associated with a simple three-element model. Adapted with permission from Garner et al. (2000). Copyright © 2000, American Society of Mechanical Engineers.

The *effect of density* (ρ in g/cm^3) and *strain rate* ($d\varepsilon/dt$ in sec^{-1}) upon the compressive strength (σ_{ult} in MPa) of bone is summarized in the following equation:

$$\sigma_{ult} = 68\rho^2 (d\varepsilon/dt)^{0.06}. \tag{9-2}$$

This includes compact bone and, allowing some scatter in the comparison with experiment, trabecular bone as well. The strength of bone is seen to depend strongly upon its density and weakly upon strain rate as shown in Figure 9-14.

Example 9-3

Calculate the density of the mineral phase of dried cow femur (c.f. Table 9-5) if the density of the organic phase and water is 1 g/cm^3.

Answer

Using a simple mixture rule, neglecting water since it is dried:

$$\rho = \rho_1 V_1 + \rho_2 V_2 + \cdots + \rho_n V_n,$$

$$V_1 + V_2 + \cdots V_n = 1.$$

From Example 9-2, $V_1 = V_2 = 0.5$. Therefore,

$$2.06 = 1 \times 0.5 + \rho_m \times 0.5,$$

$$\rho_m = \underline{3.12 \text{ g/cm}^3}.$$

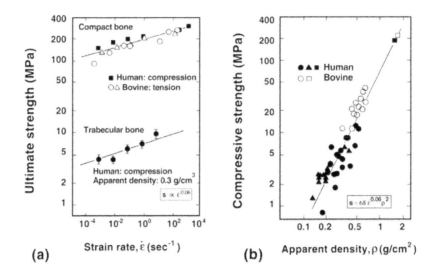

Figure 9-14. (a) Influence of strain rate on the ultimate strength of compact and trabecular bone tested without marrow in vitro. (b) Influence of density on the compressive strength of compact and trabecular bone. Reprinted with permission from Carter and Hayes (1976). Copyright © 1976, American Association for the Advancement of Science.

This value is close to the value of hydroxyapatite mineral as given in Example 6-2.

9.3.1.c. Modeling of Mechanical Properties of Bone

As mentioned earlier, bone is a composite material, therefore, many researchers have proposed composite models based on two components — the mineral and the organic phase. In this section we present several simplified composite models. They are not intended as realistic descriptions of bone. If one assumes that the load is independently borne by the two components (collagen and mineral, hydroxyapatite) then the total load (P_t) is borne by mineral (P_m) and collagen (P_c):

$$P_t = P_m + P_c. \qquad (9\text{-}3)$$

Since $\sigma = P/A = E\varepsilon$, thus,

$$P_m = A_m \times E_m \times \varepsilon_m, \qquad (9\text{-}4)$$

where A, E, and ε are area, modulus, and strain, respectively. Suppose that the strain of collagen can be assumed to be equal to that of mineral, that is,

$$P_c = P_m \frac{A_c E_c}{A_m E_m} \; ; \qquad (9\text{-}5)$$

therefore,

$$P_m = \frac{P_t A_m E_m}{A_m E_m + A_c E_c}. \qquad (9\text{-}6)$$

If we express Eq. (9-3) in terms of Young's modulus, it becomes

$$E_t = E_m V_m + E_c V_c, \tag{9-7}$$

where V is the volume fraction. This is the so-called "rule of mixtures" or Voigt model. It represents a rigorous upper bound on the stiffness of a composite. It also represents the actual stiffness of a composite with unidirectional, parallel fibers, or laminae, for uniaxial stress along the fiber direction. It is only approximately correct for bone in the longitudinal direction. The same limitations apply to Eq. (9-3). If the fibers are arranged in the perpendicular direction, then one can derive the following equation:

$$\frac{1}{E_t} = \frac{V_m}{E_m} + \frac{V_c}{E_c}. \tag{9-8}$$

Since collagenous fibers vary in their direction, one can propose another model:

$$\frac{1}{E_c} = \frac{x}{E_m V_m + E_c V_c} + (1+x)\left(\frac{V_m}{E_m} + \frac{V_c}{E_c}\right), \tag{9-9}$$

where x is the fraction of bone that conforms to the parallel direction and $(1-x)$ is the rest.

9.3.1.d. Electrical Properties of Bone

In the late 1950s it was shown that dry bone is piezoelectric in the classic sense, i.e., mechanical stress results in electric polarization, the indirect effect, and an applied electric field causes strain, the converse effect. (See the piezoelectric constitutive equations in Chapter 4.)

The piezoelectric properties of bone are of interest in view of their hypothesized role in bone remodeling. Wet collagen, however, does not exhibit a piezoelectric response. Both the dielectric properties and the piezoelectric properties of bone depend strongly upon frequency. The magnitude of the piezoelectric sensitivity coefficients of bone depends on frequency, on direction of load, and on relative humidity. Values up to 0.7 pC/N have been observed in bone, to be compared with 0.7 and 2.3 pC/N for different directions in quartz, and 600 pC/N in some piezoelectric ceramics. Piezoelectric effects occur in the kilohertz range in bone, well above the range of physiologically significant frequencies. It is, however, uncertain whether wet bone is piezoelectric in the classic sense at the relatively low frequencies that dominate in the normal loading of bone. It has been suggested that two different mechanisms are responsible for these effects: classical piezoelectricity due to the molecular asymmetry of collagen in dry bone, and fluid flow effects, possibly streaming potentials in wet bone. Streaming potentials are electrical signals resulting from flow of ionic electrolytes through channels in bone (canaliculi, Haversian canals) by the deformation of bones.

Bone exhibits additional electrical properties that are of interest. For example, the dielectric behavior (e.g., the dynamic complex permittivity, the real part of which governs the capacitance and the imaginary part of which governs the resistivity) controls the relationship between the applied electric field and the resulting electric polarization and current, as discussed in Chapter 4. Dielectric permittivity of bone has been found to increase dramatically with increasing humidity and decreasing frequency. For bone under partial hydration conditions, the dielectric permittivity (which determines the capacitance) can exceed 1,000 and the dielectric loss tangent (which determines the ratio of conductivity to capacitance) can exceed unity. Both the permittivity and the loss are greater if the electric field is aligned parallel to the bone axis. Bone under conditions of full hydration in saline behaves differently: the behavior of bovine femoral bone is essentially resistive, with very little relaxation. The resistivity is about 45–48 Ωm for the longitudinal direction, and three to four times greater in the radial direction. These values are to be compared with a resistivity of 0.72 Ωm for physiological saline alone. Since the resistivity of fully hydrated bone is about 100 times greater than that of bone under 98% relative humidity, it is suggested that at 98% humidity the larger pores are not fully filled with fluid.

Compact bone also exhibits a permanent electric polarization as well as pyroelectricity, which is a change of polarization with temperature. These phenomena are attributed to the polar structure of the collagen molecule; these molecules are oriented in bone. The orientation of permanent polarization has been mapped in various bones and has been correlated with developmental events.

The electrical properties of bone are relevant not only as a hypothesized feedback mechanism for bone remodeling (see the following section), but also in the context of external electrical stimulation of bone to aid its healing and repair. A short summary of the electrical properties of bone is given in Table 9-7.

Table 9-7. Electrical Properties of Bone

Property	Condition	Value	Reference
Dielectric permittivity	Radial, 78% rh, 37°C, 1 Hz	10^5	Lakes et al., 1977
Resistivity (Ωm)	Longitudinal, 100% hydration, 0.1–30 sec	45	Chakkkalakahl et al., 1980
	Radial, 100%hydration, 0.1–30 sec	150	
Piezoelectric Coefficients (pC/N)	75% rh, 23.5°C, 100 Hz	$d_{11} = 0.014$	Bur, 1976
		$d_{12} = 0.026$	
		$_{13} = -0.032$	
		$d_{14} = 0.105$	
		$d_{15} = -0.013$	
		$d_{16} = -0.070$	

Reprinted with permission from Lakes et al. (1977). Copyright © 1977, American Institute of Physics.

9.3.2. Bone Remodeling

9.3.2.a. Phenomenology

The relationship between the mass and form of a bone to the forces applied to it was appreciated by Galileo, who is credited with being the first to understand the balance of forces in beam bending and with applying this understanding to the mechanical analysis of bone. Julius Wolff published his seminal 1892 monograph on bone remodeling; the observation that bone is reshaped in response to the forces acting on it is presently referred to as *Wolff's law*. Indirect evidence was presented in the drawing given in Figure 9-15, in which Wolff emphasized that the remodeling of cancellous bone structure followed mathematical rules corresponding to the principal stress trajectories. Many relevant observations regarding the phenomenology of bone remodeling have been suggested (by Frost, see bibliography):

1. Remodeling is triggered not by principal stress but by "flexure."
2. Repetitive dynamic loads on bone trigger remodeling; static loads do not.
3. Dynamic flexure causes all affected bone surfaces to drift towards the concavity which arises during the act of dynamic flexure.

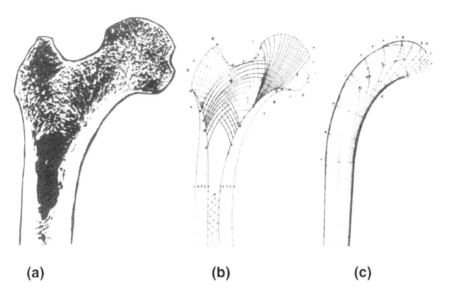

(a)   (b)   (c)

Figure 9-15. Frontal section through the proximal end of the femur (**a**); line drawing of the cancellous structure (**b**); Culmann's crane (**c**). Reprinted with permission from Wolff (1892). Copyright © 1892, Hirchwild Publishers.

These rules are essentially qualitative and do not deal with underlying causes. Additional aspects of bone remodeling may be found in the clinical literature. For example, after complete removal of a metacarpal and its replacement with graft consisting of a strut of tibial bone, the graft became remodeled to resemble a real metacarpal; the graft continued to function after 52 years. In the standards of the Swiss Association for Internal Fixation it is pointed out that severe osteoporosis can result from the use of two bone plates in the same region, as a result of the greatly reduced stress in the bone. It is suggested that as a result of bending stresses in the femur the medial and lateral aspects should be stiffer and stronger than the anterior and posterior aspects. Such a difference has actually been observed. Large cyclic stress causes more remodeling of bone than large static stress. Immobilization of humans causes loss of bone and excretion of calcium and phosphorus. Long spaceflights under zero gravity also cause loss of bone; hypergravity induced by centrifugation strengthens the bones of rats. Studies of stress-induced remodeling of living bone have been performed in vivo in pigs. In this study, strains were directly measured by strain gages before and after remodeling. Remodeling was induced by removing part of the pigs' ulna so that the radius bore all the load. Initially, the peak strain in the ulna approximately doubled. New bone was added until, after three months, the peak strain was about the same as on the normal leg bones. In-vivo experiments conducted in sheep have yielded similar results. In humans, exercise increases bone mass. It is of interest to compare the response time noted in the above experiments with the rate of bone turnover in healthy humans. The life expectancy of an individual osteon in a normal 45 year old man is 15 years, and it will have taken 100 days to produce it.

Remodeling of Haversian bone seems to influence the quantity of bone but not its quality, i.e., Young's modulus, tensile strength, and composition are not substantially changed. However, the initial remodeling of primary bone to produce Haversian bone results in a reduction in strength. As for the influence of the rate of loading on bone remodeling, there is good evidence to suggest that intermittent deformation can produce a marked adaptive response in bone, whereas static deformation has little effect. Experiments upon rabbit tibiae bear this out. In the dental field, by contrast, it is accepted that static forces of long duration move teeth in the jawbone. In this connection, the direction (as well as the type) of stresses acting on the (alveolar) bone tissue should also be considered. The forces may be static but the actual loading on the alveolar bone would be dynamic due to the mastication forces upon the jawbones. Some have pointed out that the response of different bones in the same skeleton to mechanical loads must differ; otherwise, lightly loaded bones such as the top of the human skull, or the auditory ossicles, would be resorbed.

Failure of bone remodeling to occur normally in certain disease states is of interest. For example, micropetrotic bone contains few if any viable osteocytes and usually contains a much larger number of microscopic cracks than adjacent living bone. This suggests that the osteocytes play a role in detecting and repairing the damage. In senile osteoporosis, bone tissue is removed by the body, often to such an extent that fractures occur during normal activities. Osteoporosis may be referred to as a remodeling error. Age, hormone changes, and physical inactivity contribute to osteoporosis. Osteoporosis is important in the context of biomaterials since bones made weak by this condition can fracture, resulting in treatments with bone fracture plates or joint replacement implants.

9.3.2.b. Feedback Mechanisms of Bone Remodeling

Bone remodeling appears to be governed by a feedback system in which the bone cells sense the state of strain in the bone matrix around them and either add or remove bone as needed to maintain the strain within normal limits (see Figure 13-1). The process or processes by which the cells are able to sense the strain and the important aspects of the strain field are presently unknown. It is suggested that bone is piezoelectric, i.e., that it generates electric fields in response to mechanical stress, and it is thought that the piezoelectric effect is the part of the feedback loop by which the cells sense the strain field. This hypothesis obtained support from observations of osteogenesis in response to externally applied electric fields of the same order of magnitude as those generated naturally by stress via the piezoelectric effect. The study of bone bioelectricity has received impetus from observations that externally applied electric or electromagnetic fields stimulate bone growth. The electrical hypothesis, while favored by many, has not been proven. Indeed, other investigators have advanced competing hypotheses that involve other mechanisms by which the cells are informed of the state of stress around them. Such processes may include slip between lamellae impinging on osteocyte processes; slow slip at cement lines between osteons that would align the osteons to the stress; stress-induced fluid flow in channels such as canaliculi to stimulate or nourish the osteocytes; a direct hydrostatic pressure effect on osteocytes; or an effect of stress on the crystallization kinetics of the mineral phase.

Example 9-4

Calculate using a simple rule of mixture model the percentage of load borne by mineral phase of a cow femur that is subjected to 500 N of load. Assume that the Young's modulus of mineral and collagen is about 17 and 0.1 GPa, respectively.

Answer

Since the area is proportional to the volume percentage of each component, from Example 9-2 and Eq. (9-6):

$$\frac{P_m}{P_i} = \frac{0.44 \times 17}{0.44 \times 17 + 0.40 \times 0.1} = \underline{0.9947},$$

so that 99.47% of the load is borne by the mineral phase. Actually the strength of demineralized bone is about 5–10% of that of whole bone. The rule of mixtures (Voigt model) represents an upper bound on the modulus of a composite. It is appropriate for a composite with fibers or laminae oriented in the direction of the applied load. The structure of bone is considerably more complex than these models, so that the analysis is a rough approximation.

9.3.3. Collagen-Rich Tissues

Collagen-rich tissues function mostly in a load-bearing capacity. These tissues include skin, tendon, cartilage, etc. Special functions such as transparency for the lens of the eye and shaping of the ear, tip of the nose, etc. can be carried out also by the collagenous tissues.

Table 9-8. Composition of Collagen-Rich Soft Tissues

Component	Composition (%)
Collagen	75 (dry), 30 (wet)
Mucopolysaccharides	20 (dry)
Elastin	< 5 (dry)
Water	60–70

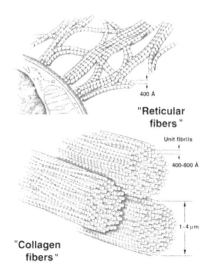

Figure 9-16. Diagram showing that the "reticular fibers" associated with the basal lamina of an epithelial cell (**above**) and the "collagen fibers" of the connective tissue in general (**below**) are both composed of unit fibrils of collagen. Those of the reticulum are somewhat smaller and interwoven in loose networks instead of in larger bundles. Reprinted with permission from Bloom and Fawcett (1968). Copyright © 1968, W.B. Saunders.

9.3.3.a. Composition and Structure

The collagen-rich tissues are mostly made up of collagen (over 75 w/o dry, see Table 9-8). The collagen is made of tropocollagen, which is made of a three-chain coiled superhelix (Figure 9-2). The collagen fibrils aggregate to form fibers, as shown in Figure 9-16. The fibrils and fibers are stabilized through intra- and intermolecular hydrogen bonding (C=O--HN) (see Figure 9-1).

The physical properties of tissues vary according to the amount and structural variations of collagen fibers. Several different fiber arrangements are found in tissues: parallel fibers, crossed fibrillar arrays, and felt-like structures in which the fibers are more randomly arranged with some fibers going through the thickness of the tissue.

9.3.3.b. Physical Properties

The collagen-rich tissues can be thought of as polymeric materials in which the highly oriented crystalline collagen fibers are embedded in the ground substance of mucopolysaccharides and amorphous elastin (a rubber-like biopolymer). When the tissue is heated in the laboratory, its specific volume increases (density decreases, exhibiting glass transition temperature, $T_g \sim 40°C$ and shrinkage, $T_s \sim 56°C$). The shrinkage temperature is considered a denaturation point for collagen.

The stress–strain curves of a collagenous structure such as tendon exhibit nonlinear behavior, as shown in Figure 9-17. Similar behavior is observed in some synthetic fibers and bundles of fibers. The initial toe region represents alignment of fibers in the direction of stress, and the steep rise in slope represents the majority of fibers stretched along their long axes. As for the later decrease in slope, individual fibers may be breaking prior to final catastrophic failure. The highest slope of the stress–strain curve is about 1.0 GPa, which is close to the modulus of individual collagen fibers. The tensile strength is, however, much lower than that of the individual fibers. Table 9-9 lists some mechanical properties of collagen and elastic fibers. The modulus of elasticity of collagen increases with the rate of loading due to the viscoelastic properties as in the case of bone discussed earlier.

Table 9-9. Elastic Properties of Elastic and Collagen Fibers

Fibers	Modulus of elasticity (MPa)	Tensile strength (MPa)	Ultimate elongation (%)
Elastic	0.6	1	100
Collagen[a]	1000	50–100	10

[a] Harkness (1968).

Unlike tendon or ligament, skin has a felt-like structure consisting of continuous fibers that are randomly arranged in layers or lamellae. The skin tissues also show mechanical anisotropy, as shown in Figure 9-18. This is the reason behind the *Langer's lines* shown in Figure 9-19. The Langer's line is determined by puncturing small circular holes on the skin of a cadaver, and following the elongated holes (which become line-like) due to the anisotropy of the skin structure, as shown in Figure 9-19. Surgeons use knowledge of this anisotropy in making incisions so that better wound healing is achieved.

Another feature of the stress–strain curve of the skin is its extensibility under small load. This compliance is observed despite the high content of collagen, a relatively stiff material, in skin. The reason is that at small extension, the fibers are straightened and aligned rather than stretched. Upon further stretching the fibrous lamellae align with respect to each other and resist further extension, as shown in Figure 9-20. When the skin is highly stretched, the modulus of elasticity approaches that of tendon.

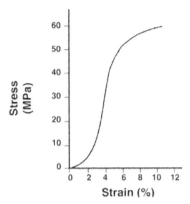

Figure 9-17. A typical stress–strain curve for tendon. Reprinted with permission from Rigby et al. (1959). Copyright © 1959, Rockefeller University Press.

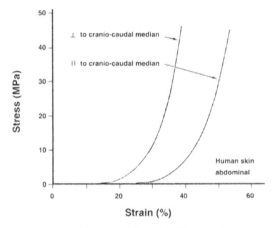

Figure 9-18. Stress–strain curves of human abdominal skin that shows more extensibility in the direction of the main axis of the body than across the abdomen. Reprinted courtesy of C.H. Daly.

Cartilage is another collagen-rich tissue that has two main physiological functions. One is the maintenance of shape (ear, tip of nose, and rings around the trachea), and the other is to provide bearing surfaces at joints. It contains very large and diffuse protein–polysaccharide molecules that form a gel in which the collagen-rich molecules are entangled (cf. Figure 9-6). They can affect the mechanical properties of the cartilage by hindering the movements through the interstices of the collagenous matrix network.

The joint cartilage has a very low coefficient of friction (< 0.01). This is largely attributed to the squeeze-film effect between cartilage and synovial fluid. The synovial fluid can be squeezed out through highly fenestrated cartilage upon compressive loading while the reverse action will take place in tension. The lubricating function is carried out in conjunction with mucopolysaccharides, especially chondroitin sulfates, as discussed previously in §4.1.2b. The modulus of elasticity (10.3–20.7 MPa) and tensile strength (3.4 MPa) are quite low. However, wherever high stress is to be supported, the cartilage is replaced by purely collagenous tissue.

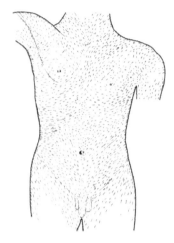

Figure 9-19. Distribution of Langer's line produced by a conical stabbing instrument. Reprinted with permission from Wilkes et al. (1973). Copyright © 1973, Chemical Rubber Co.

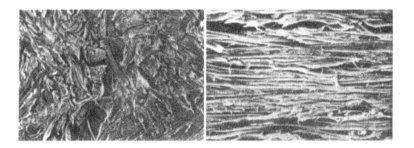

Figure 9-20. Scanning electron microphotographs of dermal skin before and after stretching. Stretching direction is horizontal (×400). Reprinted with permission from Wilkes et al. (1973). Copyright © 1973, Chemical Rubber Co.

Example 9-5

Estimate the wt% compositions of mucopolysaccharides (MPS) and elastin in Table 9-8 assuming the densities of collagen, MPS and elastin are about 1 g/cm^3 and water content is about 65%.

Answer

Based on 100 g of wet tissue.

Composition	Weight (g)	Dry (g)
Collagen	30	35 g × 0.75 = 26.25
MPS	x	35 g × 0.20 = 7.00
Elastin	y	35 g × 0.05 = 1.75
Water	65	0

$$x + y = 5 \text{ g}.$$

solving simultaneously yields:

$$x/y = 7/1.75, \quad x = 4\text{ g}, \quad \text{and} \quad y = 1\text{ g},$$

that is,

$$\text{MPS: } \underline{4\%}, \quad \text{Elastin: } \underline{1\%}.$$

This indicates that a very small amount of elastin exists in the collagen-rich tissues.

9.3.4. Elastic Tissues

The elastic tissues are compliant and therefore undergo a large deformation in response to a small load. These tissues include blood vessels, ligamentum nuchae, muscles, etc.

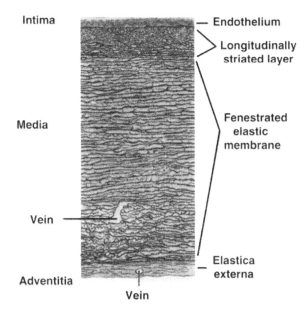

Figure 9-21. Structure of a blood vessel wall. Reprinted with permission from Bloom and Fawcett (1968). Copyright © 1968, W.B. Saunders.

9.3.4.a. Composition and Structure

Elastic tissues contain a relatively large amount of elastin, which is sometimes called "protein rubber." For example, ligamentum nuchae contains 80 w/o (dry) elastin. One of the most important elastic tissues is the blood vessel wall, which has three distinct layers when viewed in cross-section (Figure 9-21): (1) *intima*, whose structural elements are oriented longitudinally, (2) *media*, which is the thickest layer of the wall and whose components are arranged circumferentially, and (3) *adventitia*, which connects the vessels firmly to surrounding tissue via fascia. The intima and media are fenestrated by the internal elastic membrane (*elastica interna*), which is predominant in arteries of medium size. Between the media and adventitia, a thinner external elastic membrane (*elastica externa*) can be found. The smooth muscle cells are found between adjacent elastic lamellae in helical array.

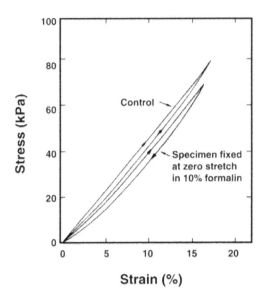

Figure 9-22. Stress–strain curve of elastin. The material is the ligamentum nuchae of cattle, which contains a small amount of collagen that was denatured by heating at 100°C for an hour. Such heating does not change the mechanical properties of elastin. The specimen is cylindrical with rectangular cross-section. Loading is uniaxial. The curve labeled "control" refers to native elastin. The curve labeled "10% formalin" refers to specimen fixed in formalin solution for a week without initial strain. Reprinted with permission from Fung (1981). Copyright © 1981, Springer-Verlag.

9.3.4.b. Properties of Elastic Tissues

The mechanical properties of elastic tissues such as elastin are similar to those of rubber. Chains in elastin are crosslinked by desmosine, isodesmosine, and lysinonorleucine (as mentioned before). Figure 9-22 shows a stress–strain curve of bovine ligamentum nuchae at low extension. As can be seen, the modulus of elasticity and amount of stored energy lost upon releasing the load are quite low. This is characteristic of elastic tissues whose primary function is the restoration of a deformed state to the original shape with minimum energy loss. Therefore, the relative amount of elastin along the blood vessel walls vary, the largest of which is the arch of the aorta, which functions as a "secondary pump," where the expelled blood from the heart is temporarily stored. The relative amount of elastin decreases with decreasing size of the vessel, as shown in Figure 9-23.

As in the case of the skin, the anisotropy of mechanical properties is prominent in the longitudinal and circumferential directions of the blood vessel, as shown in Figure 9-24. Note that the composition of the vessel walls changes along the length of the wall and, consequently, their physical properties also change (cf. Figure 9-23). Another complicating factor describing mechanical properties is the existence of involuntary smooth muscle, which is associated with arterial blood pressure regulation.

Table 9-10 illustrates that the mean pressure of the various blood vessels and the approximate tension developed at normal pressure are related as given in the Laplace equation:

$$T = P_i \times r, \tag{9-10}$$

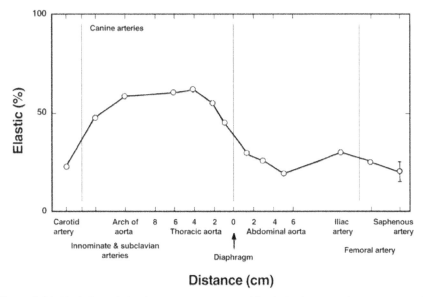

Figure 9-23. Variation of elastin percentage per combined elastin and collagen content along the major arterial tree. Reprinted with permission from Burton (1965). Copyright © 1965, Year Book Medical Publishers.

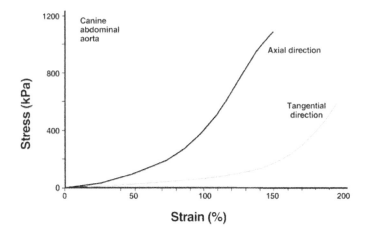

Figure 9-24. Stress–strain curves of human artery in the longitudinal and circumferential directions of the vessel. Unpublished data, J.B. Park, University of Iowa.

where T is wall tension, P_i is internal pressure, and r is the radius of the vessel. This equation is usually applied to a uniform, thin, isotropic tube in the absence of longitudinal tension. It is obvious that none of these assumptions can be met strictly by the blood vessels.

Muscle is another elastic tissue. This tissue is not discussed in detail in this book due to the "active" nature of the tissue, the importance of which is more obvious in conjunction with physiological function. As passive tissue the stress–strain curve shows nonlinear, viscoelastic behavior, as shown in Figure 9-25. Table 9-11 gives mechanical properties of some of the non-mineralized tissues for comparison.

Table 9-10. Wall Tension and Pressure Relationship of Various Sizes of Blood Vessels

Type of vessels	Mean pressure (mm Hg)	Internal pressure (dyne/cm^2)	Radius	Wall tension (dyne/cm)
Aorta, large artery	100	1.5×10^5	1.3 cm	170,000
Small artery	90	1.2×10^5	0.5 cm	60,000
Arteriole	60	0.8×10^5	62–150 µm	500–1,200
Capillaries	30	0.4×10^5	4 µm	16
Venules	20	0.26×10^5	10 µm	26
Veins	15	0.2×10^5	200 µm	400
Vena cava	10	0.13×10^5	1.6 cm	21,000

Reprinted witih permission from Burton (1965). Copyright © 1965, Year Book Medical Publishers.

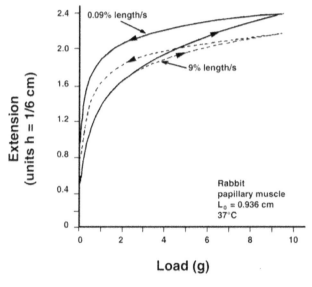

Figure 9-25. The length–tension curve of a resting papillary muscle from the right ventricle of a rabbit. Hysteresis curves at strain rates 0.09% length/s, 0.9% length/s, and 9% length/s. Length at 9 mg = 0.936 cm at 37°C. Reprinted with permission from Fung (1981). Reprinted with permission from Fung (1981). Copyright © 1981, Springer-Verlag.

Another important consideration for the physical properties of tissues is that any disturbance to the specimen may change its constituents, especially unbound water. For example, when a soft tissue specimen such as a segment of aorta is stretched in the laboratory, the unbound "free water" is squeezed out of the specimen since the laboratory preparation is an open system. It is not yet fully understood what is the role of the water in contributing to the mechanical properties of tissues.

Table 9-11. Mechanical Properties of Some of
the Non-Mineralized Human Tissues

Tissues	Tensile strength (MPa)	Ultimate elongation (%)
Skin	7.6	78
Tendon	53	9.4
Elastic cartilage	3	30
Heart valves (aortic)		
Radial	0.45	15.3
Circumferential	2.6	10.0
Aorta		
Transverse	1.1	77
Longitudinal	0.07	81
Cardiac muscle	0.11	63.8

Example 9-6

From Figure 9-18, answer.

    a.   How much stress will be developed if the abdominal skin was stretched 30% in the parallel and in the perpendicular direction to the cephalo-caudal direction of the body?

    b.   What are the strains if the skin was stressed to 1 MPa?

    c.   What are the moduli of elasticity in the two principal directions?

Answer

    a.   From Figure 9-18, the stresses perpendicular and parallel to the cephalo-caudal direction of the main body are about 1 and 0.01 MPa, respectively.

    b.   30 and 43% strain will be developed perpendicular and parallel to the cephalo-caudal direction of the body respectively.

    c.   Perpendicular direction: $E = (5.2 - 0)/(0.41 - 0.31) = \underline{52\ MPa}$,
       Parallel direction: $E = (4.6 - 0)/(0.54 - 0.44) = \underline{46\ MPa}$.

9.3.4.c. Further Considerations of the Mechanical Properties of Soft-Tissues

As with other viscoelastic materials such as polymers, soft tissues also exhibit similar behavior, which can be represented with traditional multielement Voigt and Maxwell models in series or in parallel fashion. Although this type of analysis gives a very useful handle on the mechanical behavior of tissues, it does not explicitly explain the relative importance of constitutive components in tissues. One might, therefore, try to understand the overall behavior of tissues by modeling the structure, as the example given in Figure 9-26. This model shows that the blood vessel is made up of three major components: smooth muscle fibers arranged in helical fashion with very short pitch, elastic fibers of elastin, and collagen forming a crimped network structure, which can be extended at high load.

Some researchers have explored the soft tissue behavior based on enzymolysis experiments in which a particular component is removed by an appropriate enzyme. Typical stress–strain curves obtained after removal of each major component are given in Figure 9-27. One can see that the ligamentum nuchae shows rubber-like elasticity up to 50% elongation. However, if one removes the collagen component from the tissue by an enzyme (collagenase) or autoclaving, it behaves entirely like an elastomer up to 100% elongation. Conversely, if one removes the elastin component from the same tissue, the remaining tissue behaves as collagenous tissue, except the curve is shifted toward high extension that used to be taken up by the elastin. The removal of ground substance did not alter the basic stress–strain behavior. From these experiments one can deduce a simple model, as shown in Figure 9-28.

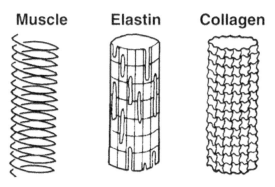

Muscle   Elastin   Collagen

Figure 9-26. Sketch of the structure of a vein. Reprinted with permission from Azuma and Hasegawa (1973). Copyright © 1973, IOS Press.

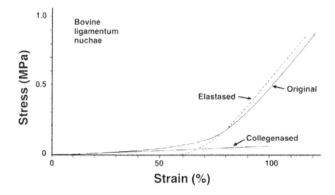

Figure 9-27. Stress–strain curves of bovine ligamentum nuchae after elastase and collagenase treatments. Reprinted with permission from Hoffman et al. (1972). Copyright © 1972, University Park Press.

In this model the elastic fiber (elastin) are stretched followed by the much stronger collagen fibers when two components are pulled together. With a little more imagination, one can envision the distribution of fiber length and thus the response can be smoother than the two fiber system as shown in Figure 9-28. From this kind of study, relating the microstructure to the macrobehavior, one can understand fully the nature of interaction between components in a connective tissue.

Example 9-7

Calculate the modulus of elasticity of the bovine ligamentum nuchae (Figure 9-27).

Answer

There are two distinct regions demarcated at about 60-70% strain.

Initial region:

$$E = \frac{\sigma}{\varepsilon} = \frac{0.06 \text{ MPa}}{100\%} = \underline{0.06 \text{ MPa}}.$$

Secondary region;

$$E = \frac{0.85 - 0 \text{ MPa}}{120\% - 72\%} = \underline{1.77 \text{ MPa}}.$$

These two values are lower and higher than the single value given in Table 9-8 for the elastic fibers.

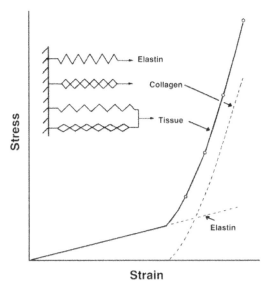

Figure 9-28. A schematic representation of the structure–property relationship of connective tissues. The collagen is represented by a loosely knitted fabric.

PROBLEMS

9-1.   Calculate the degree of polymerization of a chondroitin sulfate compound that has an average molecular weight of 100,000 g/mol.

9-2.   Calculate the volume percentage of each major component of a wet bone based on the weight percentage, i.e., 9, 69, and 22 wt% for the water, mineral, and organic phase, respectively. Assume the densities of mineral and organic phase are 2.5 and 1 g/cc, respectively.

9-3.   Calculate the density of the mineral phase of dentin if the density of the organic phase and water is 1g/cm³.

9-4.   Calculate using a simple rule of mixture model the load borne by the mineral phase of a cow femur that is subjected to 500 N of load. The Young's moduli of mineral and collagen are about 17 and 0.1GPa, respectively.

9-5.   Estimate the wt% compositions of mucopolysaccharides (MPS) and elastin in Table 9-8 assuming the densities of collagen, MPS, and elastin are about 1 g/cm³ and water content is about 67%.

9-6.   From Figure 9-18, answer.

   a. How much stress will be developed if the abdominal skin is stretched 40% in the parallel and perpendicular directions to the cephalocaudal direction of the body?

   b. What are the strains if the skin is stressed to 2 MPa?

   c. What are the moduli of elasticity in the two principal directions?

9-7.   Calculate x of Eq (9-9) for bone with a Young's modulus of 17 GPa by assuming $V_m = V_c$, $E_m = 100$ GPa, $E_c = 0.1$ GPa.

9-8.   Ligamentum nuchae is made of elastin and collagen (others do not contribute to the mechanical properties). The relative amounts excluding water are 70, 25, and 5 wt% for elastin, collagen, and others, respectively. Assume they are homogeneously distributed.

   a. Calculate the percent contribution of the elastin toward the total strength assuming elastic behavior.

   b. Compare the result with Figure 9-27.

9-9.   If a bone contains 18.9% porosity, what would be its modulus of elasticity in the absence of pores. Compare with the modulus of hydroxyapatite. Hint: $E = E_0(1 - 1.9V + V^2)$, where E is the modulus without porosity and V is the fractional of pores.

9-10.   Read de Mol BAJM, van Gaalen GL. 1996. The editor's corner: biomaterials crisis in the medical device industry: is litigation the only cause? *J Biomed Mater Res*, **33**:53–54. What are other causes?

9-11.   The force-versus-displacement curve in the following figure was obtained by tensile testing of canine skin. The skin specimen was cut by using a stamping machine that has a width of 4 mm. The thickness of the skin is about 3 mm, and the length of the sample between the grips is 20 mm.

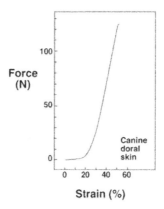

a. What is the tensile strength and fracture strain of the skin?

b. What is the modulus of elasticity in the initial and secondary regions?

c. What is the toughness of the skin?

9-12. The viscoelastic properties of compact bone have been described by using the three-element models as shown in Figure 9-12.

a. Derive a differential equation for the model.

b. Solve (a) for γ (strain) for stress relaxation tests.

c. Sketch the relaxation on a log scale similar to the one used in Figure 9-13.

d. What are the shortcomings of the model?

9-13. A piece of skin is tested for its stress relaxation properties after cutting 1 cm wide,15 cm long from the belly of a hog belly. Assume a uniform thickness of 5 mm and the properties do not change while being tested. The initial force recorded after stretching 0.1 cm between grips was 5 Newtons. The gage length was 10 cm.

a. Calculate the initial stress.

b. Calculate modulus of elasticity of the skin if the initial stretching can be considered as linear and elastic.

c. Calculate the relaxation time if the force recorded after 10 hours was 2 Newtons. Assume a single relaxation time model with exponential relaxation in time.

d. Would the relaxation of the stress be faster, slower, or the same if the initial force was 10 Newtons?

9-14. From the following tissues select the most appropriate one for the questions.

A. Blood   B. Skin   C. Bone   D. Tendon   E. Aorta

a. Which tissue contains the hydroxyapatite?

b. Which tissue contains the most elastin?

c. Which tissue has the highly oriented collagenous fibers aligned in the direction of the applied tensile force?

d. Which tissue shows the structural anisotropy that results in a Langer's line?

e. Which tissue has the highest density?

f. Which tissue has the endothelial cells?

g. Which tissue has the highest Poisson's ratio?

h. Which tissue has the highest Young's modulus?

i. Which tissue can be applied to the Laplace equation?

j. Which tissue contains the cement lines?

9-15. The compact bone is made of the following components. From the data, answer.

Component	Composition (wt%)	Density (g/cc)	Modulus (GPa)
Organic	20	1	10
Mineral	70	3.2	100
Water	10	1	0

a. What is the dry weight percentage of the organic phase?

b. What is the density of the bone? Use a simple mixture rule. $\rho = \rho_1 V_1 + \rho_2 V_2 + \dots$

c. Calculate the modulus of the bone assuming the mineral and organic phase are arranged in parallel to the loading direction: $E_c = E_1 V_1 + E_2 V_2 + \dots$

d. What is the percent (%) strain on the organic phase if the bone is strained 0.1% assuming the two phases have the same arrangement as in question ©.

e. Give at least three bone remodeling theories. Be brief.

9-16.   From the following list select the most appropriate one:

   A. Bone   B. Skin   C. Tendon   D. Artery   E. Ligament
   G. Cartilage   H. Platelet   F. Ligamentum nuchae   G. Blood

a. Low coefficient of friction

b. Elastin

c. Laplace equation

d. Langer's line

e. Haversian system

f. Media

g. Tensile force transmission

h. Thrombus

i. $Ca_{10}(PO_4)_6(OH)_2$

j. Wolff's law

9.17.   A cow femoral compact bone is cut into $1 \times 1 \times 1$ cm cubes and tested. Assume the collagen fibers are all aligned in the z-axis, which is the longitudinal direction of the cow femur.

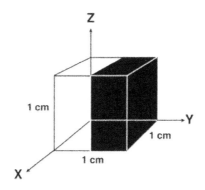

Material	Vol %	E (GPa)	σ_c (MPa)	E_{long} (%)	Density (g/cm^3)
Collagen	50	0.1	100	50	1.0
Mineral	50	300	300	1	2.2

A load of 100 Newton is applied to the bone specimen.

a. In what direction would the bone have the highest modulus?

   b. In what direction would the bone be strongest?

   c. In what direction would the bone be stretched most easily?

   d. Calculate the density of the bone.

   e. Calculate the Young's modulus of the bone in z-direction.

   f. Calculate the strain (%) in the z-direction due to the applied load of 100 N.

9-18.   The material properties of arterial prostheses change following ingrowth of tissues in vivo. A porous silicone rubber arterial prosthesis was implanted in dogs, and it was found that half of the pores became filled with tissue after 3 months of implantation. The prosthesis has a 6-mm inside diameter, a wall thickness of 1 mm, length of 10 cm, and a porosity of 30%. The solid silicone from which the prosthesis is derived has a Young's modulus and tensile strength of 10 MPa.

   a. What is the most important factor in the success of the artificial vascular graft?

   b. The silicone rubber graft leaked blood. What would you do to prevent the leakage of blood when implanted?

   c. Determine the wall tension assuming a (high) blood pressure of 200 mm Hg immediately after implantation: $T = P \times r$; 133.3 Pa = 1 mm Hg.

   d. Calculate the maximum force (N) on the arterial wall developed by the blood pressure (200 mm Hg).

   e. Would the silicone rubber be adequate for the artificial artery if a safety factor of 10 is required for it?

   f. What is the name of the anticoagulant used to prevent clotting of the prosthesis?

SYMBOLS / DEFINITIONS

Greek letters

ε: Strain

ρ: Density

σ: Stress

Latin letters

A: Area

C: Coulomb, a measure of electric charge. 1 Coulomb = 1 ampere $\times$ 1 second.

E: Young's modulus, a measure of stiffness.

P: Load (force) or pressure (force per area).

r: Radius.

T: Wall tension.

V: Volume fraction.

Words

Aspartic acid (Asp): One of the essential amino acids.

Canaliculi: Small channels (~0.3 μm in diameter) radiating from the lacunae in bone tissue.

Cementum: Calcified tissue of mesodermal origin covering the root of a tooth.

Chondroitin: One of the polysaccharides commonly found in the cornea of the eye.

Chondroitin sulfate: Sulfated mucopolysaccharides commonly found in cartilages, bones, cornea, tendon, lung, and skin.

Crown: A crown-shaped structure, especially the exposed or enamel covered portion of a tooth. It is largely (97%) made of hydroxyapatite mineral. An artificial replacement of the exposed surface or the upper part of a tooth is also called a crown.

Dentin: The chief substance of the tooth, forming the body, neck, and roots, being covered by enamel on the exposed part of the tooth and by cementum on the root. It is similar in composition and properties (but not structure) to compact bone.

Desmosine: One of the crosslinking chemicals in elastin.

Elastin: One of the proteins in connective tissue. It is highly stable at high temperatures and in chemicals. It also has rubber-like properties; hence it is nicknamed "tissue rubber."

Glutamic acid (Glu): One of the essential amino acids, much more commonly occurring in collagen than in elastin.

Glycine (Gly): One of the amino acids having the simplest structure.

Haversian system: Same as "osteon."

Histidine (His): One of the amino acids.

Hyaluronic acid: One of the polysaccharides commonly found in synovial fluid, aortic walls, etc.

Hydroxyapatite: Mineral component of bone and teeth. It is a type of calcium phosphate, with composition $Ca_{10}(PO_4)_6(OH)_2$.

Hydroxyproline (Hypro): One of the amino acids commonly occurring in collagen molecules.

Isodesmosine: Isomer of desmosine.

Lacuna: A pore (~10 × 15 × 25 μm) in Haversian bone; lacunae often contain osteocytes (bone cells).

Lysine(Lys): One of the amino acids from which hydrogen bonding takes place stabilizing collagen chains.

Lysinonorleucine: One of the crosslinking chemicals in elastin.

Mixture rule: Properties of a material made of many materials depend linearly on the amount of each material contributed.

Osteons: Large fiber-like structure (150~250 μm in diameter) in compact bone. Concentric layers or lamellae surround a central channel or Haversian canal that contains a small blood vessel. Each lamella contains smaller fibers. Osteons, also called Haversian systems, are separated by cement lines.

Piezoelectricity: Electric polarization resulting from mechanical stress upon a material; conversely, deformation resulting from an imposed electric field.

Polysaccharides: Polymerized sugar molecules found in tissues as lubricant (synovial fluid), cement (between osteons, tooth root attachment), or complexed with proteins, such as glycoproteins or mucopolysaccharides.

Pulp: Richly vascularized and innervated connective tissue inside a tooth.

Proline (Pro): One of the amino acids commonly occurring in collagen molecules.

Streaming potential: Electric potential generated in the (solid) wall of a channel when charged particles are flowing and polarization takes place in the wall.

Tropocollagen: Precursor of collagen, right-handed superhelical coil structure, which, in turn, is made of three left-handed helical peptide chains.

Valine(Val): One of the essential amino acids more commonly occurring in elastin than in collagen.

Volksmann's canal: Vascular channels in compact bone. They are not surrounded by concentric lamellae as are the Haversian canals.

Wolff's law: Remodeling of bone takes place in response to mechanical stimulation, so that the new structure becomes better adapted to the load.

BIBLIOGRAPHY

Azuma T, Hasegawa M. 1973. Distensibility of the vein: from the architectural viewpoint. *Biorheology* **10**:469–479.

Barbucci R. 2002. *Integrated biomaterials science*. New York: Kluwer Academic/Plenum.

Barker R. 1971. *Organic chemistry of biological compounds*. Englewood Cliffs, NJ: Prentice-Hall.

Black J. 1992. *Biological performance of materials: fundamentals of biocompatibility*. New York: Marcel Dekker.

Bloom W, Fawcett DW. 1968. *A textbook of histology*, 9th ed. Philadelphia: Saunders.

Bur AJ. 1976. Measurements of the dynamic piezoelectric properties of bone as a function of temperature and humidity. *J Biomech* **9**:495–507.

Burton AC. 1965. *Physiology and biophysics of circulation*. Chicago: Year Book Medical Publishers.

Carter DR, Hayes WC. 1976. Bone Compressive strength: the influence of density and strain rate. *Science* **194**:1174–1176.

Chakkalakal DA, Johnson MW, Harper RA, Katz JL. 1980. Dielectric properties of fluid-saturated bone. *IEEE Trans Biomed Eng* **BME-27**:95–100.

Chen QZ, Wong CT, Lu WW, Cheung KM, Leong JC, Luk KD. 2004. Strengthening mechanisms of bone bonding to crystalline hydroxyapatite in vivo. *Biomaterials* **25**(18):4243–4254.

Chvapil, M. 1967. *Physiology of connective tissue*. London: Butterworths.

Cima LG, Ron ES. 1991. *Tissue-inducing biomaterials*. Materials Research Society Symposium, Boston.

Cowin SC. 1981. *Mechanical properties of bone*. New York: American Society of Mechanical Engineers.

Cowin SC, Humphrey JD, eds. 2002. *Cardiovascular soft tissue mechanics*. New York: Springer.

Currey JD. 1981. What is bone for? Property–function relationships in bone. In *Mechanical properties of bone*, pp. 13–26, Ed SC Cowin. New York, ASME.

Currey JD. 1984. *Mechanical adaptations of bone*. Princeton: Princeton UP.

Daly CH. 1966. *The biomechanical characteristics of human skin*. PhD dissertation. University of Strathclyde, Scotland.

Delloye C, Cnockaert N, Cornu O. 2003. Bone substitutes in 2003: an overview. *Acta Orthop Belg* **69**(1):1–8.

Elden HR. 1971. *Biophysical properties of the skin*. New York: Wiley.

Evans FG, Lebow M. 1952. The strength of human compact bone as revealed by engineering technics. *Am J Surg* **83**:326–331.

Fung YC. 1981. *Biomechanics: mechanical properties of living tissues*. Berlin: Springer-Verlag.

Fung YC, Perrone N, Anliker M. 1972. *Biomechanics: its foundation and objectives*. Englewood Cliffs, NJ: Prentice-Hall.

Garetto LP, Turner CH, Duncan RL, Burr DB, eds. 2002. *Bridging the gap between dental and orthopaedic implants*. Indianapolis: Indiana University School of Dentistry.

Garner E, Lakes R, Lee T, Swan C, Brand R. 2000. Viscoelastic dissipation in compact bone: implications for stress-induced fluid flow in bone. *J Biomech Eng* **122**(2):166–172.

Glowacki J. 2001. Engineered cartilage, bone, joints, and menisci: potential for temporomandibular joint reconstruction. *Cells Tissues Organs* **169**(3):302–308.

Green D, Walsh D, Mann S, Oreffo RO. 2002. The potential of biomimesis in bone tissue engineering: lessons from the design and synthesis of invertebrate skeletons. *Bone* **30**(6):810–815.

Gross J. 1961. Collagen. *Sci Am* **204**:121.

Harkness RD. 1968. Mechanical properties of collagenous tissues. In *Treatise on collagen*, chapter 6. Ed BS Gould. New York: Academic Press.

Hoffman AS, Grande LA, Gibson P, Park JB, Daly CH, Ross R. 1972. Preliminary studies on mechano-chemical-structure relationships in connective tissues using enzymolysis techniques. In *Perspectives of biomedical engineering*, pp. 173–176. Ed RM Kenedi. Baltimore: University Park Press.

Hogan HA. 1992. Micromechanics modeling of Haversian cortical bone properties. *J Biomech* **25**(5):549–556.

Hohling HJ, Ashton BA, Koster HD. 1974. Quantitative electron microscope investigation of mineral nucleation in collagen. *Cell Tissue Res* **148**:11–26.

Holzapfel GA, Ogden RW, eds. 2006. *Mechanics of biological tissue.* New York: Springer.

Lakes RS. 1980. Dynamic study of couple stress effects in human compact bone. *J Mech Eng* **104**:6–11.

Lakes RS. 1988. Properties of bone and teeth. In *Encyclopedia of medical devices and instrumentation*, pp. 501–512. Ed JG Webster. New York: Wiley.

Lakes RS, Katz JL. 2006. Properties of bone and teeth. In *Encyclopedia of medical devices and instrumentation*, pp. 523–540. Ed JG Webster. New York: Wiley.

Lakes RS, Harper RA, Katz JL. 1977. Dielectric relaxation in cortical bone. *J Appl Phys* **48**:808–811.

Laurent TC. 1998. *The chemistry, biology, and medical applications of hyaluronan and its derivatives.* London: Portland Press.

Li P, Calvert P, Kokubo T, Levy R, Scheid C, ed. 2000. *Mineralization in natural and synthetic biomaterials.* Warrendale, PA: Materials Research Society.

Li S-T. 1995. Biologic materials: tissue-derived biomaterials (collagen). In *The biomedical engineering handbook*, pp. 627–647. Ed JD Bronzino. Boca Raton, FL: CRC Press.

Lieberman JR, Friedlander GE. 2005. *Bone regeneration and repair: biology and clinical applications.* Totowa, NJ: Humana Press.

Lu HH, El-Amin SF, Scott KD, Laurencin CT. 2003. Three-dimensional, bioactive, biodegradable, polymer-bioactive glass composite scaffolds with improved mechanical properties support collagen synthesis and mineralization of human osteoblast-like cells in vitro. *J Biomed Mater Res A* **64**(3):465–474.

McElhaney JH. 1966. Dynamic response of bone and muscle tissue. *J Appl Physiol* **21**:1231–1236.

Meyer U, Wiesmann HP, Meyer TE. 2006. *Bone and cartilage engineering.* New York: Springer.

Ohta Y. 1993. Comparative changes in microvasculature and bone during healing of implant and extraction sites. *J Oral Implantol* **19**(3):184–198.

Pierkarski K. 1978. Structure, properties and rheology of bone. In *Orthopaedic mechanics: procedures and devices*, p. 16. Ed D Ghista, R Roaf. New York: Academic Press.

Rammelt S, Schulze E, Witt M, Petsch E, Biewener A, Pompe W, Zwipp H. 2004. Collagen type I increases bone remodeling around hydroxyapatite implants in the rat tibia. *Cells Tissues Organs* **178**(3):146–157.

Rich A. 1999. *A new class of self-complementary beta sheet oligopeptide-based biomaterials.* Cambridge: MIT Press.

Rigby BJ, Hiraci N, Spikes JD, Eyring H. 1959. The mechanical properties of rat tail tendon. *J Gen Physiol* **43**:265–283.

Schwartz Z, Kieswetter K, Dean DD, Boyan BD. 1997. Underlying mechanisms at the bone-surface interface during regeneration. *J Periodontal Res* **32**(1 Pt 2):166–171.

Taguchi A, Sanada M, Krall E, Nakamoto T, Ohtsuka M, Suei Y, Tanimoto K, Kodama I, Tsuda M, Ohama K. 2003. Relationship between dental panoramic radiographic findings and biochemical markers of bone turnover. *J Bone Miner Res* **18**(9):1689–1694.

Tiffit JT. 1980. The organic matrix of bone tissue. In *Fundamental and Clinical Bone Physiology*, chapter 3. Ed MR Urist. Philadelphia: J.B. Lippincott.

Urist MR. 1980. *Fundamental and clinical bone physiology.* Philadelphia: J.B. Lippincott.

van Zuijlen PP, Vloemans JF, van Trier AJ, Suijker MH, van Unen E, Groenevelt F, Kreis RW, Middelkoop E. 2001. Dermal substitution in acute burns and reconstructive surgery: a subjective and objective long-term follow-up. *Plast Reconstr Surg* **108**(7):1938–1946.

Viidik A. 1973. Functional properties of collagenous tissues. *Int Rev Connect Tissue Res* **6**:127–215.

Wainwright SA, Biggs WD, Currey JD. 1976. *Mechanical design in organisms*. London: Edward Arnold.

Walenkamp GHIM, Bakker FC. 1998. *Biomaterials in surgery*. Stuttgart: New York.

Wilkes GL, Brown IA, Wildnauer RH. 1973. The biochemical properties of skin. *CRC Crit Rev Bioeng* **1**(4):453–495.

Wolff J. (1892). *Das gesetz der transformation der krochen*. Berlin: Hirchwild.

Woo SLY. 1981. The relationships of changes in stress levels on long bone remodeling. In *Mechanical properties of bone*, pp. 107–129. Ed SC Cowin. New York: American Society of Mechanical Engineers.

Woo SLY, Akeson WH, Coutts RD, Rutherford L, Doty D, Jemmott GF, Amiel D. 1976. A comparison of cortical bone atrophy secondary to fixation with plates with large differences in bending stiffness. *J Bone Joint Surg Am* **58A**:190–195.

Yamada H. 1970. *Strength of biological materials*. Baltimore: Williams & Wilkins.

Young MF. 2003. Bone matrix proteins: their function, regulation, and relationship to osteoporosis. *Osteoporos Int* **14**(Suppl 3):S35–S42.

10

TISSUE RESPONSE TO IMPLANTS

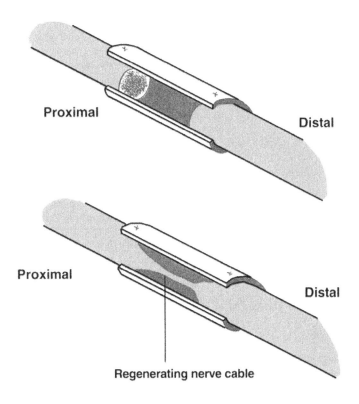

Nerve guidance channel for nerve regeneration. It is critical to have biocompatible and bioresorbable materials for this type of application.

In order to implant any prostheses, a surgeon has to first injure the tissue. The injured or diseased tissues should then be removed to some extent in the process of implantation. The success of the entire operation depends on the kind and degree of tissue response to the surgical procedure and any interactions between tissues and implants. The local and systemic response of the tissues toward implants comprise an aspect of biocompatibility. Biocompatibility entails mechanical, chemical, pharmacological, and surface compatibility (as mentioned in Chapter 1). Some examples of tissue response to various implants are summarized in Table 10-1. The study of tissue response is critical for the success of implants, yet we do not have good "quantitative methods" of measuring tissue responses. The tissue response toward injury may vary widely according to site, species, contamination, etc. However, the inflammation and cellular response to the wound for both intentional and accidental injuries are the same regardless of site.

Table 10-1. Examples of Tissue Response to Various Implants

Implants	Histology	Infection/complication
Breast	T cells, foreign body giant cells (FBGC), macrophages	Dense fibrovascular connective tissue
Heart valves and pacemakers	Endothelial cell ingrowth, macropharges, FBGC	Tissue valves: calcification, disintegration, mechanical: coagulations, formed elements of blood damages
Hernia	Fibroblasts, macropharges, rare FBGC, collagenous membrane to e-PTFE	Seroma, fistula, infection
Intraocular lens	FBGC	Posterior capsular opacification
Joints	FBGC, macropharges,	Bone cement: granulomatosis, stems and cement: particles, wear debris, osteolysis
Ossicular (ear)	Multinucleated FBGC	Degradation of UHMWPE (ultrahigh-Molecular-weight polyethylene)
Penile	Biofilm formation	Infection, partial extraction, progressive neuropathy

Modified with permission from Ravi and Aliyar (2005). Copyright © 2005, Humana Press.

10.1. NORMAL WOUND-HEALING PROCESS

10.1.1. Inflammation

Whenever tissues are injured or destroyed, the adjacent cells respond to repair them. An immediate response to any injury is the inflammatory reaction. Soon after injury, constriction of capillaries occurs (stopping blood leakage); then dilatation. Simultaneously there is greatly increased activity in the endothelial cells lining the capillaries. The capillaries become covered by adjacent leukocytes, erythrocytes, and platelets (formed elements of blood). Concurrently with vasodilatation, leakage of plasma from capillaries occurs. The leaked fluid combined with the migrating leukocytes and dead tissues will constitute *exudate*. Once enough cells (see Table 10-2 for definitions of types of cells) are accumulated by lysis, the exudate becomes pus. It is important to know that pus can sometimes occur in nonbacterial (aseptic) inflammation.

Table 10-2. Definitions of Cells Appearing in this Chapter

Types of cell	Description
Chondroblast:	an immature collagen (cartilage) producing cell
Endothelial:	a cell lining the cavities of the heart and the blood and lymph vessels
Erythrocyte:	a formed element of the blood containing hemoglobin (red blood cell)
Fibroblast:	a common fixed cell of connective tissue that elaborates the precursors of the extracellular fibrous and amorphous components
Giant Cell	
Foreign body giant cell:	a large cell derived from a macrophage in the presence of a foreign body.
Multinucleated giant cell:	a large cell having many nuclei
Granulocyte:	any blood cell containing specific granules; included are neutrophils, basophils, and eosinophils
Leukocytes:	a colorless blood corpuscle capable of ameboid movement, protects body from microorganisms and can be of five types: lymphocytes, monocytes, neutrophils, eosinophils, and basophils
Macrophage:	large phagocytic mononuclear cell; free macrophage is an ameboid phagocyte and present at the site of inflammation
Mesenchymal:	undifferentiated cell having similar role as fibroblasts but often smaller and can develop into new cell types by certain stimuli
Mononuclear:	any cell having one nucleus
Osteoblast:	an immature bone-producing cell
Phagocyte:	any cell that destroys microorganisms or harmful cells.
Platelet:	a small circular or oval disk shaped cell(3 µm dia.) precursor of a blood clot

At the time of damage to the capillaries, the local lymphatics are also damaged since they are more fragile than the capillaries. However, the leakage of fluids from capillaries will provide fibrinogen and other formed elements of the blood, which will quickly plug the damaged lymphatics, thus localizing the inflammatory reaction.

All of the reactions mentioned above — vasodilatation of capillaries, leakage of fluid into the extravascular space, and plugging of lymphatics — will provide the classic inflammatory signs: redness, swelling, and heat, which can lead to local pain.

When the tissue injury is extensive or the wound contains either irritants (foreign materials including prostheses) or bacteria, the inflammation may lead to a prolonged reaction causing extensive tissue destruction. The tissue destruction is done by *collagenase*, which is a proteolytic enzyme capable of digesting collagen. The collagenase is released from *granulocytes*, which in turn are lysed by the lower pH at the wound site. Local pH can drop from the normal values of 7.4–7.6 to below 5.2 at the injured site. If there is no drainage for the necrotic debris, lysed granulocytes, formed blood elements, etc., then the site becomes a severely destructive inflammation, resulting in a *necrotic abscess*.

If the severely destructive inflammation persists and no healing process occurs within three to five days, a *chronic* inflammatory process commences. This is marked by the presence of mononuclear cells called *macrophages*, which can coalesce to form multinuclear giant cells (Figure 10-1). The macrophages are phagocytic and remove bacteria or foreign materials if they are small enough. Sometimes the mononuclear cells evolve into *histiocytes*, which regenerate collagen. This regenerated collagen is used to close the wound or to wall-off unremovable foreign materials such as prostheses by encapsulation with a thin membrane.

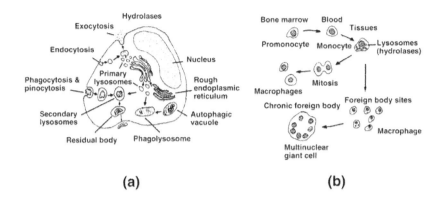

(a)   **(b)**

Figure 10-1. (**a**) Activated macrophage; (**b**) development of the multinuclear foreign body giant cell. Reprinted with permission from Black (1981). Copyright © 1981, Marcel Dekker.

In chronic inflammatory reaction, lymphocytes occur as clumps or foci. These cells are a primary source of immunogenic agents, which become active if foreign proteins are not removed by the body's primary defense.

10.1.2. Cellular Response to Repair

Soon after injury the mesenchymal cells evolve into migratory fibroblasts that move into the injured site while the necrotic debris, blood clots, etc. are removed by the granulocytes and macrophages. The inflammatory exudate contains fibrinogen, which is converted into *fibrin* by enzymes released through blood and tissue cells (see §10.3). The fibrin scaffolds the injured site. The migrating fibroblasts use the fibrin scaffold as a framework onto which the collagen is deposited. New capillaries are formed following the migration of fibroblasts, and the fibrin scaffold is removed by the fibrinolytic enzymes activated by the endothelial cells. The endothelial cells together with the fibroblasts liberate collagenase, which limits the collagen content of the wound.

After 2 to 4 weeks of fibroblastic activities the wound undergoes remodeling, during which the glycoprotein and polysaccharide content of the scar tissue decreases and the number of synthesizing fibroblasts also decreases. A new balance of collagen synthesis and dissolution is reached, and the maturation phase of the wound begins. The time required for the wound-healing process varies for various tissues, although the basic steps described here can be applied in all connective tissue wound-healing processes.

The healing of soft tissues — especially the healing of skin wounds — has been studied intensively since this is germane to all surgery. The degree of healing can be determined by histochemical or physical parameters. A combined method will give a better understanding of the overall healing process. Figure 10-2 shows a schematic diagram of sequential events of the cellular response of soft tissues after injury. The wound strength is not directly proportional to the amount of collagen deposited in the injured site, as shown in Figure 10-3. This indicates that there is a latent period for the collagen molecules (procollagen is deposited by fibroblasts) to polymerize to their maturity. It may take additional time to align the fibers in the direction of stress and to crosslink procollagen molecules to increase the physical strength closer to that of normal tissue. This collagen restructuring process requires more than 6 months to complete, although the wound strength never reaches its original value. The wound strength can be affected by many variables, i.e., severe malnutrition resulting in protein depletion, temperature, presence of other wounds, and oxygen tension. Other such factors as drugs, hormones, irradiation, and electrical and magnetic field stimulation all affect the normal wound-healing process. It is also noted that scar tissues lack elastin, resulting in noncompliant stiff collagenous tissue. The scar tissues also contract during healing due to the same reason.

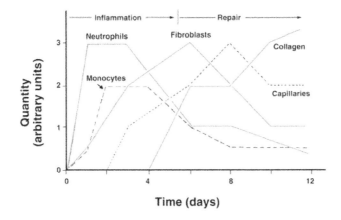

Figure 10-2. Soft tissue wound healing sequence. Reprinted with permission from Hench and Erthridge (1975). Copyright © 1975, Wiley Interscience.

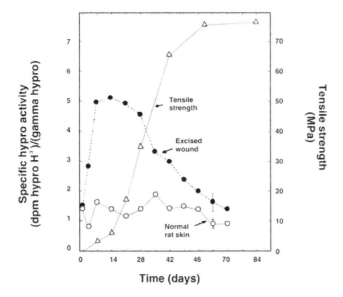

Figure 10-3. Tensile strength and rate of collagen synthesis of rat skin wounds. Reprinted with permission from Peacock and Van Winkle (1970). Copyright © 1970, W.B. Saunders.

The healing of bone fracture is *regenerative* rather than simple repair. The only other tissue that truly regenerates in humans is liver. However, the extent of regeneration is limited in humans. The cellular events following fracture of bone are illustrated in Figure 10-4. When a bone is fractured, many blood vessels (including those in the adjacent soft tissues) hemorrhage and form a blood clot around the fracture site. Shortly after fracture the *fibroblasts* in the outer layer of the *periosteum* and the osteogenic cells in the inner layer of periosteum migrate and proliferate toward the injured site. These cells lay down a fibrous collagen matrix called a *callus. Osteoblasts* evolved from the osteogenic cells near the bone surfaces start to calcify the callus into trabeculae, which are the structural elements of spongy bone. The osteogenic cells migrating further away from an established blood supply become chondroblasts, which lay down cartilage. Thus, after about 2 to 4 weeks the periosteal callus is made of three parts, as shown in Figure 10-5.

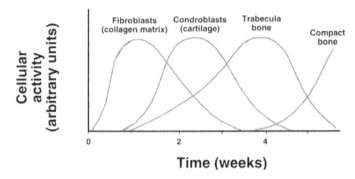

Figure 10-4. Sequence of events followed by bone fracture. Reprinted with permission from Hench and Erthridge (1975). Copyright © 1975, Wiley Interscience.

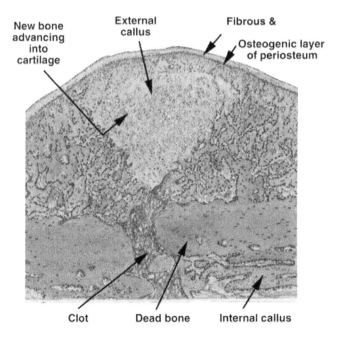

Figure 10-5. Drawing of a longitudinal section of fractured rib of a rabbit after two weeks (H & E stain). Reprinted with permission from Ham and Harris (1971). Copyright © 1971, Academic Press.

Simultaneous with external callus formation a similar repair process occurs in the marrow cavity. Since there is an abundant supply of blood, the cavity turns into callus rather quickly and becomes fibrous or spongy bone.

New trabeculae develop in the fracture site by *appositional growth*, and the spongy bone turns into compact bone. This maturation process begins after about 4 weeks.

There are many factors contributing to the healing and remodeling of the fractured bone, including energy input (mechanical, thermal, electrical, etc.), pharmacology, and the condition and location of the bone.

Some other interesting observations have been made on the healing of bone fractures in relation to the synthesis of polysaccharide on collagen. It is believed that the amount of collagen and polysaccharides is closely related to the cellular events following fracture. When the amount of collagen starts to increase, this marks the onset of the remodeling process. This occurs after about 1 week. Another interesting observation is the electrical potential (or biopotential) measured in the long bone before and after fracture, as shown in Figure 10-6. The large *electronegativity* in the vicinity of fracture marks the presence of increased cellular activities in the tissues. Thus, there is a maximum negative potential in the epiphysis in the normal bone since this zone is more active (the growth plate is in the epiphysis).

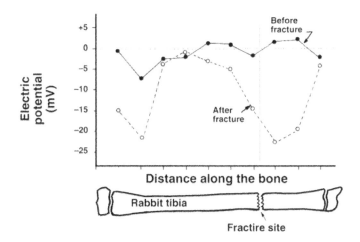

Figure 10-6. Skin surface of rabbit limb before and after fracture. Note that the fracture site has increased electronegative potential. Reprinted with permission from Friedenberg and Brighton (1966). Copyright © 1966, Charles C. Thomas.

Example 10.1

The healing process of wounds in the skin has often been investigated since such healing is relevant to every surgery. In one study electrical stimulation was used to accelerate the healing of wounds in rabbit skin, as shown in the following figure. The mean current flow was 21 μA and the mean current density was 8.4 μA/cm^2. After 7 days the load to fracture of the skin (removed from the dead animals) on the control samples was 797 g, and on the stimulated experimental side it was 1,224 g on average (Konikoff, 1976).

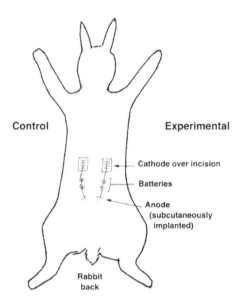

a. Calculate the percent increase of strength by stimulation.

b. The width of the testing sample was 1.6 cm. Assuming a 1.8-mm thickness of skin, calculate the tensile stress for both the control and experimental samples.

c. Compared with the strength of normal skin (about 8 MPa), what percentages of the control and experimental skin wound strengths were recovered?

d. Compare the results of (c) with the result of Figure 10-3.

e. Calculate the corrosion rate in mm/year if a platinum electrode was used.

Answer

a.
$$\frac{1.224 - 797}{797} = \underline{0.536\ (53.6\%)},$$

b.
$$\text{Stress} = \frac{797 \times 10^{-3} \times 9.8\ \text{N}}{1.8 \times 16 \times 10^{-6}\ \text{m}^2} = \underline{0.27\ \text{MPa}},$$

$$\text{Stress} = \frac{1,224 \times 10^{-3} \times 9.8\ \text{N}}{1.8 \times 16 \times 10^{-6}\ \text{m}^2} = \underline{0.42\ \text{MPa}}.$$

c.
$$\frac{0.27}{8} = \underline{0.034\ (\text{or } 3.4\%)}, \quad \frac{0.42}{8} = \underline{0.052\ (\text{or } 5.2\%)}.$$

d. About the same recovery.

e.
$$\text{Pt} \rightarrow \text{Pt}^{3+} + 3e^-, \quad \text{m.w. of Pt} = 195\ \text{g/mol}, \quad \rho_{\text{Pt}} = 21.45\ \text{g/cm}^3.$$

The ion flow is represented as a current density J:

$$J = 8.4\ \mu\text{A/cm}^2 \times [(6.25 \times 10^{18} e^-)/(A \times \text{sec})] \times 3.15 \times 10^7\ \text{sec/yr} \times 0.288\ \text{cm}^2$$

$$= 4.76 \times 1020 e^- / \text{yr}.$$

Depth of loss = $4.76 \times 10^{20} e^-$/yr $\times$ (1 ion/$3e^-$) $\times$ (195 g/mol) $\times$ (mol/6.02×10^{23} ions)

$$\times \ (cm^3/21.45 \ g) \times (1/0.288 \ cm^2)$$

$$= \underline{0.008 cm/yr} \ or \ \underline{80 \ \mu m/yr.}$$

This assumes that the current is entirely an ionic current.

10.2. BODY RESPONSE TO IMPLANTS

The response of the body toward implants varies widely according to host site and species, the degree of trauma imposed during implantation, and all the variables associated with a normal wound-healing process. The chemical composition and micro- and macrostructures of the implants induce different body responses. The response has been studied in two different areas — local (cellular) and systemic — although a single implant should be examined for both aspects. In practice, testing was not done simultaneously except in a few cases (such as bone cement).

10.2.1. Cellular Response to Implants

Generally the body reacts to foreign materials in ways to get rid of them. The foreign material could be extruded from the body if it can be moved (as in the case of a wood splinter), or walled-off if it cannot be extracted. If the material is particulate or fluid, it will be ingested by giant cells (macrophages) and removed.

These responses are related to the healing process of a wound where an implant is added as an additional factor. A typical tissue response is that the polymorphonuclear leukocytes appear near the implant followed by the *macrophages*, called *foreign body giant cells*. However, if the implant is chemically and physically inert to the tissue, the foreign body giant cells may not form. Instead, only a thin layer of collagenous tissue encapsulates the implant. If the implant is either a chemical or a physical irritant to the surrounding tissue, inflammation occurs in the implant site. The inflammation (both acute and chronic) will delay the normal healing process, resulting in granular tissues. Some implants may cause necrosis of tissues by chemical, mechanical, and thermal trauma.

It is generally very difficult to assess tissue responses to various implants due to the wide variation in experimental protocols. This is exemplified by Table 10-3, which gives the tissue reactions for various suture materials in a descriptive way. It is interesting to note that the monofilament nylon suture maintains its tensile strength and elicits minimum tissue reaction. The multifilament suture loses its strength (disintegrates) and provokes a strong tissue reaction.

The degree of the tissue responses varies according to both the physical and chemical nature of the implants. Pure metals (except the noble metals) tend to evoke a severe tissue reaction. This may be related to the high-energy state or large free energy of pure metals, which tends to lower the metal's free energy by oxidation or corrosion. In fact, titanium exhibits the minimum tissue reaction of all the metals commonly used (except gold in dentistry) in implants if its oxide layer is intact. This is due to the tenacious oxide layer that resists further diffusion of metal ions and gas (O_2) at the interface, as discussed earlier in Chapter 6. In fact, this oxide layer is a ceramic-like material that is very inert. Corrosion-resistant metal alloys such as cobalt–chromium and 316L stainless steel have a similar effect on tissue once they are passivated.

Table 10-3. Effect of Implantation on the Properties of Suture Materials

Material	Wound tensile strength	Suture tensile strength	Tissue reaction
Absorbable			
Plain catgut	Impaired	Zero by 3–6 days	Very severe
Chromic catgut	Impaired	Variable	Moderate (less severe than plain catgut, but more than nonabsorbable materials)
Nonabsorbable			
Silk	No effect	Well maintained	Slight
Nylon			
Multifilament	No effect	Very low at 6 months	Moderately severe and prolonged
Monofilament	No effect	Well maintained	Slight
Polyethylene terephthalate	No effect	Well maintained	Very slight
PTFE (Teflon®)	No effect	Well maintained	Almost none

Reprinted with permission from Newcombe (1972). Copyright © 1972, J. & A. Churchill.

Most ceramic materials investigated for their tissue compatibility are oxides such as TiO_2, Al_2O_3, ZrO_2, $BaTiO_3$, and multiphase ceramics of $CaO–Al_2O_3$, $CaO–ZrO_2$ and $CaO–TiO_2$. These materials show minimal tissue reactions with only a thin layer of encapsulation, as shown in Figure 10-7. Similar reactions were seen for carbon implants.

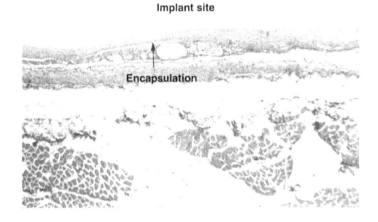

Figure 10-7. Optical micrograph of soft connect tissue adjacent to $BaTiO_3$ implants after 20 weeks (H & E stain; arrow → indicates encapsulation of dense collagenous tissue, ×400). Courtesy of G.H. Kenner and J.B. Park.

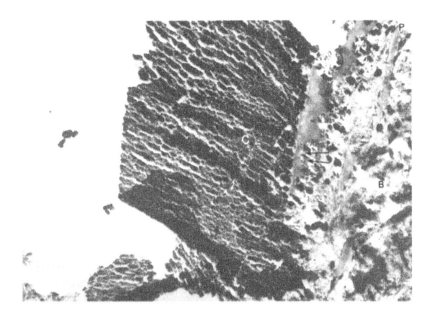

Figure 10-8. Electron micrograph of the interface between 45S5 Bioglass-ceramic (C) and bone (B). Arrows indicate region of gel formation, undecalcified ($\times 10,000$). Courtesy of L.L. Hench.

Some glass-ceramics (e.g., 45 w/o SiO_2, 24.5 w/o Ca_2O, 24.5 w/o CaO, and 6.0 w/o P_2O_5) showed a direct bonding between implant and bone. Dissolution of the silica rich gel film at the interface permits such a bond to form, as shown in Figure 10-8.

The polymers per se are quite inert toward tissue if there are no additives such as antioxidants, fillers, antidiscoloring agents, plasticizers, etc. On the other hand, the monomers can evoke an adverse tissue reaction since they are reactive. Thus, the degree of polymerization is somewhat related to the tissue reaction. Since 100% polymerization is almost impossible to achieve, a range of different-size polymer molecules exists even after complete polymerization, and small molecules can be leached out of the polymer. The particulate form of very inert polymeric materials can cause severe tissue reaction. This was amply demonstrated by polytetrafluoroethylene (Teflon®), which is quite inert in bulk form, such as rods or woven fabrics, but is very reactive in tissue when made into powder form; powder is generated as debris during wear processes. This is most likely due to the increased surface area and free radical generation by mechanical breakdown into smaller particulates. A schematic summary of tissue responses to implants is given in Figure 10-9.

There has been some concern about the possibility of tumor formation by the wide range of materials used in implantation. Although many implant materials are carcinogenic in rats, there are few well-documented cases of tumors in humans directly related to implants. It may be premature to pass final judgment since the latency time for tumor formation in humans may be longer than 20 to 30 years. In this case, we have to wait longer for a final assessment. However, the number of implants being placed in the body and the lack of strong direct evidence for carcinogenicity tend to support the conjecture that carcinogenesis is species-specific and that no tumors (or an insignificant number) will be formed in man by the implants. Carcinogenicity and its testing are discussed further in §10.4.1.

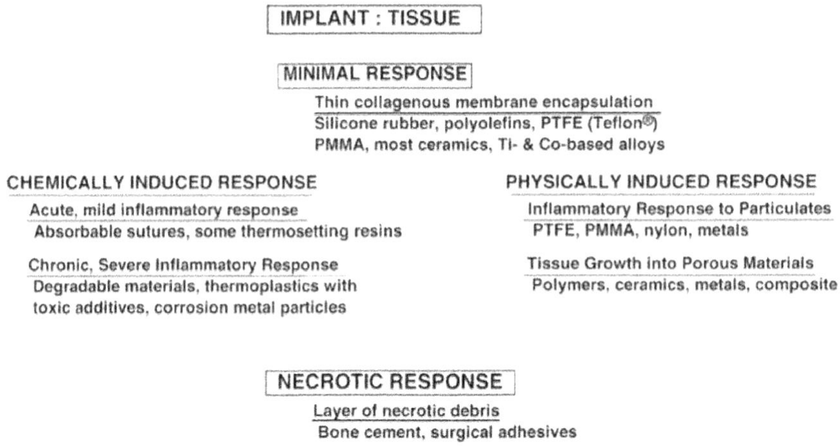

IMPLANT : TISSUE

MINIMAL RESPONSE
Thin collagenous membrane encapsulation
Silicone rubber, polyolefins, PTFE (Teflon®)
PMMA, most ceramics, Ti- & Co-based alloys

CHEMICALLY INDUCED RESPONSE

Acute, mild inflammatory response
Absorbable sutures, some thermosetting resins

Chronic, Severe Inflammatory Response
Degradable materials, thermoplastics with
toxic additives, corrosion metal particles

PHYSICALLY INDUCED RESPONSE

Inflammatory Response to Particulates
PTFE, PMMA, nylon, metals

Tissue Growth Into Porous Materials
Polymers, ceramics, metals, composite

NECROTIC RESPONSE
Layer of necrotic debris
Bone cement, surgical adhesives

Figure 10-9. Brief summary of tissue response to implants. Reprinted with permission from Williams and Roaf (1973). Copyright © 1973, W.B. Saunders.

Example 10-2

Describe the major differences between normal wound healing and the tissue responses to "inert" and "irritant" materials. What factors aside from the choice of material can affect the local tissue response to an implant?

Answer

The tissue response to an "inert" material is very much like normal wound healing. No foreign body giant cells appear, and a thin fibrous capsule is formed. The tissue in this capsule differs very little from normal scar tissue. In response to irritant materials, foreign body giant cells appear and an inflammatory response is evoked. There is an abundance of leukocytes, macrophages, and granulocytes. Granular tissue will be formed, serving the functions of phagocytosis and organization, and appears only under circumstances of irritation or infection. Healing is slow and a thick capsule forms. If the material is chemically reactive or mechanically irritating, necrosis of surrounding tissue may result. It has been suggested that the size and shape of an implant should be important factors to consider for what type of tissue reaction it could elicit.

10.2.2. Systemic Effects by Implants

The systemic effect by implants is well documented in hip joint replacement surgery. The polymethylmethacrylate bone cement applied in the femoral shaft in dough state is known to lower blood pressure significantly due to the reaction toward (liquid) methylmethacrylate monomers. This is a transient effect during surgery. There is a concern regarding the systemic effect of biodegradable implants such as absorbable suture and surgical adhesives, biodegradable polymer scaffolds (Chapter 16), and the large number of wear and corrosion particles released by metallic and polymeric implants. The latter fact is especially important in view of the fact that the period of implantation is becoming longer as prostheses are implanted in younger and more active people.

Table 10-4. Concentration of metals in tissues and organs of the rabbit after implantion

	Surr. Muscle		Liver		Kidney		Spleen		Lung		Control muscle	
	6 wk	16 wk	6 wk	16 wk	6 wk	16 wk	6 wk	16 wk	6 wk	16 wk	6 wk	16 wk
Vitallium® (61.9% Co, 28–34% Cr, 4.73% Mo, 1.52% Ni, 0.61% Fe) (2 or 3 specimens per experiment)												
Cr	30, 25, 0	0, 0, 15	0, 5, 5	5, 5, 10	0, 5, 5	5, 5, 10	0, 0, 5	5, 5, 10	0, 0, 10	5, 5, 10	0, 5, 5	0, 10, 0
Co	25,45	0, 30, 0	5, 10	5, 5, 10	0, 0, 105	5, 10,70	5, 5, 20	5, 5,300	0, 0, 300	10, 5, 10	0, 0, 50	0, 10, 0
Ni	20,35	10	5,10	5, 5, 80	5, 110	5, 5, 30	5, 205	5, 5, 1000	5,40	5, 0, 45	5,5	5
Ti	5, 5, 10	0, 10,0	0, 0, 5	0, 0, 10	0, 0, 10	0, 0, 10	0, 0, 5	0, 0, 15	0, 0, 5	0, 0, 15	0, 0, 10	0, 0, 5
Mo	0, 5, 5	0, 5,0	0, 15,80	10, 10, 70	0, 20, 50	20, 20, 75	0, 0, 10	0, 0, 5	0, 0, 10	5, 5, 5	0, 0, 0	0, 0, 0
Fe	0,40,90	0, 80,0	0,200,600	100, 160, 80	0,90, 180	110, 120, 90	0,250,270	300, 290, 110	0,90,420	110, 100, 50	0, 10,30	0,20,0
316 stainless steel (17.8% Cr, 13.4% Ni, 2.3% Mo, 0.23% Cu, 66.27% Fe)												
Cr	145,65	115,295	5,10	5.5	5.5	5.5	5.5	5,20	5.5	20,10	0,20	5.5
Co	5.5	0,10	5,20	0,15	5,10	5,115	5.5	15,600	5.5	0,250	0,0	0,90
Ni	15,50	200,70	20,20	0,95	0,10	10,10	5,10	1000,65	20,220	20,45	10,0	5.5
Ti	20,10	25,10	0,10	5.5	0,5	5,65	0,5	10,50	0,10	5.5	5,10	10,10
Mo	5,10	10,35	10,75	50,65	10,85	75,80	5,10	15,150	0,10	5.5	0,0	0,0
Fe	80,70	90,190	220,590	550,520	110,120	180,200	180,420	580,220	120,220	150,300	30,10	15,10

Amounts given in ppm dry ash, to nearest 5 ppm).

Reprinted with permission from Ferguson et al. (1962). Copyright © 1962, Journal of Bone and Joint Surgery.

Table 10-4 indicates that the various organs have different affinities for different metallic elements. These results also indicate that the corrosion-resistant metal alloys are not completely stable chemically, and some elements are released into the body. This raised another concern that the elevated ion concentrations in various organs may interfere with normal physiologic activities. The divalent metal ions may also inhibit various enzyme activities.

As mentioned before, the polymeric materials contain additives that may cause cellular as well as systemic reactions to a greater degree than the (pure) polymer itself. The silicone polymer (see Chapter 9) dimethylsiloxane (Silastic® rubber, Dow Corning Co.) has been traditionally considered to be well tolerated by the body. As discussed in Chapter 11, silicones have long been used in soft tissue reconstruction of the ears, nose, and chin. Mammary augmentation has been done with prostheses consisting of silicone gel within a silicone rubber envelope. The implants can leak or rupture after time, and they typically last a median of about 16 years. Also, some patients have been injured (via a sensitizing effect) by sterilizing agent ethylene oxide (ETO) residues in implants; rash, edema, pain, and even skin necrosis have been reported. Such tissue reactions as burns can be prevented by properly ventilating the implants to allow evaporation of the sterilizing agent. Beginning in 1982, published anecdotal reports began to appear of illness in patients with silicone breast implants. Disease suggestive of auto-immune conditions such as systemic sclerosis and chronic connective tissue inflammation were claimed to have developed after mammary augmentation with silicon. Considerable controversy surrounded such reports, and the manufacturer of silicone (Dow Corning) was subject to many lawsuits. The American Food and Drug Administration restricted the use of silicone gel-filled mammary implants. The theory that silicone breast implants cause immunological disorders has not been proven, though there have been some clinical reports of sensitivity to silicone. Silicone rubber contains a filler, silica (SiO_2) powder, to enhance its physical (mostly mechanical) properties. Silica powder itself can cause problems if inhaled in large quantities as dust, and it is an irritant when implanted in a concentrated area. Silica is considered to be relatively inert when used as a filler in a polymer matrix. However, it is not certain whether there will be later complications if a large amount of the silica is released into the tissue and retained in various organs. Indeed, immune cell response, T cell memory to silica, was observed in the offspring of women after, but not before, receiving silicone breast implants. The issue of chronic disease associated with implants is complicated by the fact that autoimmune conditions occur in the general population. It can be difficult to draw definitive conclusions unless one can compare population groups entirely equivalent except for the presence or absence of implants. Nevertheless, a review of the clinical outcome of more than 15,000 silicone finger joint implants disclosed no cases of immune reactions, autoimmune disease, or other systemic effects. Complications were rare: 2% of the finger implants fractured and 0.06% of them generated local irritation due to wear debris.

Example 10-3

A biodegradable suture will have a strength 1 MPa after 6 weeks of implantation. The strength of the implanted suture decreases according to

$$\sigma = \sigma_0 + b \ln t/a,$$

as determined by curve fitting to experimental data, where σ_0 is the original strength, $b = -2$ MPa, t = time in weeks, and a is a characteristic time, 1 week.

Determine the original strength of the suture.

Answer

$$\sigma_0 = \sigma - b \ln t/a = 1 + 2 \ln 6/1 = \underline{4.58 \text{ MPa}}.$$

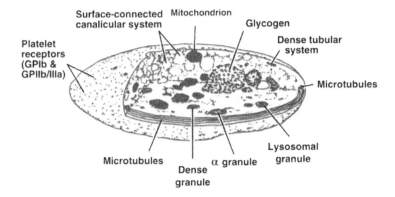

Figure 10-10. Structure of a platelet. Reprinted with permission from Hanson and Harker (1996). Copyright © 1996, W.B. Saunders.

10.3. BLOOD COMPATIBILITY

The single most important requirement for blood-interfacing implants is blood compatibility. Although blood coagulation is the most important factor for blood compatibility, the implants should not damage the proteins, enzymes, and formed elements of blood. The latter includes hemolysis (red blood cell rupture) and initiation of the platelet release reaction. The structure of the platelet is shown in Figure 10-10. The platelets adhere to a surface using pseudopods. Once adhered, platelets release α granule contents, including platelet factor 4 (PF4) and β-thromboglobulin (βGB), and dense granule contents, including adenosine diphosphate (ADP). Thrombin is generated locally through factor XIIa and platelet procoagulant activity. Thromboxine A2 (TxA2) is synthesized. ADP, TxA2, and thrombin recruit additional circulating platelets into enlarging platelet aggregate. The thrombin-generated fibrin stabilizes the platelet mass, as shown in Figure 10-11. Table 10-5 gives the properties of human clotting factors.

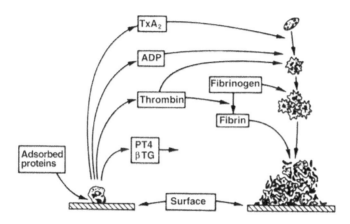

Figure 10-11. Schematic representation of platelet deposition on surfaces. Reprinted with permission from Hanson and Harker (1996). Copyright © 1996, W.B. Saunders.

The mechanism and route of blood coagulation (thrombi or emboli depending on the mobility) are quite complex. Cascading events can be triggered in blood and tissue, as shown in Figure 10-12. The kinin system is activated by the Hageman factor (Hfa, Factor XII) in contact with collagen, basement membrane, foreign bodies such as bacteria, metal, etc. Blood coagulation pathways can be depicted as shown in Figure 10-13. Again the Hageman factor triggers the cascade in intrinsic pathway while the extrinsic pathways activate Factor VII together with lipoproteins. As discussed earlier, immediately after injury the blood vessels constrict to minimize the flow of blood. Platelets adhere to the vessel walls by contacting the exposed collagen. The aggregation of platelets is achieved through release of adenosine diphosphate (ADP) from damaged red blood cells, vessel walls, and from adherent platelets.

10.3.1. Factors Affecting Blood Compatibility

The *surface roughness* is an important factor since the rougher the surface the more area is exposed to blood. Therefore, a rough surface promotes faster blood coagulation than the highly polished surfaces of glass, polymethylmethacrylate, polyethylene, and stainless steel. Sometimes thrombogenic (clot-producing) materials with rough surfaces are used to promote clotting in porous interstices to prevent initial leaking of blood and to allow later tissue ingrowth through the pores of vascular implants.

Table 10-5. Properties of Human Clotting Factors

Factors	Molecular weight (no. of chains)	Plasma concentration (µg/l)	Active form
Intrinsic system			
Factor XII	80,000 (1)	30	Serine protease
Prekallikreine	80,000 (1)	50	Serine protease
High-m.w. kininogen	105,000 (1)	70	Cofactor
Factor XI	160,000 (2)	4	Serine protease
Factor IX	68,000 (1)	6	Serine protease
Factor VIII	265,000 (1)	0.1	Cofactor
VWF	1–15,000,000ᵃ	7	Cofactor for platelet adhesion
Extrinsic system			
Factor VII	47,000 (1)	0.5	Serine protease
Tissue factor	46,000 (1)	0	Cofactor
Common pathway			
Factor X	56,000 (2)	10	Serine protease
Factor V	330,000 (1)	7	Cofactor
Prothrombin	72,000 (1)	100	Serine protease
Fibrinogen	340,000 (6)	2500	Clot structure
Factor XIII	320,000 (4)	15	Transglutaminase

ᵃ Subunit molecular weight of factor VIII/vWF is around 220,000 g/mol with a series of multimers found in circulation.
Reprinted with permission from Hanson and Harker (1996). Copyright © 1996, Elsevier.

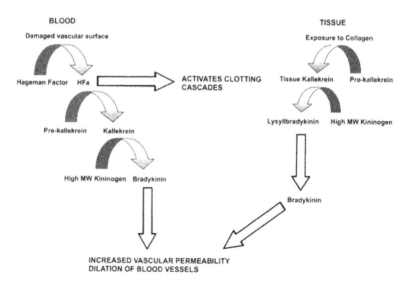

Figure 10-12. The kinin system starts with the damaged blood vessel surface. The high-molecular-weight kininogen and pre-kallikrein circulated in the plasma with Hfa trigger the cascade. Reprinted with permission from Ravi and Aliyar (2005). Copyright © 2005, Humana Press.

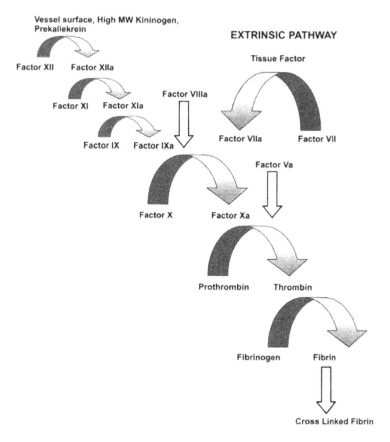

Figure 10-13. Blood coagulation pathways by intrinsic and extrinsic route. Reprinted with permission from Ravi and Aliyar (2005). Copyright © 2005, Humana Press.

Surface wettability — i.e., hydrophilic (wettable) or hydrophobic (non-wettable) — was thought of as an important factor. However, the wettability parameter, the contact angle with liquids, does not correlate consistently with blood-clotting time.

The surface of the intima of blood vessels is negatively charged (1–5 mV) with respect to the adventitia. This phenomenon is associated partially with the nonthrombogenic or thromboresistant character of the intima since the formed elements of blood are also negatively charged and hence are repelled from the surface of the intima. This was demonstrated experimentally in canine experiments by using a copper tube that is a thrombogenic material implanted as an arterial replacement. When the tube was negatively charged, clot formation was delayed to a few days when compared with the control, which clotted within a few minutes. In connection with this phenomenon, the streaming potential or zeta potential has been investigated since the formed elements of blood are flowing particles in vivo. However, it was not possible to establish a direct one-to-one relationship between clotting time and the zeta potential.

The chemical nature of a material surface interfacing with blood is closely related to the electrical nature of the surface since the type of functional groups of polymer determines the type and magnitude of the surface charge. No intrinsic surface charge exists for metals and ceramics, although some ceramics and polymers can be made piezoelectric. The surface of intima is negatively charged largely due to the presence of polysaccharides, especially chondroitin sulfate and heparin sulfate.

10.3.2. Nonthrombogenic Surfaces

There have been many efforts directed at obtaining nonthrombogenic materials. The empirical approach has been used often. These approaches can be categorized as (1) heparinized or biological surfaces, (2) surfaces with anionic radicals for negative electric charges, (3) inert surfaces, and (4) solution-perfused surfaces. An early approach to nonthrombogenic surfaces is shown in Figure 10-14.

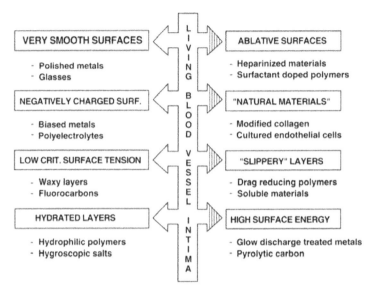

Figure 10-14. Approaches to producing thromboresistant surfaces. Reprinted with permission from Baier (1972). Copyright © 1972, New York Academy of Medicine.

Heparin is a polysaccharide with negative charges due to the sulfate groups as shown:

$$(10\text{-}1)$$

Initially, the heparin was attached to a graphite surface treated with quaternary salt, benzalkonium chloride (GBH process). Later, simpler heparinization was accomplished by exposing the polymer surface to a quarternary salt, such as tridodecylmethylammonium chloride (TDMAC). This method was further simplified by making a TDMAC and heparin solution in which the implant could be immersed followed by drying.

The heparinized materials showed a significant increase in thromboresistance compared to untreated control materials. In an interesting application, a polyester fabric graft was heparinized. This reduced the tendency of initial bleeding through the fabric, and a thin *neointima* was later formed. Many polymers were tried for heparinization, including polyethylene, silicone rubber, etc. Leaching of the heparin into the fluid medium is a drawback, although some improvement was seen by crosslinking of the heparin with gluteraldehyde and directly covalent bonding it onto the surface.

Some studies were carried out to coat the cardiovascular implant surface with other biological molecules such as albumin, gelatin (denatured collagen), and heparin. Some reported that the albumin alone can be thromboresistant and decrease platelet adhesion. Also, the vascular grafts were coated by inert materials such as carbon by depositing ULTI (ultralow-temperature isotropic) pyrolytic carbon. The pyrolytic carbon showed an excellent blood compatibility and is currently most widely used to make artificial heart valve discs.

Negatively charged surfaces with anionic radicals (acrylic acid derivatives) were made by copolymerization or grafting. Negatively charged *electrets* on the surface of a polymer enhance its thromboresistance.

Hydrogels of both hydroxyethylmethacrylate (poly-HEMA) and acrylamide are classified as inert materials since they contain neither highly negative anionic radical groups nor are negatively charged. These coatings tend to be washed away when exposed to the bloodstream, as was also seen with heparin coatings. Segmental or block polyurethanes also showed some thromboresistance without surface modification.

Another method of making surfaces nonthrombogenic is perfusion of water (saline solution) through the interstices of a porous material that interfaces with blood. This new approach to a nonthrombogenic surface has the advantage of avoiding damage to formed elements of the blood. The disadvantage is the dilution of blood plasma. This is not a serious problem since saline solution is deliberately injected for the kidney and heart/lung machine; the method can be used only in such temporary blood-interfacing applications.

Example 10-4

In the circulation of the blood, the formed elements are being destroyed by the blood pump and tube wall contacts. A bioengineer measured the rate of hemolysis (red blood cell lysis) as 0.1 g/liter pumped. If the normal cardiac output for a dog is 0.1 liter/kg/min, what is the hemolysis rate? If the animal weighs 20 kg and the critical amount of hemolysis is 0.1 g/kg of body weight, how long can the bioengineer circulate the blood before reaching a critical condition? Assume a negligible amount of new blood formation.

Answer

$$\text{Hemolysis rate} = 0.1 \text{ g/liter} \times 0.1 \text{ liter/kg/min} \times 20 \text{ kg} = \underline{2 \text{ mg/min.}}$$

$$\text{Critical hemolysis} = 0.1 \text{ g/kg} \times 20 \text{ kg} = \underline{2\text{g.}}$$

$$\frac{2 \text{ g}}{2 \text{ mg/min}} = \underline{1,000 \text{ min (or 16 hr 40 min)}}$$

Example 10-5

A bioengineer designed and made a catheter tip constructed of porous outer layer and a solid inner layer, as shown schematically. Only the porous part would be in the blood vessels. In a canine test, a complete prevention of blood clot formation was possible by perfusing 0.1 ml of saline solution per minute. (a) Calculate the total amount of saline solution to be perfused in one day? (b) What effect will dilution by the saline solution have on the normal function of the blood and other organs?

Answer

a.   Total amount of saline solution

$$0.1 \text{ ml/min} \times 60 \text{ min/hr} \times 24 \text{ hr/day} = \underline{144 \text{ ml/day or } 0.144 \text{ liter/day}}.$$

b.   The amount of saline solution perfused at this level would not cause a serious problem for renal function. For this situation, up to 10 times of saline solution may be tolerable.

10.4. CARCINOGENICITY

A variety of chemical substances are known to induce the onset of cancerous disease in human beings and are known as carcinogens. Carcinogenic agents may act upon the body by skin contact, by ingestion, by inhalation, or by direct contact with tissues. It is the last possibility that is of primary concern within the context of biomaterials.

Early studies showed that sheets or films of many polymers produced cancer when implanted in animals, especially rats. It was later found that the physical form of the implant was important, and that fibers and fabrics produced fewer tumors than sheets of the same material, and powders produced almost no tumors. Other materials, by contrast, are carcinogenic by virtue of their chemical constitution.

10.4.1. Testing of Carcinogenicity

New materials to which people may be exposed should be tested for possible carcinogenicity. Assessment of materials for possible carcinogenic effect is done as follows.

Chemical structure or function. If a material is similar structurally or pharmacologically to known carcinogenic agents, it may be suspected and evaluated further. Known carcinogens include aromatic amines, polynuclear aromatic hydrocarbons with multiple ring structures, alkylating agents including urethanes (ethylcarbamate, e.g., $H_2NCOOC_2H_5$), aflatoxins, halogenated hydrocarbons, including vinyl chloride monomer, chloroform, polychlorinated biphenyls (PCBs), and certain pesticides; metallic nickel, cadmium, and cobalt.

In-vitro tests. These tests involve exposing cultured cells to the agent in question. Their usefulness is predicated on the fact that carcinogenicity is correlated with mutagenicity. Following in-vitro testing, the cells are examined for gene mutations, chromosomal aberrations, and/or deoxyribonucleic acid (DNA) damage and repair. The most well-known and widely used of these tests is the *Ames test*, which involves exposing bacteria of the strain *Salmonella typhimurium* to the suspect agent, and looking for reverse mutations. Some mammalian metabolic capacity is provided by incubating the culture with mitochondrial extracts from rat liver. In-vitro tests have the advantages of being quick and relatively inexpensive, but they are not sensitive to all carcinogenic agents (e.g., asbestos), and they do not reflect the complexities of uptake, organ specificity, distribution, and excretion found in whole animals and in humans.

Long-term animal bioassay. Rats and mice are the animals of choice in view of their relatively low cost and short lifespan. The short lifespan permits follow-up in whole-life exposure studies; moreover, since the latency of tumor onset is a specified fraction of an animal's lifetime, the waiting period is not excessive. In a typical bioassay, there are four groups of animals: control (no exposure), maximum tolerated dose, and two intermediate doses. A minimum of 70 rodents per dose group per sex is used. The maximum tolerated dose is that which generates no overt toxicity, which does not reduce survival for reasons other than cancer. Following 2 years, animals that die are examined, and those surviving are subjected to necropsy and histopathologic study. Control and dose groups are then compared statistically. As for the relevance to human beings, virtually all materials known to be carcinogenic in humans are also carcinogenic in animals, but not *vice versa*.

Animal assays for solids. Cancer may be associated with implanted solids such as shrapnel, rifle bullets, and prosthetic implants: "foreign body" carcinogenesis. This type of situation is the most relevant to the biomaterials area. In animal assays, the suspect material is implanted in the flanks of rodents. In addition to examining for tumors, the investigators look for precancerous changes in the cells around the implant. The rationale is to increase the sensitivity of the test, since it is not feasible to increase the 'dose' as done above.

Epidemiology. This approach is clearly the most relevant since it deals directly with humans, but there are difficulties. For example, the latency period between exposure to a carcinogenic agent and development of disease is from 5 to 40 years in adult humans. The risk of a newly introduced agent, therefore, will not be apparent until many years have elapsed. Epidemiologic methods also are relatively insensitive unless a large fraction of the population has been exposed, as in the case of cigarette smoking.

10.4.2. Risk Assessment

Prediction of risk is complicated by the fact that many tests are conducted at high dose levels, but human exposure is ordinarily at very low dose levels. Therefore, in carcinogenicity testing, the *linearity hypothesis* is often referred to in interpretation of results. According to this hypothesis, the carcinogenic response is a linear function of the dose.

Example 10-6
One wishes to identify materials that will cause cancer in 1 out of 100,000 human subjects, by experiments upon rats. Is it feasible to use the same dose rate the humans would be exposed to? How many rats would one need under those conditions? Suggest an alternative experiment design based on the linearity hypothesis.

Answer
If the carcinogenic potential of the material is the same in humans as in rats, then, out of 100,000 rats, 1 would get cancer from the material, and perhaps 30,000 (30%) would get cancer by old age from other causes. A control group would contain an identical number of rats, of which about 30% would also ultimately suffer cancer. There are two problems with this experiment: 200,000 rats is an excessive number representing an excessive expense, and it would be impossible to pick out the one extra cancer case from the 30,000 "naturally" occurring ones. A better approach would be to increase the dosage of the material by a factor of 10,000, so that 10% of the rats would suffer cancer from the exposure. A smaller number of rats could then be used. This approach is warrantable if the linearity hypothesis is valid.

Other dose–response curves are conceivable, including a "threshold" or *sublinear* response in which the low-dose risk is less than that predicted by the linear model, and a *supralinear* model in which it is greater. Prediction of risk at low dose can vary by orders of magnitude depending on the model chosen. It is a standard procedure to use a linear dose–response curve.

As for the potential risk associated with implant materials, pure metallic nickel, cadmium, and cobalt are known carcinogens when injected in solution into rat muscles. Nickel is a known industrial carcinogen, and cobalt is a suspect one. Implants in animals often induce tumors, but the epidemiological evidence for human implants suggests a rather small risk.

PROBLEMS

10-1. What makes it so difficult to evaluate the tissue and blood compatibility of implants?

10-2. Calculate the corrosion rate in mm/year if a platinum electrode was used in Example 10-1.

10-3. Some materials do not normally induce tissue reaction when implanted in bulk form. However, when implanted in powder form, they become non-biocompatible. Explain why.

10-4. Most metals elicit tissue reactions in pure form (such as Al), but minimum or no reactions take place when oxidized (Al_2O_3). Why? Explain in terms of surface and reaction potential energy to other materials.

10-5. Sometimes the degree of tissue reaction toward an implant is represented by the thickness of the collagenous capsule (e.g., Figure 10-7).

    a. State what fallacy these experimental results may contain when deciding biocompatibility.

    b. What factors affect the thickness of the encapsulation?

10-6. Explain why the metals are generally less biocompatible than ceramics or polymers. What can you do to improve this disadvantage of metals as implant materials?

10-7. The temperature changes due to the heat of polymerization of bone cement (polymethylmethacrylate–polystyrene copolymer powder plus methylmethacrylate monomer liquid) was monitored at the interface between bone and cement. The mixed cement was placed in the canine femur as a 9-mm diameter plug, and the temperature was measured with a thermocouple, and the results are shown in the following figure.

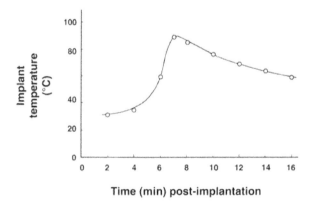

Time (min) post-implantation

    a. What will happen to the adjacent tissues due to the heat generated by polymerization?

    b. Would the temperature rise or decrease by putting a metal cylinder in the middle, in a way similar to the situation with femoral hip replacement?

    c. What problems will arise if the cement shrinks when it reaches ambient temperature?

10-8.   Proplast® is a composite of PTFE and carbon (graphite). If it is made of 50% by volume each and has 20% porosity. What is its density? Estimate its Young's modulus.

10-9.   Explain why a nylon monofilament suture is less prone to lose its strength than multifilament suture material in vivo. Also explain why monofilament suture causes less tissue reaction.

10-10.   Experience has shown that the silicone membrane used in breast implants leaks the silicone fluid contained within to the surrounding tissue. Calculate the amount of leakage in a year. Assume that the leakage is entirely by diffusion rather than by macroscopic pores. Assume the silicone oil has a molecular weight of 740 amu. Assume the membrane is 1 mm thick and the surface area is 400 cm^2. The membrane has a diffusion constant of $D = 5 \times 10^{-17}$ cm^2/sec. Assume, moreover, that the implant has a volume of 500 cm^3 and a density of $\rho = 1.2$ g/ cm^3. Discuss other ways that silicone fluid or gel could escape. Discuss the implications.

DEFINITIONS

α **granule**: Released through the vessel walls after injury, in turn releasing coagulation agents.

Adenosine diphosphate (ADP): A compound consisting of an adenosine molecule bonded to three phosphate groups, present in all living tissue. The breakage of one phosphate linkage (to form adenosine diphosphate, ADP) provides energy for physiological processes such as muscular contraction

Aflatoxin: A naturally occurring carcinogenic material. It is formed by mold infestation of food crops such as corn or peanuts.

Ames test: A screening test for carcinogenic potential of a material. Genetic mutations are observed in bacteria; carcinogenicity is correlated with mutagenicity.

Biocompatibility: Acceptance of an artificial implant by surrounding tissues and by the body as a whole. The implant should be compatible with tissues in terms of mechanical, chemical, surface, and pharmacological properties.

Callus: Unorganized fibrous collagenous tissue formed during the healing process of bone fracture. It is usually replaced with compact bone.

Carcinogen: Any substance that produces cancer.

Embolus: Any foreign matter, as a blood clot or air bubble, carried in the bloodstream.

Epidemiology: The study of disease incidence in a population of humans. Epidemiologic study constitutes the final test of the potential of a causative agent, such as a biomaterial, in inducing disease.

Formed elements of blood: This refers to the solid components of blood — red and white blood cells and platelets.

Hageman Factor (Hfa, Factor XII): The first factor of the cascading events of blood vessel wall damage control.

Heparin: A substance found in various body tissues, especially in the liver, that prevents the clotting of blood.

Hydrogel: Highly hydrated (over 30% by weight) polymer gel that is used to make soft contact lenses. Acrylamide and poly-HEMA (hydroxyethylmethacrylate) are two common hydrogels.

Kinin system: The consequences of blood vessel wall damage described in a cascading manner from triggering Hfa to the dilatation of blood vessels and increased vascular permeability.

Linearity hypothesis: This is the assumption that the incidence of cancer produced by a carcinogen is linearly proportional to the dose.

Macrophage: Any of various phagocytic cells in connective tissue, lymphatic tissue, bone marrow, etc. Sometimes called foreign body giant cell, and is associated with the presence of implants. Becomes multinucleated if the implant is not biocompatible.

Neointima: Sometimes called pseudointima. It is a new lining formed in the inner surface of porous arterial grafts, and it has similar nonthrombogenic properties as the intima (natural lining) of the arteries.

PCB: Polychlorinated biphenyl, an industrial carcinogen. It is not used in biomaterials.

Platelet: A small colorless disk-shaped cell fragment without a nucleus, found in large numbers in blood and involved in clotting. Also called a **thrombocyte**.

Pseudopod: A temporary protrusion of the surface of an ameboid cell for movement and feeding.

Silica: Silicon dioxide, SiO_2. It is chemically identical to the mineral quartz and is used in particle form as a filler in polymeric and rubbery materials.

Silicon: Chemical element, Si.

Silicone: Polydimethyl siloxane, a polymer that contains the element silicon. It is available in rubber and gel forms. It is used in soft tissue reconstruction in the face, as well as in breast implants.

Supralinear model: This is the assumption of a nonlinear relation between the dose of a carcinogen and the number of cancers produced. More commonly a linear relation is assumed. See linearity hypothesis.

Thrombin: An enzyme in blood plasma that causes the clotting of blood by converting fibrinogen to fibrin.

Thrombus: The fibrinous clot attached at the site of thrombosis.

BIBLIOGRAPHY

Ames BN. 1983. Dietary carcinogens and anticarcinogens. *Science* **221**(4617):1256–1264.

Baier RE. 1972. The role of surface energy in thrombosis. *Bull NY Acad Med* **48**:257–272.

Bechtol CO, Ferguson AB, Liang PG. 1959. *Metals and engineering in bone and joint surgery*. London: Balliere, Tindall, & Cox.

Black J. 1981. *Biological performance of materials*. New York: Dekker.

Bruck SD. 1974. *Blood compatible synthetic polymers: an introduction*. Springfield, IL: Thomas.

Cardenas-Camarena L. 1998. Ethylene oxide burns from improperly sterilized mammary implants. *Ann Plast Surg* **41**(4):361–366.

Charnley J. 1970. *Acrylic cement in orthopaedic surgery*. London: Churchill/Livingstone.

Clarke DR, Park JB. 1981. Prevention of erythrocyte adhesion onto porous surfaces by fluid perfusion. *Biomaterials* **2**:9–13.

Dawids S, ed. 1989. *Polymers: their properties and blood compatibility*. London: Kluwer Academic.

Dawids SG. 1993. *Test procedures for the blood compatibility of biomaterials.* Dordrecht: Kluwer Academic.

Elberg JJ, Kjoller KH, Krag C. 1993. Silicone mammary implants and connective tissue disease. *Scand J Plastic Reconstruct Surg Hand Surg* **27**(4):243–248.

Ferguson AB, Akahoshi Y, Laing PG, Hodge ES. 1962. Characteristics of trace ions released from embedded metal implants in the rabbit. *J Bone Joint Surg Am* **44A**:323–336.

Foliart DE. 1995. Swanson silicone finger joint implants: a review of the literature regarding long-term complications. *J Hand Surg* **20**(3):445–449.

Friedenberg ZB, Brighton CT. 1966. Bioelectric potentials in bone. *J Bone Joint Surg Am* **48A**:915–923.

Garetto LP, Turner CH, Duncan RL, Burr DB, eds. 2002. *Bridging the gap between dental and orthopaedic implants.* Indianapolis: Indiana University School of Dentistry.

Goodman CM, Cohen V, Thornby J, Netscher D. 1998. The life span of silicone gel breast implants and a comparison of mammography, ultrasonography, and magnetic resonance imaging in detecting implant rupture: a meta-analysis. *Ann Plast Surg* **41**(6):577–585.

Greco RS. 1994. *Implantation biology: the host response and biomedical devices.* Boca Raton, FL: CRC Press.

Ham AH, Harris WR. 1971. Repair and transplantation of bone. In *The biochemistry and physiology of bone*, pp. 337–399. Ed G Bourne. New York: Academic Press.

Hanson SR, Harker LA. 1996. Blood coagulation and blood-materials interactions. In *Biomaterials science: an introduction to materials in medicine*, pp. 193–199. Ed BD Ratner, AS Hoffman, FJ Schoen, JE Lemons. San Diego: Academic Press.

Harker LA, Rainer BD, Didisheim P. 1993. *Cardiovascular biomaterials and biocompatibility: a guide to the study of blood-tissue-material interactions.* New York: Elsevier.

Hastings GW. 1992. *Cardiovascular biomaterials.* London: Springer-Verlag.

Hench LL, Erthridge EC. 1975. Biomaterials: the interfacial problem. *Adv Biomed Eng* **5**:35–150.

Hench LL, Ethridge EC. 1982. *Biomaterials: an interfacial approach.* New York: Academic Press.

Hoffman AS. 1977. Applications of radiation processing in biomedical engineering: a review of the preparation and properties of novel biomaterials. *Radiat Phys Chem* **9**:207–219.

Houpt KR, Sontheimer RD. 1994. Autoimmune connective tissue disease and connective tissue disease-like illnesses after silicone gel augmentation mammoplasty. *J Am Acad Dermatol* **31**(4):626–642.

Hulbert SF, Levine SN, Moyle, DD, eds. 1974. *Prosthesis and tissue: the interfacial problems.* New York: Wiley.

Jenkins MEF, Friedman HI, von Recum AF. 1996. Breast implants: facts, controversy, and speculations for future research. *J Invest Surg* **9**(1):1–12.

Konikoff JJ. 1976. Electrical promotion of soft tissue repairs. *Ann Biomed Eng* **4**(1):1–5.

Levine SN. 1968. Materials in biomedical engineering. *Ann NY Acad Sci* **146**:3–10.

Maibach HI, Rovee DT, eds. 1972. *Epidermal wound healing.* Chicago: Year Book Medical Publishers.

Marcusson JA, Bjarnason B. 1999. Unusual skin reaction to silicone content in breast implants. *Acta Dermato-Venereologica* **79**(2):136–138.

Milman HA, Weisburger EK. 1985. *Handbook of carcinogen testing.* Park Ridge, NJ: Noyes Publications.

Newcombe JK. 1972. Wound healing. *Scientific basis of surgery*, pp. 371–390. Ed WT Irvin. London: J & A Churchill.

Ohta Y. 1993. Comparative changes in microvasculature and bone during healing of implant and extraction sites. *J Oral Implantol* **19**(3):184–98.

Peacock Jr EE, Van Winkle Jr W. 1970. *Surgery and biology of wound repair.* Philadelphia: W.B. Saunders.

Prendergast PJ, Lee TC, Carr AJ, eds. 2000. *Proceedings of the 12th conference of the European society of biomechanics.* Dublin: Royal Academy of Medicine in Ireland.

Rammelt S, Schulze E, Witt M, Petsch E, Biewener A, Pompe W, Zwipp H. 2004. Collagen type I increases bone remodelling around hydroxyapatite implants in the rat tibia. *Cells Tissues Organs* **178**(3):146–157.

Ravi N, Aliyar HA. 2005. Tissue reaction to prosthetic materials. In *The bionic human*, pp. 133–158. Ed FE Johnson, KS Virgo. Totowa, NJ: Humana Press.

Reif A. 1981. The causes of cancer. *Am Sci* **69**:437–447.

Rittmann WW, Perren SM, eds. 1974. *Cortical bone healing after internal fixation and infection; biomechanics and biology*. Berlin: Springer-Verlag.

Salzman EW. 1971. Nonthrombogenic surfaces: critical review. *Blood* **38**:509–523.

Sawyer PN, Srinivasan S. 1972. The role of electrochemical surface properties in thrombosis at vascular interfaces: cumulative experience of studies in animals and man. *Bull NY Acad Med* **48**:235–256.

Schwartz Z, Kieswetter, K. Dean DD, Boyan BD. 1997. Underlying mechanisms at the bone-surface interface during regeneration. *J Periodontal Res* **32**(1 Pt 2):166–171.

Silver FH, Christiansen DL. 1999. *Biomaterials science and biocompatability*. New York: Springer.

Smalley DL, Levine JJ, Shanklin DR, Hall MF, Stevens MV. 1996. Lymphocyte response to silica among offspring of silicone breast implant recipients. *Immunobiology* **196**(5):567–574.

Smith GK, Black J. 1977. Models for systemic effects of metallic implants. In *Retrieval and analysis of orthopaedic implants*, pp. 23–28. Ed A Weinstein, E Horowitz, AW Ruff. Gaithersburg, MD: National Bureau of Standards.

Szycher M. 1991. *High-performance biomaterials: a comprehensive guide to medical and pharmaceutical applications*. Lancaster, PA: Technomics Publishers.

Urist MR. 1980. *Fundamental and clinical bone physiology*. Philadelphia: J.B. Lippincott.

Vroman L. 1971. *Blood*. New York: Doubleday. New York: American Museum of National History Science Books.

Williams DF, ed. 1981. *Fundamental aspects of biocompatibility*. Boca Raton, FL: CRC Press.

Williams DF, Roaf R. 1973. *Implants in surgery*. Philadelphia: W.B. Saunders.

Wise DL. 1996. *Human biomaterials applications*. Totowa, NJ: Humana Press.

Wise DL, ed. 2000a. *Biomaterials engineering and devices: human applications*, Vol. 1: *Fundamentals and vascular and carrier applications*. Totowa, NJ: Humana Press.

Wise DL, ed. 2000b. *Biomaterials engineering and devices: human applications*, Vol. 2: *Orthopedic, dental, and bone graft applications*. Totowa, NJ: Humana Press.

11

SOFT TISSUE REPLACEMENT — I:
SUTURES, SKIN, AND
MAXILLOFACIAL IMPLANTS

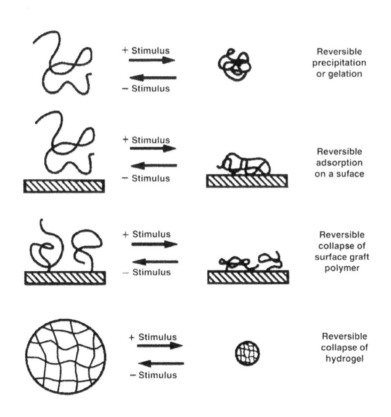

Schematic illustration of "smart," "intelligent," and "reversible memory" polymer systems. The stimulating energy can be mechanical, thermal, or chemical. Reprinted with permission from Hoffman (2004). Copyright © 2004, Elsevier.

In soft tissue implants, as in other applications that involve engineering, the performance of an implanted device depends upon both the materials used and the design of the device or implant. The initial selection of material should be based on sound materials engineering practice. The final judgment on the suitability of a material depends upon observation of the in-vivo clinical performance of the implant. Such observations may require many years or decades. This requirement of in-vivo observation represents one of the major problems in the selection of appropriate materials for use in the human body. Another problem is that the performance of an implant may also depend on the design rather than the materials themselves. Even though one may have an ideal material and design, the actual performance also greatly depends on the skill of the surgeons and the prior condition of patients.

The success of soft tissue implants has primarily been due to the development of synthetic polymers. This is mainly because the polymers can be tailor made to match the properties of soft tissues. In addition, polymers can be made into various physical forms, such as liquid for filling spaces, fibers for suture materials, films for catheter balloons, knitted fabrics for blood vessel prostheses, and solid forms for cosmetic and weight-bearing applications.

It should be recognized that different applications require different materials with specific properties. The following are minimal requirements for all soft tissue implant materials.

1. They should achieve a reasonably close approximation of the physical properties, especially flexibility and texture.
2. They should not deteriorate or change properties after implantation over time.
3. If materials are designed for degradation, rate and modes of degradation should follow the intended pathway.
4. They should not cause adverse tissue reaction.
5. They should be non-carcinogenic, non-toxic, non-allergenic, and non-immunogenic.
6. They should be sterilizable.
7. They should be low cost.

Other important factors include the feasibility of mass production and aesthetic qualities.

11.1. SUTURES, SURGICAL TAPES, AND ADHESIVES

One of the most common soft tissue implants is suture. Sutures are used to close wounds due to injury or surgery. In recent years, many new surgical tapes and tissue adhesives have been added to the surgeon's armamentarium. Although their use in actual surgery is limited to some surgical procedures, they are indispensable.

11.1.1. Sutures

There are two types of sutures according to their physical in-vivo integrity with time: absorbable (biodegradable) and nonabsorbable. They may be also distinguished according to their source of raw materials, that is, natural sutures (catgut, silk, and cotton), and synthetic sutures (nylon, polyethylene, polypropylene, stainless steel, and tantalum). Sutures may also be classified according to their physical form — monofilament and multifilament.

The various types of sutures are summarized in Table 11-1. The absorbable suture, catgut, is made of collagen derived from sheep intestinal submucosa. It is usually treated with a chromic salt to increase its strength and is crosslinked to retard resorption. Such treatment extends the life of catgut suture from 3–7 days up to 20–40 days. Synthetic sutures absorb more slowly than the catgut. Most synthetic sutures are made from PGA and its copolymer with PLLA to control absorption and flexibility for handling, as given in Table 11-2. The weight loss is directly related to the strength change, as shown in Figure 11-1. Time for essentially complete absorption is depicted in Figure 11-2. The catgut absorbs the fastest, while PDSII suture is the slowest. Table 11-3 gives initial strength data for catgut sutures according to their size. The catgut sutures are stored with needles in a physiological solution in order to prevent drying, which would make the sutures very stiff and hard and thus not easily usable.

Table 11-1. Various Types of Sutures Quoted by Roby and Kennedy

Suture type	Generic structure	Major clinical application	Representative Type[a]	Representative product	Representative manufacturer
Natural materials					
Catgut	Protein	Plain: subcutaneous, rapid-healing tissues, ophthalmic	T	Surgical gut	Ethicon
			T	Surgical Gut	Ethicon
		Chromic: Slower-healing tissues	T	Chromic, plain gut	Syneture
Silk	Protein	General suturing, ligation	B	Perma-Hand	Ethicon
			B	Softsilk	Syneture
Synthetic nonabsorbable materials					
Polyester	PET	Heart valves, vascular prostheses, general	B	Ethibond Excel	Ethicon
			B	Surgidac	Syneture
			B	Ti-Cron	Syneture
			B	Tevdek	Teleflex
	Polybutester	Plastic, cuticular	M	Novafil	Syneture
		Cardiovascular	M	Vascufil	Syneture
Polypropylene PP		General, vascular anastomosis	M	Prolene	Ethicon
			M	Surgipro	Syneture
			M	Surgipro II	Syneture
			M	Deklene II	Teleflex
Polyamide	Nylon 6, 6,6	Skin, microsurgery, tendon	M	Ethilon	Ethicon
			M	Monsof	Syneture
			M	Dermalon	Syneture
		B	Nurolon	Ethicon	
			B	Surgilon	Syneture
Stainless steel	CrNiFe alloy	Abdominal and sternal closures, tendon	M, T	Ethisteel	Ethicon
			M, T	Steel	Syneture
			M, T	Flexon	Syneture
Fluoropolymers	ePTFE	General, vascular anastomosis	M	Gore-Tex	W.L. Gore
	PVF/PHFP		M	Pronova	Ethicon
Synthetic absorbable materials					
Braids	PGA/PLLA	Peritoneal, fascial, subcutaneous	B	Vicryl	Ethicon
	PGA/PLLA		B	Vicryl Rapide	Ethicon
	PGA/PLLA		B	Panacryl	Ethicon
	PGA/PLLA		B	Polysorb	Syneture
	PGA		B	Dexon	Syneture
	PGA		B	Bondek	Teleflex
Monofilaments					
	PDO	Application dependent on tensile strength loss profile required	M	PDS II	Ethicon
	PGAIPCL		M	Monocryl	Ethicon
	PGA/PTMC/PDO		M	Biosyn	Syneture
	PGA/PTMC		M	Maxon	Syneture
	PGA/PCL/ PTMC/PLLA		M	Caprosyn	Syneture

[a]: T, Twisted monofilament; M, monofilamene; B, multifilament braid

Table 11-2. Polymer Composition of Synthetic Absorbable Sutures

Suture	Block structure	Polymer composition (%)
Multifilament braids		
Dexon	PGA homopolymer	
Vicryl	PGA/PLLA random copolymer	90/10
Polysorb	PGA/PLLA random copolymer	90/10
Panacryl	PGA/PLLA random copolymer	3/97
Monofilaments		
PDS II	PDO homopolymer	–
Maxon	PGA–PTMC/PGA-PGA	100-85/15-100
Monocryl	PGA–PCL/PGA–PGA	100-45/55-100
Biosyn	PGA/PDO–PTMC/PDO–PGA/PDO	92/8-65/35-92/8
Caprosyn	PGAIPCL/PTMC/PLLA random copolymer	70/16/8/5

Reprinted with permission from Roby and Kennedy (2004). Copyright © 2004, Elsevier.

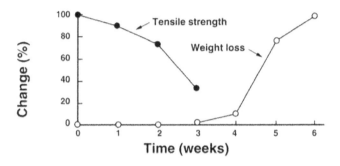

Figure 11-1. Weight and tensile strength loss after implantation of Vicryl® suture. Reprinted with permission from Fredericks et al. (1984). Copyright © 1984, Interscience Publishers.

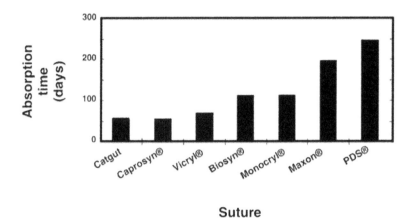

Figure 11-2. Complete absorption times for various sutures. Reprinted with permission from Roby (1998). Copyright © 1998, Sage Publications.

Table 11-3. Minimum Breaking Loads for British-Made Catgut

	Diameter (mm)		Minimum breaking load (lbf)	
Size	Minimum	Maximum	Straight pull	Over knot
7/0	0.025	0.064	0.25	0.125
6/0	0.064	0.113	0.5	0.25
5/0	0.113	0.179	1	0.5
4/0	0.179	0.241	2	1
3/0	0.241	0.318	3	1.5
2/0	0.318	0.406	5	2.5
0	0.406	0.495	7	3.5
1	0.495	0.584	10	5
2	0.584	0.673	13	6.5
3	0.673	0.762	16	8
4	0.762	0.864	20	10
5	0.864	0.978	25	12.5
6	0.978	1.105	30	15
7	1.105	1.219	35	17.5

It is interesting to note that the stress concentration at a surgical knot decreases the suture strength of catgut by half, no matter what kind of knotting technique is used. It is suggested that the most effective knotting technique is the square knot with three ties to prevent loosening. According to one study, there is no measurable difference in the rate of wound healing whether the suture is tied loosely or tightly. Therefore, loose suturing is recommended because it lessens pain and reduces cutting into soft tissues.

Catgut and other absorbable sutures [e.g., copolymer of poly (glycolic acid) and (lactic acid)] induce tissue reactions, although the effect diminishes as they are being absorbed. This is true of other natural, nonabsorbable sutures like silk and cotton, which *showed more* reaction than synthetic sutures like polyester, nylon, polyacrylonitrile, etc., as shown in Figure 11-3. As is the case of the wound-healing process (discussed in Chapter 10), the cellular response is most intensive one day after suturing and subsides in about a week.

As for the risk of infection, if the suture is contaminated even slightly, the incidence of infection increases many fold. The most significant factor in infection is the chemical structure, not the geometric configuration of the suture. Polypropylene, nylon, and PGA/PLA sutures develop lesser degrees of infection than sutures made of stainless steel, plain, and chromic catgut, and polyester. The ultimate cause of infection is a pathogenic microorganism not the biomaterial. The role of suture in infection is to provide a conduit for ingress of bacteria, to chemically or physically modify the body's immune response, or to provide an environment favorable to bacterial growth.

Example 11-1

Compare the breaking strength of catgut sutures (Table 11-3) of sizes between 7/0 and 7. What conclusion can you draw?

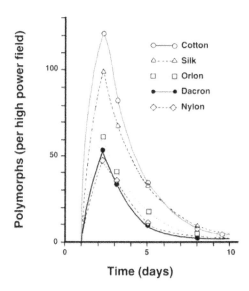

Figure 11-3. Cellular response to sutured materials. Reprinted with permission from Prostlethwait (1970). Copyright © 1970, Martin Memorial Foundation.

Answer

The breaking strength of catgut sutures 7/0 and 7 is calculated by dividing the breaking load with the cross-sectional area:

$$\sigma_{70} = 0.25 \times 4.448 \text{ N}/(\pi \times 0.125 \text{ mm}^2) = \underline{2.3 \text{ GPa}} \text{ (remember that 1 lbf = 4.448 N)},$$

$$\sigma_7 = 35 \times 4.448 \text{ N}/(\pi \times 0.52 \text{ mm}^2)$$

$$= \underline{180 \text{ MPa}}.$$

There is a tremendous increase in strength (more than tenfold) by making smaller-diameter suture, mainly due to orientation of polymer chains in the draw direction and decreased defect size in the smaller-diameter suture. Why do we not make multifilament catgut sutures?

11.1.2. Surgical Tapes and Staples

Surgical tapes are intended to offer a means of closing surgical incisions while avoiding pressure necrosis, scar tissue formation, problems of stitch abscesses, and weakened tissues. The problems with surgical tapes are similar to those experienced with band-aids, that is, (1) misaligned wound edges, (2) poor adhesion due to moisture or dirty wounds, (3) late separation of tapes when hematoma, wound drainage, etc. occur.

The wound strength and scar formation in the skin may depend on the type of incision made. If the subcutaneous muscles in the fatty tissue are cut and the overlying skin is closed with tape, the muscles retract. This in turn increases the scar area, resulting in poor cosmetic appearance when compared to a suture closure. However, due to the higher strength of scar tissue, the taped wound has higher wound strength than the sutured wounds only if the muscle is not cut. Because of this, tapes have not enjoyed the success that was anticipated when they were first introduced.

Tapes have been used successfully for assembling scraps of donor skin for skin graft, connecting nerve tissues for neural regeneration, etc.

Staples made of metals (Ta, stainless steel, and Ti–Ni alloy) can be used to facilitate closure of large surgical incisions produced in procedures such as Caesarean sections, intestinal surgery, and surgery for bone fractures. The tissue response to the staples is the same as that of synthetic sutures, but they are not used in places where aesthetic outlook is important. A self-compression NiTi alloy uses the shape memory effect, which retracts back to its austenitic phase after insertion into the broken bone and is heat to body temperature, as illustrated in Figure 11-4.

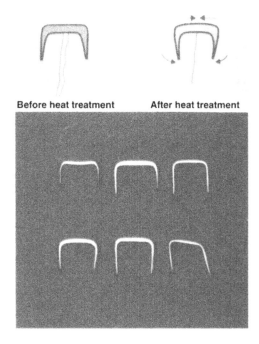

Before heat treatment　　**After heat treatment**

Figure 11-4. Staple for fractured bone before and after implant. Reprinted with permission from Haaster et al. (1990). Copyright © 1990, Butterworths-Heinemann.

11.1.3. Tissue Adhesives

The special environment of tissues and their regenerative capacity make the development of an ideal tissue adhesive difficult. Past experience indicates that the ideal tissue adhesive should be able to be wet and bond to tissues, should have adequate bond strength, be capable of rapid polymerization without producing excessive heat or toxic byproducts, be resorbable as the wounds heal, not interfere with the normal healing process, offer ease of application during surgery, be sterilizable, have an adequate shelf life, and allow ease of large-scale production.

The main strength of tissue adhesion comes from the covalent bonding between amine, carboxylic acid, and hydroxyl groups of tissues, and the functional groups of adhesives such as

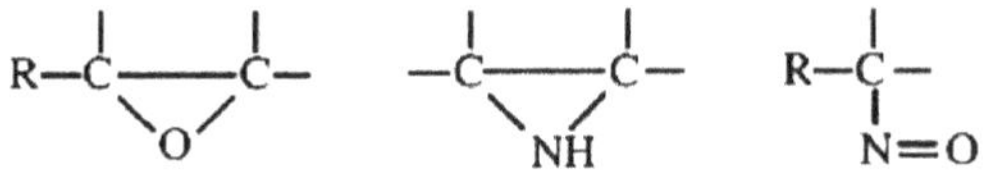

$$R-\overset{|}{\underset{O}{C}}\diagdown\overset{|}{\underset{}{C}}-\qquad -\overset{|}{\underset{NH}{C}}\diagdown\overset{|}{\underset{}{C}}-\qquad R-\overset{|}{\underset{N=O}{C}}- \qquad (11\text{-}1)$$

There are several adhesives available, of which alkyl-α-cyanoacrylate is best known. Among the homologs of alkyl-cyanoacrylate, the methyl- and ethyl-2-cyanoacylate are most promising. With the addition of some plasticizers and fillers, they are commercially known as Eastman 910®, Crazy Glue®, Super Glue®, etc. The methylcyanoacrylate has a similar chemical composition as methyl methacrylate (MMA), as shown in Figure 11-5. An interesting comparison is illustrated in Figure 11-6, which shows that the bond strength of adhesive treated wounds is about half that of a sutured wound after 10 days. Because of the lower strength and lesser predictability of the in-vivo performance of adhesives, the application is limited to use after trauma on fragile tissues such as spleen, liver, and kidney or after an extensive surgery on soft tissues such as lung. The topical use of adhesives in plastic surgery and fractured teeth has been moderately successful. As with many other adhesives, the end results of the bond depend on many variables — such as thickness, open porosity, and flexibility of the adhesive film, as well as rate of degradation. Some have tried to use adhesives derived from fibrinogen, which is one of the clotting elements of blood. Figure 11-7 shows the relationship between fibrin concentration and adhesive shear strength. The fibrin-based adhesives have sufficient strength (0.1 MPa) and elastic modulus (0.15 MPa) to sustain adhesiveness for the anastomoses of nerve, microvascular surgery, dural closing, skin and bone graft fixation, and other soft tissue fixation.

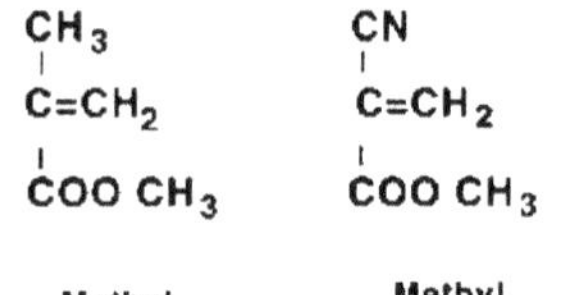

$$\begin{array}{cc}
\underset{|}{CH_3} & \underset{|}{CN} \\
\underset{|}{C=CH_2} & \underset{|}{C=CH_2} \\
COO\,CH_3 & COO\,CH_3 \\[4pt]
\textbf{Methyl} & \textbf{Methyl} \\
\textbf{methacrylate} & \textbf{cyanoacrylate}
\end{array}$$

Figure 11-5. Chemical structure of methyl methacrylate (MMA) and methyl cyanoacrylate. Polymerization takes place along double bonds ($C=CH_2$) similar to vinyl polymerization.

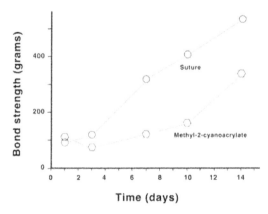

Figure 11-6. Bond strength of wounds with different closure materials. Reprinted with permission from Houston et al. (1969). Copyright © 1969, Wiley.

Table 11-4. Mechanical Properties of Dental Cements and Sealants

Materials	Compressive strength (MPa)	tensile strength (MPa)	Modulus (GPa)	Toughness K_{1C} (MPa m$^{1/2}$)
Zinc phosphate	80–100	5–7	13	~0.2
Zinc polycarboxylate	55–85	8–12	5–6	0.4–0.5
Glass ionomer	70–200	6–7	7–8	0.3–0.4
Resin sealant unfilled	90–100	20–25	2	0.3–0.4
Resin sealant filled	150	30	5	–
Resin cement	100–200	30–40	4–6	–
Composite resin filling material	350–400	45–70	15–20	1.6

Reprinted with permission from Smith (2004). Copyright © 2004, Elsevier.

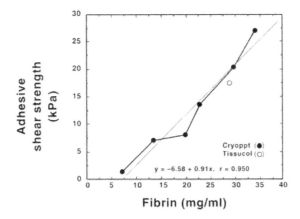

Fibrin (mg/ml)

Figure 11-7. Relationship between fibrin concentration and adhesive shear strength. Reprinted with permission from Feldman et al. (2000). Copyright © 2000, Marcel Dekker.

Table 11-4 summarizes the tissue adhesives used in both soft and hard tissues. Dental adhesives have been developed for their importance of sealing or adhering fissures, sealing after pulpectomy, cavity implants sealing, etc. The potential debonding pathway or failure modes within the dentin–adhesive–resin resin composite bonded joint is shown in Figure 11-8. As with other adhesives, bond strength decreases with aging, as depicted in Figure 11-9 for 1 and 6 months. The initial tight adhesion between the dentin surfaces–adhesive–dental composite can be compromised with microcrack formations due to trapped voids, monomer vapor, etc. The mechanical properties of dental cements and sealants are given in Table 11-4. The composite resin filling materials are discussed in §8.3.1. As in soft tissue adhesion, adhesion to hard tissues is primarily via calcium ions (Ca^{++}), as shown in Figure 11-10. Similar bonding mechanisms may take place if one uses similar cements and adhesives for bone bonding. However, this type of tight bonding was not successful in attaching for broken long bones. One of the main reasons for failure of bonding is the viability of tissues by the bonding due to almost complete separation by an adhesive or sealant layer. If one could make these cements or adhesives with cell-communicating capability through the adhesive layer, it might become a viable solution.

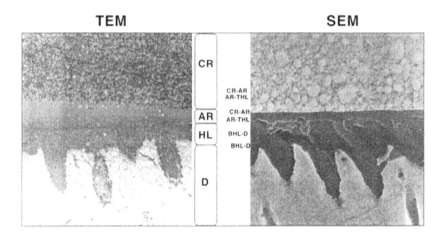

Figure 11-8. TEM and SEM pictures of joint after dentin–adhesive resin–resin composite bonding. Courtesy of S.R. Armstrong, University of Iowa, 1998.

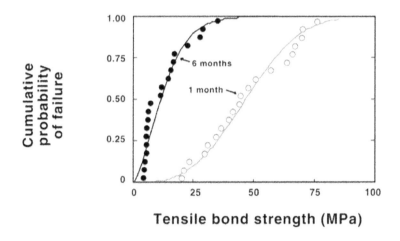

Figure 11-9. Cumulative probability of failure distributions for dentin–adhesive resin–resin composite bonds after 1 and 6 months in water storage. Courtesy of S.R. Armstrong, University of Iowa, 1998.

Example 11-2

A nylon suture was implanted in the abdominal cavity of a dog. The suture was removed after 10 days, and a second piece of the same suture was removed after 20 days, and its average tensile strength was measured. The strength decreased by 40 and 50%, respectively. How long will it take for the strength to decay 60% of its original strength? Assume an exponential decay of strength.

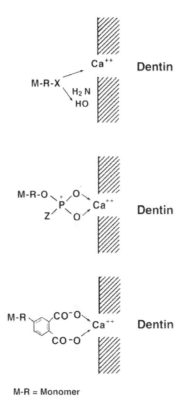

M-R = Monomer

Figure 11-10. Bonding of dental resin monomers with bone via Ca". Reprinted with permission from Asmussen et al. (1989). Copyright © 1989, Elsevier Science.

Answer

Since the strength decreases exponentially, we can assume

$$\frac{\sigma_t}{\sigma_0} = A\exp(-Bt),$$

where A and B are constants, t is time (days), σ_t is the strength at time t, and σ_0 is the original strength.

Therefore,

$$0.6 = A\exp(-10B),$$

$$0.5 = A\exp(-20B).$$

By solving simultaneously

$$\frac{\sigma}{\sigma_0} = 0.72\exp(-0.018t),$$

$$0.4 = 0.72\exp(-0.018t),$$

$$t = \underline{33\ days}.$$

11.2. PERCUTANEOUS AND SKIN IMPLANTS

Percutaneous (trans, or through the skin) implants are used in the context of artificial kidneys and hearts, and to allow prolonged injection of drugs and nutrients. Artificial skin (or dressing) can be used to maintain the hydration and body temperature of severely burned patients. Actual permanent replacement of skin by biomaterials is beyond the capability of today's technology.

11.2.1. Percutaneous Devices

The problem of obtaining a functional and a viable interface between the tissue (skin) and an implant (percutaneous) device is primarily due to the following factors. First, although initial attachment of the tissue into the interstices of the implant surface occurs, it cannot be maintained for a long period of time, since the dermal tissue cells turn over continuously and dynamically. Furthermore, downgrowth of epithelium around the implant (extrusion) or overgrowth of implant (invagination) occurs. Second, any openings large enough for bacteria to infiltrate may result in *infection* even though initially a complete sealing between skin and implant is achieved.

Many variables and factors are involved in the development of percutaneous devices. These are:

1. End-use factors
   a. Transmission of information: biopotentials, temperature, pressure, blood flow rate, etc.
   b. Energy: electrical and electromagnetic stimulation, power for heart assist devices, cochlear implants, etc.
   c. Matter: cannula for kidney dialysis and blood infusion or exchange, etc.
   d. Load : attachment of prosthesis.
2. Engineering factors
   a. Materials selection: polymers, ceramics, metals, and composites.
   b. Design variations: button, tube with and without skirt, porous or smooth surface, etc.
   c. Mechanical stresses: soft and hard interface, porous or smooth interface.
3. Biological factors
   a. Implant host: man, dog, hog, rabbit, sheep, etc.
   b. Implant location: abdominal, dorsal, forearm, etc.
4. Human factors
   a. Postsurgical care.
   b. Implantation technique.
   c. Aesthetic outlook.

Figure 11-11 shows a simplified cross-sectional view of a generalized percutaneous device (PD), which can be broken down into five regions:

A. Interface between the epidermis and PD, which should be completely sealed against invasion by foreign organisms.
B. Interface between the dermis and PD, which should reinforce the sealing of (A), as well as resist mechanical stresses. Due to the relatively large thickness of the dermis, the mechanical aspect is more important at this interface.
C. Interface between the hypodermis and PD should reinforce the function of (B). Immobilization of the PD against piston action is a primary function of (C).

D. Implant material per se should meet all the requirements of an implant for soft tissue replacement.

E. The line where epidermis, air, and PD meet is called a three-phase line, similar to (A).

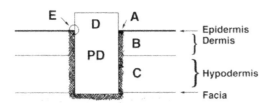

Figure 11-11. Simplified cross-sectional view of PD–skin interfaces. Reprinted with permission from von Recum and Park (1979). Copyright © 1979, Chemical Rubber Co.

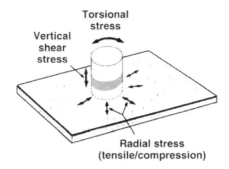

Figure 11-12. Various mechanical stresses acting at the PD–skin interface. Reprinted with permission from von Recum and Park (1979). Copyright © 1979, Chemical Rubber Co.

The stresses generated between a cylindrical percutaneous device and skin tissue can be simplified, as shown in Figure 11-12. The relative motion of the skin and implant results in shear stresses that can be avoided if the implant floats (or moves) freely with movement of the skin. For this reason PDs without connected leads or catheters function longer. There have been many different PD designs to minimize shear stresses. All designs have centered around creating a good skin tissue/implant attachment in order to stabilize the implant. This is done by providing felts, velours, and other porous materials at the interface. Figure 11-13 shows a design to minimize the transfer of stresses and strains to the skin. The device includes making an air chamber made of a rubber balloon (a) interposed between skin and PD, and firmer fixation of the cannula by providing a large surface for tissue ingrowth (b and c). Some designs have tried to minimize the trauma imposed by the external tubes and wires by providing a pin connector with good provision for firm tissue attachment subcutaneously.

There have been no percutaneous devices that are completely satisfactory. Nevertheless, some researchers believe that hydroxyapatite may be a solution to the problem. In one experimental trial, hydroxyapatite based PDs showed very little epidermal downgrowth (1 mm after

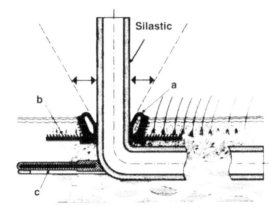

Figure 11-13. Schematic drawing of a Grosse-Siestrup PD. Courtesy C. Grosse-Siestrup.

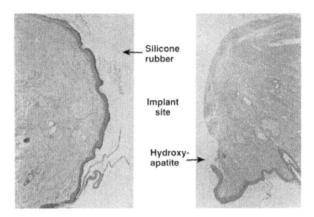

Figure 11-14. Histological view of the canine dermal tissues adjacent to the percutaneous device made of hydroxyapatite (left) and silicone rubber (right) 3 months after implantation (100× magnification). Reprinted with permission from Aoki et al. (1972). Copyright © 1972, Institute of Electrical and Electronics Engineers.

17 months versus 4.6 mm after 3 months for the silicone rubber control specimens in dorsal skin of canines, see Figure 11-14) and a high level of success rate (over 80% versus less than 50% for the control). In this context, success refers to patency of the device and freedom from infection, not to any improvement in health due to the device. The amino acid contents of the tissue capsules formed over the subcutaneous implants of the same materials showed that the hydroxyapatite site had the same composition as the periosteum of the femur, while the control site showed a similar composition to that found in pathological tissues. Some researchers have tried to switch to subcutaneous implants, which can be accessed by a needle for peritoneal dialysis, as illustrated in Figure 11-15.

11.2.2. Artificial Skins and Burn Dressing

Artificial skin can be thought of as a percutaneous implant, so that the problems are similar to those described in the previous section. Most useful for this application is a material that can adhere to a large (burned) surface and thus prevent loss of fluids, electrolytes, and other biomolecules until the wound has healed. Figure 11-16 illustrates the degree of burns and their depths. First- and second-degree burns can be treated with temporary burn covering membranes or dressing, while third-degree burns can be treated with autografts at present. Although a permanent skin implant could benefit those who have lost skin from burns or injury, this is a long way from being realized for the same reasons given in the case of percutaneous implants proper. Presently, autografting and homografting (skin transplants) are available as a permanent solution. Table 11-5 summarizes some commonly used wound membranes and their principal characteristics. Figure 11-17 illustrates various synthetic membranes.

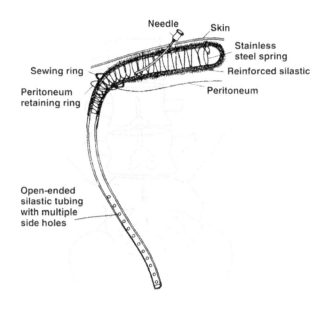

Figure 11-15. Subcutaneous peritoneal dialysis assist device. Reprinted with permission from Kablitz et al. (1979). Copyright © 1979, Blackwell Science.

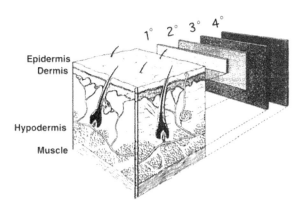

Figure 11-16. Degrees of burn relative to depth of skin. Reprinted with permission from Morgan et al. (2004). Copyright © 2004, Elsevier Science.

Table 11-5. Some Commonly Used Wound Membranes and Their Principal Characteristics

Membrane	Selected characteristics
Temporary	
Porcine xenograft	Adheres to coagulum, excellent pain control
Biobrane[a]	Bilaminate, fibrovascular ingrowth into inner layer
Split-thickness allograft	Vascularizes and provides durable temporary closure
Various semipermeable membranes	Provides vapor and bacterial barrier
Various hydrocolloid dressings	Provides vapor and bacterial barrier, absorbs exudate
Various impregnated gauzes	Provides barrier while allowing drainage
Allogeneic dressings	Provides temporary cover while supplementing growth factors
Permanent	
Epicel[b]	Provides autologous epithelial layer
Integra[c]	Provides scaffold for neodermis, requires delayed thin autograft grafting
AlloDermd	Consists of cell-free human dermal scaffold, requires immediate thin autograft

[a]Mylan Laboratories, Inc. [b]Genzyme Biosurgery Inc., Cambridge, MA. [c]Integra Life Sciences Corporation, Plainsboro, NJ. [d]LifeCell Inc.. Branchburg, NJ.

Figure 11-17. Various skin burn membranes from bottom left-hand corners: meshed split-thickness autograft, TransCyte, Epicel, cryopreserved cadaver allograft, Biobrane, split-thickness autograft, EZ Derm, and Integra Dermal Regeneration Template. Reprinted with permission from Morgan et al. (2004). Copyright © 2004, Elsevier Science.

In one study, wound closure (burn dressing) was achieved by controlling the physico-chemical properties of the wound-covering material (membrane). Six ways were suggested to improve certain physicochemical and mechanical requirements necessary in the design of artificial skin. These are shown schematically in Figure 11-18. Biomechanical and chemical analysis conducted in this study led to the design of a crosslinked collagen–polysaccharide (chondroitin 6-sulfate) composite membrane chosen for ease in controlling porosity (5–150 μm in diameter), flexibility (by varying crosslink density), and moisture flux rate.

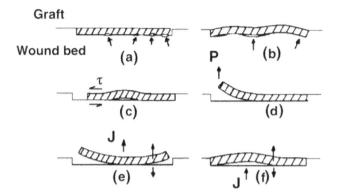

Figure 11-18. Schematic representation (not drawn according to scale) of certain physico-chemical and mechanical requirements in the design of an effective wound closure. (**a**) Skin graft (cross-hatched) does not displace air pockets (arrows) efficiently from the graft–wound bed interface. (**b**) Flexural rigidity of graft is excessive; graft does not deform sufficiently under its own weight to make contact with depressions in wound bed surface, resulting in air pockets (arrows). (**c**) Shear stresses (arrows) cause buckling of graft, ruptures of graft–wound bed bond, and formation of air pocket. (**d**) Peeling force P lifts graft away from wound bed. (**e**) Excessively high moisture flux rate through graft causes dehydration and development of shrinkage stresses at edges (arrows), which cause lift-off from wound bed. (**f**) Very low moisture flux J causes accumulation (edema) at graft–wound bed interface and peeling off (arrows). Reprinted with permission from Yannas and Burke (1980). Copyright © 1980, Wiley.

Several polymeric materials, including reconstituted collagen, have also been tried as burn dressings. Among them are the copolymers of vinyl chloride and acetate and methyl-2-cyanoacrylate. Methyl-2-cyanoacrylate was found to be too brittle and histotoxic for use as a burn dressing. The ingrowth of tissue into the pores of sponge (Ivalon®, polyvinyl alcohol) and woven fabric (nylon and silicone rubber velour) was also attempted without much success. Sometimes plastic tapes have been used to hold skin grafts during microtoming (ultrathin sectioning) and grafting procedures. For severe burns, immersion of the patient into silicone fluid was found to be beneficial for prevention of early fluid loss, decubitus ulcers, and reduction of pain.

Rapid epithelial (epidermal) layer growth by culturing cells in vitro from the skin of the burn patient for covering the wound area may offer a better solution. However, the multiplication of epidermal layer through tissue engineering has been available but not as popular as first hoped. Methods for growing and implanting whole dermis are not available as of yet.

11.3. MAXILLOFACIAL AND OTHER SOFT-TISSUE AUGMENTATION

In the previous section we dealt with problems associated with wound closing and wound/tissue interfacial implants. In this section we study (cosmetic) reconstructive implants. Although soft-tissue implants can be divided into (1) space filler, (2) mechanical support, and (3) fluid carrier or storage device, most have two or more combined functions. For example, breast implants fill space and provide mechanical support.

11.3.1. Maxillofacial Implants

There are two types of maxillofacial implant (often called prosthetics, which implies extracorporeal attachment) materials: extraoral and intraoral. The latter is defined as "the art and science of anatomic, functional or cosmetic reconstruction by means of artificial substitutes of those regions in the maxilla, mandible, and face that are missing or defective because of surgical intervention, trauma, etc."

There are many polymeric materials available for extraoral implants, which require: (1) color and texture should be matched with that of patients, (2) it should be mechanically and chemically stable, i.e., should not creep or change color or irritate the skin, and (3) it should be easily fabricated. Polyvinyl chloride and acetate (5–20%) copolymers, polymethylmethacrylate, silicone, and polyurethane rubbers are currently used.

The requirements for intraoral implants are the same as for other implant materials since they are in fact implanted. For maxilla, mandibular, and facial bone defects, metallic materials such as tantalum, titanium, and Co–Cr alloys, etc. are used. For soft tissues like gum and chin, polymers such as silicone rubber, polymethylmethacrylate, etc. are used for augmentation.

The use of injectable silicones that polymerize in situ has proven partially successful for correcting facial deformities. Although this is obviously a better approach in terms of minimal initial surgical damage, this procedure was not accepted due to the tissue reaction and the eventual displacement or migration of the implant. The use of collagen paste as a space-filling material for cosmetic purposes has a similar drawback: the collagen can be resorbed by the body or it can migrate. This leads many such patients to seek repeated collagen injections.

11.3.2. Ear and Eye Implants

The external ear serves to gather sound, but replacement of a damaged or diseased external ear is done principally for cosmetic reasons. Such replacement is considered a maxillofacial reconstruction, considered above.

As for the middle ear, conduction of sound depends on the ossicular chain of small bones (malleus, incus, and stapes). The use of implants can restore the conductive hearing loss associated with partial or complete impairment of the ossicles. Such impairment can result from otosclerosis (a hereditary defect that involves a change in the bony tissue of the ear) and chronic otitis media (inflammation of the middle ear). Many different prostheses are available to correct the defects, some of which are shown in Figure 11-19. The porous polyethylene total ossicular replacement implant is used to obtain a firm fixation of the implant by tissue ingrowth. The tilt-top implant is designed to retard tissue ingrowth into the section of the shaft, which may diminish sound conduction. More modern versions of ossicular implants are illustrated in Figure 11-20.

As for the inner ear, the cochlea is a fluid-filled spiral structure in which sound of different frequency excites nerve endings linked to the brain. Deafness due to disease of the inner ear has been treated with cochlear implants. These ear implants, which enable the user to hear speech, have been developed and have become quite widely used worldwide. The inner, middle, and outer ear are schematically shown in Figure 11-21. The cutaway view of the cochlea showing the fluid-filled chambers of the inner ear, scala tympani, scala media, and scala vestibuli is shown in Figure 11-22. In cochlear implants, sound waves are collected, amplified, and processed before the signals are conducted to the cochlear nerve endings. Placement of the stimulating electrodes is shown in the enlarged view in Figure 11-22. There are more than 30,000 nerve fibers in each ear. In implant designs, one typically uses only 20 or so stimulating conductors, made mostly of platinum and other noble metal alloys. Figure 11-23 shows an

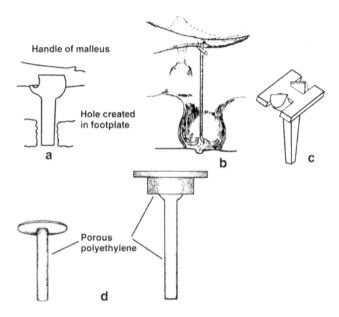

Figure 11-19. Prostheses for the reconstruction of ossicles. (**a**) PTFE 'piston' stapes prosthesis of Shea. [Reprinted with permission from Shea et al. (1962). Copyright © 1962, American Medical Association.] (**b**) Incus replacement prosthesis of Sheehy. [Reprinted with permission from Sheehy (1969). Copyright © 1969, W.B. Saunders.] (**c**) Tabor prosthesis for replacement of whole ossicular chain. [Reprinted with permission from Tabor (1970). Copyright © 1970, American Medical Association.] (**d**) Porous polyethylene total ossicular replacement prosthesis. [Reprinted with permission from Yannas and Burke (1980). Copyright © 1980, Wiley.] (**e**) Same as (**d**) except the stem can be tilted. [Reprinted with permission from Yannas and Burke (1980). Copyright © 1980, Wiley.]

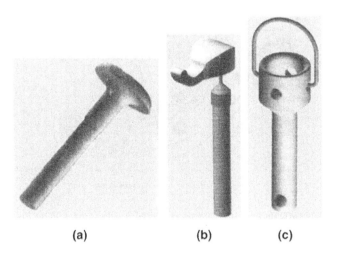

Figure 11-20. Examples of (**a**) partial ossicular prosthesis (PORP), (**b**) incus stapes replacement prosthesis (ISRP), and (**c**) stapes prosthesis. Courtesy of Medtronic ENT.

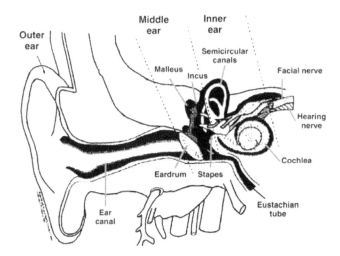

Figure 11-21. Schematic representation of the inner, middle, and outer ear. Courtesy of V.M. Vrockel, Hearing Research Center, University of Washington, Seattle.

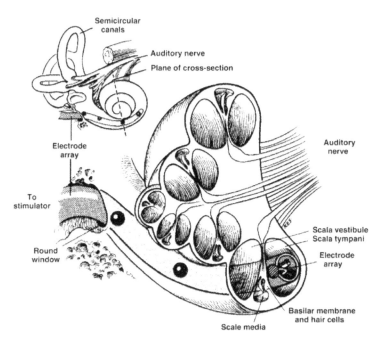

Figure 11-22. Cutaway view of the cochlea showing the fluid-filled chambers of the inner ear: scala tympani, scala media, and scala vestibuli. Also, the placement of electrode is shown in the enlarged view. Reprinted with permission from Spelman (1988). Copyright © 1988, Wiley.

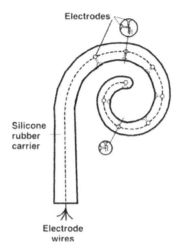

Electrodes

Silicone
rubber
carrier

Electrode
wires

Figure 11-23. Schematic view of a cochlear-stimulating electrode. Reprinted with permission from Spelman (1988). Copyright © 1988, Wiley.

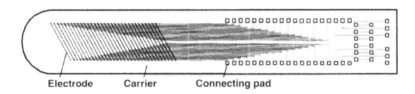

Electrode   Carrier   Connecting pad

Figure 11-24. Layout of blade electrodes of a cochlear implant. Reprinted with permission from Spelman (1988). Copyright © 1988, Wiley.

electrode resembling the shape of the cochlea. The blade array of electrodes is arranged as shown schematically in Figure 11-24. Figure 11-25 shows the field produced by current (0.1 mA) driven through a conducting strip that is mounted on an insulator that faces a conductive, homogenous, isotropic medium. Figure 11-26 shows potential versus strip-line position referring to Figure 11-25. The strip-line lies on an insulator whose resistivity is 160 times that of the conducting medium and produces a current of 0.1 mA.

It is estimated that about 43,000 cochlear implants were implanted by 2001, and the number is growing fast. As with other implants the success rate varies between 40 to 90% according to the test noise level, brand, and who collected the data. These types of cochlear implants have electrodes that stimulate the cochlear nerve cells. The electrodes are insulated individually and connected to nerve endings. The electrodes are made of noble metals and their alloys (Pt, Pt–10%Ir), which would have the most corrosion resistance, least tissue reaction, and threshold potential elevation. However, the release of metal ions, however minute, may affect the performance of implants, but not much is yet known. The implant also has a transducer that transforms sound into electrical impulses and a frequency-selective amplifier tuned to the frequency range associated with speech. Electrical impulses can be conducted through coupled external and internal coils, as shown in Figure 11-27. The electrical impulses can also be transmitted directly by means of a percutaneous device (PD). Most widely used is the LTI pyrolytic carbon-coated graphite substrate similar to the heart valve disc. The center of the PD has a hole, which permits one to pass electrodes through. Usually the ends of electrodes are embedded on the PD, and another connector is attached to minimize stress on the PD and making it possible to detach electrodes from the PD.

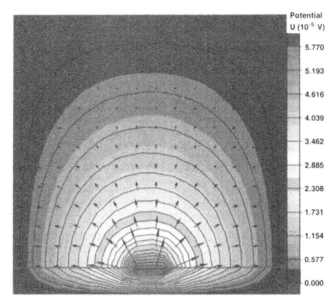

Figure 11-25. The field produced by current (0.1 mA) driven through a superconducting strip that is mounted on an insulator that faces a conductive, homogenous, isotropic medium. Reprinted with permission from Spelman (1988). Copyright © 1988, Wiley.

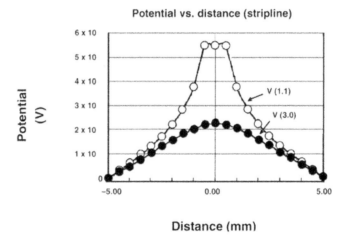

Figure 11-26. Potential vs. stripline of Figure 11-23. The stripline lies on an insulator whose resistivity is 160 times that of the conducting medium and produces a current of 0.1 mA. Reprinted with permission from Spelman (1988). Copyright © 1988, Wiley.

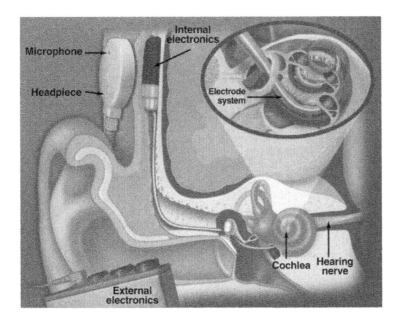

Figure 11-27. Basic components of cochlear implants. Courtesy of Advanced Bionics, Sylmar, CA.

Ear implants have been fabricated using many different materials: polymethylmethacrylate, polytetrafluoroethylene, polyethylene, silicone rubber, stainless steel, and tantalum. A polytetrafluoroethylene–carbon composite (Proplast®), porous polyethylene (Plastipore®), and pyrolytic carbon (Pyrolite®) have been shown to be suitable materials for cochlear (inner ear) implants.

Eye implants are used to restore the functionality of the cornea and lens when they are damaged or diseased. The basic structure of the eye is shown in Figure 11-28. Contact lenses, both soft and hard, are not implants and were discussed earlier in the context of oxygen permeability of elastomers (§7.3.6).

The cornea, if diseased or damaged, is usually transplanted from a suitable donor rather than implanted since the longevity of the cornea implant is uncertain due to fixation problems and infection. Figure 11-29 shows some of the eye implants tried clinically. They are made from "transparent" acrylics, especially polymethylmethacrylate, which has a comparatively high refractive index (1.50). The epikeratphakia procedure and epikeratoprosthesis (artificial epithelium) can be utilized, as shown in Figure 11-30. Figure 11-31a shows a schematic representation of intracorneal implants used to change the curvature of the cornea in refractive keratoplasty and an intrastromal hydrogel intracorneal implant. Figure 11-31b shows an intrastromal corneal ring. Surgery on the cornea is also done in treatments for refractive errors, typically myopia. For example, the Lasik and Visx methods use a laser to reshape the cornea. A flap is cut in the cornea and is folded back, revealing the stroma, the middle section of the cornea. Pulses from a laser vaporize a portion of the stroma, and the flap is replaced. The related Ladar approach uses an eye tracker to stabilize the laser beam. The objective is to reduce a person's dependence upon eyeglasses or contact lenses.

In cataracts, the lens of the eye becomes cloudy; the lens can then be removed surgically. The lost optical power can be restored with thick-lens spectacles, but these cause distortion and

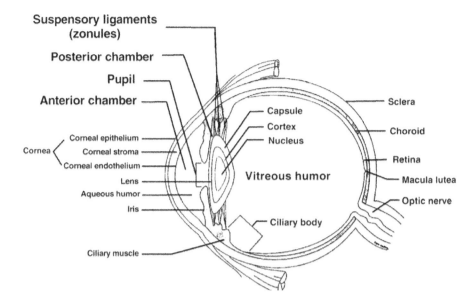

Figure 11-28. Structure of the eye. Reprinted with permission from Refojo (2004). Copyright © 2004, Elsevier.

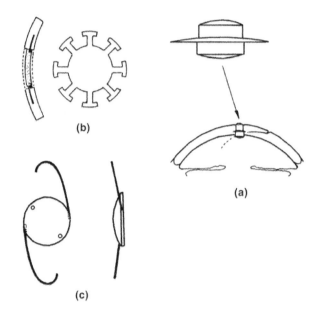

Figure 11-29. (**a**) Corneal implant of McPherson and Anderson. [Reprinted with permission from McPherson and Anderson (1953). Copyright © 1953, British Medical Association.] (**b**) Corneal implant of Cardona. [Reprinted with permission from Cardona (1962). Copyright © 1962, Elsevier Science.] (**c**) Intraocular lens. [Courtesy of Intra-Intermedics Inc., Pasadena, CA]

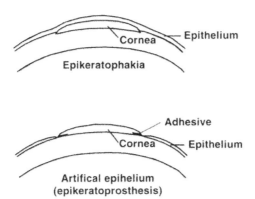

Figure 11-30. Schematic representation of corneal implants. Reprinted with permission from Refojo (2004). Copyright © 2004, Elsevier.

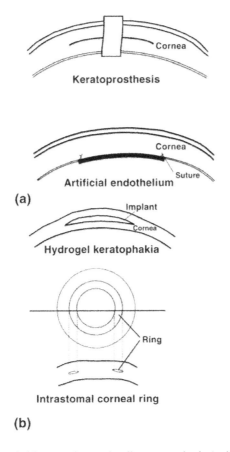

Figure 11-31. Epikeratphakia procedure and epikeratoprosthesis (artificial epithelium) (**a**) and schematic representation of intracorneal implants to correct the curvature of the cornea (**b**). Reprinted with permission from Refojo (2004). Copyright © 2004, Elsevier.

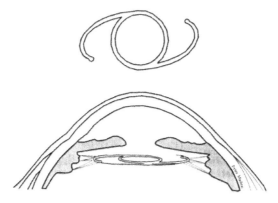

Figure 11-32. Schematic diagram showing placement of the IOL in the anterior segment of the eye. Reprinted with permission from Refojo (2004). Copyright © 2004, Elsevier.

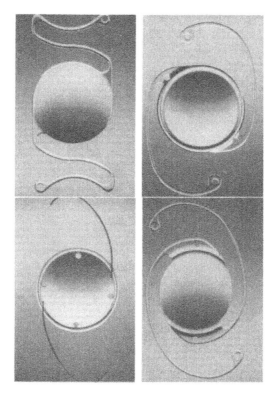

Figure 11-33. Various types of IOLs. Reprinted with permission from Obstbaum (1996). Copyright © 1996, Elsevier.

restriction of the field of view, and some people object to their appearance. Intraocular lenses (IOLs) are implanted surgically to replace the original eye lens, and they restore function without the problems associated with thick spectacles. Figure 11-32 shows placement of an IOL in the anterior segment of the eye. The various types of IOLs are given in Figure 11-33. The IOLs can be monofocal, multifocal, and foldable. The lens materials are polymethylmethacrylate (PMMA) and its copolymers, polysiloxane, and polyHEMA (hydroxy ethyl methacrylate, hydrogel, ~30% water content) with UV absorbers. The haptics can be PMMA, polyamide, and polypropylene with or without the UV absorbers. Problems of infection and fixation of the lens to the tissues can occur but have been substantially reduced by refinement of the technique. The intraocular lens can damage the soft structures to which it is attached, and it can become dislodged. Nevertheless, this type of cataract surgery has become commonplace and successful; many such implantation procedures are successfully conducted. As for the nerves of the eye, some researchers have tried to develop an artificial eye for people who have lost all the conductive functions of the optic nerve or of the retina. One such device provides stimulation to the brain cells, as shown in Figure 11-34. One of the major problems with this type of total organ replacement is the development of suitable electrode materials that will last a long time in vivo without changing their characteristics electrochemically. Another difficulty with an artificial eye is that significant image processing goes on in the retina. Consequently, simple electrical stimulation of the visual cortex of the brain yields a very poor image. It would be possible to have a prosthesis that could process different light waves that in turn can be interfaced directly to the optic nerves. However, such materials and technology of interfacing with the nerves are yet to be developed.

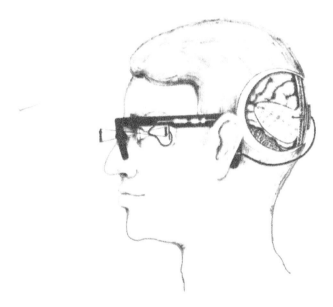

Figure 11-34. Diagram of the concept of an artificial eye. Television cameras in the glasses relate the message via microcomputers with radio waves to the array of electrodes on the visual cortex of the brain. Reprinted with permission from Dobelle et al. (1974). Copyright © 1974, American Association for the Advancement of Science.

11.3.3. Fluid Transfer Implants

Fluid transfer implants are used for cases such as hydrocephalus, urinary incontinence, and chronic ear infection. Hydrocephalus, caused by abnormally high pressure of the cerebrospinal

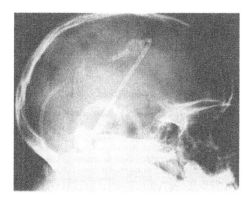

Figure 11-35. Postoperative x-ray picture of hydrocephalus shunt implantation.

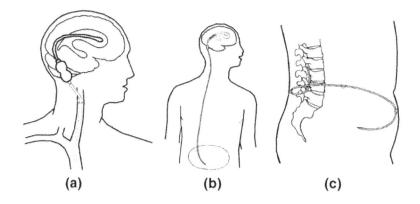

Figure 11-36. Various locations for emptying fluid from the brain: (**a**) ventriculoatrial, (**b**) ventriculoperitoneal, and (**c**) lumboperitoneal shunt system. Reprinted with permission from Rustamzadeh and Lam (2005). Copyright © 2005, Humana Press.

fluid in the brain, can be treated by draining the fluid (essentially an ultrafiltrate of blood) through a cannula. The postoperative x-ray radiograph is shown in as Figure 11-35. The fluid can be emptied in various locations, as shown in Figure 11-36. Various designs of shunt valves are shown in Figure 11-37. Programmable and differential pressure valves are available, as depicted in Figure 11-38. The drainage tubes for chronic ear infection can be made from polytetrafluoroethylene (Teflon®) or other inert materials. These are not permanent implants. Penile implants have been used to treat cases of impotence due to biological causes and which cannot be treated by drugs. Figure 11-39 shows examples of penile implants. The simple malleable rod type does not collapse, while the inflatable one does. Placement of the cylinders is shown in Figure 11-40. Table 11-6 shows the clinical use of these implants in Europe during 2001. Penile implants are made of silicone rubber, which can be made to various hardness depending on the amount of crosslinking and filler materials (SiO_2).

The use of implants for correcting problems in the urinary system has been difficult because of the difficulty of joining a prosthesis to a living system to achieve fluid tightness. In

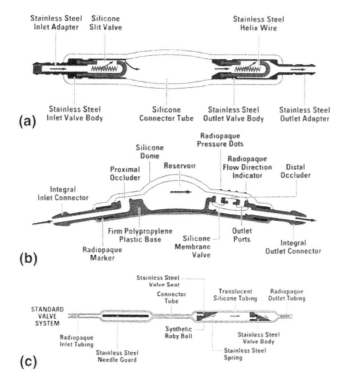

Figure 11-37. Various designs of shunt valves: (**a**) silicone slit (Codman® Holter valve, courtesy of Johnson & Johnson). (**b**) silicone membrane (CSF-Flow Control Valve, Contoured) (Courtesy of Medtronic PS Medical), and (**c**) stainless steel needle valve (Codman® Hakim Valve System, courtesy of Johnson & Johnson).

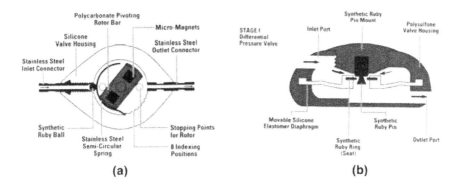

Figure 11-38. Programmable and differential pressure valves: (**a**) Sophy® programmable (courtesy of Sophysa, Costa Mesa, CA); (**b**) Orbis-Sigma® differential pressure valve (courtesy of Cordis Corporation, Miami Lakes, FL).

addition, blockage of the passage by deposits from urine and constant danger of infection have been problematical. Many materials have been tried — including glass, rubber, silver, tantalum, Vitallium®, polyethylene, Dacron®, Teflon®, polyvinyl alcohol, etc. — without much long-term success. Some have tried to develop a balloon filled with polymer gel that could be placed around the urethral opening of the bladder to aid in closing the urethra. A one-way valve is used to prevent leaking after filling the balloon with a gel. However, a hydrogel solution, block copolymer of poly(ethylene)-*b*-(propylene) oxide [PEO-*b*-PPO-*b*-PEO] and sodium hyaluronate (SH), which become a gel at body temperature from liquid solution at room temperature, has been developed. This would have the benefit of not requiring a one-way valve, and even if the hydrogel is diffused out of the membrane, it will be resorbed by the body.

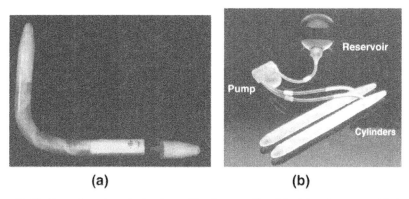

(a)   (b)

Figure 11-39. Examples of penile implants. Simple noncollapsible (**a**) and more sophisticated inflatable with a pump (**b**) are shown. (**a**) Reprinted with permission from Lynch (1982). Copyright © 1982, Van Nostrand Reinhold. (**b**) Reprinted with permission from Mulcahy (2005). Copyright © 2005, Humana Press.

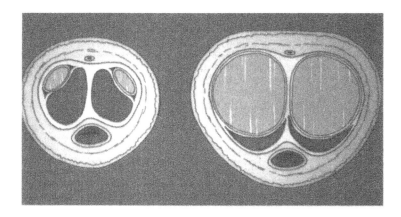

Figure 11-40. Cross-section of penis before and after inflation of the implant. Reprinted with permission from Mulcahy (2005). Copyright © 2005, Humana Press.

An artificial urethral sphincter (AUS) to control urethral incontinency has been around since early 1970s. Urinary incontinence occurs in men after transurethral resection of the prostate (TURP) or radical prostatectomy and in women with postpartum incontinence. The AUS devices are shown in Figure 11-41. These devices are similar to the penile implant with a pump. Table 11-7, plotted in Figure 11-42, shows the success rate of AUS devices for 5- and 10-year follow-up for males and females. In this context, the success rate is usually defined as patency of the device and absence of infection. Some devices continued to function but had to be removed largely due to infection and erosion; this indicates a much lower rate of success of these devices compared to others, such as total hip joint prostheses.

Table 11-6, Penile Implantation Cases in Europe in 2001

	Inflatable	Malleable
Benelux (Belgium, Holland, and Luxembourg)	46	8
France	350	50
Germany	420	35
United Kingdom	150	95
Italy	70	60
Spain	300	90
Switzerland	20	5
Czech Republic	50	0

Reprinted with permission from Evans (2005). Copyright © 2005, Humana Press.

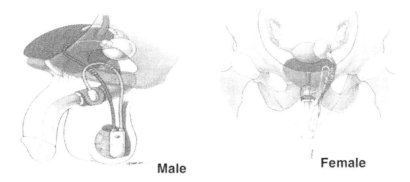

Male   Female

Figure 11-41. Artificial urethral sphincter (AUS) for males and females. Courtesy of American Medical Systems Inc., Minnetonka, MN.

Example 11-3

A bioengineer is trying to make a cochlear nerve stimulating implant using a piezoelectric ceramic. This will possibly eliminate the use of a speech processor as well as a power source. The piezoelectric sensitivity coefficient can vary from 0.7 for bone, 2.3 for quartz and up to 600 pC/N for some piezoelectric ceramics (pp. 65–67, Park and Lakes, 1992). Assume an RMS sound level of 100 dB, which produces 2 Pa pressure (p. 86, Gorga and Neely, 1994), and a 1-mm thick ceramic implant with a piezoelectric sensitivity of 600 pC/N, and then calculate the potential output for use in cochlear nerve cell stimulation.

Answer

The charge density q/A is the product of the piezoelectric sensitivity coefficient and stress:

$$q/A = 600 \text{ pC/N} \times 2 \text{ Pa} = 1.2 \times 10^{-9} \text{ C/m}^2. \tag{11-1}$$

Table 11-7. Statistics on the Success Rate of Artificial Urethral Sphincter (AUS) in All Patients, with Bladder Neck and Bulbar Cuffs for Male and Female

| | | Implant period (years) | |
		5	10
Types of AUS			
All patients	Overall	60	39
	Medical	68	59
	Mechanical	90	66
Female bladder neck AUS	Overall	40	18
	Medical	48	39
	Mechanical	85	39
Male bladder neck AUS	Overall	68	42
	Medical	74	63
	Mechanical	92	68
Male bladder cuff AUS	Overall	71	52
	Medical	79	72
	Mechanical	90	70

Reprinted with permission from Vern et al. (2000). Copyright © 2000, Elsevier.

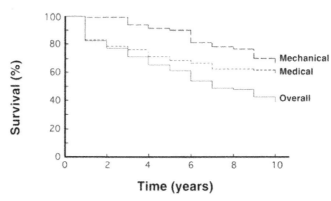

Figure 11-42. Plot of the survival rate of AUS devices. Medical: survival of the original device with adequate performance excluding mechanical failure. Mechanical: survival of the original device with adequate performance excluding failure due to infection and erosion into the urethra. Reprinted with permission from Vern et al. (2000). Copyright © 2000, Lippincott, Williams & Wilkins.

Under the assumptions given, the implant behaves as a capacitor of capacitance C, for which the charge q is $q = CV$, in which V is voltage. $V = q/[\kappa\varepsilon_0 A/t]$, with k being the dielectric constant, ε_0 is the permittivity of space, A is the cross-sectional area, and t is the thickness. Using the charge density give in Eq. (11.1) with 1-mm thickness,

$$V = [1.2 \times 10^{-9} \text{ C/m}^2][10^{-3} \text{ m}]/[1000 \times 8.85 \times 10^{-12} \text{ C/Vm}]$$

$$= 1.4 \times 10^{-4} \text{ V} = 0.14 \text{ mV}. \tag{11-2}$$

The amount of sound energy reduction when the sound waves hit the eardrum, pass through the tissues, and arrive on the surface of the implant can be calculated. The acoustic impedances

($Z = \rho v$, where ρ is density and v is the velocity of sound in the material) of air, tissue (average), and piezoelectric ceramic such as barium titanate are 0.04, 163, and 2408 kRayl, respectively (Park and Lakes, 1992, p. 74). The amplitude reflection coefficient associated with an interface between material 1, containing the incident wave, and material 2 is given by

$$R_A = (Z_2 - Z_1)/(Z_2 + Z_1) = (163 - 0.04)/(163 + 0.04)$$

$$= 0.9995. \tag{11-3}$$

The amount of sound wave reflected at the soft tissue is 99.95%, that is, only 0.05% of the wave passes through the soft tissue. For the soft tissue and implant,

$$R_B = (2408 - 163)/(2408 + 163) = 0.873. \tag{11-4}$$

The net reflection of sound waves is 87.3%, and only 12.7% is transmitted through. Therefore, the net fraction of sound reaching the nerve cells would be 6.2×10^{-5}. Recalling that the potential generation in Eq. (11.2) is 0.14 mV; hence, 0.14 mV $\times$ 6.2×10^{-5} = 8.4×10^{-9} V (8.4 nV). This potential is about 1/100 of the recorded maximum potential since the amplitude of the auditory evoked potential is on the order of 1 mV maximum in all frequency ranges (Stapells, 1994, p. 257). The theoretical calculations give a much smaller value than that obtained from the auditory nerve cell signals. One may consider a piezoelectric polymer, which offers a better match of acoustic impedance than the ceramic. In that case, one could stimulate the cochlear nerve cells without the use of a tuned amplifier. It is also possible to stimulate the cochlear nerve cells without the use of a tuned amplifier if one were to connect the electrode directly to the piezoelectric "implant," which would be placed outside the skin. In that case, more than enough sound energy could be delivered to the nerve tissues. The electric current or voltage output may depend on the size, shape, and angles made with the direction of sound. It is also conceivable that this technique could be used to grow nerve tissues since electrical energy is known to stimulate regeneration of hard and soft tissues.

11.3.4. Space-Filling Implants

Breast implants are quite common space-filling implants. At one time, enlargement of breasts was done with various materials such as paraffin wax, bee's wax, silicone fluids, etc. by direct injection or by enclosure in a rubber balloon. There have been several problems associated with directly injected implants, including progressive instability and ultimate loss of original shape and texture, as well as infection, pain, etc. In the 1960s the FDA banned such practices by classifying injectable implants such as silicone gel as drugs.

Another of the early efforts in breast augmentation was to implant a sponge made of polyvinyl alcohol. However, soft tissues grew into the pores and then calcified with time, and the so-called marble breast resulted. Although the enlargement or replacement of breast for cosmetic reasons alone is not recommended, prostheses have been developed for the patient who has undergone radical mastectomy or who has nonsymmetrical deformities. In the case of cancer surgery, the implants are considered beneficial for psychological reasons. In this case a silicone rubber bag filled with silicone gel and backed with polyester mesh to permit tissue ingrowth for fixation, has been a widely used prosthesis, as shown in Figure 11-43a. The silicone gel was replaced with saline solution due to litigation associated with silicone breast implants (see Figure 11-43b). This type of breast implant has certain advantages and disadvantages with respect to the silicone gel-filled ones. The main advantages are that the saline can be reintroduced if it has leaked, its lower density, and that it is easy to implant the membrane via a tube. However, it lacks the adequate feel of natural breast, and a shift of saline can result in collapse of the membrane on one side. The same filling material used for urethral incontinence can be utilized for a breast implant. These temperature-sensitive hydrogels with sodium hyarulonate can be inserted into the membrane at room temperature in liquid form, as with saline, yet they become gel once warmed up at body temperature. An artificial penis, testicles, and vagina fall into the same category as breast implants in that they make use of silicones and are implanted for psychological reasons rather than to improve physical health.

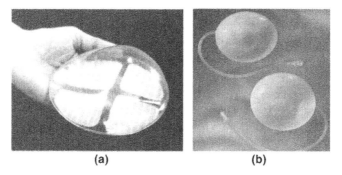

(a)   (b)

Figure 11-43. Example of an artificial breast filled with silicone gel (**a**) and with a saline solution filling tube (**b**). Courtesy of Mentor Corp., Santa Barbara, CA.

Example 11-4

Experience has shown that the silicone membrane used in breast implants leaked the silicone fluid into surrounding tissue. Calculate the amount of leakage during a year. Assume that the leakage is entirely by diffusion rather than by macroscopic pores. Assume the silicone oil has a molecular weight of 740 amu. Assume the membrane is 1 mm thick and the surface area 400 cm^2. The membrane has a diffusion constant of $D = 5 \times 10^{-17}$ cm^2/sec. Assume, moreover, that the implant has a volume of 1000 cm^3 and a density of $\rho = 1.5$ g/cm^3.

Answer

From Fick's first law for diffusion, the flux is written as

$$F = -D \frac{dc}{dx}.$$

in which D is the diffusion coefficient and c is concentration. The flux is in units of mass per unit area per time, so that if the concentration is initially zero in the tissue,

$$\text{mass/time} = \text{flux} \times \text{area} = FA = D \frac{dc}{dx} 400 \text{ cm}^2$$

$$= 5 \times 10^{-17} \text{ cm}^2 / \text{sec} \frac{1.5 \text{ g/cm}^3}{0.1 \text{ cm}} 400 \text{ cm}^2$$

$$= \underline{3 \times 10^{-13} \text{ g/sec}} \text{ or } \underline{9.6 \times 10^{-6} \text{ g/yr}}.$$

The body normally tolerates silicones well; problems do not usually arise unless gross amounts are lost and migrate through the tissues.

We remark that the given volume corresponds to a mass $m = \rho V = 1.5$ kg, corresponding to a weight of about 3.3 pounds for each breast. If the shape is hemispherical, the volume is $2\pi r^3/3 = 1000$ cm^3, so that the diameter (twice the radius) is 15.6 cm. The area of the curved surface is $2\pi r^2 = 384$ cm^2. Commercially available implants are not quite hemispherical. The largest one manufactured is 18 cm in diameter with a volume of 600 cm^3.

PROBLEMS

11-1.   A bioengineer is trying to understand the biomechanics of a hole created in the skin for a transcutaneous implant. The engineer made a hole using a circular biopsy drill in the dorsal skin of a dog. The diameter of the drill is 5 mm. If the hole becomes an ellipse with a minor and major axis of 3 and 7 mm, answer the following questions.

   a.   In which direction is the internal stress in the skin greater?
   b.   In which direction are the collagen fibers more oriented?
   c.   How can the bioengineer obtain a circular rather than elliptical hole for the implant?
   d.   Assuming the implant is non-deformable compared to the skin, what problems will arise between skin and implant when a load or force is applied to the skin or implant by handling accidentally?

11-2.   Calculate the breaking strength of the size 00 suture wire given in Table 13-1. Compare the result with the tensile strength of fully annealed 316L stainless steel (refer to Table 5-2).

11-3.   Design a blood access device for kidney dialysis or other long-term use and give specific materials selected for each part. In addition, explain why you chose the particular material.

11-4.   Draw the anatomy of the eye and label salient features.

11-5.   A breast implant is made of silicone rubber membrane filled with silicone rubber foam. Discuss the advantages and disadvantages of this design in comparison with an oil-filled implant.

11-6.   Proplast$^®$ is a composite of PTFE fiber and carbon (graphite). If it is made up of 50 vol% of each and has 20% porosity, what is its density? Estimate its Young's modulus.

11-7.   Design a penile implant that can carry out erectile function for a person who has lost that capability due to disease or injury. What kind of materials would you need for its construction?

11.8.   The retention of tensile breaking strength of absorbable sutures for chromic catgut and PGA suture is shown in the following illustration:

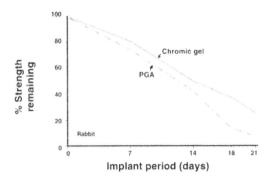

a. Express the rate of strength decrease mathematically for both sutures.

b. From the mathematical expression, calculate the zero strength times.

11-9. An ideal suture is defined as on that "handles comfortably and naturally, minimum tissue reaction, adequate tensile strength and knot security" and is "not favorable for bacterial growth and easily sterilizable, nonelectrolytic, noncapillary, nonallergenic and noncarcinogenic" (Chu, 1983).

11-10. Explain why the nylon monofilament suture is less prone to lose its strength than multifilament suture material in vivo. Also explain why the monofilament suture causes less tissue reaction.

11-11. The telephone has two piezoelectric ceramics to transform sound wave energy to electricity and vice versa for talking and listening. Could you use these ceramics for stimulating the cochlear nerve cells as in Ex. 11-3? If one could make such an implant, how could the deaf person be trained to recognize a voice?

DEFINITIONS

Catgut: A material used for the strings of musical instruments and for surgical absorbable sutures. It is made of the dried twisted intestines of sheep or horses, but not cats.

Chromic salt: Chemical compound that is used to treat collagen to achieve crosslinking between molecular chains of collagen. Such treatment increases its strength but decreases its flexibility.

Cochlea: The spiral cavity of the inner ear containing the organ of Corti, which produces nerve impulses in response to sound vibrations.

Cyanoacrylate: A polymer used as a tissue adhesive since it can polymerize fast in the presence of water.

Dacron®: Polyethylene terephthalate polyester that is made into fibers. If the same polymer is made into a film, it is called Mylar®.

FDA: Food and Drug Administration, which regulates the use of medical devices in the United States.

Fibrin: An insoluble protein formed from fibrinogen during the clotting of blood. It forms a fibrous mesh that impedes the flow of blood.

Fibrinogen: A plasma protein of high molecular weight that is converted to fibrin through the action of thrombin. This material is used to make (absorbable) tissue adhesives.

Hydrocephalus: A condition in which fluid accumulates in the brain, typically in young children, enlarging the head and sometimes causing brain damage.

Hydrogel: A gel in which the liquid component is water (>30% by weight).

Keratoplasty: Surgery carried out on the cornea, especially corneal transplantation.

LADAR: Similar to lasik and uses an active radar eye tracking system, which compensates for involuntary eye movements.

LASIK (laser in-situ keratomileusis): Surgical correction of the curvature of the cornea using a laser.

Ossicle (ossicular, adj.): A small bone, especially one of those in the middle ear.

Percutaneous device (PD): An implant designed to transfer matter, information, etc. from the body to the outside of the body transcutaneously.

Plastipore®: Porous polyethylene.

Polyglycolic acid (PGA): Polymer made from glycolic acid and used to make absorbable sutures or other products.

Polylactic acid (PLA): Polymer made from lactic acid and used to make absorbable sutures or other products.

Proplast®: A composite material made of fibrous polytetrafluoroethylene and carbon. It is usually porous and has a low modulus and low strength.

Pulpectomy: Removal of dental pulp.

Pyrolite®: Pyrolytic carbon.

Scala media: The central duct of the cochlea in the inner ear, containing the sensory cells and separated from the scala tympani and scala vestibuli by membranes.

Silicone: A polymer containing the element silicon. Depending on molecular weight, it may be a gel or a rubber.

Suture: Material used in closing a wound with stitches.

Teflon®: Polytetrafluoroethylene.

Vitallium®: Co-Cr alloy.

VISX: A commercial method using an excimer laser to treat myopia.

BIBLIOGRAPHY

Aoki H, Kato K, Shibata M, Naganuma A. 1972. Thermal transition of calcium phosphates and observation by a scanning electron microscope [in Japanese]. *Rep Inst Med Dent Eng* **6**:50–61.

Armstrong SR. 1999. *Mechanical testing and failure analysis of the dentin–adhesive resin–resin composite bonded joint*. PhD dissertation. Iowa City: The University of Iowa.

Asmussen E, Araújo PA, Peutzfeldt A. 1989. In vitro bonding of resins to enamel and dentin: an update. *Trans Acad Dent Mater* **2**:36.

Barbucci R. 2002. *Integrated biomaterials science*. New York: Kluwer Academic/Plenum.

Black J. 1992. *Biological performance of materials: fundamentals of biocompatibility*. New York: Marcel Dekker.

Black J, Hastings GW. 1998. *Handbook of biomaterial properties*. London: Chapman & Hall.

Bulbulian AH. 1973. *Facial prosthetics*. Springfield, IL: Thomas.

Cardona H. 1962. Keratoprosthesis. *Am J Opthalmol* **54**:284.

Christel P. 1994. Synthetic ligament replacement and augmentation. *Clin Mater* **15**(1):75.

Chu CC. 1983. Survey of clinically important wound closure biomaterials. In *Biocompatible polymers, metals, and composites*, pp. 477–523. Ed M Szycher. Lancaster, PA: Technomics Publishers.

Chu CC. 2000. Biodegradable polymeric biomaterials: an updated overview. In *The biomedical engineering handbook*, 2d ed., pp. 41-1–22. Boca Raton, FL, CRC Press.

Chu CC, Von Fraunhofer JA, Greisler HP. 1997. *Wound closure biomaterials and devices*. Boca Raton, FL: CRC Press.

Chvapil M. 1982. Considerations on manufacturing principles of a synthetic burn dressing: a review. *J Biomed Mater Res* **16**:245–263.

Dagalakis N, Flink J, Stasikelis P, Burke JF, Yannas IV. 1980. Design of an artificial skin, III: control of pore structure. *J Biomed Mater Res* **14**:511–528.

Davis JR. 2003. *Handbook of materials for medical devices*. Materials Park, OH: ASM International.

Dobelle WH, Mladejovsky MG, Girvin JP. 1974. Artificial vision for the blind: Electrical stimulation of visual cortex offers hope for a functional prosthesis. *Science* **183**:440–444.

Dumitriu S. 1994. *Polymeric biomaterials*. New York: Marcel Dekker.

Edwards WS. 1965. *Plastic arterial grafts*. Springfield, IL: Thomas.

Evans CM. 2005. European counterpoint to chapter 13. In *The bionic human*, pp. 302–312. Ed FE Johnson, KS Virgo. Totowa, NJ: Humana Press.

Feldman D, Barker T, Blum B, Bowman J, Kilpadi D, Redden R. 2000. Biomaterial-enhanced regeneration for skin wounds. In *Biomaterials and bioengineering handbook*, pp. 807–842. Ed DL Wise. New York: Marcel Dekker.

Fredericks RJ, Melveger AJ, Dolegiewitz LJ. 1984. Morphological and structural changes in a copolymer of glicolide and lactide occurring as a result of hydrolysis. *J Polym Sci* **22**:57–66.

Gantz BJ. 1987. Cochlear implants: an overview. *Acta Otolaryngol Head Neck Surg* **1**:171–200.

Gorga MP, Neely ST, eds. 1994. *Stimulus calibration in auditory evoked potential measurements: principles and applications in auditory evoked potentials*. Needham Heights, MA: Allyn and Bacon.

Grosse-Siestrup C. 1978. *Entwicklung und klinische erprobung von hautdurchleitungen veterinaermedizin*. PhD dissertation. Free University, Berlin.

Haaster J, von Salis-Solico G, Bensmann G. 1990. The use of Ni–Ti as an implant material in orthopedics. In *engineering aspects of shape memory alloys*, pp. 426–444. Ed TW Duerig, KN Melton, D Stocker, CM Wayman. London: Butterworths-Heinemann.

Hastings GW. 1976. Adhesives and tissues. In *Biocompatibility of implant materials*, chapter 17. Ed DF Williams. London: Sector Publishing.

Hoffman AS. 2004. Application of "smart polymers" as biomaterials. In *Biomaterials science: an introduction to materials in medicine*, pp. 107–115. Ed BD Ratner, AS Hoffman, FJ Schoen, JE Lemons. Amsterdam: Elsevier Academic Press.

Houston S, Hodge Jr JW, Ousterhout DK, Leonard F. 1969. The effect of alpha-cyanoacrylate on would healing. *J Biomed Mater Res* **3**:281–289.

Kablitz C, Kessler T, Dew PA, Stephen RL, Kolff WJ. 1979. Subcutaneous peritoneal catheter: 2-1/2 years experience. *Artif Organs* **3**:210–217.

Lee H, Neville K. 1971. *Handbook of biomedical plastics*. Pasadena, CA: Pasadena Technology Press (chapters 3–5, 13).

Lynch W. 1982. *Implants: reconstructing human body*. New York: Van Nostrand Reinhold.

Maniglia AJ, Proops DW. 2001. *Implantable electronic otologic devices: state of the art*. Philadelphia: W.B. Saunders.

McPherson DG, Anderson JM. 1953. Keratoplasty with acrylic implant. *Brit Med J* **1**:330.

Miloro M, Ghali GE, Larsen P, Waite P, eds. 2003. *Peterson's principles of oral and maxillofacial surgery*. London: B.C. Decker.

Morgan JR, Sheridan RL, Tompkins RG, Yarmush ML, Burke JF. 2004. Burn dressings and skin substitutes. In *Biomaterials science: an introduction to materials in medicine*, pp. 602–614. Amsterdam: Elsevier Academic.

Mukherjee N, Roseman RD, Willging JP. 2000. The piezoelectric cochlear implant: concept, feasibility, challenges, and issues. *J Biomed Mater Res* **53**(2):181–7.

Mulcahy JJ. 2005. Penile prostheses. In *The bionic human*, pp. 289–299. Ed FE Johnson, KS Virgo. Totowa, NJ: Humana Press.

Nulsen FE, Spitz EB. 1951. Treatment of hydrocephalus by direct shunt from ventricle to jugular vein. *Surg Forum* **2**:399–403.

Obstbaum SA. 1996. Ophthalmic implantation. In *Biomaterials science: an introduction to materials in medicine*, pp. 435–443. Ed BD Ratner, AS Hoffman, FJ Schoen, JE Lemons. Amsterdam: Elsevier.

Park JB, Lakes RS. 1992. *Biomaterials: an introduction*. New York: Plenum.

Park JB, DeVries KL, Statton WO. 1978. Chain rupture during tensile deformation of nylon fibers. *J Macromol Sci Phys* **15**:205–227.

Plastic-pore material technical information. 1980. Memphis, TN: Richard Manufacturing.

Poitout DG. 2004. *Biomechanics and biomaterials in orthopedics*. New York: Springer.

Prostlethwait RW, Schaube JF, Dillon ML, Morgan J. 1959. Wound healing, II: an evaluation of surgical suture material. *Surg Gynecol Obstet* **108**:555–566.

Refojo MF. 1996. Ophthalmological applications. In *Biomaterials science: an introduction to materials in medicine*, pp. 328–335. Ed BD Ratner, AS Hoffman, FJ Schoen, JE Lemons. Amsterdam: Elsevier.

Refojo MF. 2004. Ophthalmological applications. In *Biomaterials science: an introduction to materials in medicine*, pp. 583–591. Ed BD Ratner, AS Hoffman, FJ Schoen, JE Lemons. Amsterdam: Elsevier.

Roby MS, ed. 1998. *Recent advances in absorbable sutures.* International Conference on Advances in Biomaterials and Tissue Engineering.

Roby MS, Kennedy J. 2004. Sutures. In *Biomaterials science: an introduction to materials in medicine*, pp. 614–627. Ed BD Ratner, AS Hoffman, FJ Schoen, JE Lemons. Amsterdam: Elsevier.

Rustamzadeh E, Lam CH. 2005. Cerebrospinal fluid shunts. In *The bionic human*, pp. 333–358. Ed FE Johnson, KS Virgo. Totowa, NJ: Humana Press.

Rutter LAG. 1958. Natural materials. In *Modern trends in surgical materials*, p. 208. Ed L Gills. London: Butterworths.

Shea JJ, Sanabria F, Smyth GDL. 1962. Teflon piston operation of otosclerosis. *Arch Otolaryngol* **76**:516–521.

Sheehy JL. 1969. Stapes surgery when the incus is missing. In *Hearing loss-problems in diagnosis and treatment*, p. 141. Ed LR Boies. Philadelphia: W.B. Saunders.

Silastic. 1972. Hospital-Surgical Products Bulletin 51-051A. Midland, MI: Dow Corning.

Smith AM, Callow JA, eds. 2006. *Biological adhesives.* New York: Springer.

Smith DC. 2004. Adhesives and sealants. In *Biomaterials science: an introduction to materials in medicine*, pp. 572–583. Ed BD Ratner, AS Hoffman, FJ Schoen, JE Lemons. Amsterdam: Elsevier.

Spelman FA. 1988. Cochlear prosthesis. In *Encyclopedia of medical devices and instrumentation*, pp. 720–727. Ed JG Webster. New York: Wiley.

Spelman FA. 2004. Cochlear prosthesis. In *Biomaterials science: an introduction to materials in medicine*, pp. 657–669. Ed BD Ratner, AS Hoffman, FJ Schoen, JE Lemons. Amsterdam: Elsevier.

Tabor JR. 1970. Reconstruction of the ossicular chain. *Arch Otolaryngol* **92**:141–146.

Thacker JG, Rodeheaver G, Moore JW, Kauzlarich JJ, Kurtz L, Edgerton MT, Edlich RF. 1975. Mechanical performance of surgical sutures. *Am J Surg* **130**:374–380.

Turunen T, Peltola J, Kangasniemi I, Jussila J, Uusipaikka E, Yli-Urpo A, Happonen R-P, eds. 1995. *Augmentation of the maxillary sinus wall using bioactive glass and autologous bone.* Oxford: Pergamon/Elsevier Science.

Vern SN, Greenwell TJ, Mundy AR. 2000. The long-term outcome of artificial urinary sphincters. *J Urol* **164**:702–707.

von Recum AF, Park JB. 1979. Percutaneous devices. *CRC Crit Rev Bioeng* **5**:37–77.

Wang PY. 1976. Performance of tissue adhesives in vivo. In *Biocompatibility of implant materials*, chapter 18. Ed DF Williams. London: Sector Publishing.

Williams DF, ed. 1981. *Fundamental aspects of biocompatibility.* Boca Raton, FL: CRC Press.

Williams DF, ed. 1982a. *Biocompatibility in clinical practice.* Boca Raton, FL: CRC Press.

Williams DF. 1982b. Biodegradation of surgical polymers. *J Mater Sci* **17**:1233–1246.

Wise DL. 1996. *Human biomaterials applications.* Totowa, NJ: Humana Press.

Wong JY. 2004. *Architecture and application of biomaterials and biomolecular materials.* Materials Research Society Symposium Proceedings. Boston: Materials Research Society.

Yannas IV, Burke JF. 1980. Design of an artificial skin, I: basic design principles. *J Biomed Mater Res* **14**:65–81.

Yannas IV, Burke JF, Gordon PL, Huang G, Rubenstein RH. 1980. Design of an artificial skin, II: control of chemical composition. *J Biomed Mater Res* **14**:107–131.

12

SOFT TISSUE REPLACEMENT — II:
BLOOD INTERFACING IMPLANTS

**Glass cylinder to hold
liquid to be injected**

Injection syringe, c. 1865

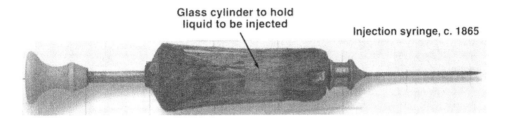

The injection syringe. One of the important inventions in medicine. Reprinted with permission from Platt (1994). Copyright © 1994, Marcel Dekker.

Blood-interfacing materials can be divided into two categories: (1) short-term extracorporeal devices such as membranes for artificial organs (kidney and heart/lung machine), tubes, and catheters for the transport of blood, and (2) long-term in-situ implants such as vascular implants and implantable artificial organs. Although pacemakers for the heart are not interfaced with blood directly, they are considered here since they are devices that help the heart to circulate blood throughout the body.

The single most important requirement for blood-interfacing implants is blood compatibility (review §10.3). Blood coagulation is the most important aspect of blood compatibility: the implant should not cause the blood to clot. In addition, implants should not damage proteins, enzymes, and the formed elements of blood (red blood cells, white blood cells, and platelets). The implants should not cause hemolysis (red blood cell rupture) or initiation of the platelet release reaction.

Blood is circulated throughout the body according to the sequence shown in Figure 12-1. Implants are usually used to replace or patch large arteries and veins as well as the heart and its valves. Surgical treatment without using implants is usually preferred. However, there are many clinical situations in which the surgeon in consultation with the patient chooses to anastomose or replace a large segment with implants.

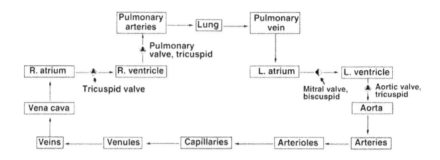

Figure 12-1. Schematic diagram of blood circulation in the body.

Table 12-1. Comparison of Cell Properties

Components	Normal concentration in blood (/liter)	Cell diameter (µm)	Average density (g/ml)	Sedimentation coefficient ($S \times 10^7$)
Red blood cell	$4.2–6.2 \times 10^{12}$	8	1.098	12.0
White blood cell	$4.0–11.0 \times 10^9$	–	–	–
Lymphocyte	$1.5–3.5 \times 10^9$	7–18	1.072	1.2
Granulocyte	$2.5–8.0 \times 10^9$	10–15	1.082	–
Monocyte	$0.2–0.8 \times 10^9$	12–20	1.062	–
Platelet	$150–400 \times 10^9$	2–4	1.058	0.032
Plasma	–	–	1.027	–

Modified with permission from Malchesky (2004). Copyright © 2004, Elsevier.

12.1. BLOOD SUBSTITUTES AND ACCESS CATHETERS

Table 12-1 gives some properties of blood for reference. If much blood is lost in an injury or in surgery, there may be insufficient transport of oxygen to tissues. Blood transfusions can make up for lost blood, but such blood may not always be available in sufficient amounts for emergencies. Also, some people object to blood transfusions on religious grounds. Therefore, blood substitutes have been explored. Artificial blood substitutes are in clinical trials. Red blood cell substitutes based on perfuloroctyl bromide ($C_8F_{17}Br$) or perfluorodichloroctane ($C_8F_{16}Cl_2$) make use of the solubility of oxygen in some materials. The other formed elements of blood —white blood cells and platelets — cannot be substituted at this time. The hemoglobin of the red cells can be crosslinked by glutaraldehyde in the presence of red cell enzyme, catalase to improve the flexibility of the crosslinked cells, and, more importantly, solubility of oxygen.

Blood is circulated throughout the body according to the sequence shown in Figure 12-1. Blood access catheters have been developed for transient and more permanent indwelling devices, as shown in Figures 12-2 and 12-3. Implantable venous access ports of various sizes and septum diameters made of silicone rubber are shown in Figure 12-4. These are used to administer drugs and draw blood for diagnosis.

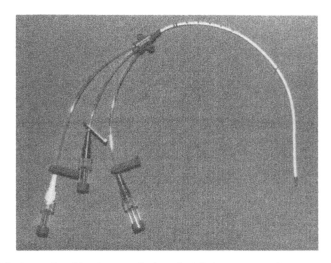

Figure 12-2. A transient blood access device of triple lumen; central venous catheter made of polyurethane. Reprinted with permission from Raaf and Vinton (1993). Copyright © 1993, Churchill Livingstone.

12.2. CARDIOVASCULAR GRAFTS AND STENTS

The blood vessels walls in disease states may become thinner in aneurysms or thicker in the intima in the case of atherosclerosis (see §9.3.4). Replacement of the artery or reinforcing the wall can be made by autograft of vein, artificial prosthesis, reinforcement by stent, and opening of completely blocked arteries using balloons with or without an embedded stent. These remedies can be transient (short term) or long term. Ideal stents should have:

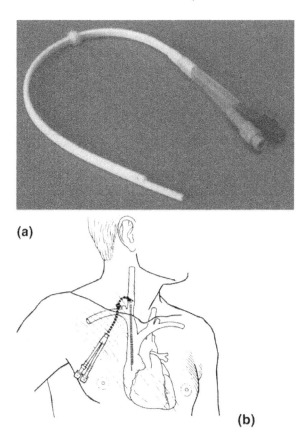

Figure 12-3. Semipermanent indwelling dual-lumen atrial catheter having a large bore and staggered tip (**a**) and placement (**b**). Reprinted with permission from Raaf (1991). Copyright © 1991, Williams and Wilkins.

Figure 12-4. Implantable venous access ports of various sizes and septum diameters, made of silicone rubber. Reprinted with permission from Raaf and Vinton (1993). Copyright © 1993, Churchill Livingstone.

1. Ease of visualization with x-rays,
2. High hoop strength to resist arterial recoil,
3. Longitudinal flexibility to pass tortuous vessels and bifurcation with a contralateral approach,
4. Radial elasticity under external compression,
5. Minimal or no change after implantation,
6. High expansion ratio and low profile for passage through small introducers or guiding catheters or through tight stenosis,
7. Retrievability,
8. Side-branch accessibility,
9. Minimal induction of hyperplasia of intima,
10. Thromboresistance,
11. Fatigue resistance,
12. Biocompatibility.

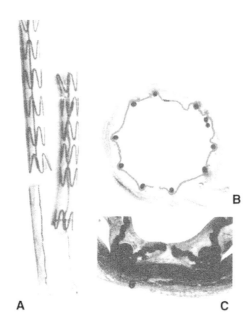

Figure 12-5. Stent grafts. (**a**) Configuration of device showing composite metal and fabric portions. (**b**) Low-power photomicrograph of well-healed experimental device explanted from a dog aorta. The lumen is widely patent and the fabric and metal components are visible. (**c**) High-power photomicrograph of stent graft interaction with the vascular wall, demonstrating mild intimal thickening. Reprinted with permission from Padera and Schoen (2004). Copyright © 2004, Elsevier.

Angioplasty employs balloons made of a polymer (polyethylene terephthalate) sheet wrapped around a tube for insertion and expands after placement. The expanded artery may not remain patent for long times. In some instances stents are deployed to prevent reoccurrence of stenosis. A typical stent is shown in Figure 12-5 before after implantation in canine aorta. The stent is a composite of balloon and metal wire, as can be seen in the histology. The self-expandable stents are made of shape memory alloys (see §5.4.3), while others are made from stainless steels. Many variations of stents could be made to achieve drug delivery (anticoagulant, anti-tissue growth, etc.), be covered (ePTFE, polyester), or contain holes on the coil surfaces to promote delivery of drugs. Figure 12-6 shows some promising results using covered and drug impregnated stents. Longer-term clinical study results are needed before one can draw conclusions about the success of such methods.

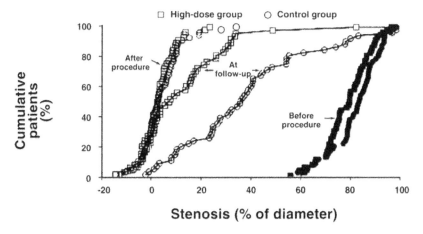

Figure 12-6. Cumulative distribution of percent stenosis in the high-dose and control groups. The distributions were similar at baseline (about 80%) and immediately after stent placement (about 0%). At 6 months follow-up, the distribution in the high-dose group remained similar to the distribution immediately after stenting, whereas the control group suffered greater restenosis (about 40%). Reprinted with permission from Padera and Schoen (2004). Copyright © 2004, Elsevier. Based on the work presented by Park (2003).

12.3. BLOOD VESSEL IMPLANTS

Atherosclerotic artery disease is a leading cause of mortality (death) and sickness (morbidity) in industrial nations, and these trends are increasing in developing nations. Blood vessels can become narrowed from atherosclerosis, or they can undergo an abnormal bulge, called an aneurysm. In an attempt to deal with these conditions, implants have been used in various circumstances to treat vascular maladies. Examples of implants include simple sutures for anastomoses after removal of vessel segments, vessel patches for aneurysms, as well as total replacements for large arteries. An anastomosis is a connection made surgically between blood vessels or segments of vessels, parts of the intestine, or other channels of the body, or the operation in which this is constructed. Vein implants have encountered more difficulties than arterial ones because of the collapse of an adjacent vein or clot formation, which in turn is due to low blood pressure and stagnant blood flow in veins as compared with arteries. Vein replacements have not been a major concern since autografting can be performed for the majority of cases. Nonetheless, many materials — including nylon, polytetrafluoroethylene, polyester — have been fabricated for clinical applications.

Early trials for arterial replacements used solid-wall tubes made of glass, aluminum, gold, silver, and polymethylmethacrylate. All implants developed clots and became useless. In the

1940s aneurysms were dealt with by attempting to reinforce them to delay rupture. Indeed, renowned physicist Albert Einstein had an abdominal aorta aneurysm wrapped with cellophane in 1948 in the hope of inducing a fibrous capsule to strengthen the bulging aorta. Einstein suffered a bout of abdominal pain in 1955; surgeons suggested the leaking aneurysm be replaced with a cadaver graft. Einstein refused further surgery and died of a ruptured aneurysm soon after at the age of 76. In the early 1950s porous implants made of fabrics were introduced. These allowed tissue growth into the interstices, as shown in Figure 12-7. The new tissues interface well with blood and thus minimize clotting. Ironically, for this type of application thrombogenic materials were found to be more satisfactory. Another advantage of tissue ingrowth is fixation of the implant by the ingrown tissues, which constitute a viable anchor. The initial leakage through pores is disadvantageous, but this can be prevented by pre-clotting the outside surface of the implant prior to placement. Crimping of the prosthesis, as shown in Figure 12-8, is done to prevent kinking when the implant is flexed. Also, the crimping allows expansion of the graft in the longitudinal direction, which reduces strain on the prosthesis wall. Natural arteries can expand circumferentially and longitudinally to accommodate the pulsatile flow of the blood. Artificial arteries are used only to replace large arteries. For clogged small arteries, as in coronary artery bypass surgery, natural vessels are harvested from other places in the patient's body.

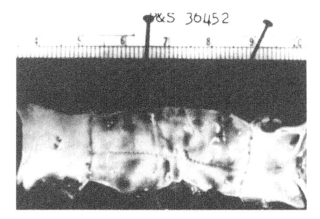

Figure 12-7. The first arterial graft made by stitching fabrics together by hand. Reprinted with permission from Hufnagel (1983). Copyright © 1983, C.W. Wright.

Although the exact sequence of tissue formation in implants in humans is not fully documented, quite a bit is known about reactions in animals. Generally, soon after implantation the inner and outer surfaces of the implant are covered with fibrin and fibrous tissue, respectively. A layer of fibroblasts replaces the fibrin, becoming *neointima*, which is sometimes called *pseudointima* or *pseudoneointima*. The long-term fate of the neointima varies with the species of animal; in dogs it stabilizes into a constant thickness, while for a pig it will grow until it occludes the vessel. In humans the initial phase of healing is the same as for animals, but in later stages the inner surface is covered by both fibrin and a cellular layer of fibroblasts. The sequence for healing of arterial implants in humans, dogs, and pigs is given Figure 12-9. Figure 12-10 depicts the healing process of an arterial wall with an implant. Some clinical results comparing ePTFE, autologous vein, and umbilical vein grafts are shown in Figure 12-11. The relatively short-term results (up to 5 years) show about 50% patency for synthetic artery compared to over 75% for autologous vein transplants.

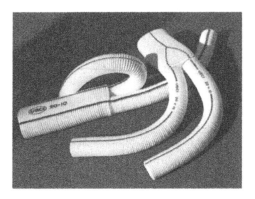

Figure 12-8. Modern arterial graft. Note the crimping.

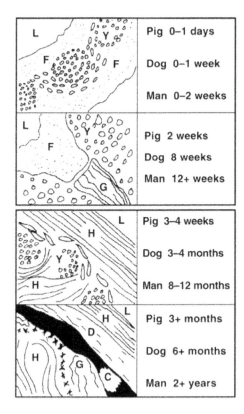

Figure 12-9. Basic healing pattern of arterial prosthesis. L = lumen of prosthesis, F = fibrin, Y = yarn bundle, G = organizing granulation tissue, H = healed fibrous capsular tissue, D = degenerative fibrous capsular tissue, C = calcified capsular tissue. Reprinted with permission from Wesolowski et al. (1968). Copyright © 1968, Chemical Rubber Company.

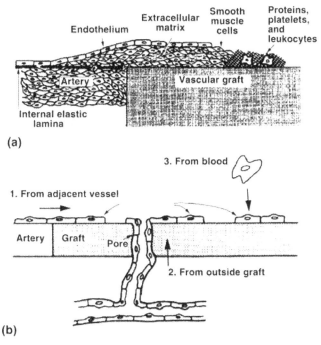

Figure 12-10. Vascular graft healing. (a) Schematic diagram of pannus formation. The major mode of graft healing with currently available vascular grafts. Smooth muscle cells migrate from the media to the intima of the adjacent artery and extend over and proliferate on the graft surface; this smooth muscle cell layer is covered by a proliferating layer of endothelial cells. (b) Possible sources of endothelium on the blood-contacting surface of the vascular graft. Reprinted with permission from Schoen (1989). Copyright © 1989, W.B. Saunders.

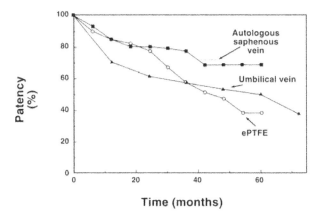

Figure 12-11. Comparison of patency rates in a six-year prospective multicenter randomized comparison of autologous saphenous vein transplant and expanded polytetrafluoroethylene grafts in infrainguinal arterial reconstructions. Reprinted with permission from Padera and Schoen (2004). Copyright © 2004, Elsevier. Based on the work presented by Park (2003).

The types of materials and the geometry of the implant influence the rate and nature of tissue ingrowth. A number of polymer materials have been used to fabricate implants, including nylon, polyester, polytetrafluoroethylene, polypropylene, polyacrylonitrile, and silicone rubber. However, polytetrafluoroethylene, polyester, polypropylene, and silicone rubber are the most favorable materials due to minimal deterioration of their physical properties in vivo, as discussed in Chapter 8. Polyester (particularly polyethyleneterephthalate, Dacron®) is usually preferred because of its superior handling properties.

A pyrolytic carbon coated arterial graft was developed using the techniques of ultralow-temperature isotropic (ULTI) deposition. The nonthrombogenic properties of the pyrolytic carbon may enhance the patency of a graft made from this material and decrease prescriptions for postsurgical anticoagulant drugs used to prevent clots in grafts.

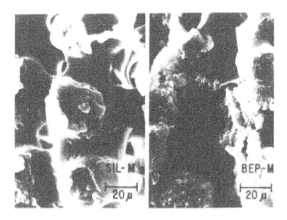

Figure 12-12. Scanning electron microscopic view of the replamineform silastic arterial graft. Reprinted with permission from Hiratzka et al. (1979). Copyright © 1979, American Medical Association.

Another interesting arterial graft is made by pressure-injecting silastic rubber into premachined molds made of the tentacles of sea urchins. The objective is to achieve a microporous structure for the tissues to grow into. After the silastic rubber is cured, the mold is dissolved away by acid treatment, leaving a replamineform of the ultrastructure, as shown in Figure 12-12. Animal experiments showed promising results.

The geometry of fabrics and porosity have a great influence on healing characteristics. The preferred porosity is such that 5,000 to 10,000 ml of water is passed per square centimeter of fabric per minute at a pressure of 120 mm Hg. The fluid permeability depends not only on the porosity (volume fraction of pores) but also on the size, shape, and connectivity of pores. The lower limit will prevent excessive leakage of blood, and the higher limit is better for tissue ingrowth and healing characteristics. The thickness of the implant is directly related to the amount of thrombus formation: the thinner the wall, the smaller or thinner the thrombus deposited. Less thrombus results in faster organization of the neointima. Also, a smaller-caliber (< 5 mm in diameter) prostheses can be made more easily with thinner walls.

The long-term testing of vascular prostheses is as important as it is with any other implants. A simple in-vivo testing machine is shown in Figure 12-13 in which the pseudoextracellular fluid is drawn through valve "A" and pushed out through valve "B" of the graft 96 cycles per minute with a peak pressure at 150 mm Hg at 37°C. Various grafts were tested and compared with in-vivo implantation results, as shown in Figure 12-14. It can be seen that the Teflon® knit graft did not lose its tenacity (a measure of normalized strength). The initial values of tenacity for Teflon® and Dacron® knit prostheses are about 1.3 and 3.0 g/denier, respectively. The Dacron® grafts showed initial decreases and stabilized after 6 months under both in-vivo and in-vitro conditions.

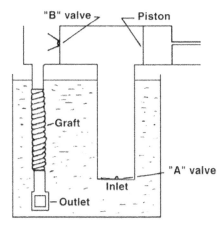

Figure 12-13. Schematic diagram of arterial graft life tester. Reprinted with permission from Botzko et al. (1979). Copyright © 1979, American Society for Testing and Materials.

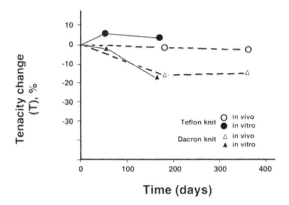

Figure 12-14. Percent change in tenacity for three types of prostheses after life testing of canine implant. Reprinted with permission from Botzko et al. (1979). Copyright © 1979, American Society for Testing and Materials.

Such testing, while important, is insufficient to evaluate the success of biomaterials, which, after all, are intended to improve the health of humans. Clinical evaluation is essential. For example, vascular surgery can be highly invasive and traumatic. Aneurysm repair surgery can cause kidney failure, sexual dysfunction, nerve paralysis, ischemia, and necrosis of the bowel, pelvis, rectum, spine, or buttock, stroke, leg ischemia and paraplegic paralysis. In this context, it is a further testimony to the perceptive qualities of Albert Einstein that he refused surgery for a leaking aneurysm, saying: "I want to go when I want. It is tasteless to prolong life artificially. I have done my share; it is time to go. I will do it elegantly." In an effort to reduce problems due to surgical trauma, endovascular aortic aneurysm repair has been introduced. In this approach, a graft consisting of a nitinol frame covered with a woven polyester fabric is inserted via the femoral artery, and secured within the aneurysm by self-expanding stents or hooks. Complications are nevertheless serious and frequent.

Example 12-1

The material properties of arterial prostheses change following ingrowth of tissues in vivo. A porous silicone rubber arterial prosthesis was implanted in dogs, and it was found that half the pores became filled with tissue after 3 months of implantation. The prosthesis has a 5-mm inside diameter, a wall thickness of 1 mm, and a porosity of 30%. The solid silicone from which the prosthesis is derived has a Young's modulus of 10 MPa. Answer the following.

a. Determine the Young's modulus of the porous silicone.

b. Find the elastic modulus of the prosthesis following tissue ingrowth. Assume that the ingrown tissue is similar to the natural arterial wall ($E = 0.1$ MPa).

c. Determine the wall tension assuming a (high) blood pressure of 200 mm Hg.

Answer

a. There are several relationships, which can be used for determination of the properties of porous materials. For example, we may consider the empirical relationship

$$E = E_{solid}(1 - 1.9V + 0.9V^2),$$

in which V is the porosity (volume fraction of pores) and E_{solid} is the elastic modulus of the solid without porosity. So

$$E = 10(1 - 1.9 \times 0.3 + 0.9 \times 0.32) = \underline{5.1 \text{ MPa}}.$$

Alternatively, we may consider the model of Gibson and Ashby, which has both empirical and theoretical justification:

$$E = E_{solid}(\rho/\rho_{solid})^2.$$

The density ratio is

$$\frac{\rho}{\rho_{solid}} = 1 - V,$$

so that

$$E = 10(0.49) = \underline{4.9 \text{ MPa}}.$$

The actual elastic modulus will depend on the shape of the pores and their orientation.

b. This is a rather complicated system. Obviously, the maximum elastic modulus for this problem would be 10 MPa if the pores were filled with an identical silicone rubber. We may approximate the modulus (using the second value above) as

$$E = 4.90 + 0.1 \times 0.3 = \underline{4.93 \text{ MPa}}.$$

This is under the assumption that there is an intimate attachment between tissue and implant.

c. $\quad T = P \times r = 200 \text{ mmHg} \times \dfrac{1333 \text{ dyne/cm}^2}{\text{mmHg}} \times \dfrac{6 \text{ mm} + 1 \text{ mm}}{2} = \underline{80,000 \text{ dyn/cm}}.$

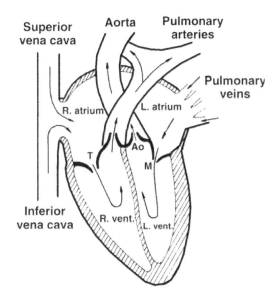

Figure 12-15. Circulation of blood in the heart. Compare with Figure 12-1.

12.4. HEART VALVE IMPLANTS

There are four valves in the ventricles of the human heart, as shown in Figure 12-15 (cf. Figure 12-1). In the majority of cases of valve problems, the left-ventricular valves (mitral and aortic) become incompetent more frequently than the right-ventricular valves as the result of higher left-ventricular pressure. Most important and frequently critical is the aortic valve, since it is the last gate the blood has to go through before being circulated into the body. In some cases the surgeon repairs the valve rather than replaces it. Heart valve replacement as practiced today is commonly done and is considered to have a high success rate.

There have been many different types of valve implants. The early ones in the 1960s were made of flexible leaflets, which mimicked natural valves. Invariably the leaflets could not withstand fatigue for more than 3 years. In addition to hemolysis, regurgitation and incompetence were major problems. Butterfly leaflets and ball- or disk-in-the-cage valves were later introduced. A summary of implants used through 1994 is given in Table 12-2. Some of the valves are shown in Figures 12-16 and 12-17. The material requirements for valve implants are the same as for vascular implants. Some additional requirements are related to the blood flow and pressure, i.e., the formed elements of blood should not be damaged and the blood pressure should not drop below a clinically significant value. Also, valve noise should be minimal, for psychological reasons.

Table 12-2. Summary of Various Heart Valve Implantations up to 1994

Types	Model	Year introduced	Total implanted up to 1994
Mechanical			
Ball-in-cage	Starr-Edwards	1965	200,000
Tilting disc			
	Bjork-Shiley		360,000
	Medtronic Hall	1977	178,000
	Omniscience	1978	48,000
	Monostrut	1982	94,000
Bileaflet	St. Jude	1977	580,000
	Carbomedic	1986	110,000
Tissue			
Porcine	Hancock	1970	177,000
	Hancock Modified Orifice	1978	32,000
	Carpentier Edwards (CE)	1971	400,000
	CE Supra Annular	1982	45,000
Porcine (stentless)	Toronto Stentless	1991	5,000
	Medtronic Freestyle	1992	5,000
Pericardial	Carpentier Edwards	1982	35,000
Homograft	Various	1962	28,000
Autogenous	Pulmonary	1967	2,000

Reprinted with permission from Bonow et al. (1998). Copyright © 1998, Elsevier.

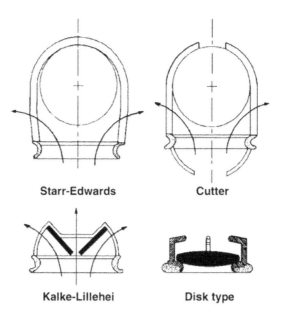

Figure 12-16. Schematic diagram of various types of heart valves.

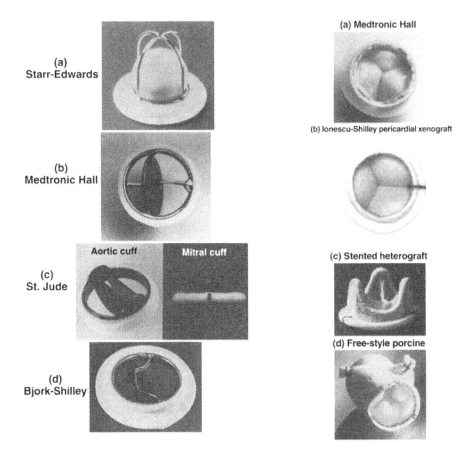

Figure 12-17. Pictures of mechanical heart valves: (a) Starr-Edwards, (**b**) Medtronic Hall, (**c**) St. Jude, (**d**) Bjork-Shiley. Modified with permission from Arya and Labovitz (2005). Copyright © 2005, Humana Press.

Figure 12-18. Tissue heart valves. (**a**) Medtronic Hall, (**b**) Ionescu-Shilley pericardial xenograft, (**c**) Stented heterograft, (**d**) Free-style porcine. Modified with permission from Arya and Labovitz (2005).

Figure 12-18 shows tissue valves made from such collagen-rich material as pericardial tissues, and porcine heart valves with and without stent. Basically, the pericardium is made up of three layers of collagen fibers oriented 60° from each layer and about 0.5 mm thick in the case of bovine pericardium. It can be crosslinked by glutaraldehyde. During this treatment, the cell viability is destroyed and the proteins denatured. Therefore, the implant does not provoke immunological reactions. Porcine xenograft valves have also been used. They are also treated with a chemical process, which denatures the proteins and kills any living cells.

All mechanical valves have a sewing ring covered with various polymeric fabrics. This helps during initial fixation of the implant. Later, the ingrown tissue will render the fixation viable in a manner similar to the porous vascular implants. The cage itself is usually made of metals and covered with fabrics, to reduce noise, or with pyrolytic carbons for a nonthrombogenic surface (the disk or ball is also coated with pyrolytic carbon at the same time). The practice of covering the struts with fabrics has been abandoned since the fabric fatigued and broke into pieces.

The ball (or disk) is usually made as a hollow structure composed of solid (nonporous) polymers (polypropylene, polyoxymethylene, polychlorotrifluoroethylene, etc.), metals (titanium, Co–Cr alloy), or pyrolytic carbon deposited on a graphite substrate. The early use of a silicone rubber poppet was found undesirable due to lipid absorption and subsequent swelling and dimensional changes. This problem has been corrected in modern valves. Although this was an unfortunate episode (some results were fatal), it helped to reinforce the realization that the in-vitro experiment alone is not sufficient to predict all circumstances that arise during in-vivo use, no matter how carefully one tries to predict. This is true of any implant research, even with very simple devices.

The clinical trial outcome for the aortic valve replacement is compared with the natural history of aortic valve disease population in Figure 12-19. The graph shows clearly the far better survival percentage with the prosthesis, but it does not reveal the health of the surviving patients.

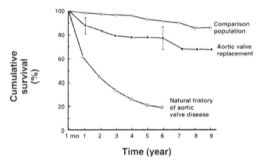

Figure 12-19. Cumulative survival percentage of comparable healthy, natural valve diseases, and valve replaced population. Reprinted with permission from Padera and Schoen (2004). Copyright © 2004, Elsevier. Based on the work presented by Park (2003).

Example 12-2

The bovine pericardium has been tested for its mechanical properties. The stress–strain curve is shown in in the following illustration. Answer the following questions.

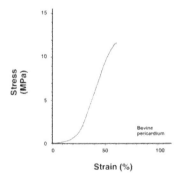

Stress–strain curve of bovine pericardium.

a.   Calculate the initial modulus.

b.   Calculate the secondary modulus.

c.   What is the toughness?

Answer

a.   The initial modulus is, from the slope of the graph,

$$E_i = \frac{15-0}{0.75-0} = \underline{2.0 \text{ MPa}}.$$

b.   The secondary modulus is, from the slope of the graph,

$$E_S = \frac{15-0}{0.62-0.21} = \underline{37 \text{ MPa}}.$$

c.   The toughness is the area under the curve up to the failure strain. It can be approximated by a triangle:

$$\text{Toughness} = 15 \text{ MPa} \times \frac{0.62\text{-}0.21}{2} = 3.1 \text{ MPa} \frac{m}{m}$$

$$= \underline{3.1 \times 10^6} \frac{\text{Nm}}{\text{m}^3} \text{ or } \underline{3.1 \text{ MPa} \cdot \text{m}}.$$

12.5. HEART AND LUNG ASSIST DEVICES

The function of the heart in pumping blood can be temporarily taken over by a mechanical pump. This procedure is most useful in cardiac surgery, in which a surgical field free of blood is required. The pump must propel the blood at the correct pressure and flow rate, and its internal parts must be compatible with blood. Moreover, damage to red blood corpuscles should be minimized. Use of artificial devices to take over the function of the lungs is also common in thoracic surgery. The human heart actually contains two pumps: one for systemic circulation and one for pulmonary (lung) circulation. Therefore, even if a cardiac surgery patient has normal lung function, it is usual to assume lung function by a machine, to simplify connections between the pump and the patient's circulatory system. The combination of a blood pump and an oxygenator is known as the *heart-and-lung* machine, a schematic diagram of which is shown in Figure 12-20.

There are basically three types of oxygenators, which are the core of the artificial lung, as shown in Figure 12-21. In all cases oxygen gas is allowed to diffuse into the blood and simultaneously waste gas (CO_2) is removed. In order to increase the rate of gas exchanges at the blood/gas interface of the bubble oxygenator, the gas is broken into small bubbles (about 1 mm in diameter; if smaller, it is difficult to remove them from blood) to increase the surface contact area. Sometimes the blood is spread thinly as a film and exposed to the oxygen. This is called a film oxygenator. A membrane oxygenator is similar to the membrane-type artificial kidney to be discussed later. The main difference is that the oxygenator membrane is permeable to gases only, while the kidney membrane is also permeable to liquids.

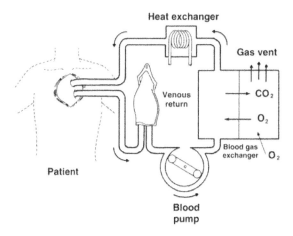

Figure 12-20. Schematic diagram showing heart and lung bypass. Reprinted with permission from Wagner et al. (2004). Copyright © 2004, Elsevier.

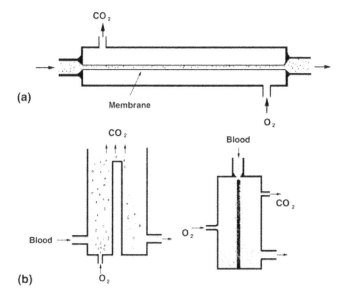

Figure 12-21. Schematic diagrams of different oxygenators: (**a**) membrane, (**b**) bubble, and (**c**) film.

Membrane and bubble oxygenators each have advantages. The membrane oxygenator is considered physiologically superior in view of the fact that there is no blood–gas interface, and as a result there is a less turbulence-induced hemolysis and better platelet function than with the bubble oxygenator. Moreover, the membrane oxygenator does not introduce any micro-bubbles or microemboli in the blood. Control of exchange of oxygen and carbon dioxide according to the needs of the patient is readily achieved, and there is no need for antifoam agents such as with the bubble oxygenator. Membrane oxygenators are used for prolonged procedures; however, for short surgeries bubble oxygenators are preferred since they are simpler to operate and consequently less expensive.

Table 12-3. Physical Characteristics of Natural versus Artificial Lung

Characteristics	Natural lung	Artificial lung
Pulmonary flow (liter/min)	5	5
Head of pressure (mm Hg)	12	0–200
Pulmonary blood volume (liter)	1	1–4
Blood transit time (sec)	0.1–0.3	3–30
Blood film thickness (mm)	0.005–0.010	0.1–3.0
Length of capillary (mm)	0.1	20–200
Pulmonary ventilation (liter/min)	7	2–10
Exchange surface (m²)	50–100	2–10
Veno-alveolar O_2 gradient (mm Hg)	40–50	650
Veno-alveolar CO_2 gradient (mm Hg)	3–5	30–50

Reprinted with permission from Cooney (1976). Copyright © 1976, Marcel Dekker.

Table 12-4. Gas Permeability of Teflon and Silicone Rubber Membranes*

Materials	Thickness (mil)	Oxygen	Carbon dioxide	Nitrogen	Helium
Teflon	1/8	239	645	106	1425
	1/4	117	302	56	730
	3/8	77	181	35	430
	1/2	61	126	30	345
	3/4	41	86	23	240
	1	29			
Silicone rubber	3	391	2072	184	224
	4	306	1605	159	187
	5	206	1112	105	133
	7	159	802	81	94
	12	93	425	48	51
	20	59	279	31	43

*Permeation rates of oxygen, carbon dioxide, nitrogen, and helium across Teflon and silicone rubber membranes of a given thickness, in ml/min-m²-atm (STP). 1 mil = 0.001 inch.

Reprinted with permission from Cooney (1976). Copyright © 1976, Marcel Dekker.

Some of the mechanical and chemical characteristics of the natural and artificial lung (oxygenator) are compared in Table 12-3. The surface area of the artificial membrane is about ten times larger than the natural lung, since the amount of oxygen transfer through a membrane is proportional to the surface area, pressure, and transit time but inversely proportional to the (blood) film thickness. The blood film thickness for the artificial membrane is about 30 times larger than in the natural lung. This has to be compensated for by increased transit time and higher pressure (650 mm Hg) to achieve the same amount of oxygen transfer through the artificial lung.

The membranes are usually made of silicone rubber or polytetrafluoroethylene. The gas permeability of these materials is given in Table 12-4. Silicone rubber is 40 and 80 times more permeable to O_2 and CO_2, respectively, than polytetrafluoroethylene, but the latter can be made about 20 times thinner due to its higher strength. Therefore, silicone rubber is only two and four times better than polytetrafluoroethylene for O_2 and CO_2 transfer, respectively.

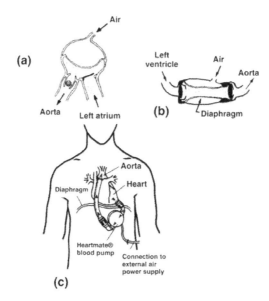

Figure 12-22. Schematic diagram of heart assist devices: (a) DeBakey left ventricular bypass, (b) Bernard-Teco assist pump [Reprinted with permission from Lee and Neville (1971). Copyright © 1971, Pasadena Technology Press.], and (c) depiction of the use in vivo. [Reprinted with permission from Wagner et al. (2004). Copyright © 2004, Elsevier.]

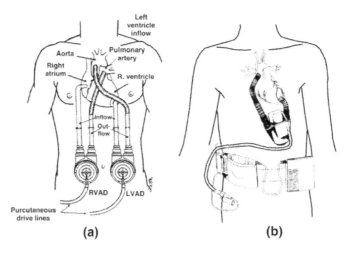

Figure 12-23. Depiction of (a) both right- and left-ventricular assist devices (RVAD, LVAD) and (b) a wearable LVAD. Reprinted with permission from Wagner et al. (2004). Copyright © 2004, Elsevier.

Polyurethane, natural, and silicone rubbers have been used for constructing balloon-type assist devices such as the left-ventricular assist device (LVAD) shown in Figure 12-22. This is because these materials are thromboresistant. Some use both right- and left-ventricular assist devices, as shown in Figure 12-23, as well as a wearable LVAD. The inner surfaces of the pumps can be coated by Ti microspheres or textured in order to induce tissue growth, resulting in neointima, as depicted in Figure 12-24. Figure 12-25 shows neointima (pseudointima) formation on the textured surfaces of the pumps after about 8 months in vivo. Sometimes the surfaces are coated with heparin and other nonthrombogenic molecules. Infection and thrombus formations are the major problems causing failure.

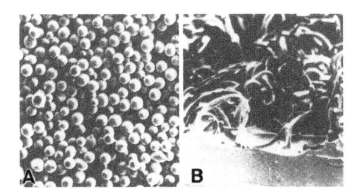

Figure 12-24. Coating the inner surface of the heart-assist devices with (**a**) Ti microspheres (50–75 μm diameter) and (**b**) polyurethane felt (18-μm diameter fibrils). Reprinted with permission from Wagner et al. (2004). Copyright © 2004, Elsevier.

Figure 12-25. Neo(pseudo)intima formation on the textured surface of Ti spheres coated (**a**) and polyurethane fibrils coated (**b**). Reprinted with permission from Wagner et al. (2004). Copyright © 2004, Elsevier.

12.6. ARTIFICIAL ORGANS

The ultimate triumph of biomaterials science and technology would be to make implants behave or function the same way as the organs or tissues they replace without affecting other tissues or organs, and without any negative effect on the patient's mental condition. True regeneration of the natural organ would of course be superior to any artificial one, but that is beyond the scope of biomaterials. However, some organs such as the liver could be tissue engineered, as discussed in Chapter 16. As mentioned in the introduction, most implants are designed to substitute mechanical functions, passive optical functions (eye lens), or passive diffusive functions. The electrical functions can be taken over by some implants (pacemakers), and some primitive yet vital chemical functions (passive diffusion) can also be delegated to implants (kidney dialysis machine and oxygenator). Most of the *artificial heart* and *heart assist devices* use a simple balloon and valve system to circulate the blood. In all cases a balloon or membrane is used to displace blood. A simpler heart device is the intraaortic balloon, which is placed in the descending aorta. During the diastolic phase of the heart the balloon is inflated to prevent backflow.

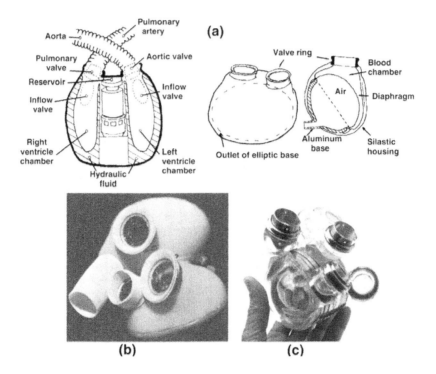

Figure 12-26. Artificial hearts: (**a**) Schumacker-Burns electrohydraulic heart [Reprinted with permission from Lee and Neville (1971). Copyright © 1971, Pasadena Technology Press.]; (**b**) Jarvik heart and its actual photograph [Reprinted with permission from Kolff (1975). Copyright © 1975, Springer.]; (**c**) totally implantable artificial heart [courtesy of Abiomed Inc., Danvers, MA, 2006].

12.6.1. Artificial Hearts

Heart disease afflicts many people in industrial societies. Several artificial hearts are shown in Figure 12-26. Although the design principle and material requirements are the same as those for assist devices, the power consumption (about 6 watts) is quite high for the device to be completely implanted for some early devices. In some devices the power is introduced through a percutaneous device (§11.2.1) in the form of compressed air or electricity. Such an external power unit was used with the first heart replacement done for Dr. Barney Clark at the University of Utah in 1982. The artificial heart kept the patient alive for 112 days, but he suffered seizures, and the refrigerator-sized external power unit restricted his movements. More recently, developments in battery and electrical methods have enabled an artificial heart powered by an internal battery and external battery pack the size of a book to be implanted in a human volunteer. The patient's quality of life was so poor he said "This is nothing, nothing like I thought it would be. ... If I had to do it over again, I wouldn't do it. No ma'am. I would take my chances on life." He suffered strokes, breathing problems, and excruciating pain. Bills were sent to the family in error. The patient's wife asked the surgeon to find whatever was left of Mr. Quinn's original heart, and to put it back inside his chest. In death, she wanted her husband to have a human heart. Some researchers believe artificial hearts may be useful for short-term use in keeping patients with end-stage heart disease alive until a transplant heart becomes available. The small, totally implantable hearts are now being used on a small-scale experimental basis until heart transplantation. The social and economic issues prevent wider use of this and other medical devices.

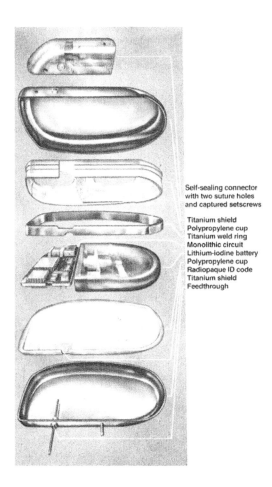

Self-sealing connector
with two suture holes
and captured setscrews

Titanium shield
Polypropylene cup
Titanium weld ring
Monolithic circuit
Lithium-Iodine battery
Polypropylene cup
Radiopaque ID code
Titanium shield
Feedthrough

Figure 12-27. A typical pacemaker consists of a power source and electronic circuitry encased in solid plastic. The electrical wires are coated with a flexible polymer, usually a silicone rubber. Courtesy of Medtronic Inc., Minneapolis, MN.

12.6.2. Cardiac Pacemaker

A *cardiac pacemaker* is used to assist the regular contractile rhythm of heart muscles. The sino-atrial (SA) node of the heart originates the electrical impulses which pass through the bundle of His to the atrio-ventricular (AV) node. In the majority of cases of disease in natural heart rhythm control, pacemakers are used to correct the conduction problem in the bundle of His. Basically, the pacemakers should deliver an exact amount of electrical stimulation to the heart at varying heart rates. The pacemaker consists of conducting electrodes attached to a stimulator, as shown in Figure 12-27. Single-chamber, dual-chamber, and adaptive-rate pacemakers are available. Similar technology could be used for such other purposes as brain pacemakers for Parkinson's disease or other tremor sufferers. The basic components are similar to cardiac pacemakers. The stimulating tip is placed in the thalamus, as illustrated in Figure 12-28. In contrast to cardiac pacemkers, which are established therapy, brain stimulators are currently a research topic.

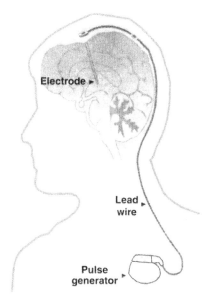

Figure 12-28. Brain pacemaker.

The electrodes are well insulated with rubber (usually silicone or polyurethane), except for the tips, which are sutured or directly embedded into the cardiac wall, as depicted in Figure 12-29. The tip is usually made of a noncorrosive noble metal with reasonable mechanical strength such as Pt–10%Ir alloy. The most significant problems are the fatigue of the electrodes (they are coiled like springs to prevent this) and the formation of collagenous scar tissue at the tip, which increases threshold electrical resistance at the point of tissue contact. The battery and electronic components are sealed hermetically by a titanium case, while the electrode outlets are sealed by a polypropylene cuff.

Pacemakers are usually changed after 2–5 years due to the limitation of the power source. A nuclear energy powered pacemaker is commercially available. Although this and other new power packs (such as lithium battery) may lengthen the life of the power source, the fatigue of the wires and diminishing conductivity due to tissue thickening limits the maximum life of a pacemaker to less than 10 years. A porous electrode at the tip of the wires may be fixed to the cardiac muscles by tissue ingrowth, as in the case of a vascular prosthesis. This may diminish the interfacial problems, as shown in Figure 12-30.

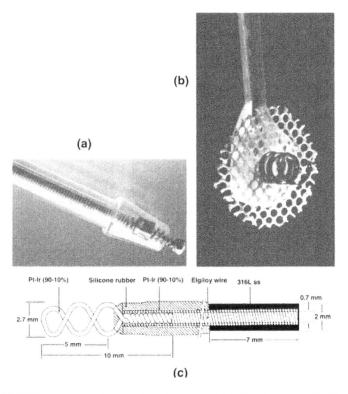

(b)

(a)

(c)

Figure 12-29. Different types of pacemaker electrodes: (**a**) ball-point electrode (Cordis, Miami Lakes, FL). The ball has a 1-mm diameter and the surface area is 8 mm²; (**b**) screw-in electrode (Medtronic, Minneapolis, MN); (**c**) details of an arterial electrode (Medtronic 6991). Reprinted with permission from Greatbatch (1981). Copyright © 1981, Chemical Rubber Company.

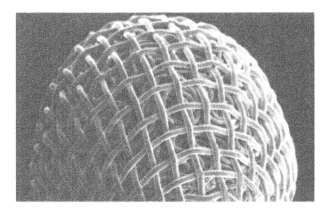

Figure 12-30. Amundsen/CPI porous electrodes. Reprinted with permission from Amundsen et al. (1979). Copyright © 1979, Futura Publishing.

Example 12-3

A bioengineer is designing an arterial stent from NiTi alloy and polyester cloth to enlarge an atherosclerotic artery. Calculate the hoop stress in an 8-mm diameter artery with a thickness of 1 mm due to a blood pressure of 90 mm Hg. Assume the artery to be a uniform tube. Discuss the relevance of the stress in the design process.

Answer

Assuming uniform thickness and diameter, the Laplace Eq. (9-10) can be applied,

$$T = P_i\, r\, /\, t,$$

where T is the wall stress, P_i is the internal pressure, and r is the radius of the vessel. If one assumes about 90 mm Hg for P_i, then

$$P = 9 \text{ cm Hg } (1333 \text{ Pa/cm Hg}) = \underline{12 \text{ kPa}},$$

$$T = P_i\, r\, /\, t = 12 \text{ kPa } (8 \text{ mm/2})/1 \text{ mm} = \underline{50 \text{ kPa}}.$$

This stress is relatively low. Expansion of the artery depends upon the stress–strain relation for the artery. This relation is nonlinear, and the effective modulus for physiological pressure is low. To design a stent to expand an artery beyond the diameter due to normal blood pressure, the engineer needs more detail about the properties of the artery. Since an atherosclerotic vessel has a higher stiffness than a natural one, a higher hoop stress must be applied to expand it.

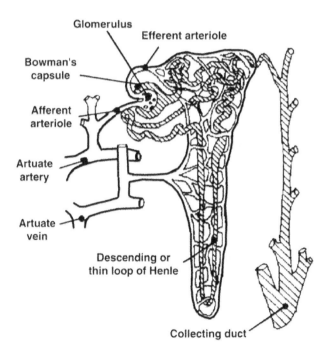

Figure 12-31. Schematic diagram of a kidney nephron.

12.6.3. Artificial Kidney Dialysis Membrane

The primary function of the kidney is to remove metabolic waste products. This is accomplished by passing blood through the glomerulus (Figure 12-31) under a pressure of about 75 mm Hg. The glomerulus contains up to 10 primary branches and 50 secondary loops to filter the load. The glomeruli are contained in the Bowman capsule, which in turn is a part of the nephron. The nephron is the functional unit of the kidney (see Figure 12-31). The main filtrate is urea (70 times the urea content of normal blood), followed by sodium, chloride, bicarbonate, potassium, glucose, creatinine, and uronic acid.

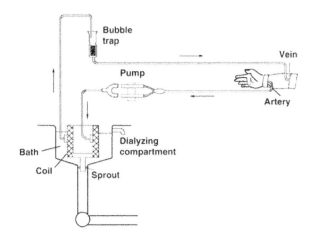

Figure 12-32. Schematic diagram of a typical dialyzer.

The artificial kidney uses a synthetic semipermeable membrane to perform the filtering action in a way similar to that of a natural kidney. The membrane is the key component of the artificial kidney machine. In fact, the first attempt to filter or dialyze blood with a machine failed due to an inadequate membrane. In addition to a membrane filter, the kidney dialyzer consists of a bath of saline fluid into which the waste products diffuse out from the blood, and a pump to circulate blood from an artery and return the filtered blood to the vein, as shown in Figure 12-32. Hemofiltration is similar to hemodialysis, except that the fluid is removed from whole blood, as shown in Figure 12-33. Solute and water are removed by convection. The membrane materials are similar to hemodyalysis but use a higher-molecular-weight (up to 50,000, but usually 20,000 g/mol) filtering membrane. This technique is not popular due to the difficulty of balancing the filtration rate, balancing electrolyte concentrations, and requiring high sensitivity of the monitoring machine.

There are basically three types of membranes for the kidney dialyzer, shown in Figure 12-34. The flat-plate type membrane, historically was developed first and can have two or four layers. The blood passes through the spaces between the membrane layers while the dialysate is passed through the spaces between the membrane and the restraining boards. The second and most widely used type is the coil membrane, in which two cellophane tubes (each with 9 cm in circumference and 108 cm long) are flattened and coiled with an open-mesh spacer material made of nylon. The newest type of kidney is made of hollow fibers. Each fiber has dimensions of 255 and 285 mm inside and outside diameter, respectively, and are 13.5 cm long. Each unit contains up to 11,000 hollow fibers. The blood flows through the fibers while dialysate is passed through the outside of the fibers. The operational characteristics of the various dialyzers are given in Table 12-5. The fibers can also be made from (soda-lime) glass, which is made porous by phase separation techniques. This type of glass fiber has an advantage over the organic fiber in that it can be reused after cleaning and sterilizing. However, one should be careful in supposing that it is always good to reuse the medical products since they are quite expensive, and some products may be completely safe when reused. It is not practical to reuse them, since the patients have some reservations and the manufacturing companies always object to such practice since they cannot sell as many. Also, there are concerns about the possible litigation arising from such practices.

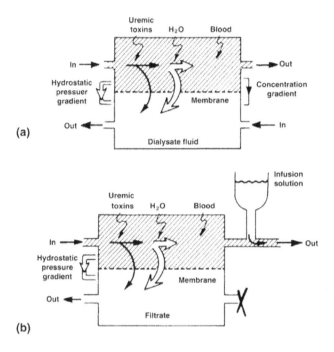

Figure 12-33. Schematic representation of hemodialysis (**a**) and hemofiltration (**b**). Reprinted with permission from Malchesky (2004). Copyright © 2004, Elsevier.

There have been some efforts to improve dialyzers by using charcoals. The blood can be circulated directly over the charcoal or the charcoal can be made into microcapsulates incorporating enzymes or other drugs. One drawback of activated carbon filtering is its ineffective absorption of urea, which is one of the major byproducts to be eliminated by dialysis.

The majority of dialysis membranes are made from cellophane, which is derived from cellulose. Ideally, the membrane should selectively remove all the metabolic wastes as does the normal kidney. Specifically, the membrane should not selectively sequester materials from dialyzing fluid, should be blood compatible so that an anticoagulant is not needed, and should have sufficient wet strength to permit ultrafiltration without significant dimensional changes. It should allow passages of low-molecular-weight waste products while preventing passage of plasma proteins.

There are two clinical grade cellophanes available: Cuprophane® (Bemberg Co., Wuppertal, Germany) and Visking® (American Viscose Co., Fredricksberg, VA). The cellophane films contain 2.5-mm diameter pores that can filter molecules smaller than 4,000 g/mol.

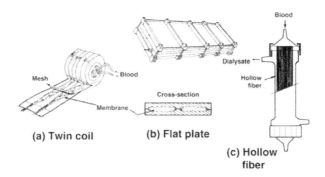

Figure 12-34. Three types of artificial kidney dialyzer: (**a**) twin coil, (**b**) flat plate, and (**c**) hollow fibers.

Table 12-5. Comparison of the plate and Coil Artificial Kidneys

Characteristics	Flat plate (2 layers)	Coil (twin)
Membrane area (m²)	1.15	1.9
Priming volume (liter)	130	1000
Pump needed?	No	Yes
Blood flow rate (milliliter/min)	140–200	200–300
Dialysate flow rate (liter/min)	2.0	20–30
Blood channel thickness (mm)	0.2	1.2
Treatment time (hr)	6–8	6–8

Reprinted with permission from Cooney (1976). Copyright © 1976, Marcel Dekker.

There have been many attempts to improve cellophane membrane wet strength by crosslinking, copolymerization, and reinforcement with other polymers such as nylon fibers. Also, the surface has been coated with heparin in order to prevent clotting. Other membranes such as copolymers of polyethylene glycol and polyethyleneterephthalate (PET) can filter selectively due to their alternate hydrophilic and hydrophobic segments. High-strength amorphous polymers such as polysulfone, polycarbonate, and polymethylmethacrylate are also being used to make membranes. Besides improving the membrane for better dialysis, the main thrust of kidney research is to make the kidney machine more compact (portable or wearable kidney, see Figure 12-35a) and less costly (home dialysis, reusable or disposable filters, etc.).

The other important factor in dialysis is the use of a cannula, which is used to gain access to blood vessels. In order to minimize repeated trauma on the blood vessels, the cannula is sometimes implanted for a long period for chronic kidney patients. Also for the same reason, a single-needle dialysis technique has been developed (depicted in Figure 12-35b).

In addition to hemodialysis by a machine, dialysis can also be carried out by using the patient's own peritoneum, which is a semipermeable membrane. The blood is brought to the membrane through the microcirculation of the peritoneum while dialysate is introduced into the peritoneal cavity through a catheter, one of which is shown in Figure 11-15 implanted in the abdominal wall. The dialysate is drained through the same catheter after solute exchange takes place and is replaced by a fresh bottle. Glucose is added to the dialysate to increase its osmotic pressure gradient for ultrafiltration since it is impossible to obtain a high hydrostatic pressure gradient.

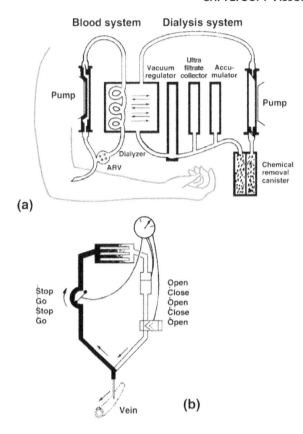

Figure 12-35. (**a**) Diagram of wearable artificial kidney. (**b**) Schematic arrangement of the single-needle dialysis of Dr. Klaus Kopp. The pump operates continuously or is intermittently synchronized with the inflow and outflow of blood. Reprinted with permission from Kolff (1976). Copyright © 1976, Springer-Verlag.

A sorbent cartridge that regenerates the dialysis fluid for reuse has been tried. The cartridge was originally developed for hemodialysis and could reduce water requirements substantially, making dialysis more portable. A schematic diagram of the sorbent regenerated cartridge is shown in Figure 12-36. It is interesting to note that in order to remove urea, enzymolysis is carried out by urease since carbon cannot absorb urea effectively, as mentioned before.

Example 12-4

Calculate the number of ions released in a year from a platinum pacemaker tip. Assume that the average current flow is 10 μA and that the surface area is 1 cm^2.

Answer

$Pt \rightarrow Pt^{++} + 2e^-$, so that the number of atoms per year is

$$10 \times 10^{-6} \frac{coul}{s} \frac{3.15 \times 10^7 \text{ s/yr}}{1.6 \times 10^{-19} \text{ coul/electron}}$$

1.97×10^{21} electrons/yr $= \underline{9.9 \times 10^{20}}$ Pt atoms/yr.

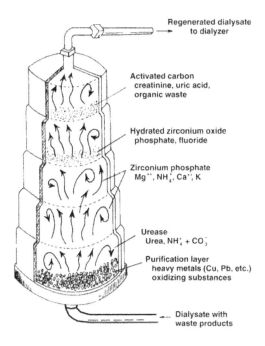

Regenerated dialysate
to dialyzer

Activated carbon
creatinine, uric acid,
organic waste

Hydrated zirconium oxide
phosphate, fluoride

Zirconium phosphate
Mg^{++}, NH_4^+, Ca^{++}; K

Urease
Urea, NH_4^+ + CO_3^-

Purification layer
heavy metals (Cu, Pb, etc.)
oxidizing substances

Dialysate with
waste products

Figure 12-36. Schematic diagram of a sorbent regeneration cartridge. Reprinted courtesy of R.A. Ward.

Example 12-5

Is atherosclerosis irreversible? Survey the literature to find an answer.

Answer

Atherosclerosis, or hardening of the arteries, is commonly viewed as a consequence of aging. Recently it has been observed that in human volunteers 40 to 60 years old who practiced calorie restriction (Fontana et al., 2004), blood pressure normalized at childhood levels (99 mm Hg systolic, 61 mm Hg diastolic), blood lipid levels improved substantially, and carotid artery thickening (as evaluated by ultrasound) reversed itself.

PROBLEMS

12.1. What will be the blood urea nitrogen concentration after 5 and 10 hours of dialysis if the initial concentration is 100 mg% (where mg% means concentration in mg per 100 ml). The concentration after dialysis can be expressed exponentially,

$$C_t = C_0 \exp\left(\frac{Q_b(b-1)t}{V}\right),$$

in which C_0 is the original dialysate concentration, Q_b is the blood flow rate, t is time, V is the volume of body fluid (60% of body weight), and b is a constant determined by mass transfer coefficient (K), Q_b, and membrane surface area (A) by exp (KA/Q_b) according to Cooney (1976).

12-2.   A porous polyurethane rubber arterial prosthesis was implanted in dogs, and it was found that half pores were filled with tissue after 3 months of implantation. The prosthesis had a 5-mm inside diameter and was 1 mm thick, 10 cm long, and had a porosity of 30%.

   a.   Calculate the wall tension assuming a (high) blood pressure of 200 mm Hg immediately after implantation: $T = P \times r$; 133.3 Pa = 1 mm Hg.

   b.   Calculate the maximum force (N) on the arterial wall developed by the blood pressure (200 mm Hg).

   c.   Calculate the maximum stress (MPa) on the arterial wall developed by the blood pressure (200 mm Hg).

   d.   What is the most important factor related to success of the artificial vascular graft?

   e.   The porous polyurethane rubber graft leaked blood. What would you do to prevent leakage of blood after implantation?

   f.   What is the name of the anticoagulant used to prevent clotting of the prosthesis?

12-3.   Select the most related one.

   a.   low coefficient of friction   1. bone ( )
   b.   tensile force transmission   2. skin ( )
   c.   Laplace equation   3. tendon ( )
   d.   Haversian system   4. artery ( )
   e.   elastin   5. ligamentum nuchae ( )
   f.   Langer's line   6. cartilage ( )

   a. Al_2O_3 1. fluidized bed ( )
   b.   hydroxy-apatite   2. resorbable ceramic ( )
   c.   pyrolytic carbon   3. sapphire ( )
   d.   calcium phosphate   4. mineral phase of bone ( )
   e.   Bioglass   5. piezoelectric ( )
   f.   barium titanate   6. SiO_2–CaO–Na_2O–P_2O_5 ( )
   g.   graphite   7. substrate of heart valve disk ( )

   a.   PMMA   1. polyolefin ( )
   b.   polyamide   2. styrene/butadiene copolymer ( )
   c.   poly HEMA   3. natural rubber ( )
   d.   polyacrylamide   4. Teflon ( )
   e.   polytetrafluoroethylene   5. nylon ( )
   f.   SBR   6. soft contact lens ( )
   g.   polyisoprene   7. bone cement ( )
   h.   polypropylene   8. hydrogel ( )

12-4.   Answer the following questions.

   a.   Calculate the maximum wall tension (N/mm) due to internal pressure developed for a 0.5-cm-diameter artery. Assume the maximum pressure will be 250 mm Hg and the artery is uniform in the longitudinal direction.

   b.   If one wishes to replace a 5 cm long section of artery, what is the maximum force exerted on the wall?

   c.   Can one use silicone rubber for the replacement material if the wall thickness is 1 mm and the safety factor is 10?

12-5.   Design a heart valve and give specific materials selected for each part, and explain why you selected the particular material for each part.

12-6.   Is it practical to use porous glass tubules for the hollow-fiber type kidney dialysis machine? Suppose that the porous glass has filtering characteristics similar to those of currently used materials, by virtue of appropriate pore size and tubule thickness.

12-7.   The surface of a kidney dialysis membrane is coated with polyHEMA (hydroxyethyl methacrylate). Discuss the advantages and disadvantages.

12-8.   Why are pacemaker wires coiled?

12-9.   A bioengineer is trying to make an arterial stent from NiTi alloy and polyester cloth to enlarge an atherosclerotic artery. Calculate the minimum hoop stress needed to expand an 8-mm-diameter artery with a thickness of 1 mm. Assume the artery to be a uniform tube.

12-10.   Some researchers tried to correct arrhythmic heart with embryonic stem cells from pigs. They were partially successful in restoring normal function in about half the animals. Discuss the major obstacles of further advancement in such a method for human use. Discuss how you propose to overcome the obstacles.

12-11.   Discuss the ethical and economic aspects of procedures that have a high rate of severe complications. As an example, consider surgery for aortic aneurysms as reviewed in §12.3 and in the references given in the bibliography. Suppose that the cost of the likely complications may be equivalent to the four-year college expenses of several students. If you were 80 years old and had such an aneurysm, would you choose surgery? If you are a 20-year-old student, would you (and several friends) give up your college education to pay for the complications?

DEFINITIONS

Anastomosis: Interconnection between two blood vessels.

Aneurysm: Abnormal dilatation or bulging of a segment of a blood vessel, often involving the aorta or pulmonary artery.

Angioplasty: Surgical repair or unblocking of a blood vessel, especially a coronary artery.

Artificial blood: Use of synthetic polymers [perfloroctyl bromide ($C_8F_{17}Br$) or perfluorodichloroctane ($C_8F_{16}Cl_2$)] or regenerated blood cells to substitute blood.

Autograft: A transplanted tissue or organ transferred from one part of a body to another part of the same body.

Cannula: A thin tube inserted into a vein or body cavity to administer medicine, drain off fluid, or insert a surgical instrument.

Denier: A unit of weight for measuring the fineness of threads of silk, rayon, nylon, etc., equal to 1 gram per 9000 meters.

Dialysate: The part of a mixture that passes through the membrane in dialysis.

> • the solution this forms with the fluid on the other side of the membrane.

> • the fluid used on the other side of the membrane during dialysis to remove impurities.

Dialysis: The separation of particles in a liquid on the basis of differences in their ability to pass through a membrane.

Fibroblast: A cell in connective tissue that produces collagen and other fibers.

Glomerulus: A small tuft or cluster, as of blood vessels or nerve fibers; applied especially to the coils of blood vessels, one projecting into the expanded end or capsule of each of the uriniferous tubules of the kidney.

Incompetence: Incomplete closure of a heart valve.

Left-ventricular assist device (LVAD): Balloon-type heart assist device connected to left ventricle of the heat.

Neointima: New lining, intima (blood vessel lining) formed by ingrowth of tissues through pores of a vascular graft.

Oxygenator: An apparatus by which oxygen is introduced into the blood during circulation outside the body, as during open-heart surgery.

Pacemaker (cardiac): A device designed to stimulate, by electrical impulses, contraction of the heart muscle at a certain rate.

Pericardium: The membrane enclosing the heart, consisting of an outer fibrous layer and an inner double layer of serous membrane.

Pseudointima: See neointima.

Pyrolysis (pyrolytic, adj.): Decomposition brought about by high temperatures.

Regurgitation: Backflow of blood of a heart valve.

Replamineform: Devices or materials made by replicating natural tissues such as sea urchin tentacles. The objective is to achieve better tissue ingrowth into the pores.

Sorbent: A substance that has the property of collecting molecules of another substance by sorption.

Sorption: The ability of some substances to soak up or attract materials such as contaminants and hold them.

Stent: A tubular support placed temporarily inside a blood vessel, canal, or duct to aid healing or relieve an obstruction.

Tenacity: The normalized strength of a fiber; strength per unit size (e.g., expressed in terms of denier). Tensile strength = tenacity $\times$ density $\times$ constant, in appropriate units.

Xenograft: A transplanted tissue or organ transferred from an individual of another species.

BIBLIOGRAPHY

Akutzu T, ed. 1986. *Artificial heart*, Vol. 1. Berlin: Springer-Verlag.

Amundsen D, McArthur W, Mosharrafa M. 1979. A new porous electrode for endocardial stimulation. *Pace* **2**:40.

Antunes MJ, Franco CG. 1996. Advances in surgical treatment of acquired valve disease. *Curr Opin Cardiol* **11**(2):139–154.

Arya MD, Labovitz AJ. 2005. Cardiac valves. In *The bionic human*, pp. 489–522. Ed FE Johnson, KS Virgo. Totowa, NJ: Humana Press.

Atala A. 2001. Bladder regeneration by tissue engineering. *BJU Int* **88**(7):765–770.

Blum U, Voshage G, Lammer J, Beyersdorf F, Tollner D, Kretschmer G, Spillner G, Polterauer P, Nagel G, Holzenbein T. 1997. Endoluminal stent-grafts for infrarenal abdominal aortic aneurysms. *N Engl J Med* **336**(1):13–20.

Bonow RO, Carabello B, de Leon AC, Edmunds LH Jr, Fedderly BJ, Freed MD, Gaasch WH, McKay CR, Nishimura RA, O'Gara PT, O'Rourke RA, Rahimtoola SH, Ritchie JL, Cheitlin MD, Eagle KA, Gardner TJ, Garson A Jr, Gibbons RJ, Russell RO, Ryan TJ, Smith SC Jr. 1998. ACC/AHA task force report: ACC/AHA guidelines for the management of patients with valvular heart disease. *J Am Coll Cardiol* **32**(5):1486–1588. Full text available at http://en.wikipedia.org/wiki/Mitral_stenosis.

Botzko K, Snyder R, Larkin J, Edwards WS. 1979. In vivo/in vitro life testing of vascular prostheses. In *Corrosion and degradation of implant materials*, pp. 76–88. Ed BC Syrett, A Acharya. Philadelphia: ASTM (STP 684:76–88).

Bown MJ, Fishwick G, Sayers RD. 2004. The post-operative complications of endovascular aortic aneurysm repair. *J Cardiovasc Surg (Torino)* **45**(4):335–347.

Bruck SD. 1974. *Blood compatible synthetic polymers: an introduction*. Springfield, IL: Thomas.

Bulbulian AH. 1973. *Facial prosthetics*. Springfield, IL: Thomas.

Chang TMS, ed. 1998. *Blood substitutes: principles, methods, products, and clinical trials*. Basel: Karger.

Cohen JR, Graver LM. 1990. The ruptured abdominal aortic aneurysm of Albert Einstein. *Surg Gynecol Obstet* **170**(5):455–458.

Cooney DO. 1976. *Biomedical engineering principles*. New York: Marcel Dekker.

Coselli JS, LeMaire SA, Conklin LD, Koksoy C, Schmittling ZC. 2002. Morbidity and mortality after extent II thoracoabdominal aortic aneurysm repair. *Ann Thorac Surg* **73**(4):1107–1115; discussion 1115–1116.

Cruz CP, Drouilhet JC, Southern FN, Eidt JF, Barnes RW, Moursi MM. 2001. Abdominal aortic aneurysm repair. *Vasc Surg* **35**(5):335–344.

Dardik H, ed. 1978. *Graft materials in vascular surgery*. Chicago: IL, Year Book Medical Publishers.

Fielder JH. 2005. Social issues. In *The bionic human*, pp. 89–114. Ed FE Johnson, KS Virgo. Totowa, NJ: Humana Press.

Fontana L, Meyer TE, Klein S, Holloszy JO. 2004. Long-term calorie restriction is highly effective in reducing the risk for atherosclerosis in humans. *Proc Natl Acad Sci USA* **101**(17):6659–6663.

Greatbatch W. 1981. Metal electrodes in bioengineering. *CRC Crit Rev Bioeng* **5**:1–36.

Gyers GH, Parsonnet V. 1969. *Engineering in the heart and blood vessels*. New York: Wiley.

Haubold AD, Shim HS, Bokros JC. 1979. Carbon cardiovascular devices. In *Assisted circulation*, pp. 520–532. Ed F Unger. New York: Academic Press.

Haubold AD, Yapp RA, Bokros JC. 1986. Carbons for biomedical applications. In *Encyclopedia of materials science and engineering*, pp. 514–520. Ed MB Beaver. Oxford: Pergamon/MIT Press.

Heijkants RG, van Calck RV, De Groot JH, Pennings AJ, Schouten AJ, van Tienen TG, Ramrattan N, Buma P, Veth RP. 2004. Design, synthesis and properties of a degradable polyurethane scaffold for meniscus regeneration. *J Mater Sci Mater Med* **15**(4):423–427.

Hiratzka LF, Goeken JA, White RA, Wright CB. 1979. In vivo comparison of replamineform, silastic and bioelectric polyurethane arterial grafts. *Arch Surg* **114**:698–702.

Homsy CA, Armeniades CD, eds. 1972. *Biomaterials for skeletal and cardiovascular applications*, Vol. 3. New York: Wiley.

Hufnagel CA. 1983. History of vascular grafting. In *Vascular grafting: clinical applications and techniques*, pp. 1–12. Ed CB Wright. Boston: J. Wright.

Ishii K, Adachi H, Tsubaki K, Ohta Y, Yamamoto M, Ino T. 2004. Evaluation of recurrent nerve paralysis due to thoracic aortic aneurysm and aneurysm repair. *Laryngoscope* **114**(12):2176–2181.

Jimenez JC, Smith MM, E. S. Wilson 2004. Sexual dysfunction in men after open or endovascular repair of abdominal aortic aneurysms. *Vascular* **12**(3):186–191.

Kolff WJ. 1975. Artificial organs and their impact. In *Polymers in medicine and surgery*, pp. 1–28. RL Kronenthal, Z Oser, E Martin. New York: Plenum.

Kolff WJ. 1976. *Artificial organs.* New York: Wiley.

Korossis SA, Fisher J, Ingham E. 2000. Cardiac valve replacement: a bioengineering approach. *Biomed Mater Eng* **10**(2):83–124.

Lee H, Neville K. 1971. *Handbook of biomedical plastics.* Pasadena, CA: Pasadena Technology Press (chapters 3–5, 13).

Liddicoat JE, Bekassy SM, Beall AC, Glaeser DH, DeBakey ME. 1975. Membrane vs bubble oxygenator: clinical comparison. *Ann Surg* **184**:747–753.

Lynch W. 1982. *Implants: reconstructing human body.* New York: Van Nostrand Reinhold.

Malchesky PS. 2004. Extracorporeal artificial organs. In *Biomaterials science: an introduction to materials in medicine*, pp. 514–526. Ed BD Ratner, AS Hoffman, FJ Schoen, JE Lemons. Amsterdam: Elsevier.

Noguchi M, Hirashima S, Eishi K, Takahashi H, Hazama S, Takai H, Koga S. 2004. Surgical treatment of abdominal aortic aneurysm associated with horseshoe kidney and coagulopathy case report. *J Cardiovasc Surg (Torino)* **45**(5):505–509.

Padera RF, Schoen FJ. 2004. Cardiovascular medical devices. In *Biomaterials science: an introduction to materials in medicine*, pp. 470–494. BD Ratner, AS Hoffman, FJ Schoen, JE Lemons. Amsterdam: Elsevier.

Park SJ, Shim WH, Ho DS, Raizner AE, Park SW, Hong MK, Lee CW, Choi D, Jang Y, Lam R, Weissman NJ, Mintz GS. 2003. A paclitaxel-elutin stent for the prevention of coronary restenosis. *N Engl J Med* **348**:1537–1545.

Platt R. 1994. *Smithsonian visual timeline of inventions.* New York: Marcel Dekker.

Presnall JJ. 1996. *Artificial organs.* San Diego, CA: Lucent Books.

Raaf JH. 1991. Vascular access, pumps, and infusion. In *Comprehensive textbook of oncology*, 2nd ed. pp. 583–589. Ed AR Moossa, SC Schimpff, MC Robson. Baltimore: Williams & Wilkins.

Raaf JH, Vinton D. 1993. Vascular access: central venous catheters. In *Current practice of surgery*, pp. 1–11. Ed BA Levine, EMI Copeland, RJ Howard, H Sugerman, AL Warshaw. London: Churchill Livingstone.

Raaf JH, Heil D, Rollins DL. 1994. Vascular access, pumps, and infusion. In *Cancer surgery*, pp. 47–62. Ed RJ McKenna, GP Murphy. Philadelphia: J.B. Lippincott.

Sawyer PN, Kaplitt MH. 1978. *Vascular grafts.* New York: Appleton-Century-Crofts.

Schoen FJ. 1989. *Interventional and surgical cardiovascular pathology: clinical correlations and basic principles.* Philadelphia: W.B. Saunders.

Silver FH. 1994. *Biomaterials, medical devices, and tissue engineering: an integrated approach.* London: Chapman & Hall.

Stanley C, Burkel WE, Lindenauer SM, Bartlett RH, Turcotte JG, eds. 1972. *Biologic and synthetic vascular prostheses.* New York: Grune & Stratton.

Starr A, Fessler CL, Grunkemeier G, He GW. 2002. Heart valve replacement surgery: past, present and future. *Clin Exp Pharmacol Physiol* **29**(8):735–738.

Stolberg SG. 2002. On medicine's frontier: the last journey of James Quinn. *New York Times*, Oct 8.

Voorhees AB, Jaretski A, Blackmore AH. 1952. Use of tubes constructed from Vinyon-N cloth in bridging arterial defects. *Ann Surg* **135**:332–336.

Wagner WR, Borovetz HS, Griffith BP. 2004. Implantable cardiac assist devices. In *Biomaterials science: an introduction to materials in medicine*, pp. 494–507. Ed BD Ratner, AS Hoffman, FJ Schoen, JE Lemons. Amsterdam: Elsevier.

Ward RA. 1982. *Investigation of the risk and hazards with devices associated with peritoneal dialysis and sorbent regenerated dialysate delivery systems.* Revised draft report, FDA Contract No. 223-81-5001.

Welborn MB, Seeger JM. 2001. Prevention and management of sigmoid and pelvic ischemia associated with aortic surgery. *Semin Vasc Surg* **14**(4):255–265.

Wesolowski SA, Fries CC, Martinez A, McMahon JD. 1968. Arterial prosthetic materials. *Ann NY Acad Sci* **146**:325–344.

Williams DF, ed. 1981. *Fundamental aspects of biocompatibility*. Boca Raton, FL: CRC Press.

Williams DF. 1993. *Biomaterials for tomorrow*. London: Institute of Materials.

Wise DL. 2000. *Biomaterials engineering and devices: human applications*, Vol. 2: *Orthopedic, dental, and bone graft applications*. Totowa, NJ: Humana Press.

13

HARD TISSUE REPLACEMENT — I: LONG BONE REPAIR

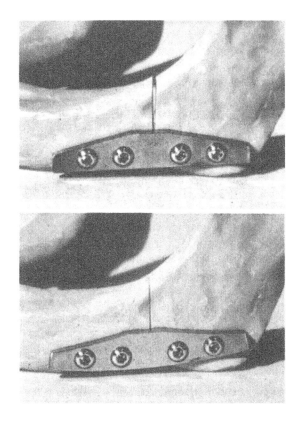

Use of shape memory alloy in closing the jaw bone: (**top**) before applying heat; (**bottom**) after applying heat. Reprinted with permission from Duerig (1990). Copyright © 1990, Butterworths-Heinemann.

Biomaterials used in the surgical treatment of bone fractures are discussed in this chapter. For a fracture to heal properly, the bone ends must be aligned and stabilized. Stabilization may be accomplished externally, as with a cast, or internally (surgically), with screws, plates, pins, or rods.

The design principles, selection of materials, and manufacturing criteria for orthopedic implants are the same as for any other engineering products undergoing dynamic loading. Although it is tempting to duplicate the natural tissues with materials having the same strength and shape, this has been neither practical nor desirable since the natural tissues and organs have one major advantage over man-made implants, that is, their ability to adjust to a new set of circumstances by remodeling their micro- and macrostructure in response to prevailing stress conditions, and to repair damage. Consequently, the mechanical fatigue of tissues is minimal unless a disease hinders the natural healing processes or unless they are overloaded beyond their ability to heal.

When we try to replace a joint or heal a fractured bone, it is logical that bone repairs should be made according to the best repair course that the tissues themselves follow. Therefore, if the bone heals faster when a compressive force (or strain) is exerted, we should provide compression through an appropriate implant design. Likewise, if the compression is detrimental for wound healing, the opposite approach should be taken. Unfortunately, the effects of compressive or tensile forces on the repair of long bones are not fully understood. Moreover, many experimental results thus far are contradictory.

It is believed that the osteogenic (bone generating) and osteoclastic (bone removing) processes are normal activities of the bone in vivo. Thus, the equilibrium between osteogenic and osteoclastic activities is governed according to the static and dynamic force applied in vivo, that is, if more load is applied the equilibrium tilts toward more osteogenic activity to better support the load and vice versa (*Wolff's law*), as shown in Figure 13-1. Of course, excessive load should not be imposed by the implant; too much force can damage the cells rather than enhance their activities.

The causal mechanism underlying bone remodeling has not been definitively identified. Some scientists consider it to be related to the *piezoelectric* phenomenon of bone in which the *strain-generated electric potentials* occur due to asymmetry in the collagen molecules. The strain-generated potentials in living bone are also thought to be a result of streaming potentials from fluid flow in the channels in bone. Strain-generated potentials may stimulate cells to trigger the bone remodeling response. This is the rationale for the electrically stimulated fracture repair of clinical *nonunions*. Bone remodeling also may be stimulated by pressure upon the cells as mediated by tissue fluid, or by flow of the fluid in response to deformation of the bone.

Historically speaking, until Dr. J. Lister's aseptic surgical technique was developed in the 1860s, various metal devices such as wires and pins constructed of iron, gold, silver, platinum, etc., were not successful largely due to infection after implantation. Most of the modern implant developments have been centered around repairing long bones and joints. Lane of England designed a fracture plate in the early 1900s using steel, as shown in Figure 13-2a. Sherman of Pittsburgh modified the Lane plate to reduce stress concentration by eliminating sharp corners (Figure 13-2b). He used vanadium alloy steel for its toughness, ductility, and better corrosion resistance than ordinary steel. Vanadium steel was used clinically for some years, but problems with in-vivo corrosion led to its abandonment. Subsequently, Stellite® (Co–Cr based alloy) was found to be the most inert material for skeletal implantation by Zierold in 1924. Soon 18/8 (18 w/o Cr, 8 w/o Ni) and 18-8sMo (2–4 w/o Mo) stainless steels were introduced for their corrosion resistance, with 18-8sMo being especially resistant in saline solution. Later, another stainless steel (19 w/o Cr, 9 w/o Ni) named Vitallium® was introduced into medicine. (The name Vitallium® is now used for the Co-based alloy.) Another metal, tantalum, was introduced in 1939, but its poor mechanical properties made it unpopular in orthopedics, yet it found wide use in neurological and plastic surgery. During the post-Lister period, all the various designs and materials could not be related specifically to the success of an implant despite many studies. It became customary to remove any metal implant as soon as possible after its initial function was served.

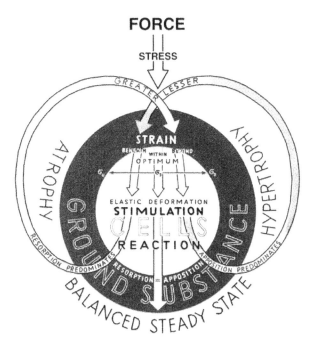

Figure 13-1. Proposed model for the feedback mechanism of the bone remodeling process due to mechanical energy input. Reprinted with permission from Kummer (1976). Copyright © 1976, Springer-Verlag.

Newer additions to the bone plate materials are the biodegradable polymers, especially poly-glycolideide (PGA), as shown in Figure 13-3. This material is self-reinforced with PGA fiber to improve its mechanical properties, as shown schematically in Figure 13-4.

13.1. WIRES, PINS, AND SCREWS

Although the exact mechanism of bone fracture repair by the body is not known at this time, stability of the implant with respect to the wound surfaces is clinically known to be an important factor in facilitating the healing process. Whether fixation is accomplished by compressive or tensile force, the reduction (i.e., the placement of the broken bone ends near their former positions) should be anatomical, and the bone ends should be firmly fixed so that the healing processes cannot be disturbed by unnecessary micro- and macromovement. Bone fractures may be conservatively managed (fixed by external means such as a cast), or surgically reduced and fixed depending upon the extent of the fracture and the experience of the surgeon. Surgical techniques usually involve the use of metallic fixation devices. These vary in complexity from simple pins to complex multicomponent hip nails. Almost all of these devices are made from metal alloys, which were discussed in Chapter 5.

(a)

(b)

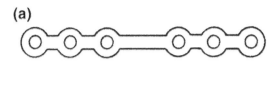

Figure 13-2. Early design of bone fracture plate: (**a**) Lane, (**b**) Sherman.

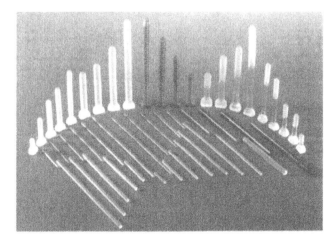

Figure 13-3. Absorbable polymers are used to make pins, screws, and rods for orthopedic applications. Reprinted with permission from van der Elst et al. (2000). Copyright © 2000, Marcel Dekker.

13.1.1. Wires

The simplest but most versatile implants are the various metal wires (called Kirschner wires when the diameter is less than 3/32 inch (2.38 mm) and Steinmann pins for those of larger diameter), which can be used to hold fragments of bones together. Wires can be monofilament or multifilament. Wires are also used to reattach the greater trochanter in hip joint replacements or for long oblique or spiral fracture of long bones. The common problems of fatigue combined with corrosion of metals may weaken the wires in vivo. The added necessity of twisting and knotting the wires for fastening aggravates the problem since strength can be reduced by 25% or more due to the stress concentration effect. The deformed region of the wire will be more prone to corrosion than the undeformed region due to the higher strain energy. The wires are classified as given in Table 13-1.

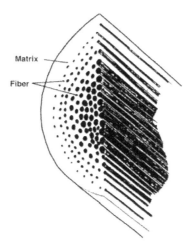

Figure 13-4. Schematic illustration of self-reinforcing biodegradable PGA (P. Tormala et al., US Pat # 4,743,257, 1988). Reprinted with permission from van der Elst et al. (2000). Copyright © 2000, Marcel Dekker.

Table 13-1. Nomenclature and Specifications of Surgical Wires

Suture size	Wire gauge no.	American wire gage (diam) in.	American wire gage (diam) mm	Standard wire gage (diam) in.	Standard wire gage (diam) mm	Diameter (mm) Min	Diameter (mm) Max	Knot-pull tensile str. (kgf), class 3
10-0						0.013	0.025	0.05
9-0						0.025	0.038	0.06
8-0						0.038	0.051	0.11
7-0						0.051	0.076	0.16
6-0	40	0.0031	0.079	0.0048	0.1222			
						0.076	0.102	0.27
6-0	38	0.0040	0.102	0.0060	0.152			
5-0	35	0.0056	0.142	0.0084	0.213	0.102	0.152	0.54
4-0	34	0.0063	0.160	0.0092	0.234			
					0.152	0.203	0.82	
4-0	32	0.0080	0.203	0.0108	0.274			
000	30	0.0100	0.254	0.0124	0.315	0.203	0.254	1.36
00	28	0.0126	0.320	0.0148	0.376	0.254	0.330	1.80
0	26	0.0159	0.404	0.0180	0.457	0.330	0.406	3.40
1	25	0.0179	0.455	0.0200	0.508	0.406	0.483	4.76
2	24	0.0201	0.511	0.0220	0.559	0.483	0.559	5.90
3	23	0.0226	0.574	0.0240	0.610	0.559	0.635	7.26
4	22	0.0254	0.643	0.0280	0.712	0.635	0.711	9.11
5	20	0.0320	0.813	0.0360	0.915	0.711	0.813	11.40
6	19	0.0359	0.912	0.0400	1.020	0.914	1.016	13.60
7	18	0.0403	1.061	0.0480	1.220	0.914	1.016	15.90
	10	0.0109	2.590	0.1280	3.260			
	1	0.2893	7.340	0.3000	7.630			
	0	0.3249	8.250	0.3240	8.230			
	6-0	0.5800	14.700	0.4640	11.800			
	7-0			0.5000	12.700			

13.1.2. Pins

The Steinmann pin is also a versatile implant often used for internal fixation in cases when it is difficult to use a plate (see §13.2) or when adequate stability cannot be obtained by other means. The tip of the pin is designed to penetrate bone easily when the pin end is screwed into the bone. The fluting of the pin end differs from that of screws in that the flute angles of the pin are opposite to those of the screws. The reason for this is, in contrast to a screw or a drill, there is a lack of space between the hole created and the pin. Three types of tip designs are shown in Figure 13-5. The trochar tip is the most efficient in cutting; hence, it is often used for cortical bone insertion. The fractured bones can be held together by two or more pins inserted percutaneously away from the fracture site, and the pins are fixed by a device such as the Hoffmann external fixator shwn in Figure 13-6.

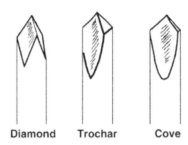

Diamond   Trochar   Cove

Figure 13-5. Types of Steinman pin tips.

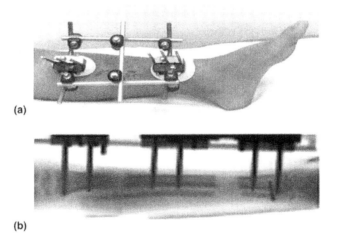

(a)

(b)

Figure 13-6. Hoffman external long-bone fracture fixation device: (**a**) external view, (**b**) x-ray view (http://www.rch.org.au/limbrecon/prof.cfm?doc_id=4873#Orthofix_fixator). Similar external ring fixators such as Ilizarov and Taylor can be used for limb-lengthening operations.

13.1.3. Screws

Screws are some of the most widely used devices for fixation of bone fragments to each other or in conjunction with fracture plates. Figure 13-7 illustrates the various parts of a screw with various head designs. There are basically two types of screws: one is *self-tapping* and the other *non-self-tapping* (Figure 13-8). As the name indicates, a self-tapping screw cuts its own threads as it is screwed in. The extra step of tapping (i.e., cutting threads in the bone) required of the non-self-tapping screws make them less favorable, although the holding power (or pull-out strength) of the two types of screws is about the same. The variations in thread design (Figure 13-9) do not influence holding power. However, the radial stress transfer between the screw thread and bone is slightly less for V-shaped thread than buttress thread, indicating that the latter can withstand longitudinal load slightly better.

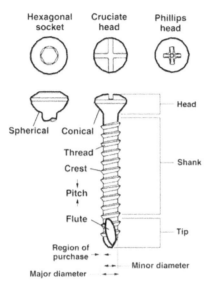

Figure 13-7. Illustration showing the various parts of a self-tapping bone screw, including various head designs. Reprinted with permission from Mears (1979). Copyright © 1979, Williams & Wilkins.

Figure 13-8. Photographs of the points of a self-tapping and a non-self-tapping screw.

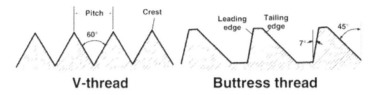

Figure 13-9. Types of thread in screws.

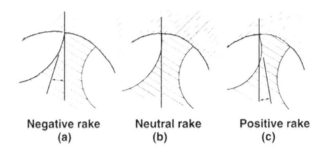

Figure 13-10. Illustration showing the relationship of various rake angles to the outer edge of the cross-section of a cutting flute.

The rake angle of the cutting edge is also an important factor in screw design (see Figure 13-10). A positive rake angle requires higher cutting force yet results in lower cutting temperatures due to less drag, while the opposite effect is obtained with a negative rake angle. Almost all bone screws are made with positive rake angles, while hard metal cutting drills that can withstand larger cutting loads are made with negative rake angles.

The pull-out strength or holding strength of screws is an important factor in the selection of a particular screw design. However, regardless of the differences in design, the pull-out strength depends only on the size (diameter) of the screw, as illustrated in Figure 13-11. The larger screw, of course, has a higher pull-out strength.

The tissues immediately adjacent to the screw often necrose and reabsorb initially, but if the screws are firmly fixed then the dead tissues are replaced by living tissues. When micro- or macromovement exists, a collagenous fibrous tissue encapsulates the screw. This is the reason why loading of the repaired bone by the patient should be delayed until a firm fixation takes place between screw and bone.

Example 13.1

Calculate the breaking strength of the size 00 suture wire given in Table 13-1. Compare the result with the tensile strength of fully annealed 316L stainless steel (Table 6-5).

Answer

Since the knot pulling strength (3.40 kgf) is given in Table 13-1,

$$\text{Tensile strength} = \frac{\text{force}}{\text{area}} = \frac{3.4 \text{ kgf}}{\pi(0.404 \text{ mm})} = 26.52 \text{ kgf/mm}^2.$$

But 1 kgf = 1 kg $\times$ 9.81 m/sec^2 = 9.81 N. So

$$\text{Tensile strength} = 26.52 \times 9.8 \times 106 \text{ Pa} = \underline{260 \text{ MPa}}.$$

The ultimate tensile strength of annealed 316L stainless steel is 73,000 psi or 505 MPa, which is almost twice the knot pulling strength. One may recall that other sutures have a similar strength reduction after knotting, as a result of stress concentration.

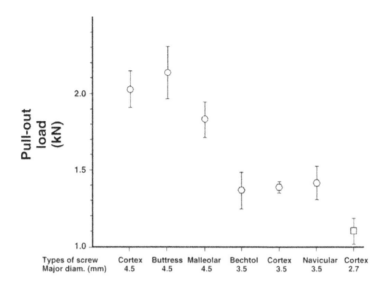

Figure 13-11. Average pull-out strength with 95% confidence limits for various Richards and Richards Osteo bone screws. Reprinted with permission from *Bone screw technical information*. Memphis, TN: Richards Manufacturing Company (1980). Copyright © 1980, Richards Manufacturing Company.

13.2. FRACTURE PLATES

13.2.1. Cortical Bone Plates

There are many different types and sizes of fracture plates, as shown in Figure 13-12. Since the forces generated by the muscles in the limbs are very large, generating large bending moments (see Table 13-2), the plates must be strong. This is especially true for the femoral and tibial plates. The bending moment versus bend angle (rotation) of various devices are plotted in Figure 13-13. In comparison with the bending moment at the proximal end of the femur during normal activities (cf. Table 13-2), one can see that the plates cannot withstand the maximum bending moment applied. Therefore, some type of restriction on the patient's movement is essential in the early stages of healing.

Equally important is the adequate fixation of the plate to the bone with the screws, as mentioned previously. However, overtightening may result in necrosed bone as well as deformed screws which may fail later due to the corrosion process at the deformed region (see §6.3.3).

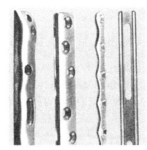

Figure 13-12. Bone plates. From left: Richard-Hirschorn plate (stress is evenly distributed throughout the length of the plate); AO compression plate; Sherman plate (stress is evenly distributed throughout the length of the plate, but the plate is much weaker than the two plates on the left); Egger's plate. Reprinted with permission from Albright et al. (1978). Copyright © 1978, Academic Press.

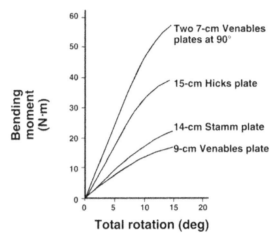

Figure 13-13. Bending moment versus total rotation of various bone plates. Reprinted with permission from Laurence et al. (1969). Copyright © 1969, Charles C. Thomas.

Table 13-2. Greatest Resistable Bending Moment at Proximal End of Femur

| | No. of subjects | | Bending moment (Nm) | |
Muscle group	Men	Women	Range	Mean
Hamstring	11		54–93	72
		17	26–54	35
Quadriceps	6		42–60	51
Hip abductors	6		38–108	63
		3	24–48	39
Hip adductors	6		60–126	81
		3	32–40	30

Reprinted with permission from Laurence et al. (1969). Copyright © 1969, British Editorial Society of Bone and Joint Surgery.

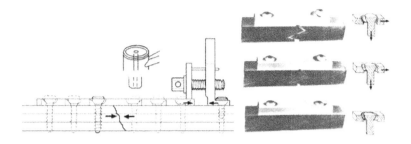

Figure 13-14. Principle of a dynamic compression plate (DCP): method of compression with a device (upper) and illustration of the principle by a model (lower). Reprinted with permission from Allgower et al. (1973). Copyright © 1973, Springer-Verlag.

A bone plate device to compress the ends of fractured bones together is shown in Figure 13-14. A similar effect can be achieved by using a self-compression plate and screw system. The added complexity of the devices and the controversy as to whether the compressive forces or strain is beneficial had hindered early acceptance of the devices. It is interesting to note that traditionally a large amount of *callus formation* has been considered to be a favorable sign and even essential for good healing. However, with the use of the compression plate, the opposite is thought to be a more favorable sign of healing. In this situation the amount of callus formed is proportional to the amount of motion between plate and bone. One major drawback of healing by rigid plate fixation is weakening of the underlying bone such that refracture may occur following removal of the plate. This is largely due the *stress-shield effect* upon the bone underneath the rigid plate. The stiff plate can carry so much of the load that the bone is understressed, so that it is reabsorbed by the body according to Wolff's law. In addition, the tightening of the screws that hold the bone plate creates a concentrated stress upon the bone; this can have a deleterious effect if the screws are in cancellous bone. In order to alleviate osteopenia due to stress shielding, resorbable bone plates made of poly-l-glycolic acid (PLGA) have been tried experimentally. Such plates are available for small bones, as shown in Figure 13-2. It can be difficult to achieve uniform resorption of this kind of implant and to achieve the correct rate of resorption. In addition, it has been difficult to make sufficiently strong screws from this type of material. This problem is partially solved by reinforcing the PGA with fibrous PGA, as shown in Figure 13-3. If conventional metal screws are used instead, they are usually removed surgically after healing, as is the case with conventional metal bone plates. Moreover, manipulation of the resorbable plate in situ by the surgeon is difficult, if not impossible; additional equipment such as an oven may be required to heat and shape these thermoplastic materials to fit the bone.

It would be desirable to reduce the problems of refracturing of bone due to screw holes (Figure 13-15) and weakened osteoporotic bones under the fixation plate. These problems may be reduced if one were to design the plate as shown in Figure 13-16, where the polymer insert (UHMWPE) can be made to creep (slowly deform with time), transferring load more toward to bone, which will result in a reduced osteoporotic effect (see Figure 13-17).

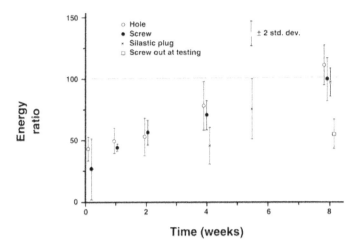

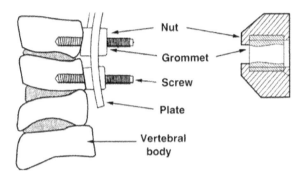

Figure 13-15. Effect of holes on the healing of cortical rabbit bones. Reprinted with permission from Curry and Frankel (1972). Copyright © 1972. Charles C. Thomas.

Figure 13-16. Bone plate with a polymer insert (grommet) to reduce osteoporosis by transferring load from the plate-screw system to the bone as the bone heals and the polymer deforms applied to spinal disc fusion. Reprinted with permission from Park et al. (1999). Copyright © 1999, University of Iowa Research Foundation.

13.2.2. Cancellous Bone Plate

A considerable amount of care must be exercised when fixing cancellous bone since this kind of bone has lower density and much lower stiffness and strength than cortical bone. An example of the fixation of the ends of a long bone is shown in Figure 13-18, in which the fractured bones are fixed with a combination of screws, plates, bolts, and nuts. The bulk necessary for adequate stabilization of the fracture increases the chance of infection near the site of implants.

Oftentimes small bones and cancellous bone can be fixed with absorbable screws, as shown in Figure 13-19. Sometimes one can fix a cancellous bone fracture by using a simple nail, as shown in Figure 13-20. This is a special case since the patient was a young child (who incidentally had two to three times the trabecular bone mass normally present in the adult cancellous femoral head and neck region), and since the epiphyseal plate lies close to the hip joint the loading is essentially normal to the fracture surface. Also, the freely mobile hip joint relieves stress except in the compression cycle. Obviously, a wide range of choices is available. The choice is largely determined by the surgeon(s), not the patients, and biomedical engineers.

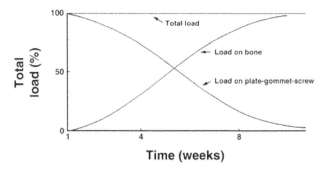

Figure 13-17. Schematic illustration of load transfer from the plate-screw system to the bone depicted in Figure 13-16. Modified with permission from Goel et al. (1991). Copyright © 1991, American Association of Neurological Surgeons.

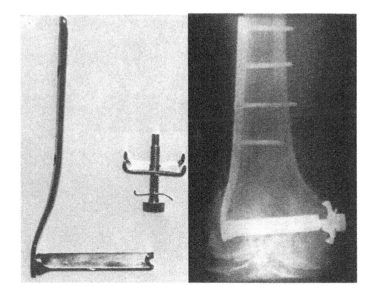

Figure 13-18. Devices to fix cancellous bone in a supracondylar fracture of the femur. Reprinted with permission from Brown and D'Arcy (1971). Copyright © 1971, Charles C. Thomas.

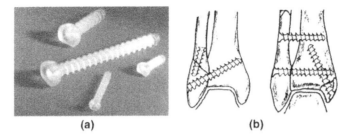

(a)   (b)

Figure 13-19. Biodegradable PGA bone screws (**a**) and depiction of the application in cancellous bone fractures (**b**). Reprinted with permission from van der Elst et al. (2000). Copyright © 2000, Marcel Dekker.

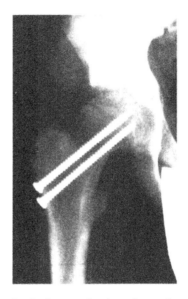

Figure 13-20. Example of a simple fracture fixation of cancellous bone. Reprinted with permission from Frost (1973). Copyright © 1973, Charles C. Thomas.

Example 13-2

A bioengineer is asked to make a composite material from carbon fiber and PMMA resin (as a matrix) for a bone fracture plate. The following data are given.

Material	E (GPa)	Density (g/cm^3)	Strength (MPa)
Carbon fiber	250	1.95	5,000
PMMA	3	1.20	70

a.   What is the amount of carbon fiber required to make the modulus of the plate 100 GPa? Assume the fibers are aligned in the direction of the test.

b.   How many fibers are required if the diameter of the fibers is 10 μm and the cross-sectional area of the specimen is 1 cm^2?

c.   How much stress in the fiber direction can the composite take?

Answer

a. Using a simple mixture rule, since the fibers are aligned,

$$E_c = E_f V_f + E_m V_m,$$ (1)

$$V_f + V_m = 1.$$ (2)

Substituting (2) into (1) and rearranging, we have

$$V_f = \frac{E_c - E_m}{E_f - E_m} = \frac{97}{247} = \underline{0.39 \ (39 \ v/o)}.$$

b. Volume of fibers = $\pi r2NL$, where N is the total number of fibers and L is the length of the composite.

Volume of composite = 1 cm^{2}L:

$$\frac{V_f}{V_{total}} = \frac{\pi r^2 N \ L}{1 \ L} = 0.39,$$

$$N = \frac{0.39 - 1}{\pi r^2} = \frac{0.39}{\pi (10 \times 10^{-4})^2} = \underline{1.24 \times 10^5}.$$

c.

$$\sigma_c = \sigma_f V_f + \sigma_m V_m,$$

$$\sigma_c = 5000 \times 0.39 + 3000 \times 0.02 \times 0.61 \ = \underline{1.99 \ GPa}.$$

The fiber can only stretch 2% (5/250) at its maximum load, while the PMMA can stretch 2.3% (70/300) at its maximum load. Actually, in this geometry both the fibers and matrix undergo the same strain when the composite is loaded along the fibers, so that the actual strength will be somewhat less than the value predicted by the simple rule of mixtures.

13.3. INTRAMEDULLARY DEVICES

Intramedullary devices are used to fix fractures of long bones. The devices are snugly inserted into the medullary cavity (Figure 13-21). This type of implant should have some spring in it to exert some elastic force inside the bone cavity to prevent rotation of the device and to fix the fracture firmly.

Compared to plate fixation, the intramedullary device does not involve placement of multiple screws. However, its torsional resistance is much less than that of the plate. It is also believed that the intramedullary device destroys the intramedullary blood supply. Even so, it does not disturb the periosteal blood supply, in contrast to the case of plate fixation. Another advantage of the intramedullary device is that it does not require opening of a large area to operate, and the device can be nailed through a small incision.

There have been many studies concerned with the medullary blood supply and on fracture healing in view of the extensive damage to the medullary canal by insertion of the device. The long bone blood supply comes from three sources: the nutrient arteries and their intramedullary branches, the metaphyseal arteries, and the periosteal arteries. If fracture occurs, the extraosseous circulation from the surrounding soft tissues becomes active and forms the fourth source of blood supply. When the intramedullary canal is reamed and a device is inserted, the nutrient artery and its branches are destroyed. Nevertheless, this procedure does not significantly damage to the viability of the bone. Thus, a solid reunion can be achieved with this method of treatment. Other studies have demonstrated that a tight-fitting nail delays healing due to the time necessary to reestablish intracortical (or intramedullary) circulation.

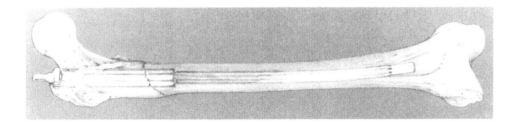

Figure 13-21. Illustration of an intramedullary device used in the femur.

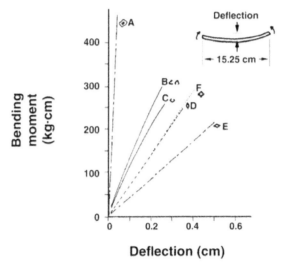

Figure 13-22. Bending deflections of the femur and of three intramedullary nails (9-mm cloverleaf, Schneider, and diamond-shaped) are compared as they would appear under identical loading. The length of each structure is taken as 15.25 cm. Curve A for the femur shows the bone to be more rigid than any of the nails. The cloverleaf nail is stiffer with the slot in tension (curve B) than in compression (curve C). The diamond-shaped nail is 50% more rigid when bent in its major plane (curve D) than in its minor plane (curve E). The Schneider nail has the same rigidity as the diamond-shaped nail (curve F), but even in its most unfavorable orientation the Schneider nail has a higher ultimate bending strength (not shown). Reprinted with permission from Soto-Hall and McCloy (1953). Copyright © 1953, Lippincott, Williams & Wilkins.

There are many different types of intramedullary devices, varying to a large extent only in their cross-sectional shapes. For a given size of device, the resistance to bending and torque is different for different devices, as shown in Figures 13-22 and 13-23. The closed (solid) designs of the four-flanged nail (Schneider) and the diamond nail showed higher resistance to torsion than the open cloverleaf nail. However, in bending, the cloverleaf nail showed the highest resistance due to its larger bending moment of inertia.

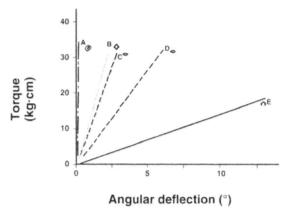

Angular deflection (°)

Figure 13-23. Curves of torque versus angular deflection for the femur and three intramedullary nails. The femur (curve A) is more rigid than any of the nails. The Schneider nail (curve B) is about one tenth as rigid as the femur. As the length of the diamond shape is doubled, the rigidity is halved (curves C and D). The cloverleaf nail is the least rigid (curve E). The length of each nail is 20.25 cm except for curve D. Reprinted with permission from Soto-Hall and McCloy (1953). Copyright © 1953, Lippincott, Williams & Wilkins.

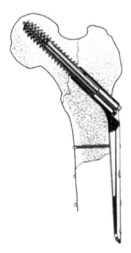

Figure 13-24. Illustration of hip nail on a fix fractured femoral head. Courtesy of DePuy, Division of Bio-Dynamics Inc., Warsaw, IN.

The intramedullary nail is usually plated to the proximal femur for fixation of a femoral neck and intertrochanteric bone fracture, as shown in Figure 13-24. There are many different types of cross-sectional area designs to prevent rotation after fixation, as shown in Figure 13-25. Note that all of them except the V-shaped devices have a guide hole in the center. A femoral neck fracture fixation is usually made to compress the broken bones together by tightening a screw, which also helps to stabilize the fracture.

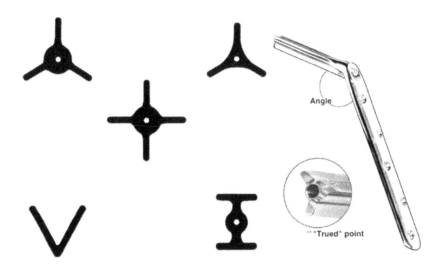

Figure 13-25. Cross-section of various hip nails and a typical implant and its insertion tip. Courtesy of Smith & Nephew Richards Inc., Memphis, TN.

13.4. ACCELERATION OF BONE HEALING

13.4.1. Electrical, Electromagnetic, Ultrasound and Mechanical Stimulation

It has been recognized that some of the most challenging problems in orthopedic surgery are the clinical nonunions or delayed unions of fractures and congenital and acquired pseudo-arthroses. The latter term means false joint; it is a gap in the bone allowing abnormal movement. Until recently, the usual methods of treatments were grafting, plating, and nailing. When one or a combination of any of the methods failed to repair the gap in the bone, the patient was left with a nonfunctional limb. Amputation was done to allow the patient to walk with an artificial leg. Many investigators demonstrated that the application of electromagnetic energy by means of *direct* (or *alternating*) *current* or a pulsing *electromagnetic field* may enhance and stimulate osteogenic activities. Stimulation by vibration or ultrasound has also been tried. The exact mechanism of the stimulation is not yet well understood. It is thought that the electrical and mechanical stimulation causes increases in ionic and metabolic activity and changes the balance in cell differentiation and proliferation (Figure 10-6). The extra electrical, mechanical, or thermal energy input into the wound area seems to trigger more osteogenic cellular activities. Bone tissue responds favorably only over appropriate ranges of input signal in the case of electrical stimulation: 10–40 μWatts, 5–20 μAmperes, and less than 1 Volt. The response to the stimulation is also closely related to the nature of the electrode material, the surface area, and location.

One commercial electrical stimulator is shown in Figure 13-26. The four electrodes (cathodes) are inserted into the fracture site transcutaneously after drilling the bone, and the positive electrode (anode) is placed over the skin using a conducting pad. The electrode can be used with or without internal fixation devices, as shown in Figure 13-27. This device is not currently available.

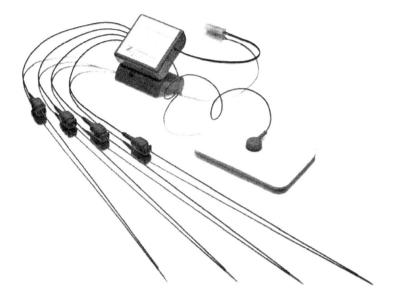

Figure 13-26. Commercial electrical stimulator for bone fracture repair. Courtesy of Zimmer USA, Warsaw, IN.

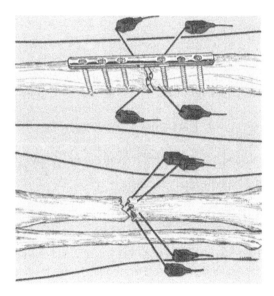

Figure 13-27. Schematic illustration of the use of an electrical stimulator with or without fracture fixation devices.

The electromagnetic stimulators use a pair of Helmholtz coils, which are aligned across the wound site. Electrical current in the coils generates a magnetic field that penetrates the tissues; the changing magnetic field induces an electric field, hence also an electric current in the bone. The coils are excited with a monophasic waveform with a 150-msec period, and with a repetition rate of 75 Hz. The stimulation is applied as shown in Figure 13-28. This pulse input induces 1–1.6 mV/cm of potential gradient in the bone. The magnetic stimulation has one big advantage over the direct current stimulation: it is a noninvasive technique. The efficacy of both types of stimulation is about the same: an over 70% success rate.

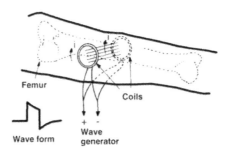

Figure 13-28. Representation of pulsed electromagnetic field bone stimulation.

As for mechanical stimulation with vibration or ultrasound, little is known about the dose–response relation. It is, however, known that a relatively small number of cycles at high intensity will stimulate formation of bone. Weightlifters, for example, have dense bone. Weightlifting by older people can prevent or reverse osteoporosis (an age-related thinning of bones), but a patient with a nonunion could not exercise the affected limb. A schematic diagram showing mechanical stimulation is given in Figure 13-1.

Example 13-3

A bioengineer is studying the effect of electrical stimulation on the healing of bone fractures. In an experimental model, a platinum wire is inserted into a fracture site and a small electric current passed through it. As a control, an identical platinum wire is inserted into a similar fracture but no current is applied. The rationale is to isolate the effect of the electric current from the effect of the wire alone.

a. What are the variables which will affect the results of this study?

b. Write a critique of the experimental procedure.

Answer

a. Any study of this nature will be influenced by many factors including the following.

    i. The nature of the injury that occurs during fracture and during surgery, the location of the fracture in the skeleton, and the type of bone (cortical or cancellous).

    ii. The patient's physical condition, age, sex, and state of health.

    iii. The type of metal of which the electrode is made, its geometry and surface area.

    iv. The electrical power source voltage, current, frequency content, DC signal content, and duty cycle (Jingade and Sangur, 2005).

    v. The length of the observation period. The effect may occur only during a limited time interval following stimulation.

b. To some extent the experiment can discriminate the active from the inactive electrodes provided the fracture sites are matched for factors listed under (i), (ii) and (iii) above.

However, it is difficult to isolate the effect of electric current alone since *ions* are released from both electrodes. The inactive electrode releases ions via corrosion. The active electrode provides an electron current and an ion current. Additional experiments such as trace metal analysis of tissues and urine would be needed to evaluate how much of each is present.

The issue of metal ion currents is circumvented by the use of pulsing electromagnetic fields (PEMF) for stimulation. This technique is noninvasive. A possible confounding variable in the study of this method is heating of the tissue caused by the induced currents as well as heat from the coils used to generate the fields. This heating effect is small. Nevertheless, a proper scientific study should include collateral experiments to isolate the effect of heating from the effect of the electromagnetic fields.

13.4.2. Chemical and Pharmacological Stimulation: BMPs, Growth Factors, and Apatites

Bone morphogenetic proteins (BMPs), transforming growth factor (TGF-β), and, to a lesser extent, the hydroxyapatites have been used to accelerate healing of bone fractures and to augment bone. It is not obvious whether the quality of bone growth stimulated by BMPs, TGF-β, and apatites is any better than that of bone grown by electrical and electromagnetic stimulation. If functional bone tissue is grown by any method, it becomes subject to the remodeling that occurs in any bone.

13.4.3. Bone Augmentation

There have been many studies to substitute bone in orthopedics as well as in dentistry. A relatively small defect (less than a 1-cm gap) in bone can be healed to its original state by use of an autograft (see Chapter 15). Defects may also be treated by bone paste made of collagen mixed with calcium phosphate paste, as given in Table 13-3. Polymers and ceramics have been used for bone substitutes. For long-term success, these materials should be resorbable so that new bone tissue can replace them with time. Natural or synthetic polymers are used in addition to the osteoconductive or osteoinductive materials (BMPs, TGF-β) and such ceramic materials as (mono-, di-, tri-, and tetra-) calcium phosphates and hydroxyapatite. Glass ceramics such as Bioglass® have been utilized due to surface film formation that resembles hydroxyapatite (review §6.4).

PROBLEMS

13-1. A bioengineer is asked by a supervisor to construct a bone fracture plate from the following materials: 316L SS, Ti6Al4V, cast CoCrMo, wrought CoNiCrMo, Al$_2$O$_3$, carbon fiber–carbon composite.

    a. Which two would be chosen for their biocompatibility?

    b. Which one would be chosen for its strength?

    c. Which one would be chosen for its specific strength (σ/ρ).

    d. Which ones should be chosen for FDA approval reasons alone?

Table 13-3. Classifications for Bone Substitutes

A. Source of raw materials
    1.   Natural
        a. Human and bovine bone, reprocessed whole bone
        b. Anorganic mineral, hydroxyapatite
        c. Organic collagen
    2.   Synthetic
        a. Ceramics, hydroxyapatite, calcium phosphate (mono-, di-, tri-, and tetra-CaP), glass-ceramic
           (Bioglass®), calcium sulfate
        b. Polymer-ceramic composite (calcium layered polymethylmethacrylate with hydroxyethylmethacrylate)

B. Physical form
    1. Granular or large porous sponge-like
    2. Injectable paste

All the formulations could use the BMPs and TGF-β to enhance the performance.

Reprinted with permission from Hollinger et al. (1996). Copyright © 1996, J.B. Lipincott.

13-2.   If Ti6Al4V were chosen, discuss its advantages and disadvantages for making a fracture plate. Explain possible degradation mechanisms of polymers in vitro and in vivo (review §7.5).

13-3.   What is the Oppenheimer effect? Read Oppenehimer et al. (1961).

13-4.   After bone screws have been removed from healed bone, holes remain that act as stress concentrators. Estimate the value of the stress concentration factor. Discuss the short-term consequences for the patient. What do you think will happen over a period of years? [Read Brooks (1970) and Burstein (1972).]

13-5.   A bioengineer has designed a bone plate that has a rubber washer between the plate and screws. What advantages and disadvantages would this insert have in comparison with conventional screws and plate? What would be the effect of such a washer in conjunction with a compression plate?

13-6.   An intramedullary fixation device is designed to be of tubular shape and made of homogeneous metal. Discuss the advantages and disadvantages. Suppose the design included two orthogonal holes on each end, perpendicular to the tube axis. Would this modification be advantageous?

13-7.   Design a self-compression bone plate that forces the bone fragments together as the screws are tightened. Illustrate its principle in the form of a two-hole plate.

13-8.   A bioengineer is asked to make a composite material from carbon fiber and PMMA resin (as a matrix) for a bone fracture plate. The following data are given:

Material	E (GPa)	Density (g/cm^3)	Strength (MPa)
Kevlar® fiber	130	1.45	2700
PMMA	3	1.2	70

    a. What is the amount of carbon fiber required to make the modulus of the plate 100 GPa?

    b. How many fibers are required if the diameter of the fibers is 10 μm and the cross-sectional area of the specimens is 1 cm^2?

    c. How much stress in the fiber direction can the composite take?

13-9. Bioglass® is coated on the surface of Ti6Al4V alloy for making implants that are more biocompatible. The 10-mm-diameter Ti6Al4V is coated with 1-mm-thick Bioglass. Use the following data to answer (a)–(d).

Material	Young's modulus(GPa)	Strength (MPa)	Density (g/cm³)
Ti6Al4V	79	795 (yield)	4.54
Bioglass	300	300 (fracture)	4.5

   a. What is the maximum load the composite can carry in the axial direction?

   b. What is Young's modulus of the composite in the axial direction?

   c. What is the density of the composite?

   d. Give two reasons why you would not use this composite to make orthopedic implants over plain stainless steel.

13-10. A bioengineer is asked by a boss to construct a bone fracture plate from the following materials:

   A. 316L SS   B. Ti6Al4V   C. cast CoCrMo

   D. Al₂O₃   E. wrought CoNiCrMo   F. carbon fiber–carbon composite

   a. Which one would be chosen for its biocompatibility alone?

   b. Which one would be chosen for strength reasons alone?

   c. Which one would be chosen for its specific strength [σ/ρ]?

   d. Which one should be chosen for its Young's modulus (the highest) alone?

   e. Which one should be chosen for its lost wax investment method?

   f. Which one should be chosen to prevent stress-shielding osteoporosis alone?

DEFINITIONS

BMP: Bone morphogenic protein. See morphogen below. This is a protein used to stimulate bone growth.

Bone inductions: Interactions among pluripotential cells and bone morphogenetic proteins (BMPs), converting these cells to osteoblasts.

Cancellous bone: Spongy bone.

Cortical bone: Dense compact bone found in the cortex (outer layer) of a bone.

Compression plate: Bone plate designed to give compression on the fractured site of a broken bone for fast healing. A minimum amount of callus is observed; this is considered to be a better sign of healing in comparison with the conventional plate.

Electrical stimulation: Tissues can be stimulated by applying a small amount of electrical energy in either alternating or direct current at low electrical potential so as not to hydrolyze the water medium. Dosage–response relationships are not fully established yet for various tissues.

18-8sMo: One of the stainless steels made of 18w/o Cr, 8 w/o Ni and a small (2–4 w/o) amount of Mo for saltwater corrosion resistance.

Fracture plate: Plate used to fix broken bones by open (surgical) reduction. Screws are used to fix it to the bone.

Helmholtz coil: A wire-wound coil to generate (pulsed) electromagnetic field to induce electric current in the tissues to stimulate their growth.

Hoffmann external fixator: A mechanical device used to reattach broken long bones using percutaneously inserted pins to compress the bones together for faster healing.

Intramedullary device: A rod-like device inserted into the intramedullary marrow cavity to promote healing of long bone (spiral, non-comminutive) fractures.

Kirschner wire: Metal surgical wires with diameters less than 3/32 inches (2.38 mm).

Lane plate: First fracture plate used extensively for repair of broken long bones designed by Dr. Lane in the first years of the twentieth century.

Lordosis: Abnormally increased concavity in the curvature of the lumbar spine.

Morphogens (BMPs): Cartilage-derived morphogenetic proteins-1 and -2 (CDMP-1 and -2) to regenerate form for skeletal insufficiencies caused by oncologic resection, trauma, disease, or developmental deficiency.

Osteopenia: Condition of insufficient bone mass or density.

Osteoporosis: Condition of abnormally high porosity of bone. Osteoporosis is associated with aging and with disuse atrophy due to inactivity. It can also be caused locally by shielding of stress by an implant or deficiency of calcium or vitamin D.

Schneider nail: One of the intramedullary rods used to fix broken long bones.

Scoliosis: Lateral curvature of vertebral column; it is always abnormal.

Steinmann pin: Metal pins with diameters greater than 3/32 inches (2.38 mm).

Stellite®: Co–Cr alloy.

Strain-generated electric potential: Electric potential generated due to strains during deformation of the bone. They may be related to the remodeling processes of bone.

Stress-shield effect: Prolonged reduction of stress on a bone may result in porotic bone (osteoporosis), which may weaken it. The process can be reversed if the natural state of stress can be restored to its original state.

Vitallium®: Originally the name was given to 19-9 stainless steel, but it was changed to designate a Co–Cr alloy.

BIBLIOGRAPHY

Bone screw technical information. 1980. Memphis, TN: Richards Manufacturing.

Albright JA, Johnson TR, Saha S. 1978. Principles of internal fixation. In *Orthopedic mechanics: procedures and devices*, pp. 123–229. Ed DN Ghista, R Roaf. New York: Academic Press.

Allgower M, Matter P, Perren SM, Ruedi T. 1973. *The dynamic compression plate, DCP*. Berlin: Springer-Verlag.

An HS, Cotler JM. 1999. *Spinal instrumentation*. Philadelphia: Lippincott Williams & Wilkins.

Bassett CAL. 1984. The development and application of pulsed electromagnetic (PEMFs) for ununited fractures and arthrodeses. *Orthop Clin North Am* **15**:61–88.

Bassett CAL, Becker RO. 1962. Generation of electric potentials in bone in response to mechanical stress. *Science* **137**:1063–1064.

Bassett CAL, Pilla AA, Pawluk RJ. 1977. A non-operative salvage of surgically resistant pseudoarthroses and nonunions by pulsing electromagnetic fields. *Clin Orthop Relat Res* **124**:128–143.

Brighton CT, Friedenberg ZB, Mitchell EI, Booth RE. 1977. Treatment of nonunion with constant direct current. *Clin Orthop Relat Res* **124**:106–123.

Brooks DB, Burstein AH, Frankel VHJ. 1970. The biomechanics of torsional fractures: the stress concentration effect of a drill hole. *J Bone Joint Surg Am* **52A**:507–514.

Brown A, D'Arcy JC. 1971. Internal fixation for supro-condylar fractures of the femur in the elderly patients. *J Bone Joint Surg Am* **53B**:420–424.

Burstein AH, Currey JD, Frankel VH, Heiple KG, Lunseth P, Vessely JC. 1972. Bone strength: the effect of screw holes. *J Bone Joint Surg Am* **54A**:1143–1156.

Curry J, Frankel VH. 1972. Bone strength: the effect of screw holes. *J Bone Joint Surg Am* **54A**:1151.

Deng M, Johnson R, Latour R, Shalaby S. 1996. Effects of gamma-ray irradiation on thermal and tensile properties of ultrahigh-molecular-weight polyethylene systems. *Irradiat Polym* **620**:293–301.

Duerig TW, Melton KN, Stöckel D, Wayman CM. 1990. *Engineering aspects of shape memory alloys.* London: Butterworths-Heinemann.

Dumbleton JH, Black J. 1975. *An introduction to orthopedic materials.* Springfield, IL: Thomas.

Edidin A, Jewett C, Kalinowski A, Kwarteng K, Kurtz S. 2000. Degradation of mechanical behavior in UHMWPE after natural and accelerated aging. *Biomaterials* **21**(14):1451–1460.

Einhorn TA, Lane JM, Physick PS. 1998. *Association of bone and joint surgeon workshop supplement: fracture healing enhancement.* Hagerstown, MD: Lippincott Williams & Wilkins.

Frost HM. 1973. *Orthopedic biomechanics.* Springfield, IL: Thomas.

Goel VK, Lim TH, Gwon JK, Chen J-Y, Winterbottom JM, Park JB, Weinstein JN, Ahn J-Y. 1991. Effects of an internal fixation device: A comprehensive biomechanical investigation. *Spine* **16**:s155–s161.

Hassler CR, Rybicki EF, Diegle RB, Clark LC. 1977. Studies of enhanced bone healing via electrical stimuli. *Clin Orthop Relat Res* **124**:9–19.

Hollinger JO, Breike J, Gruskin E, Lee D. 1996. Role of bone substitutes. *Clin Orthop Relat Res* **324**:55–56.

Jingade RRK, Sangur RR. 2005. Biomechanics of dental implants: an FEM study. *J Indian Prosthodont Soc* **5**(1):18–22.

Kim YK, Yeo HH, Lim SC. 1997. Tissue response to titanium plates: a transmitted electron microscopic study. *J Oral Maxillofac Surg* **55**(4):322–326.

Kummer B. 1976. Biomechanics of the hip and knee joint. In *Advances in hip and knee joint technology,* pp. 24–52. Ed M Schaldach, D Hohmann. Berlin: Springer-Verlag.

Kuntscher G. 1967. *Practice of intramedullary nailing.* Springfield, IL: Thomas.

Landes CA, Kriener S. 2003. Resorbable plate osteosynthesis of sagittal split osteotomies with major bone movement. *Plast Reconstr Surg* **111**(6):1828–1840.

Laurence M, Freeman MAR, Swanson SAV. 1969. Engineering considerations in the internal fixation of fractures of the tibial shaft. *J Bone Joint Surg Am* **51B**:754–768.

Lieberman JR, Friedlander GE. 2005. *Bone regeneration and repair: biology and clinical applications.* Totowa, NJ: Humana Press.

Martinez SA. 1999. *Fracture management and bone healing.* Philadelphia: W.B. Saunders.

Martz EO, Goel VK, Pope MH, Park JB. 1997. Materials and design of spinal implants: a review. *J Biomed Mater Res* **38**(3):267–288.

Martz EO, Lakes RS, Goel VK, Park JB. 2005. Design of an artificial intervertebral disc exhibiting a negative Poisson's ratio. *Cell Polym* **24**:127–138.

McKellop H, Shen F, DiMaio W, Lancaster J. 1999. Wear of gamma-crosslinked polyethylene acetabular cups against roughened femoral balls. *Clin Orthop Relat Res* **369**:73–82.

Mears DC. 1979. *Materials and orthopaedic surgery.* Baltimore: Williams & Wilkins.

Mow VC, Huiskes R. 2005. *Basic orthopaedic biomechanics and mechano-biology.* Philadelphia: Lippincott Williams & Wilkins.

Muratoglu O, O'Connor D, Bragdon C, Delaney J, Jasty M, Harris W, Merrill E, Venugopalan P. 2002. Gradient crosslinking of UHMWPE using irradiation in molten state for total joint arthroplasty. *Biomaterials* **23**(3):717–724.

Ohura K, Hamanishi C, Tanaka S, Matsuda N. 1999. Healing of segmental bone defects in rats induced by a beta-TCP-MCPM cement combined with rhBMP-2. *J Biomed Mater Res* **44**(2):168–75.

Oppenheimer ET, Willhite M, Danishefsky I, Stout AP. 1961. Observations on the effects of powdered polymer in the carcinogenic process. *Cancer Res* **21**:132–134.

Park JB, Goel VK, Pope MH, Weinstein JN. 1999. *Surgically implantable fastening system*. University of Iowa Research Foundation: USPTO 6004323.

Park KD, Khang GS, Lee HB, Park JB. 2001. Characterization of compression-molded UHMWPE, PMMA and PMMA/MMA treated UHMWPE: density measurement, FTIR-ATR, and DSC. *Biomed Mater Eng* **11**(4):311–323.

Park KD, Kim JK, Yang SJ, Yao A, Park JB. 2003. Preliminary study of interfacial shear strength between PMMA precoated UHMWPE acetabular cup and PMMA bone cement. *J Biomed Mater Res B, Appl Biomat* **65B**(2):272–279.

Park KD, Park JB. 2002. Investigation of interfacial strength and its structure on the development of a new design of UHMWPE acetabular component. *J Biomed Mater Res* **63**(3):363–372.

Rockwood CA, Green DP, Heckman JD, Bucholz RW. 2001. *Rockwood and Green's fractures in adults*. Philadelphia: Lippincott Williams & Wilkins.

Rogers LF. 2002. *Radiology of skeletal trauma*. Philadelphia: Churchill Livingstone.

Schlag G, Redl H. 1994. *Wound healing*. Berlin: Springer-Verlag.

Schnall SB, Osmon DR. 2003. Proceedings of the musculoskeletal infection society 2002. *Clin Orthop Relat Res* (special issue), no. 414.

Soto-Hall R, McCloy NP. 1953. Cause and treatment of angulation of femoral intramedullary nails. *Clin Orthop Relat Res* **2**:66–74.

Sumner-Smith G, Fackelman GE. 2002. *Bone in clinical orthopedics*. Stuttgart: Thieme.

Uhthoff HK, ed. 1980. *Current concepts of internal fixation of fractures*. Berlin: Springer-Verlag.

van der Elst M, Klein CPAT, Patka P, Haarman HJTM. 2000. Biodegradable fracture fixation devices. In *Biomaterials and bioengineering handbook*, pp. 509–524. Ed DL Wise. New York: Marcel Dekker.

Venable CS, Stuck WC. 1947. *The internal fixation of fractures*. Springfield, IL: Thomas.

Wang S, Lewallen D, Bolander M, Chao E, Ilstrup D, Greenleaf J. 1994. Low-intensity ultrasound treatment increases strength in a rat femoral fracture model. *J Orthop Res* **12**:40–47.

Williams DF, ed. 1981. *Fundamental aspects of biocompatibility*. Boca Raton, FL: CRC Press.

Williams DF, Roaf R. 1973. *Implants in surgery*. Philadelphia: W.B. Saunders.

Wood MB, Gilbert A. 1997. *Microvascular bone reconstruction*. St. Louis: Mosby.

Yang K, Paravizi J, Wang S, Lewallen D, Kinnick R, Greenleaf J, Bolander M. 1996. Exposure to low-intensity ultrasound increases aggrecan gene expression in a rat femur fracture model. *J Orthop Res* **14**:802–809.

14

HARD TISSUE REPLACEMENT — II: JOINTS AND TEETH

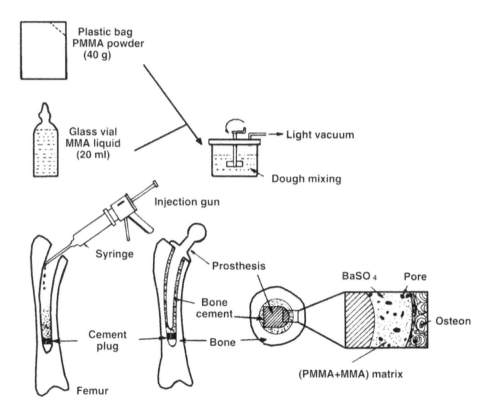

Schematic illustration of hip joint implantation procedure. Courtesy of J.B. Park, Univeristy of Iowa.

The articulation of joints poses some additional problems for the bioengineer as compared with long bone fracture repairs. These include wear and corrosion and their products, as well as complicated load transfer dynamics. In addition, the massive nature of (total) joint replacements, such as the knee and elbow, and their proximity to the skin also makes for a greater possibility of infection. More importantly, if the replacement fails for any reason, it is much more difficult to replace the joint a second time since a large portion of the natural tissue has already been destroyed.

For these reasons orthopedic surgeons try to salvage the existing joint if possible and make use of implants as a last resort. However, the hip prosthesis has shown favorable acceptance in recent years by older patients. The four primary indicators for any joint replacement are (1) pain, (2) instability, (3) stiffness, and (4) deformity. The conditions that lead patients to seek joint replacement include severe trauma and arthritis. Recently joint replacements of many types have become more popular, as shown in Figure 14-1. It is estimated that close to one million hip replacement surgeries, both primary and revision, are being performed worldwide each year.

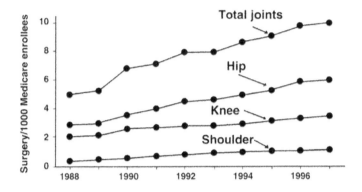

Figure 14-1. Some statistics of joint replacements in the 1990s. Reprinted with permission from Weinstein (2000). Copyright © 2000, American Hospital Association Press.

The interface between a tooth and the surrounding bone is actually a joint. In the dental setting, the dentist's first choice in treating tooth decay is to replace only the decayed portion with a filling. If much of the tooth structure is damaged or lost, the upper portion may be replaced by a crown. If a tooth is lost entirely, dentures (false teeth) or dental implants may be used. Modern dentistry is successful in preventing a great deal of misery. For example, if tooth decay proceeds without intervention, the tooth begins to ache. A painful abscess may then form in the jawbone. Good dental health is also associated with a lower incidence of cardiovascular disease and stroke. Dental infection including periodontal disease causes a systemic microbial and inflammatory burden that has system-wide consequences.

The use of dental implants for missing or extracted teeth has not been as successful as the use of joint replacements; the difference is due mainly to the *percutaneous* nature of dental implants. Exposure of the interface between implants and tissues to the hostile oral environment makes them easily infected. In fact, it is very difficult if not impossible to get rid of a low-grade infection even for a successful and functional dental implant. In addition, the extremely large and varying direction of the forces applied during mastication limit the selection of implant materials.

14.1. JOINT REPLACEMENTS

The hip and shoulder joints have a *ball-and-socket* articulation, while other joints such as the knee and the elbow have a *hinge-type* articulation. However, they all possess two opposing smooth cartilaginous articular surfaces that are lubricated by viscous synovial fluid. This fluid is made of polysaccharides that adhere to the cartilage and upon loading can be permeated out onto the surface to reduce friction. The cartilage is not vascularized, and nutrition of the tissues appears to be a diffusional process.

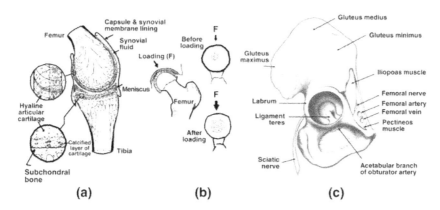

Figure 14-2. Structural arrangement of a knee joint and hip joint before (**a**) and after loading (**b**). Reprinted with permission from Frost (1973). Copyright © 1973, Charles C. Thomas. (**c**) Anatomy of the hip. Reprinted with permission from Eftekhar (1978). Copyright © 1978, Mosby.

For the hip and knee joints, nature provided the surface of the joint with a large area to distribute load, as shown in Figure 14-2. The shock of loading is reduced by cartilage compliance and compliance of the trabecular subchondral bone underlying the cartilaginous tissue. These tissues also have viscoelastic properties, beneficial in reducing shock loads. The anatomy of the hip socket is also illustrated in Figure 14-2. A ligament in the middle of the socket is attached to the femoral head, preventing dislocation of the joint. The actual articulation of the joint is stabilized by the body's coordination of the ligaments, tendons, and muscles.

An analysis of the forces acting on the various tendons and ligaments is very complicated. Even the center of the knee joint cannot be determined with any great precision; in fact, it shifts position with each movement. The eccentric joint movement helps to distribute the load throughout the entire joint surface.

Some joints such as the knee have fibrous, cartilaginous menisci-shaped wedges located between the sliding surfaces (Figure 14-2). It is believed that the main function of the menisci is to transfer load over a larger area than is possible without them.

The joint forces applied during a range of activities are given in Table 14-1. Of course, the forces applied during walking vary considerably with each motion, as shown in Figure 14-3. It should not be surprising that the forces are up to 8 times the body weight as a result of the leverage geometry of the muscles and the dynamic nature of human activity. Biomechanical analysis can be of use in designing a better implant since it is necessary to know the load that will be applied to the implant in order to design sufficient strength and stiffness into it.

Table 14-1. Average Maximum Values of Forces at Hip and
Tibio-Femoral Joints during a Range of Activities

	Maximum joint force (multiples of body weight)	
Activity	Hip	Knee
Level walking		
Slow	4.9	2.7
Normal	4.9	2.8
Fast	7.6	4.3
Stairs		
Up	7.2	4.4
Down	7.1	4.4
Ramp		
Up	5.9	3.7
Down	5.1	4.4

Reprinted with permission from Paul (1976). Copyright © 1976, Springer-Verlag.

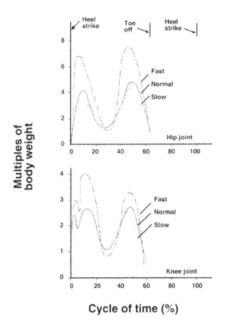

Figure 14-3. Variation of forces with time in the hip and knee joint during walking. Reprinted with permission from Paul (1976). Copyright © 1976, Springer-Verlag.

14.1.1. Lower Extremity Implants

14.1.1.a. Hip Joint Replacements

The early methods of correcting hip joint malfunctions involved only the acetabular cup or femoral head. One technique of restoring the hip joint function is to place a cup over the femoral head while the surface of acetabulum is also resected to fit the cup. The implant serves as a *mold* interposing between the articulating surfaces, which eventually adapt according to the function of the joint. Today, both the acetabulum and femoral head surfaces are replaced as shown in Figure 14-4.

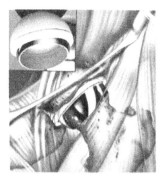

Figure 14-4. Example of double cup arthroplasty. In the early days only the femoral cup was placed (mold arthroplasty) in order to obtain a movable joint. Courtesy of Howmedica Inc., Rutherford, NJ.

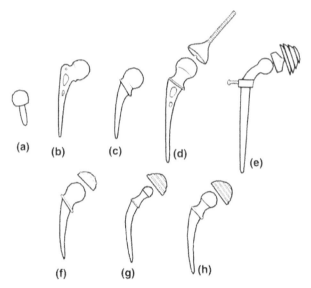

Figure 14-5. Hemiarthroplasty prostheses: (**a**) Acrylic Judet; (**b**) self-locking Moore; (**c**) Thompson, total arthroplasty (bipolar) prosthesis; (**d**) Ring; (**e**) Sivash; (**f**) McKee-Farrar; (**g**) Charnley; (**h**) Muller. Note: no cement was used in (**d**) and (**e**), but acrylic cement was employed in (**f**), (**g**), and (**h**). Reprinted with permission from Eftekhar (1978). Copyright © 1978, Mosby.

Some surgeons have tried to replace the femoral head with implants of various design after resection, as shown in Figure 14-5. The wide variety of implants reflects the limited knowledge of the function of joints and the ability of the joint to accommodate insult imposed upon it by various implants. Most femoral head replacements are made with installation of an acetabular cup. This is the so-called *total hip joint replacement* (arthroplasty). Ever since

Dr. J. Charnley introduced the use of bone cement in the fixation of artificial hip joints on the advice of Dr. D. Smith in the late 1950s, it helped to popularize the procedure throughout the world. The initial success of the procedure has been tempered by the problems related not only to the bone cement but also the implants per se, surgical techniques, patient selection, and so on.

Surgical insertion of a total hip replacement is done as follows. The diseased femoral head is cut off, and the medullary canal of the femur is drilled and reamed to prepare it for the stem of the prosthesis. The cartilage of the acetabulum is also reamed. The PMMA (polymethylmethacrylate) bone cement is prepared from polymer powder and monomer liquid. When the cement reaches the correct "doughy" consistency, it is packed into the medullary canal of the femur and the femoral stem is inserted. The acetabular component is similarly cemented. The alignment and articulation of the artificial ball-and-socket joint are then verified. In cases where both joints are severely arthritic, the operations are performed bilaterally. The various types of hip implants can be grouped into ball and socket, retained ball and socket, trunnion bearing, floating acetabulum, and double cup, as illustrated in Figure 14-6.

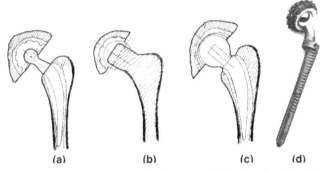

(a)   **(b)**   **(c)** **(d)**

Figure 14-6. Different types of total hip implants: (**a**) ball and socket, (**b**) double cup, (**c**) trunnion, and (**d**) retained ball and socket. Sivash Design, Courtesy of U.S. Surgical Corporation, New York, NY.

Traditionally, the "Charnley type" prosthesis has served well with a polyethylene acetabular cup and metallic femoral head, as shown in Figure 14-7. (See the x-ray radiograph on the cover page of this chapter.) Over the years, the materials have changed from high-density polyethylene (HDPE), to ultrahigh-molecular-weight polyethylene (UHMWPE), and crosslinked UHMWPE for the cup. Originally stainless steel (presumably 18-8sMo, i.e., 316L), CoCr-, and Ti-based alloys were used for the femoral head and stem. The ceramic acetabular cup showed promise early on with alumina (Al_2O_3), which had lower wear and friction when combined with itself, polymer, and metal. Due to its brittleness, the femoral stem is not made of ceramic, although it was tried during an early development stage. The relatively newer material, zirconia (ZrO_2), has proven not as suitable as originally thought due to phase transformation in vivo under stress in contrast to experiments in many in vitro and some animal tests. Figure 14-8 shows a zirconia-based acetabular cup and femoral head with a metal stem. Coating the surfaces of the head and cup with diamond was attempted, as shown in Figure 14-9. Ultrahigh pressure (6–7 GPa) and temperature (1300–1500°C) were achieved inside a chamber with six cylinders through which the electrical power was also introduced. The coating is polycrystalline in micron-sized randomly oriented crystals of diamond (95–99+% pure with a metal catalyst). This synthetic (industrial grade) diamond is made with a similar apparatus.

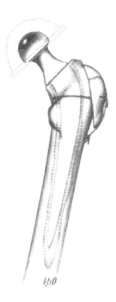

Figure 14-7. Charnley hip joint prosthesis. Polymeric cup and femoral stem with small-diameter head claimed to be low friction with less contact surface. Note the reattachment of greater trochanter bone with a wire. Reprinted with permission from Eftekhar (1978). Copyright © 1978, Mosby.

Figure 14-8. Zirconia-based acetabular cup and femoral head with metal stem. Courtesy of Norton Advanced Ceramics, Export, PA.

The acetabular cup was originally made by Charnley of polytetrafluoroethylene (PTFE) due to its low friction and tissue compatibility as a bulk material. The wear particles, however, caused a tremendous tissue reaction by free radical formation in vivo, which is quite toxic to tissues. These free radicals are similar to peroxides and are intentionally used for sterilization of wounds. The free radicals disappear rapidly by reacting with air in vitro. Similar problems with free radicals may cause a tissue reaction to polyethylene (UHMWPE); however, in this case the material is much stronger and less wear particulates are generated. This is one of the main reasons the ceramic acetabular cup is used. The wear of the polyethylene (UHMWPE) cup can be reduced by crosslinking while the outer surfaces are precoated by bone cement. The method for preparing the cup is depicted in Figure 14-10. A graph of the reduction in wear by physically crosslinking (using radiation as in sterilization) is shown in Figure 14-11. The wear testing was carried out on biaxial pin-on-disk matching UHMWPE pins with CoCr metal. The crosslinked polyethylene tends to be more noncrystalline, has a higher molecular weight, and has lower mechanical strength in comparison to the original polymer. The proper crosslink procedure involves control of crosslink density and the use of an inert atmosphere. Long-term clinical trials are needed before this crosslinked polyethylene can be evaluated for its effectiveness.

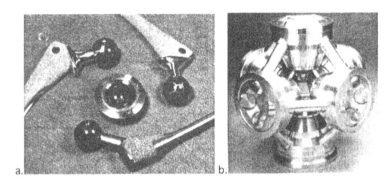

Figure 14-9. Coating the surfaces of the head and acetabular cup (**a**) with compact diamond (**b**). Reprinted with permission from Taylor and Pope (2001). Copyright © 2001, Butterworths.

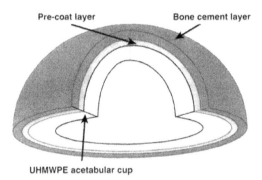

Figure 14-10. Pre-coating all-poly acetabular cup outer surface with bone cement layer combined with crosslinking of UHMWPE by γ-radiation. Modified with permission from Park et al. (1999). Copyright © 1999, Begell House.

The single most difficult problem of the hip joint as well as other joint replacements is fixation of the implants. This is due to the fact that the implant has an interface with the cancellous bone, which is much weaker than compact bone. The bone may have insufficient density of trabeculae to support the increased load imposed. Also, the stress concentration of the implant at points of sharp contact, such as the *calcar region* and the (lateral) end of the femoral stem, causes the already weakened bone to resorb. In fact, the first wide acceptance of total hip replacement was achieved by providing an acceptable fixation using an acrylic bone cement (see §8.2.3).

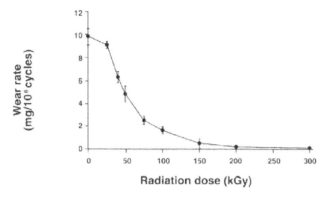

Figure 4-11. Wear rate versus degree of crosslinks (radiation dose) of UHMWPE. Reprinted with permission from Muratoglu et al. (1999). Copyright © 1999, Marcel Dekker.

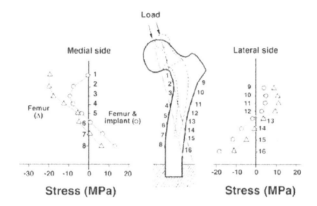

Figure 14-12. Stresses on the surface of the femoral stem by a load of 4,000 Newtons. Numbers indicate the location of strain gauges to measure deformations. Note that there is no stress in position 1 (calcar region) after insertion of the implant. Reprinted with permission from Swanson and Freeman (1977). Copyright © 1977, Wiley.

The cement serves not only as the initial attachment of the implant with bone, but it also acts as a shock absorber since it is a viscoelastic polymer. The bone cement also helps to spread the load more evenly over a large area and reduces the stress concentration on the bone by the prosthesis. However, the stress in the bone in the region of the distal stem is much higher than normal. In contrast, the bone stress in the proximal region is reduced by the presence of the implant, as shown in Figure 14-12. This stress-shielding effect causes bone resorption of the proximal region due to the reduced stress upon this region, which, in turn, will lead to either loosening or fracture of the stem. Nevertheless, total hip joint prostheses commonly last ten years in older patients.

To obviate the problem of calcar resorption, one can beneficially increase the load in the proximal calcar region by making the neck portion of the stem longer. However, this arrangement increases the moment applied in the mid-stem, predisposing it to fracture.

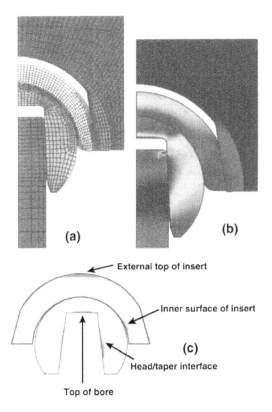

Figure 14-13. Finite-element analysis or model (FEA or FEM) for stress distribution (ceramic–ceramic and metal femoral stem): (**a**) finite-element grids, (**b**) calculated stress in each element expressed in graphic form, (**c**) overall view of the model. Reprinted with permission from Cales et al. (2000). Copyright © 2000, Marcel Dekker.

Some authors advocate the use of a passive fixation method, in which the femoral stem is allowed to act as a wedge jammed into the intramedullary canal. Moore and Ring prostheses use this technique for fixation.

Analysis of the stress distribution using finite-element analysis or modeling (FEA or FEM) can provide a detailed stress distribution before and after implantation on the implant and tissues. One such model is given in Figure 14-13 for the head and cup contact region similar to the zirconia ceramic implants shown in Figure 14-8. Three-dimensional FEA can be quite expensive due to the large number of elements and lengthy computing time.

The cement itself sometimes involves problems such as monomer vapors interfering with the body's systemic function, thereby briefly decreasing the blood pressure. The highly exothermic polymerization reaction can cause a local temperature rise that can result in cell necrosis, as mentioned in §8.2.3. Also, the extensive intramedullary cavity preparation for the cement space can block bone sinusoids, causing tissue necrosis and fat embolism.

Another problem is difficulty of removal and the extent of tissues destroyed if the implant has to be removed for any reason. Thus, the *replaceability* of an implant is an important aspect of the design. In this regard, the original Sivash implant has an inherent weakness due to its design, that is, the whole prosthesis had to be replaced even if only one component failed. It is now designed so that it is possible to replace one part without removing the other implant.

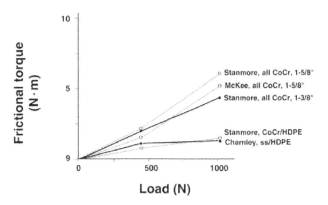

Figure 14-14. Frictional torque versus applied load for various hip prostheses. Reprinted with permission from Wilson and Scales (1970). Copyright © 1970, Lippincott, Williams & Wilkins.

Friction between the ball and cup of the hip joint is problematic when it creates excessive torque. Especially for large loads, the frictional torque becomes very significant for the cobalt-chromium alloy hip joint, as shown in Figure 14-14. The stainless steel–polyethylene and cobalt–chromium alloy–polyethylene combinations are better for reducing frictional torque and wear than all-metal systems. The high frictional torque of an all-metal system may also be due to the larger surface contact area since the femoral head is much larger than in the metal–polymer prosthesis. In actual use, the all-metal system works well without exerting high frictional torque. This is mainly due to lubrication of the surfaces by tissue fluids. More recently, all-metal prostheses perform better than earlier ones due to better manufacturing, especially the congruence of the head and socket and better metallurgical control.

The problems due to the cement/prosthesis interface can be diminished by precoating the prosthesis with bone cement at the factory; such implants are commercially available. Precoating not only increases the interfacial strength but also helps to eliminate bare metal exposure, and reduces the amount of bone cement used at the time of surgery, thus reducing the amount of heat and monomers released at the time of surgery, etc. The precoating technique can be used for any orthopedic implants.

The prosthesis/bone interface generates the most significant problems in joint replacements. Specifically, the failure of hip joint replacements is usually due to *loosening* of the acetabular and femoral components. Loosening can be largely divided into mechanical and radiological loosening. Loosening may cause pain or it may not be related to any clinical symptoms. Inadequate fit of implants and the surgical and cementing techniques were thought to be the main factors in loosening. Some workers attributed loosening to the blood clots interposed at the time of surgery and shrinkage of bone cement during polymerization. Also, bone remodeling and resorption in the proximal femur due to stress-shielding by the stiff prosthesis could contribute to loosening in long-term implants. Interface problems in orthopedic implants are discussed more fully in §14.6.

Another method of fixation is biological fixation. This is the same as direct fixation of an implant into the bone with and without porous surface coating, as shown in Figure 14-15, for a porous coated acetabular cup microcontact x-ray radiograph. Clearly, the initial bone ingrowth is demonstrated as given in Table 14-2. However, the bone (bio)dynamics may dictate the rest of the implant period for fixation. Indeed, if the stress level increases too much inside the pores, then the bone will be resorbed. Also, one should be careful that the ions (such as Ca and

P) may accumulate in the pores too much, causing resorption in case of a hydroxyapatite coating on the porous surfaces. This may cause fibrous membrane formation around a porous-surfaced hip prosthesis coated with hydroxyapatite, as shown in Figure 14-16.

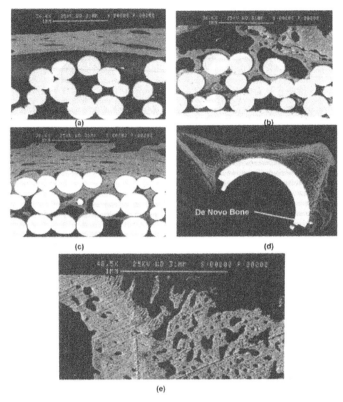

Figure 14-15. Scanning electron microscopy of the gap region of a control component shows no bone filling or ingrowth into the porous surface (**a**). The αBSM(bone substitute material)-treated component shows some new bone (**b**). The rhBMP-2/αBSM-treated component shows extensive bone ingrowth into the porous coated acetabular cup (**c**). All after 12 weeks in canine. (**d**) A contact x-ray radiograph shows de novo bone formation in the porous coating (arrow). (**e**) Scanning electron microscopy shows de novo bone with rhBMP-αBSM. Reprinted with permission from Bragdon et al. (2003). Copyright © 2003, Lippincott, Williams & Wilkins.

Based on the extensive reporting system in Scandinavian countries, especially Sweden and Norway, the failure (success) rates of joint (especially hip and knee) prostheses have been extensively analyzed and reported yearly at the AAOS (American Association of Orthopedic Surgeons) meetings. This is one of the best ways of improving the design and materials of joint implants and the system overall. If it is mandated, this reporting system could be used throughout the world, and this would aid in improving design and make it possible to select the most appropriate materials.

The recent report for the "Prognosis of total hip replacement; update of results and risk-ratio analysis for revision and re-revision from the Swedish National Hip Arthroplasty Resister 1979–2000" gives very informative results. Figure 14-17 shows the results of cemented and hybrid (cemented stem and uncemented cup) implants of Charnley-type hip prosthesis. Clearly, the choice of fixation will dictate the longevity of a prosthesis regardless of type, material, condition of patients, etc. Obviously, other factors can be analyzed, such as type of cement, acetabular cup, primary or revision, age, etc., which will help the biomaterials community to consider theses factors for final products. These large data sets show a clearly better performance by cemented rather than uncemented hip arthroplasty.

Table 14-2. SEM Quantification of Bone Ingrowth into the Porous Surface

Test group	Region of gap defect			Region of direct apposition		
	Area fraction of ingrowth into the void space of the porous layer (%)	Extent of the external surface of the porous layer occupied by new bone (% apposition)	% of depth penetration	Area fraction of ingrowth the void space of the porous layer (%)	Extent of external surface of the porous layer occupied by new bone (% apposition)	% of death penetration
Control	2.4 ± 1.9	31.0 ± 31.5	21.1 ± 10.5	18.0 ± 6.7	86.0 ± 14.1	50.2 ± 13.6
αBSM	13.6 ± 10.3	63.9 ± 42.1	40.3 ± 24.9	22.4 ± 8.6	76.2 ± 26.0	59.8 ± 18.1
rhBMP-2/αBSM	22.6 ± 8.3*	96 ± 8.9*	65.2 ± 24.8*	25.4 ± 7.7	93.2 ± 9.6	68.3 ± 24.2

*Significantly higher than the control.

Modified with permission from Bragdon et al. (2003). Copyright © 2003, J.B. Lippincott.

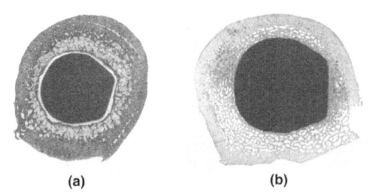

(a) **(b)**

Figure 14-16. Histology of cross-section of canine hip femoral stem with and without porous surface coated by hydroxyapatite for better interdigitation with ingrown bone. Note that no tissue directly apposes the smooth stem surface (**a**) and the integration of tissues in the porous surface (**b**). Reprinted with permission from Geesink (1993). Copyright © 1993, Raven Press.

Another interesting study of 5300 uncemented acetabular cups in young patients compared with an all-polyethylene (all-poly) acetabular cup (Charnley type) showed that the all-poly cup survived better. The investigators concluded that metal-backed acetabular cups (uncemented) do not perform as well as the cemented all-poly acetabular cup. One might also consider the cost of a metal-backed cup, which is much higher. This suggests that it is not always better to have more expensive, more complicated prostheses for the in-vivo performance. By contrast, it is always better to use a clinically proven prosthesis rather than employ a "trendy" prosthesis. This happens mainly due to a lack of clinical data on any particular design on a long-term basis, although the Swedish and Norwegian registries may provide valuable information.

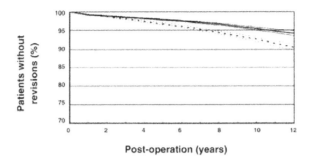

Figure 14-17. Survival data for all diagnoses; all reasons for re-vision are shown. The survival rate and 95% confidence intervals are indicated for the Charnley stem (dotted line; DePuy, Warsaw, IN) and the Ogee all-polyethylene cemented cup (solid line; DePuy). Reprinted with permission from Malchau et al. (2005). Copyright © 2005, Lippincott, Williams & Wilkins.

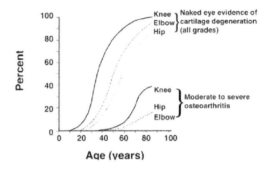

Figure 14-18. Incidence of joint degeneration. Reprinted with permission from Greenwald and Matejczyk (1980). Copyright © 1980, Appleton-Century-Crofts.

14.1.1.b. Knee Joint Replacements

The development and acceptance of knee joint prostheses have been slower than that of the hip joint due to the knee's more complicated geometry and biomechanics of movement, and lesser stability in comparison with the hip. The incidence of knee joint degeneration is higher than that of any other joints, as shown in Figure 14-18.

Knee joint implants can be classified into hinged and non-hinged types. The latter is further divided into uni- and bicompartmental. A cross-section of a natural knee is shown in Figure 14-2, and typical artificial knee joints are shown in Figure 14-19. Knee surgery with bone cement requires a complete clean-up of the cement and bone chip debris, which can cause severe damage to the articulating surfaces, especially the tibial plateau. The selection of a particular implant depends upon the health of the knee and the types of disease and range of activities required. As in the case of hip joint replacement, the major problems are loosening and infection. Sinking of the tibial plateaus can occur, and it is due to crushing of the trabecular bone under the implant. This problem can be corrected by making the prosthesis larger. A metallic backing is made under the polymer (ultrahigh-molecular-weight polyethylene), as shown in Figure 14-20. The metal in this case is in contact with cortical bone, which is much stronger than trabecular bone. The polymer insert is also much thinner than the all-poly implant even though the recommended thickness is a minimum of 8 mm.

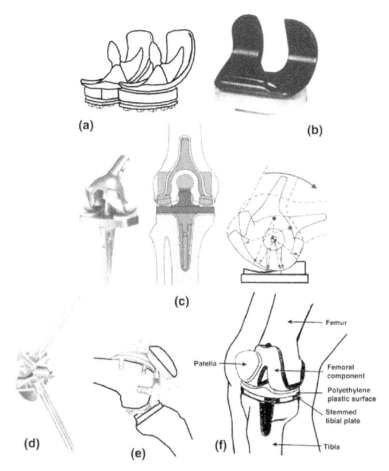

Figure 14-19. Examples of various types of knee replacements: (**a**) Marmor [Reprinted with permission from Marmor (1980). Copyright © 1980, Appleton-Century-Crofts]; (**b**) Freeman-Swanson ICLH [Reprinted with permission from Freeman (1980). Copyright © 1980, Appleton-Century-Crofts]; (**c**) Spherocentric [Reprinted with permission from Sonstegard et al. (1977). Copyright © 1977, Charles C. Thomas]; (**d**) Walldius [Reprinted with permission from Walldius (1980). Copyright © 1980, Appleton-Century-Crofts.]; (**e**) Bechtol [Reprinted with permission from Bechtol (1976). Copyright © 1976, Richards Manufacturing Company]; (**f**) Knee prosthesis components [Reprinted with permission from Davis (2003). Copyright © 2003, American Society for Microbiology].

UHMWPE has demonstrated excellent properties, such as low friction, high wear resistance, high toughness, high impact strength, excellent biocompatibility, and low cost. Despite the excellent performance in vivo, wear of the UHMWPE in contact with harder femoral heads made of metals or ceramics is still a major concern in total joint replacements since wear particulate debris may induce osteolysis, resulting in loosening of implants and weakening of the structure of bone. Wear itself is not avoidable, and the major concerns are wear debris and weakening of the matrix. Studies have shown that wear debris from UHMWPE may limit the longevity of implants. Some studies show that wear particle size, distribution, and geometry are attributable to the type of tissue responses, which could lead to cell necrosis.

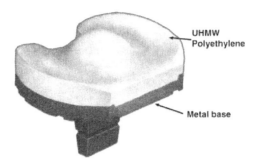

Figure 14-20. Typical metal-backed tibial implant designed to prevent sinking of the component.

The introduction of crosslinks to improve the wear resistance of orthopedic implants made of UHMWPE is relatively recent, as mentioned earlier. Crosslinking of UHMWPE is made by radicalizing the side chains by radiation (x-ray, gamma, electron beam, microwave, etc.), or by chemicals. In this process, the main chains may also be affected, resulting in degraded polymer with a lower molecular weight (see §7.2.1). Although crosslinking enhances wear properties, other mechanical properties, such as ductility, modulus, toughness, impact strength, and fatigue strength could be compromised. Furthermore, gamma radiation, which is currently most widely used for crosslinking, generates a significant amount of free radicals as a byproduct. In fact, when polymeric medical devices are exposed to gamma radiation, they undergo either crosslinking or oxidation (and thus chain scission). Crosslinking is beneficial, whereas oxidation is detrimental. Both processes are triggered by free radicals generated by gamma radiation.

It is interesting to note that the statistics from Scandinavian countries show that the all-polymer acetabular cup performed as well as the metal-backed acetabular cups, as mentioned earlier. In fact, many times the all-poly cup performed better. One reason is that the all-polymer cup has a thicker layer of polymer than the metal-backed one. All-poly cups provide some compliance to reduce impact and also more material to allow for wear or creep. Any incongruence between the liner outer surface and the inner surface of the metal will cause faster wear. Another important aspect to consider is cost: the metal-backed cup is more expensive than the all-polymer acetabular cup. The same considerations apply to the all-polymer tibial plateau in knee replacements.

UHMWPE has a similar structure as polytetrafluoroethylene, PTFE, Teflon®, and has very low surface energy. Consequently, it is very difficult to adhere to any other polymers, including acrylic bone cement, with UHMWPE. It has been shown that UHMWPE can be precoated with PMMA (polymethylmethacrylate) and MMA monomer, which have been used for bone cement to fix implants. Thus, the all-poly acetabular cup, which is made of a PMMA-precoated outermost layer for better fixation to bone cement and a crosslinked innermost layer for enhanced wear resistance against metal or ceramic femoral heads, uses direct compression molding, as shown in Figure 14-21. It would be interesting to determine the in-vivo performance of this implant if it becomes available for patients.

Porous coated implants have been developed to avoid the problems associated with cemented prostheses. For example, the femoral stem can be made porous. A porous surface layer permits bony tissue ingrowth to achieve a dynamic interface of bone and implant (Figure 14-22). This implant does not require the use of bone cement as most other orthopedic implants do. The porous coated implants should be used for relatively healthy knees since their stability is entirely dependent on ingrown tissues. It is also expected that the time required before the patient can walk (ambulate) will be much longer than for the cement-fixed case since it will take some time for the tissues to grow into the pores and premature loading may be detrimental for the ingrowth process.

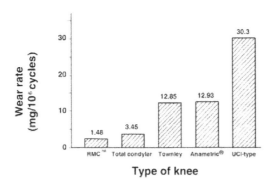

Figure 14-21. Typical all-poly tibial and patellar components with metal condylar implants. Reprinted with permission from Wilde (1989). Copyright © 1989, Springer-Verlag.

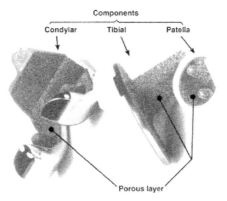

Figure 14-22. Porous metal coated total knee implant. Reprinted with permission from *Porous coated anatomic (PCA) total knee system* (1981). Copyright © 1981, Orthopaedic Division, Howmedica Inc.

Figure 14-23. Bar graph of the wear rate results. Reprinted with permission from *In vitro testing of the RMC total knee* (1978). Copyright © 1978, Richards Manufacturing Company.

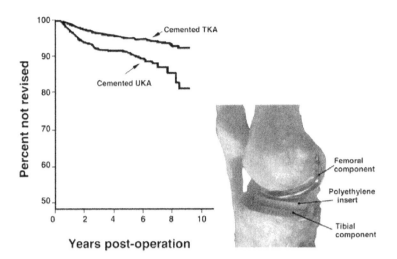

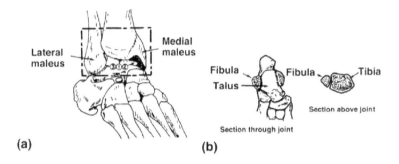

Figure 14-24. Statistical analysis of knee joint replacement in unicompartmental and tricompartmental (patella resected) cemented arthroplasty fixed with bone cement (**a**) [Reprinted with permission from Furnes et al. (2005). Copyright © 2005, Orthopedic Research Society]; (**b**) UKA inserted in vivo for a better illustration [Reprinted with permission from Suggs et al. (2006). Copyright © 2006, Wiley Interscience.

Figure 14-25. (**a**) Anatomical features of the ankle joint articulating surfaces: (1) the distal tibia and the superior surface of the talus; (2) the medial malleolus and the medial side of the talus; and (3) the lateral malleolus or fibula and the lateral side of the talus. (**b**) Sectional views through and above the ankle joint. Reprinted with permission from Mears (1979). Copyright © 1979, Williams & Wilkins.

Wear of the surface of the tibial plateaus of knee implants can be significant, as illustrated in Figure 14-23, which is an in-vitro measurement. In-vivo performance, however, may be quite different.

The Norwegian Arthroplasty Registry has also reported similar statistics on knee prosthesis. For example, Figure 14-24 shows the results with unicompartmental and tricompartmental cemented primary knee arthroplasty from 1994 to 2003. Again, there are large variations in the specific prosthesis, but one could decide that the TKA is better than the UKA in cemented primary arthroplasty.

14.1.1.c. Ankle Joint Replacement

The ankle joint consists of three articulating surfaces: the distal tibia and the superior surface of the talus, the medial malleolus and the medial side of the talus, and the lateral malleolus or fibula and the lateral side of the talus, as shown in Figure 14-25. The ankle joint has principal movements of dorsiflexion and plantar flexion. The joint, however, works as a universal joint and can rotate up to 14° in normal walking. The motion of the joint is not entirely that of a hinge; it has a slight rotating type of motion, as in the knee and has a gliding motion as shown in Figure 14-26. This freedom of motion makes it more difficult to replace the ankle in an implant.

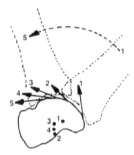

Figure 14-26. In this schematic view of the lateral ankle joint of a normal non-weight-bearing male, all of the instant centers are located within the talar body. Surface velocity shows the distraction at the beginning of motion with the subsequent sliding. Reprinted with permission from Sammarco et al. (1973). Copyright © 1973, W.B. Saunders.

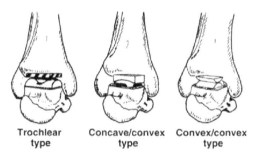

| Trochlear type | Concave/convex type | Convex/convex type |

Figure 12-27. Incongruent surface types of total ankle joint replacements. Reprinted with permission from Pappas et al. (1976). Copyright © 1976, Lippincott, Williams & Wilkins.

Ankle joint prostheses are of two types: congruent and incongruent. There are three types of incongruent ankle prostheses: trochlear, concave/convex, and convex/convex, as shown in Figure 14-27. These types of implants may suffer a high stress concentration effect and intrinsic instability.

There are four basic variations of the congruent type of implant (Figure 14-28): spherical (ball-and-socket), spheroidal (barrel-shaped), conical, and cylindrical. These designs may give greater stability of the joint and lessen the stress concentration more than the incongruent type, due to the larger contact surfaces. An actual ankle prosthesis is shown in Figure 14-29. Although improved design and materials may someday result in a "clinically satisfactory prosthesis," the long-term prospects are not as good as with the hip or knee joint.

The materials used to construct ankle joints are usually Co–Cr, Ti-alloy, and UHMWPE. Carbon fiber-reinforced UHMWPE composite has been used to fabricate the tibial component to provide higher strength and creep resistance.

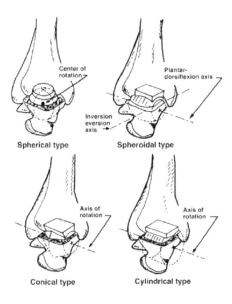

Figure 12-28. Congruent surface types of total ankle joint replacements. Reprinted with permission from Pappas et al. (1976). Copyright © 1976, Lippincott, Williams & Wilkins.

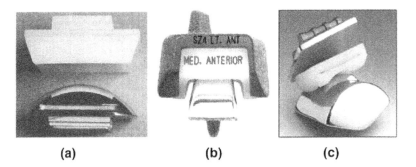

Figure 14-29. Some ankle prostheses: (**a**) original Mayo total ankle prosthesis [Reprinted with permission from Stauffer (1989). Copyright © 1989, Springer-Verlag]; b. FDA-approved prosthesis for cement implantation; it is used as an ingrowth device. (**c**) Scandinavian total ankle prosthesis [Reprinted with permission from Haddad (2006). Copyright © 2006, Orthopedic Technology Review.

14.1.1.d. Toe Joint Prosthesis

The toe joint is very difficult to replace due to the high degree of motion and stresses. There are no satisfactory materials and designs available for such a demanding prosthesis at this time. This is again due to the fixation problems and issues of long-term service life in a highly dynamic situation. Similar to finger joint replacement, only cosmetic purposes with minimum function can be achieved with silastic rubber implants. Any attempt to use other materials may result in unsatisfactory results, such as those seen in a screw toe prosthesis made of ceramic.

Example 14-1

A bioengineer is trying to determine the amount of gap developed between bone and cement when a femoral hip replacement arthroplasty is performed. It is assumed that the system is modeled as a set of concentric cylinders. Calculate the gap developed between bone and cement if the temperatures of cement, implant, and bone reached are 55, 50, and 45°C, respectively, throughout each component. Assume the thermal expansion coefficient (α) of the implant is $17 \times 10^{-6}/°C$, and no direct adhesion takes place between bone and cement, that the temperature is uniform, and that each constituent is 10 mm thick.

Answer

From Table 2-1 the linear coefficients of thermal expansion α are 8.3 and $81 \times 10^{-6}/°C$ for bone and cement. The strain is $\varepsilon = \alpha \Delta T$, and the change in temperature ΔT is computed with body temperature 37°C as a reference point. The change in length is $\Delta L = \varepsilon L = \alpha \Delta T L$.

The length change after equilibration with body temperature for each component will be (in a one-dimensional approximation)

$$\text{Implant:} \quad \Delta L = (17 \times 10^{-6}/°C \ (37\text{--}50)°C) \ 10 \text{ mm } = -2.2 \ \mu\text{m},$$

$$\text{Cement:} \quad \Delta L = (81.0 \times 10^{-6}/°C \ (37\text{--}55)°C) \ 10 \text{ mm} = -15 \ \mu\text{m},$$

$$\text{Bone:} \quad \Delta L = (8.3 \times 10^{-6}/°C \ (37\text{--}45)°C) \ 10 \text{ mm} = -0.66 \ \mu\text{m}.$$

Actually, a gap can form only if the surfaces are perfectly smooth (unlikely in the case of tissue) and if the difference in expansion gives rise to tension. Since the cement shrinks more than the metal as it cools, compression will occur at the interface and the cement will impose a *hoop stress* around the implant. If the hoop stress becomes large enough, the cement may break from its own shrinkage stress. Cement breakage and a microgap between the implant and bone have been observed in clinical situations, although the former is a rare case. Shrinkage stress is calculated for a tooth filling in Example 3-5.

Example 14-2

Consider a total hip replacement prosthesis. The proximal portion of the supporting bone cement has crumbled, so that there is an 8-inch (203 mm) distance (defined as L) between the point of application of the force at the head of the prosthesis and the undamaged distal bone cement. Assume that the stem of the prosthesis is $b = 0.8$ in (20.3 mm) wide and $h = 0.31$ in (7.87 mm) deep in the direction of bending. The patient weighs 100 lbs (mass 45.5 kg), and walks slowly with the damaged hip replacement in place. Suppose that for the hip geometry in question, the bending component of force transverse to the implant stem is three times the body weight, or 300 lbs. This is realistic since the forces in walking can be from 3 to 5 times the body weight. Neglect the bending moment due to the vertical component of load. The stem is made of a cobalt chrome with a yield strength σ_y=122,000 psi and an ultimate tensile strength of σ_{ult} = 185,000 psi. Determine whether the implant will break.

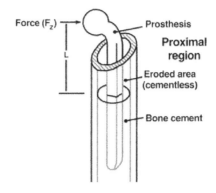

Answer

Consider cantilever bending with the load horizontal and the stem (considered as a beam) vertical. Then,

$$\sigma_{zz} = My/I = Fzy/I = 6FL/bh^2,$$

since $y = h$ at the surface where the stress is greatest, and the area moment of inertia $I = bh^3/12$, so

$$\sigma_{zz} = (300 \text{ lb} \times 6 \times 8 \text{ in})/(0.1 \text{ in} \times 2 \times 0.8 \text{ in}) = \underline{180,000 \text{ psi}} \ (\underline{1.24 \text{ GPa}}),$$

which exceeds the yield strength and is nearly as large as the ultimate strength. The stem will yield at the first step and fracture due to fatigue after a few steps. The stress is so large as a result of a lack of proximal support of the stem ($z = 8$ in). This example is an extreme example of a continuing clinical problem. Breakdown of the proximal cement causes excessive stress to occur in the stem, which can fracture in fatigue after many cycles of walking. Loosening of the implant can also cause pain to the patient.

14.1.2. Upper Extremity Implants

14.1.2.a. Shoulder Joint Replacements

The major shoulder joint motion originates from the ball-and-socket articulation of the gleno-humeral joint, as shown in Figure 14-30. The hemispherical, incongruent joint provides the largest motion in the body. As in hip joint replacements, the first shoulder joint replacement by Neer was attempted by merely replacing the humeral head, as shown in Figure 14-31. Observe the large articulating surface and two holes, which provide better fixation and resistance to rotation. There are three generations of the shoulder arthroplasty, as given in Table 14-3. It is not clear, however, whether the newer generation of implants can be as effective or better than the earlier ones. Again, a large amount of clinical data is needed in this area. One factor making this prosthesis difficult is that the gleno-humeral joint is close to spherical in shape. Figure 14-32 shows a modular design for a third-generation shoulder prosthesis.

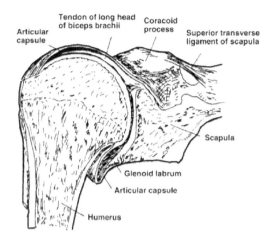

Figure 14-30. Anatomy of the glenohumeral joint. Reprinted with permission from Goss (1975). Copyright © 1975, Lippincott, Williams & Wilkins.

14.1.2.b. Elbow Joint Prosthesis

The *elbow joint* is a hinge-type joint allowing mostly flexion and extension, but having polycentric motion. The distal end of the humerus has two articulating surfaces: the trochlea and the capitulum, as shown in Figure 14-33. Most elbow joint implants are either hinge or surface replacement types. The major problems involved are loosening of implants and the limited soft tissue coverage of the implant, making it vulnerable to infection.

Most elbow joint implants are either hinge or surface replacement types, as given in Table 14-4. Major problems include loosening of the implant and limited soft tissue coverage of the implant, making it vulnerable to infection. An in-situ elbow prosthesis is shown in Figure 14-34. There has been little progress with this type of implant that offer long-term clinical satisfaction, although many designs have been attempted.

14.1.2.c. Wrist Joint Replacements

The *wrist joint* allows flexion, extension, adduction, and abduction primarily through the radiocarpal joint, as shown in Figure 14-35. Since the wrist joint arthroplasty for prosthetic replacement includes removal of the capitate, where the anatomical instantaneous center or axis of motion of a radiocarpal complex is located, it is very difficult to place the prosthesis in the correct position. The unnatural position of the implant will constrain its movement, causing an excessive generation of bending moments in adduction or abduction. This is also a major cause of complications in total wrist replacements.

Figure 14-36 illustrates some of the wrist implants now in use. The Meuli and Volz implants are ball-in-socket, while Swanson's is a space-filler type made of silicone rubber, like his finger joint prosthesis. The more recent wrist implants may not give clinically satisfactory results, as shown in Figure 14-36. However, improved materials and a better understanding of surgeons of the limitations of such implants may yield better-performing implants. Pyrolytic carbon-coated implant materials and high-performance silicone rubber incorporating more filler materials for better tear resistance have been tried. The clinical results are poorer than with other joint replacements due to more complex motion of the wrist and surgical inconsistency due to a lack of landmarks to rely on, as mentioned before.

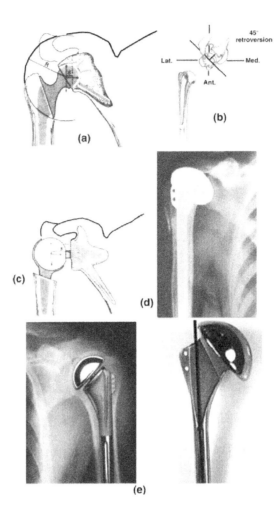

Figure 14-31. Types of shoulder joint prostheses: (**a**) Stanmore; (**b**) Bechtol; (**c**) Fenlin [Reprinted with permission from Fenlin (1975). Copyright © 1975, W.B. Saunders]; (**d**) Neer [Reprinted with permission from Neer (1974). Copyright © 1974, Charles C. Thomas]; (**e**) Global total shoulder prosthesis.

Table 14-3. Generations of Shoulder Prostheses

Generations	Characteristics	Commercial products
First	Monoblock, non-modular	Cofiels (Richards) Fenlin (Zimmer) Neer (3M)
Second	Modular	Biomodular (Biomet) Global(DePuy) Neer (3M) Select Shoulder (Intermedics Orthopedics Inc.)
Third	Modular and adaptable	Aeqalis (Tornier)

Reprinted with permission from Boileau and Walch (2006). Copyright © 2006, Maitrise Orthopédique.

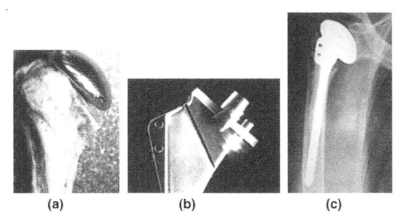

Figure 14-32. Modular design of a shoulder prosthesis. (**a**) Anatomical osteotomy of the humeral head does not always allow one to correctly position a standard prosthesis. Conversely, cutting the humeral head according to a fixed inclination may change the rational centers. (**b**) The prosthesis (Aequalis) allows adaptation to individual inclination through a system of variable necks between the stem and the prosthetic head. (**c**) X-ray radiograph of the prosthesis. Reprinted with permission from Boileau and Walch (2006). Copyright © 2006, Maitrise Orthopédique.

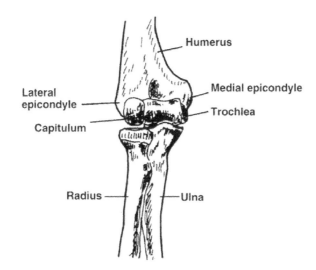

Figure 14-33. Detailed anatomical drawing of the elbow joint. Reprinted with permission from Kapandji (1970). Copyright © 1970, Livingstone.

14.1.2.d. Finger Joint Replacements

A finger contains distal, middle, and proximal joints and is provided with adequate controls by ligaments and tendons so as not to collapse under a compressive load, as shown in Figure 14-37. The exact mechanics of the joint movements are rather complex. The implants and various types of resectional arthroplasties can be divided into five major categories, as shown in Figure 14-38. Table 14-5 lists three major types of finger joint replacement. The traditional treatment for arthritic degeneration is resection of the joint, which usually relieves pain and corrects deformity but results in a loss of stability and strength. One can divide the various types of resectional and implant arthroplasties into five major categories, as depicted in Figure 14-38. Alternate approaches of joint replacement were divided into four categories: hinge type, polycentric type, space-filler type, and a combination of space filler and hemiresection arthroplasty. Some actual finger joint implants are shown in Figure 14-39. A more recent finger joint prosthesis is shown in Figure 14-40. This implant is to be fixed by tissue ingrowth over the titanium coating, which is deposited on a polymer (PEEK, polyetheretherketone) body.

Table 14-4. Types of Elbow Prostheses

Prosthesis type	Fixation	Materials
Hinge (loose, narrow, rigid) surface replacement	Stem, IMS, Cement	Co-Cr, Ti-alloy, PE

IMS: intramedullary stem, PE: UHMWPE (earlier HDPE)

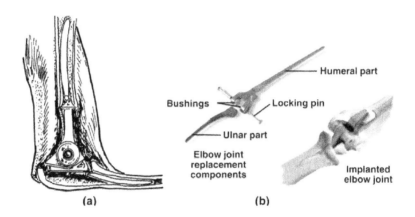

(a)   **(b)**

Figure 14-34. (a) Elbow prosthesis in situ (Dee prosthesis). Apart from a portion of the humeral intramedullary stem, the remainder of the implant is buried within the bone [Reprinted with permission from Youm et al. (1978). Copyright © 1978, Charles C. Thomas]; (b) Zimmer elbow prosthesis [Reprinted with permission from Zimmer (2006). Copyright © 2006, The Zimmer Corporation].

The concept of fixing the implant with an encapsulated collagenous membrane, which is created around the implant, is believed to be a clinically sound one. This primarily takes advantage of nature's response to any implant in the body, especially a moving implant, that is, encapsulation. The finger joint is best suited for this means of fixation since cortices of finger bones are often too thin to allow ingrowth of an adequate amount of bone tissue around a metal prosthesis.

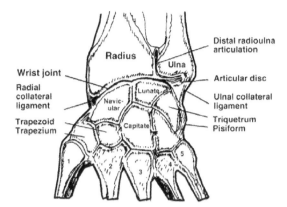

Figure 14-35. Anatomical features of the wrist joint. Reprinted with permission from Goss (1975). Copyright © 1975, Lea & Febiger.

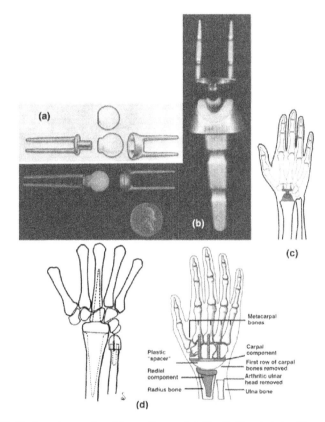

Figure 14-36. Various types of wrist prostheses: (**a**) Meuli [Reprinted with permission from Meuli (1975). Copyright © 2075, Ferdinand Enke Verlag]; (**b**) Voiz [Reprinted with permission from Volz (1977). Copyright © 1977, Lippincott, Williams & Wilkins]; (**c**) Swanson [Reprinted with permission from Swanson (1973). Copyright © 1973, Mosby]; (**d**) Recent wrist joint prosthesis [Reprinted with permission from AAOS (2002). Copyright © 2002, American Academy of Orthopaedic Surgeons].

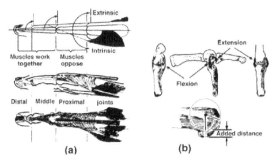

Figure 14-37. (a) This schematic diagram of a finger indicates the principal musculotendinous structures. Note the arrangement of muscles and tendons. (b) Schematic diagram of the metacarpophalangeal joint in extension and flexion. When the joint is in extension, the collateral ligaments are slack to permit abduction and adduction. The joint is stabilized in flexion because the ligaments are tightened in both the longitudinal and transverse planes. Reprinted with permission from Flatt (1972). Copyright © 1972, Mosby.

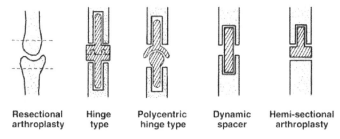

Figure 14-38. Schematic diagrams indicating the various types of resectional and implant arthroplasties available for treatment of joints in the hand. Reprinted with permission from Burton (1973). Copyright © 1973, W.B. Saunders.

Table 14-5. Types of Finger Joint Replacements

Type	Remarks	Examples
Hinge	Loosening, excessive wear, and breakage, subsidence of implants UHMWPE/Co-Cr alloy	Schultz (Figure 14-39a) St. Georg (Figure 14-39b)
Polycentric	Good duplication of anatomical motion, unstable joint UHMWPE/Co-Cr alloy	Steffee (Figure 14-39c)
Space-filler	Good clinical success, tear sensitive, strong grip not possible due to weakness of implant Silicone rubber, silicone rubber-Dacron® composite, polypropylene	Swanson (Figure 14-39d) Calnan-Nicolle (14-39e) Niebauer-Cutter (Figure 14-39f)

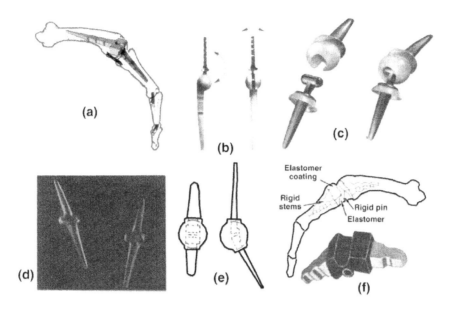

Figure 14-39. Various types of finger joint prostheses: (**a**) Schultz; (**b**) St. Georg; (**c**) Steffee; (**d**) Swanson; (**e**) Calnan-Nicolle; (**f**) Niebauer-Cutter; (**g**) Lord's Bonded Bion® elastomer titanium joint.

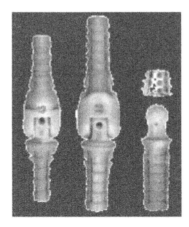

Figure 14-40. RM Finger joint prosthesis. Distal and proximal joint components are coated with titanium for osteointegration. Reprinted with permission from Mathys Medical (2006). Copyright © 2006, Mathys Medical Ltd.

Example 14-3

Bioglass® is coated on the surface of stainless steel for making implants that are more biocompatible. 10-mm diameter stainless steel is coated with 1-mm thick Bioglass®. Using the following data, answer questions a–d.

Material	Young's modulus (GPa)	Strength (MPa)	Density (g/cm^3)
Stainless Steel	200	300 (yield)	7.9
Bioglass®	300	300 (fracture)	4.5

a. What is the maximum load the composite can carry in the axial direction?

b. What is the Young's modulus of the composite in the axial direction?

c. What is the density of the composite?

d. Give two reasons why you would not use this composite to make orthopedic implants over plain stainless steel.

Answer

a. In this composite system we calculate the maximum strain for each component by using Hooke's law, and hence,

$$\text{Maximum strain for stainless steel} = \frac{300 \times 10^6 \text{ N/m}^2}{200 \times 10^9 \text{ N/m}^2} = 1.5 \times 10^{-3},$$

$$\text{Maximum strain for Bioglass}^® = \frac{300 \times 10^6 \text{ N/m}^2}{300 \times 10^9 \text{ N/m}^2} = 1.0 \times 10^{-3}.$$

Therefore, the maximum strain for the composite is the smaller of the two, that is, 1×10^{-3} (0.1%). The cross-sectional areas of each component are

$$A_{ss} = \pi(5 \times 10^{-3} \text{m})^2 = 25\pi \times 10^{-6} \text{ m}^2,$$

$$A_{\text{Bioglass}} = \pi[(6 \times 10^{-3} \text{ m})^2 - (5 \times 10^{-3} \text{m})^2] = 11\pi \times 10^{-6} \text{m}^2.$$

Therefore,

$$F_{\text{comp}} = F_{ss} + F_{BG}$$

$$= 200 \times 10^6 \text{ N/m}^2 \times 25 \times 10^{-6} \text{ m}^2 + 300 \times 10^6 \text{ N/m}^2 \times 11\pi \times 11^{-6} \text{ m}^2$$

$$= 8300\pi \text{ N}$$

$$= \underline{26.1 \text{ kN}}.$$

b. Since

$$E_{\text{comp}} = E_1 V_1 + E_2 V_2 + \dots$$

$$= E_1 A_1 + E_2 A_2 + \dots$$

where V and A are the volume and area fractions of each component; therefore,

$$E_{\text{comp}} = 200 \times 25\pi/26\pi + 300 \times 11\pi/36\pi$$

$$= \underline{231 \text{ GPa}}.$$

$$\rho_{\text{comp}} = \rho_1 V_1 + \rho_2 V_2 + \dots$$

$$= \rho_1 A_1 + \rho_2 A_2 + \dots$$

$$= 7.9 \times 2.5\pi/36\pi + 4.5 \times 11\pi/36\pi$$

$$= \underline{6.85 \text{ g/cm}^3}.$$

c.

1. It is difficult to obtain perfect bonding between the stainless steel and Bioglass® due to the differences in thermal expansion coefficient. In addition, Bioglass® is brittle; consequently, a moderate or severe bending load on the implant causes the Bioglass® to develop cracks and microcracks.

2. The bonding between the (hard) tissue and Bioglass® cannot be maintained for a prolonged time without further weakening of the Bioglass® since a continuous dissolution of the surface film is essential for the bonding to be maintained. (Remember that the tissue cells renew themselves.)

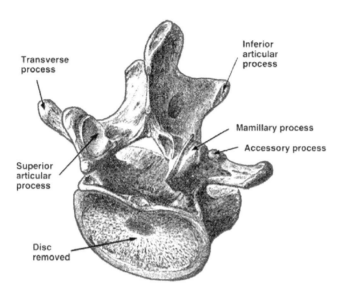

Figure 14-41. Anatomy of a spinal unit.

14.2. SPINAL IMPLANTS

Each vertebra has two sets of facet joints. These spinal joints are called facet, apophyseal, or zygapophyseal joints. One pair faces upward (superior articular facet) and one downward (inferior articular facet), as shown in Figure 14-41. There is one joint on each side (right and left). Facet joints are gliding joints and link vertebrae together; they are located at the back of the spine (posterior). Facet joints are synovial joints surrounded by a capsule of collagenous tissue. The surfaces of synovial joints are coated with cartilage allowing joints for smooth (articulate) motion against each other. These joints allow flexion (bend forward), extension (bend backward), and rotating or twisting motion. The spine is made more stable due to the interlocking nature of adjacent vertebrae restricting certain types of motions.

The spinal discs are located between pairs of vertebrae, making up about a quarter of the spinal column height. The discs act as shock absorbers and allow deformation of the spine. The disc is made up of the outer portion, the annulus fibrosus, and the inner nucleus pulposus. The annulus is made of layers of collagenous lamellae. The fibers of the lamellae are angled at 30°, and the fibers of each lamella run in a direction opposite to the adjacent layers. This creates a structure that is exceptionally strong and flexible.

The nucleus pulposus is made of a gel-like material mostly composed of collagen fibers, proteins, and water; water makes up 90% of the disc weight at birth but decreases to 70% by age 50. Excess disc bulging or complete herniation may take place by injury or aging, allowing the disc material to escape the disc. A bulging disc or nucleus material may compress the nerves or spinal cord, causing pain. When a person turns 50, the blood supply to discs decreases; some discs may degenerate. The disc begins to lose water content and shrinks. The range of motion and shock-absorbing ability of the spine are decreased. This may result in damage to the nerve and vertebrae, and the aging disc itself may generate pain.

In a young healthy person, the spinal components function in unison to provide flexibility, support of the upper body weight, and protect the spinal cord and nerve roots. Abnormal spinal components can compromise quality of life. Backaches and back pains are very common. Back pain is exacerbated by emotional stress, such as that due to the workplace, as well as to smoking and excess alcohol consumption. Most cases respond well to conservative treatment, which may involve exercise to restore flexibility and achieving a proper balance of strength in the muscles that stabilize the spine. Of people who suffer an episode of low back pain, 95% respond to conservative treatment (e.g., rest and muscle exercise) and return to work within 3 months. Surgery may be done for those who do not respond to conservative treatment. Surgical interventions include discectomy and fusion. Fusion surgery is followed by long recuperation, and it gives rise to many complications, including loss of motion and degeneration of adjacent discs. Artificial disc replacements have been considered to restore the motion of the original disc, but problems with these implants have included dislocation and subsidence. Graft materials, metals, polymers, ceramics, and composites in various forms are used to restore spinal function. More devices are being introduced in this important and yet underdeveloped field.

14.2.1. Interpositional Barriers

Interpositional barriers are used after laminectomies, foraminotomies, and discectomies, which invariably result in scar (granulation) tissue formation, also termed the "laminectomy membrane." The potential complication pertaining to the laminectomy membrane are persistent pain following surgery, increase in the technical difficulty and risk of re-operation by increasing the possibility of tearing the *dura mater* (a fibrous layer surrounding the spinal cord) and damaging the emerging nerve roots. Despite the acclaim of autogenic free fat grafts as the most effective interpositional barrier, there are drawbacks: graft necrosis and atrophy, infection at the harvest site, migration of the graft, and *cauda equina syndrome*, a serious nerve complication in which control of the bladder or bowel is lost. Other problems include a lack of graft to cover the area of operation without fibrous tissue adhering to the graft, and in thin people, the harvest site of a graft leaving a dimpling appearance. Fat grafts can also shrink by as much as 50%, and this must be taken into account when harvesting them. Several materials have been introduced as an alternative to fat graft, but with varied results (see Table 14-6).

14.2.2. Interbody (Disc) Spacers

Surgical procedures involving a total discectomy invariably lead to a loss of disc height, and an unstable spinal segment. Both allo- and autologous bone grafts, as interbody spacers, present surgeons with several complications: pain, dislodgment of the anterior bone graft, pseudoarthrosis, fracture, collapse of the bone graft, loss of alignment, etc. Both "rigid" and "nonrigid" or articulating" type alternatives have been used to enhance stabilization of the segment, prevent postoperative collapse, and supplement autologous bone grafts.

Table 14-6. Summary of the Various Materials Used as Interpositional Barriers

Materials	Classification	Examples
Natural	Natural tissue grafts	Autogenic fat, pedicle fat
	Reprocessed tissues	Collagen (foam or film) Heparinized collagen Hyaluronic acid[a]
Synthetic	Polymeric Implants	ePTFE[b], silicone, oxidized rayon fiber

[a] Can be obtained from tissues or synthesized.

[b] Expanded polytetrafluoroethylene (Teflon®).

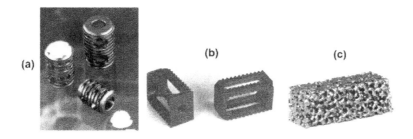

Figure 14-42. (**a**) BAK cervical spinal fixation device [Reprinted with permission from *BAK Interbody fusion system* (1996). Copyright © 1996, SpineTech Inc.]. (**b**) Carbon fiber cage made from Ultrapek® [Reprinted with permission from Brantigan et al. (1991). Copyright © 1991, American Association of Neurological Surgeons. (**c**) Co-Cr metal sponge [Reprinted with permission from Waisbrod (1988). Copyright © 1988, Springer-Verlag].

14.2.2.a. "Rigid" Spacers

Metallic cages of various shape are used to restore disc height and stability. These cages are normally packed with bone chips (allo-, auto-, with or without BMP or TGF-β) to enhance fusion for long-term stability. The BAK® interbody fusion cage is one such system, as shown in Figure 14-42a. It is a threaded, hollow, porous Ti-6Al-4V shell into which bone chips are packed. The ends of the implant are capped with an ultrahigh-molecular-weight polyethylene (UHMWPE) plug to better contain the bone graft material, as well as to minimize the likelihood of adhesions to the surrounding nerves and blood vessels.. Another cage device, the titanium Moss cage, provides a scaffolding for osteosynthesis by allowing one to pack bone grafts within it, thereby promoting fusion between vertebral bodies in patients requiring vertebral resection. A composite of PEKEKK (polyether ketone ether ketone ketone) with (pyrolytic) carbon fibers interspersed within the polymer matrix has been used to fabricate a cage with teeth located superiorly and inferiorly to resist expulsion (see Figure 14-42b). These cages can be used in the treatment of burst fractures and tumors of the lumbar and thoracic spine. The

Young's modulus of this composite matches that of cortical bone (17 GPa). The lower modulus of the material, as opposed to a metal, may decrease the likelihood of stress shielding. Postoperative visualization of the implant area by MRI is better than that of a metallic implant, due to the radiolucent and nonmagnetic nature of the implant material.

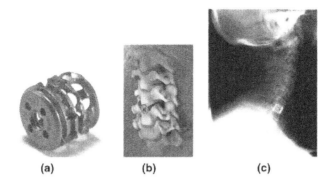

(a)   **(b)**   **(c)**

Figure 14-43. Newer version of the BAK cage (**a**), in vitro (**b**), and x-ray radiograph after implantation (**c**). Reprinted with permission from Regan (2006). Copyright © 2006, SpineUniverse.com.

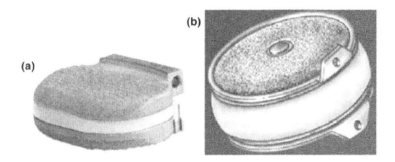

Figure 14-44. Different types of cervical spine prosthesis: (**a**) PCM® artificial cervical disc [Reprinted with permission from Spine Solutions (2006). Copyright © 2006, Spine Solutions Inc.]. (**b**) similar to lumber spine disc implant [Reprinted with permission from Sasso (2006). Copyright © 2006, SpineUniverse.com].

A cobalt–chrome metal sponge with a porous structure (pore diameter 800~1500 μm, and pore volume > 60%) similar to that of cancellous bone is shown in Figure 14-42c. Additionally, as a result of the increased porous surface area, there may be problems associated with a greater amount of metal ion release into surrounding tissues. Similarly, a titanium mesh block was developed as a substitute for an autogenous bone graft in situations that are known to respond well to anterior discectomy and interbody fusion. These titanium mesh blocks are claimed to have a mechanical compliance near that of trabecular bone. However, because of the flexibility claimed with this device, there may be problems of collapse should the fusion not take hold. A newer version of the BAK® cage is shown in Figure14-43a for an anterior approach, as shown by the x-ray radiograph in Figure 14-43c. Other designs are also either in the experimental or clinical trial stage, as shown in Figure 14-44.

Some surgeons have tried PMMA (polymethylmethacrylate) bone cement for immediate fixation of the spine, and as an alternative for external bracing. Using PMMA bone cement to stabilize the anterior column is believed to have merit only in patients with a limited life expectancy who have a loss of vertebral bone stock anteriorly due to destruction by neoplastic disease.

Ceramic prostheses to replace a lumbar vertebra in sheep have been developed. The implant, which is an apatite–wollanstonite (A-W) glass-ceramic, has excellent biocompatibility. Complete bonding of bone to implant took a year, but at 3 months early apatite bonding was observed, suggesting that if a more rigid fixation was possible then complete bonding could occur sooner. Glass-ceramic spacers made of the same A-W, in place of the bone graft, have been used in conjunction with laminoplasty in the cervical spine region with some success.

14.2.2.b. "Articulating" Spacers

It would be ideal if one could restore the original function of the damaged disc without fusing the segment, since a rigid fusion in one joint could yield excessive motion in its neighboring joints. The concept of restoring function to a damaged disc becomes feasible only when the rest of the spinal components are not affected by pathology. Both artificial and graft materials have been used to achieve this goal.

Fresh-frozen allografts, consisting of a vertebral body and its adjacent discs, were implanted to restore nearly normal biomechanical characteristics of the segment in a canine model. Histological study revealed revascularization of the osseous structures after 18 months. However, transplanted discs were not entirely normal morphologically or metabolically.

A number of artificial disc prostheses have been designed, ranging from a stainless steel ball-type ball-bearing prosthesis, to sandwiches of porous coated metal plates and elastomeric bodies to incorporate a cushioning effect. A high-density polyethylene (HDPE) pad inserted between two metal components has been developed for cervical spine. The HDPE washer rests on the HDPE pad below the upper metal plate, and this washer acts as a slide bearing for the HDPE pad, as shown in Figure 14-45a. This prosthesis is anchored in place with bone screws, but without cement.

An all-metallic artificial disc provides 15–20 degrees of flexion/extension, 3 degrees of lateral bending, but less than 1 degree of axial rotation. It also duplicates the inherent stiffness of the natural disc in flexion and extension. Two Ti6Al4V springs are pocketed between either HIPed (hot isostatic press) or forged CoCr end-plates, with CoCr alloy beads sintered or plasma sprayed onto the end-plate surfaces to facilitate long-term fixation through bony ingrowth, as shown in Figure 14-45b. The springs as well as three hot isostatic pressed (HIPed) CoCr hourglass fatigue specimens have been tested to 100 million cycles without failure. A different design with metallic articulating disc is a two-piece ball-and-socket implant, as shown in Figure 14-45c, with two sets of grooves on both the superior and inferior end-plates, and a porous coating separating them. These grooves should confer some degree of initial fixation to the implant, while over time the porous coating will allow for bony ingrowth. The ball and socket is designed for six degrees of freedom, but restricts axial compression and lateral translation. The Link® SB Charité disc prostheses are lumbar intervertebral replacements, based on low-friction principles. The implant is composed of two CoCr alloy end-plates with an UHMWPE body, as shown in Figure 14-45d, also shown in situ within the implant in Figure 14-45e.

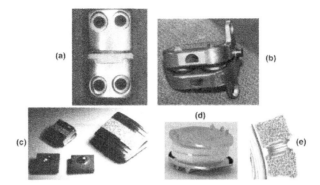

Figure 14-45. (**a**) Metal cementless prosthesis [Reprinted with permission from Solini et al. (1989). Copyright © 1989, Lippincott, Williams & Wilkins]; (**b**) Danek artificial disc [*An artificial intervertebral disc arthroplasty* (1995). Copyright © 1995, Sofamor Danek Group]; (**c**) Kostuik artificial disc [Reprinted with permission from Kostuik (1992). Copyright © 1992, Raven Press]; (**d**) SB Charité intervertebral prosthesis [Reprinted with permission from Griffith et al. (1994). Copyright © 1994, American Association of Neurological Surgeons]; (**e**) In situ position of (d).

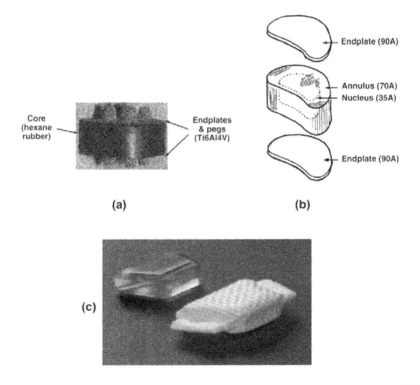

Figure 14-46. (**a**) Steffee artificial disc [Reprinted with permission from Enker et al. (1993). Copyright © 1993, American Association of Neurological Surgeons]. (**b**) Artificial disc simulating a natural disc [Reprinted with permission from Vuono-Hawkins (1991). Copyright © 1991, American Association of Neurological Surgeons]. (**c**) Prosthetic disc nucleus (PDN) [Reprinted with permission from An and Juarez (2006). Copyright © 2006, SpineUniverse.com].

Another disc is composed of a hexene-based polyolefin rubber (Hexsyn™) core that is bonded to two Ti6Al4V end-plates. These end-plates have two layers of titanium beads (average diameter ≈250 μm, c.p.–Ti) sintered onto both the superior and inferior surfaces of both titanium end-plates, as shown in Figure 14-46a. This disc has four cylindrical protrusions that bore into the superior and inferior vertebrae to provide initial fixation, while the porous coating hopes to induce bony ingrowth for long-term fixation. The Hexsyn™ core provides 2 to 3 degrees of axial rotation, accommodates bending of the back, and attenuates forces. However, in 1990 this design was discontinued because of a report that 2-mercaptobenzothiazole, a chemical used in the vulcanization process of the rubber core, was possibly carcinogenic in rats. Another polymer-based artificial disc is composed of three parts, all made from a single material, but with each part exhibiting a different stiffness: the "nucleus" portion is the least stiff, followed by the "annulus," and the "end-plates" have the highest hardness value, as shown in Figure 14-46b. It is made from a thermoplastic elastomeric material (polysiloxane-modified styrene/ethylene/butylene block copolymer, C-FLEX® TPE) without fiber reinforcement. Another artificial disc device is similar to the prosthetic disc nucleus (PDN). Like the metal implant, the purpose of it is to restore disc height and allow for normal spinal motion. However, instead of the whole disc being removed, only the inner nucleus material is removed and is then replaced with two mini "pillows," as shown in Figure 14-46c.

Another concept within the realm of the artificial disc is to implant flexible "capsules" into the nucleus space. These capsular implants are intended to restore mechanical support, with the possibility of delivering various therapeutic agents aimed at healing the disc and surrounding tissues. A semipermeable membrane can enclose a thixotropic gel such as hyaluronic acid, which would preferably be viscous, and have velocity–shear behavior similar to the normal nuclear intradiscal tissue. When the prostheses are implanted, initially they are slack and partly dehydrated, but the hygroscopic hyaluronic acid will cause the implant to swell and thus secure itself within the disc space. Some have proposed an artificial nucleus composed of a hydrogel. The polyvinyl alcohol prosthesis will preferably be composed of two parts to make insertion into the excavated disc space easier, and lessen the incision into the annulus. Table 14-7 gives a short summary of the materials used for the artificial disc spacer.

Before any artificial disc prosthesis becomes a viable solution, many issues need to be resolved — including fatigue, wear, wear debris, design of the implant, fixation to the bone, and tissue compatibility. Perhaps most importantly, though, there need to be clear definitions as to under what conditions an artificial intervertebral disc would be warranted.

14.2.3. Intersegmental Fastening Devices

The devices used for imparting stability across an abnormal spinal segment can be grouped as wires, tapes, cables, pedicle screws, hooks, plates, bands, and rods. Under extreme cases, bone cement (ordinary and bioabsorbable) may be utilized to reinforce intersegmental fixation.

14.2.3.a. Wires, Tapes and Cables

Arthrodesis of a single or multiple spinal segments is often facilitated with the aid of wires. The most commonly used wire material is 316L stainless steel. Stainless steel wire is relatively strong, ductile, and inexpensive. The material is highly resistant to corrosion, due to its extremely low carbon content (0.03% maximum, by weight versus 0.8% for other stainless steels). Stainless steel wire can cause problems (artifacts) in magnetic resonance images. If it is considered important to reduce artifacts in these medical images, titanium alloy wires have been considered. However, biomechanical testing has shown that titanium alloy such as Ti6Al4V may be a poor choice because of its decreased fatigue resistance due to notch sensitivity. The titanium alloy wires are brittle and harder to bend permanently than stainless steel. Pure titanium wires are more ductile and easy to bend; they do not possess adequate strength.

Table 14-7. Types of interbody disc spacers and materials

Type	Examples	Materials	References
Rigid	BAK™	Ti6Al4V: cage UHMWPE: caps	Spine-Tech, Inc.
	Moss	Ti	Schueller and DeWald (Schueller et al., 1994)
	Carbon fiber cage	PEKEKK (poly ether ketone ether ketone ketone) carbon fibers	McMillin and Brantigen (1993)
	Porous sponge	CoCr alloy	Waisbrod (1988)
	Bone cement	Polymethylmethacrylate (PMMA)	McAfee et al. (1986)
	Ceramic disc	Apatite-wollastonite	Yamamuro et al. (1990)
Nonrigid or articulating	Spinal allograft	Tissues with disc	Olson et al. (1991)
	Ball bearing	Stainless steel	Fernström (1972)
	Metal plate with polymer spacer	Metal: not known, spacer: high-density polyethylene	Solini et al. (1989)
	Link® SB Charité disc	CoCr alloy, UHMWPE spacer	Griffith et al. (1994)
	Kostuik disc	CoCr alloy end plates with Ti6Al4V springs	Kostuik (1992)
	Sofamor Danek	Ti6Al4V end plates with alumina ceramic ball spacer	Sofamor Danek Group (1996)
	Steffee artificial disc	Ti6Al4V end plates with hexene- based polyolefin rubber (Hexsyn™) core	Steffee (1992)
	Vuono-Hawkins disc	Thermoplastic elastomer (C-Flex TPE) and CoCr alloy (Vitallium®)	Vuono-Hawkins et al. (1991)
	Capsular disc	Polyvinyl alcohol with hydrogel	Bao and Hingman (1991, 1993) Ray (1991, 1992)

The in-situ strength of a wire can vary greatly depending on the type of knots or twist that is used to secure it irrespective of the material of the wire itself. In tension tests of stainless steel wires tied together with various types of knots, the square knot had the highest failure load, 922 N, and it failed by wire fracture. The symmetric twist was the next highest at 516 N, but this failed by untwisting, as did various other knots tested. Since the main function of a wire is to maintain tension over time, once the tension is relieved, the wire has "failed."

Alternatives to wires have been proposed primarily due to the concern that metal wires for sublaminar fixation may inflict neurologic damage or the wire may cut through bone. Polymer tapes have been introduced such as nylon and Mersilene® (strands of woven polyethylene). Another alternative to monofilament wires is to use cables, whose greatest advantage is that the significantly increased flexibility prevents repeated contusions to the spinal cord. However, this is also a disadvantage because sublaminar passage is more difficult, but this may be overcome through the use of a lead wire. Cables also are less subject to stress concentration, can conform better to irregular surfaces, and are more resistant to fatigue than wires. It was found that stainless steel (316L) cable assemblies outperform titanium ones.

14.2.3.b. Pedicle Screws and Hooks

Screws are used in several ways to augment spinal fusions. When placed in the pedicles or the vertebral bodies themselves, the screws act as anchor points for bands, rods, and plates to restore spinal stability. Strength and stability across the screw–bone interface may influence the success of the fixation. There was a significant correlation between bone mineral density and pull-out strength of Caspar cervical plating screws. The strength of the transpedicular screw–vertebra interface may be increased by increasing the depth of penetration of the screw. The use of PMMA bone cement or other grouting agents to fix the screw in place with enhanced interface strength may be hazardous because PMMA may encroach into the spinal canal. Hooks, alone or in combination with screws, are also used to enhance the in-situ performance of the device. Screws in situ are subjected to cyclic loads. The fatigue strength of screws made from 316L stainless steel can be enhanced through surface treatments such as nitrogen ion implantation.

14.2.3.c. Bands, Plates and Rods

Increasingly, pedicle screws are used in spinal constructs as a means to anchor longitudinal bands, rods or plates to a vertebra. A system consisting of bilateral "polymer" bands wrapped around the pedicular screws across a segment have been tested experimentally, as shown in Figure 14-47a. This system was found to restore a post-laminectomy motion segment to normal stiffness (i.e., pre-injury) levels.

The rigidity of a plate or rod-type device (Figure 14-47b) can be changed with the material choice [e.g., from stainless steel (316 LVM or 22-13-5 stainless steel) to titanium], or the stiffness of the plate or rod itself. The fatigue life of rod-based systems undergoing cyclical compressive flexural loading was higher than the plate based systems during in vitro cyclic fatigue tests. If the system is too rigid, it becomes difficult to bend and twist to provide the normal curvature of the spine such as the appropriate lordosis in the lumbar region, and may lead to stress-induced osteopenia of the vertebral bodies and spinal degeneration of the adjacent normal segments. However, clinical and animal-based biomechanical studies have shown that a rigid system promotes and accelerates the solid fusion process in the early stages of healing. These observations dictate that one should design a system whose rigidity decreases with time, being very rigid until fusion occurs, and thereafter allowing its rigidity to diminish. In one such design, viscoelastic polymer (UHMWPE, ultrahigh-molecular-weight polyethylene) washers are inserted between the plate and a nut of a conventional system. This concept should provide rigid fixation immediately after surgery and decreasing rigidity with time due to the viscoelastic behavior of the polymer washers. This, in turn, imposes higher loads over time on the vertebral bodies across the stabilized segments. Biodegradable washers have been proposed as a substitute to reduce the effects of polymer debris since this debris can be absorbed by the body. The rigidity of a posterior spinal fixation device and its ability to share load with the fusion mass are considered essential for fusion to occur. If the load transferred through the fusion mass, however, is increased without sacrificing the rigidity of the construct, a more favorable environment for fusion may be created. To achieve this objective, several "semi-rigid" devices have been designed and tested, as shown in Figure 14-47c-e.

It is possible that one may design biodegradable devices made from such materials as PLA or PGA, or a copolymer of the two. Currently, these materials lack the bending and torsional strength required for vertebral fixation, as given in Table 14-8. A radiotransparent biodegradable plate and rod system, whose components are made of Phusiline® (a polylactic acid), has been developed to induce anterior osteosynthesis of the cervical spine. Phusiline degrades in vivo within 18 months into lactic acid, and via the Krebs cycle into water and carbon dioxide.

The procedure made use of an interbody xenograft to assist in the fusion, and the results of an initial trial with five patients were sufficiently encouraging to warrant further investigation. Designing a biodegradable implant poses many challenges — in particular how to best match the rate of healing with the rate of degradation and a decrease in strength of the biodegradable material. Additional issues include the effects of sterilization, in-vivo loading conditions, long-term tissue, and systemic reaction to the byproducts of degradation, as well as wear and wear debris. To date, the majority of such orthopedic applications have not gained sufficient acceptance from either manufacturers or surgeons.

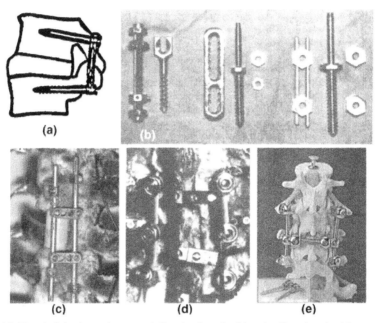

Figure 14-47. (**a**) Spinal specimens stabilized using Graf bands. [Reprinted with permission from Goel and Gilbertson (1997). Copyright © 1996, Lippincott, Williams & Wilkins]. (**b**) Pedicle screws, plates, and rods [Reprinted with permission from Goel and Gilbertson (1997). Copyright © 1997, Lippincott, Williams & Wilkins]. (**c**) Dymanized ALC device [Reprinted with permission from Goel et al. (1997). Copyright © 1997, Lippincott, Williams & Wilkins]. (**d**) DDS® cable [Reprinted with permission from Goel et al. (1997). Copyright © 1997, Lippincott, Williams & Wilkins]. (**e**) Posterior "hinged"-type pedicle screw rod device [Reprinted with permission from Scifert et al. (1997). Copyright © 1997, North American Spine Society].

Preliminary work on treating scoliotic spines with Nitinol® (see §5.4.3) rods has been carried out in goats. The shape memory alloys (SMAs) are shaped in a higher temperature phase (austenitic), and then cooled to the martensitic phase, where these materials can again be shaped, but when the material is heated it will assume its original configuration by reverting to the original austenitic phase. Therefore, the theory behind using shape memory alloy rods for the treatment of scoliosis is that one can shape it into a straight, corrective rod in the austenitic phase, cool it, and bend the rod to conform to a scoliotic spine. After it has been fitted to the spine, the rod would then be reheated (by electric current or magnetic field induction) to above its transformation temperature, which will result in the rod reverting to its original straight shape.

Table 14-8. Material Properties of Biodegradable Polyglycolic Acid (PGA) before and after 4 Weeks in Distilled Water at 37°C

Materials	Initial values		After 4 weeks in distilled water, 37°C	
	Bending strength (M Pa)	Shear strength (MPa)	Bending strength (MPa)	Shear strength (MPa)
SR-PGA (uncoated)	350 ± 50	200 ± 50	45 ± 8	137 ± 7
SR-PGA (PLLA coated)	350 ± 50	250 ± 50	176 ± 27	181 ± 5

Note: Material was shaped into cylindrical rods 3.2 mm in diameter and 50 mm long.

Adapted with permission from Vasenius et al. (1989). Copyright © 1989, Elsevier Science.

The outcome of a surgical procedure is dependent upon a complex process primarily influenced by the graft material used as well as many other factors that affect the fusion healing response. These factors may have positive or negative effects on graft healing. There are two general areas of fusion enhancement currently under consideration. Electrical stimulation [direct current or pulsed electromagnetic field (PEMF)] of bone healing has been studied and has yielded accelerated fracture repair (see the previous chapter). The electrical approach has been a subject of some controversy. The other major area for biological enhancement of spinal fusion is that of bone graft substitutes (allograft, bone marrow, xenograft, ceramics — calcium phosphate consisting of hydroxyapatite and tricalcium phosphate), and osteoconductive growth factors (BMP and TGF-ß). It seems appropriate to explore the possibility of increasing the fusion success rate through basic and applied research on enhancement of the biological fusion process. The osteoconductive growth factors may also be used to fill the space within cages previously mentioned, and thus lessen the use of autologous bone graft in that procedure.

It is also a possibility in the somewhat distant future that a tissue engineering (see Chapter 16) technique can be used to grow whole spinal discs or vertebrae for replacement. It is also tempting to use better and innovative materials, such as a hydrogel that is liquid at room temperature and becomes gel at body temperature. One may insert the gel into a balloon, which could be used as a spacer for the disc.

14.3. DENTAL RESTORATIONS AND IMPLANTS

14.3.1. Dental Restorations

Tooth decay is treated by the dentist to drill out the decayed portion of the tooth and to replace the decayed portion with a filling. Since the filling material replaces tooth enamel, which is rather stiff and strong, the filling must have adequate mechanical properties. As with other biomaterials, dental materials must be adequately stable in the hydrated environment in which they are placed, and they must not injure tissues. Thermal expansion is relevant in the dental setting, since temperatures in the mouth can vary considerably due to consumption of hot or cold foods. Metallic filling materials, including silver amalgam and gold, are discussed in Chapter 5 (§5.4). Polymers were tried for restoration of anterior teeth. The reason was to avoid objections to the cosmetic appearance of metal in the front teeth. Pure polymers were found to have inadequate strength and wear properties. Composite filling materials contain a polymer matrix and ceramic inclusions, as discussed in Chapter 8 (§§8.1, 8.3.1). These materials resemble the tooth in appearance, and they have mechanical properties adequate for many applications in dental fillings. If much of the tooth structure is damaged or lost, the upper portion may be replaced by a crown. Crowns as well as dentures are made using ceramics (§§6.3, 6.4). Crowns as well as gold restorations are held in place by dental cements, which typically contain particles of ceramic such as calcined zinc and magnesium oxides. The powder is mixed by the dentist with a liquid containing phosphoric acid and water.

14.3.2. Dental Implants

In this section we examine total tooth replacements or alveolar bone augmentation with manmade materials. Replacement of whole teeth is a very challenging task in view of the transcutaneous (or percutaneous) nature of the implant in the hostile oral environment, which continually changes its chemical composition, pH, temperature, etc. Teeth experience large forces (up to 850 Newtons) upon a small area; consequently they undergo the most severe compressive stress in the body. Satisfactory materials or techniques have not yet been found that withstand not only the compressive stress but also the added torque and shear stresses during mastication.

In an early comprehensive conference on the benefits and risks of dental implants, dental implants were classified largely into two categories; subperiosteal/staple/transosteal and endosseous tooth implants. The former is to support dentures and the latter is to restore the function of teeth with or without a supporting bridge framework.

14.3.2.a. Endosseous Tooth Implants

The endosseous dental implant is inserted into the site of missing or extracted teeth to restore original function. The ideal implant would be the tooth itself, replaced in the same socket it was lost from. Teeth that have been knocked out traumatically can sometimes be replanted. In most cases of tooth loss, however, the teeth or their supporting structures are diseased. Artificial teeth supported by the gums (false teeth) are a partial solution to the problem of tooth loss; however, they present problems of their own: lack of stability, poor esthetics, and resorption of the jaw bone. Artificial teeth fixed to the jawbone represent an attempt to achieve more natural replacements; these are known as endosseous implants or simply dental implants. There are many different types of designs in the early times for endosseous implants, as shown in Figure 14-48. The main idea behind the various root portions of self-tapping screws, spiral screwvent, and blade-vent implants, is to achieve immediate stabilization as well as long-term viable fixation. The post will be covered with an appropriate crown after the implant has been fixed firmly for about 14 months. Some implants used a more complicated system of fixation, as shown in Figure 14-49, in which the implant root is first implanted in the tooth extraction site (preferably after complete healing of the site), and completely buried; then the post is installed through a punctured hole on the mucosa, and finally the crown is made. Despite the more elaborate work and design of the implants, the success rate of this implant system did not improve over other implants such as blade vents (Linkow types). However, the dental implants are very popular recently, as shown in Figure 14-50. Most of them mimic the Brånemark type two-stage implants made of Ti6Al4V. Figure 14-51 shows three ways of connecting the root portion to extend it for the crown portion. The implant may not induce the collagenous tissue between implant surface and alveolar or maxillary bone unlike other implants. This phenomenon is termed "osseo-integration." It is not clear why this phenomenon appears in dental implants but not in others such as joint implants. Even the natural tooth is embedded in soft tissues, cementum, and ligament. Such viscoelastic materials provide damping or shock-absorbing ability, so that peaks in the mastication force are moderated. Implants lack these

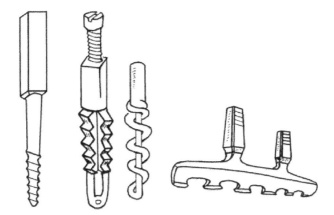

Figure 14-48. Various designs of self-tapping dental (root) implants. Reprinted with permission from Grenoble and Voss (1976). Copyright © 1976, Marcel Dekker.

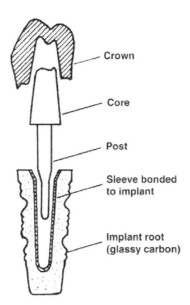

Crown

Core

Post

Sleeve bonded
to implant

Implant root
(glassy carbon)

Figure 14-49. Tooth root implant fabricated from glassy carbon for two stage implantation. Reprinted with permission from Grenoble and Voss (1976). Copyright © 1976, Marcel Dekker.

tissues. A similar concept of fixation technique has been tried in hip and knee joint implants, as mentioned earlier (§14.1.1), with somewhat limited success. These implants are installed in two stages, as shown in Figure 14-52. First, the root is implanted and allowed to heal for a few weeks; then the top potion is screwed to the root portion. Despite the popularity of these implants, care is called for in their use, specifically excellent hygiene to prevent infection and effort to minimize applied forces. Such implants tend to be more successful if used as supporters for bridges rather than standing alone.

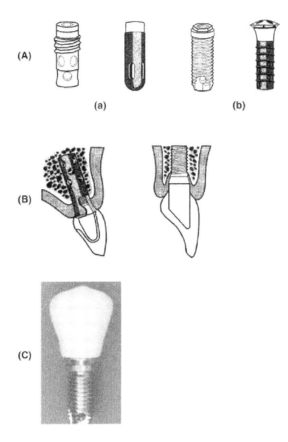

Figure 14-50. (**A**) Two types of tooth root implants: cylindrical (**a**) and screw type (**b**). (**B**) In situ depiction of (A) with alumina crown or alumina-porcelain crown for aesthetic look [Reprinted with permission from Rubenstein and Lang (2003). Copyright © 2003, Quintessence Publishing]. (**C**) Screw-shaped implant with abutment and final prosthesis in position [Reprinted with permission from Searson et al. (2005). Copyright © 2005, Quintessence Publishing].

Figure 14-51. External and internal hex and Morse taper connectors. Reprinted with permission from Searson et al. (2005). Copyright © 2005, Quintessence Publishing.

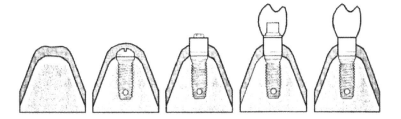

Figure 14-52. Two stages of implantation operation. The first stage is for firm fixation of the root part by submucosal implantation; the second stage is for attaching the crown portion with ceramics and porcelain. Reprinted with permission from Brånemark et al. (1977). Copyright © 1977, Swedish Society of Medicine.

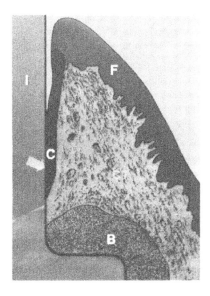

Figure 14-53. Schematic diagram of the important interface zones between the regenerated crevicular epithelium (C) and the implanted biomaterial (I). This is the area of the biological seal (arrow) that must be established if an implant is to be successful. Also shown are epithelia of the free gingival margin (F), the lamina propria (CT), and regenerated bone (B). Reprinted with permission from McKinney et al. (1985). Copyright © 1985, PSG Publishing.

The three-phase interface among gingival, tooth surface, and environment is a very critical area for survival of the tooth implant, as shown in Figure 14-53, as it is in percutaneous implants (see §11.2). This is mainly due to the ease of invasion of foreign organisms through this junction. The natural tooth surface has an affinity for collagen, which facilitates its attachment to the crevicular epithelium of the gingiva, thus tightly sealing the tooth root. However, if the patient fails to achieve thorough cleaning of the surface in every way possible, the attachment may yield to foreign materials such as plaques, which will not only hinder attachment of gingival tissue but also open the way for foreign organisms to gain entrance to the inside of the body.

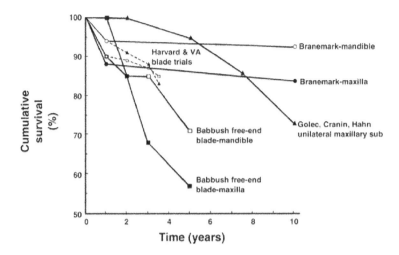

Figure 14-54. Comparative survival of encapsulated individual blade and unilateral subperiosteal implant and osseointegrated root-form implants. Reprinted with permission from Shulman (1991). Copyright © 1991, W.B. Saunders.

The survival rates of the blade-vent and Brånemark implants vary for various investigators due to many factors: surgical techniques, patient selection, location, evaluation criteria, etc. Figure 14-54 shows the comparative survival of encapsulated individual blade and unilateral subperiosteal implant and osseointegrated root-form implants. Most of the blade-vent endosseous implants are made from stainless steel, Co–Cr alloy, Ti, and Ti6A14V alloy. Brånemark implants are almost exclusively made of Ti alloys. Single-crystal alumina has been used in Japan. The main problem is its brittleness and difficulty in attaching the crown after the root is fixed. There have been efforts to coat the surfaces of the implants with ceramics (alumina, zirconia, and hydroxyapatite) and polytetrafluoroethylene composite (Proplast®), with little significant improvement in their performance. Others have used pyrolytic carbon, polycrystalline alumina, and single-crystal alumina. Recently, surface-textured implants and porous implants with electrical stimulation have been tried. Some researchers have even tried to use anorganic bone/acrylic polymer (PMMA) composite material to induce bony tissue to grow into the spaces originally filled with anorganic bone. This results in ingrown tissue and fixation of the bone to the artificial tooth root.

Example 14-4

During an experiment involving tissue growth into the pores of a porous surfaced tooth implant, a push-out test was performed. The experimenter sectioned out 2-mm disks (two for each implant) and one of the curves is given below (the diameter is 4.6 mm).

a.   Calculate the maximum interfacial shear stress between the bone and the implant.

b.   What is the stiffness of the interface?

c.   Is the interfacial strength adequate for fixation of the implant?

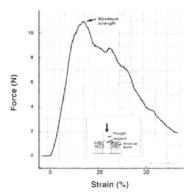

Answer

a. The "nominal" shear stress is

$$\sigma = \frac{F}{A} = \frac{F}{\pi \times D \times H} = \frac{11\ N}{28.9\ mm^2} = \underline{0.38\ MPa}.$$

b. The stiffness can be calculated:

$$S = \frac{\sigma}{\varepsilon} = \frac{5.8\ N/28.9\ mm^2}{0.05} = \underline{4\ MPa.}$$

Note that the strain is the "shear" strain.

c. The interfacial strength may not substantially contribute to the total mastication load (because the natural tooth has a conical shape, the applied load is distributed into two major components, shear and compression); therefore, the interfacial shear strength calculated in (a) is a high enough value. (The direct attachment of artificial tooth by tissue growth into the pores of an implant has certain advantages. However, the types of tissues, whether hard or soft, to be grown into to achieve a viable fixation and give the best results have not been established.)

14.3.2.b. Subperiosteal and Staple/Transosteal Implants

Implants have been successfully used to provide a framework for dentures for the edentulous alveolar ridge, as shown in Figure 14-55. Although similar functions can be duplicated by implanting osseous dental implants, periosteal and/or staple/transosteal implants have been developed to compensate the weakness of the thin alveolar ridge in many endentulous patients. The rationale is to provide better support for dentures or other bridgework. Unfortunately, these implants thus far have been subjected to problems similar to those associated with individual dental implants.

The more successful osseointegration of dental roots can provide the foundation for the framework of dentures to be anchored, as shown in Figure 14-56. Modern CAD/CAM (computer-aided design/machining) technology can custom make these prosthetics more easily and precisely ensure better fit and comfort, as well as make them longer lasting. Better understanding of the biomechanics of mastication and the role of supporting implants that in turn support the dentures is essential for overall success, as shown in Figure 14-57.

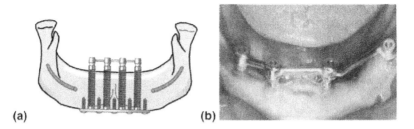

(a)   **(b)**

Figure 14-55. (**a**) Transosseous implant frame. (**b**) Intraoral view of transmandibular staple implant. Reprinted with permission from Searson et al. (2005). Copyright © 2005, Quintessence Publishing.

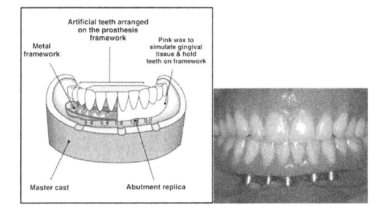

Figure 14-56. A master cast shows the prosthodontic simulation of a fixed implant-supported prosthesis for a totally edentulous mandible (**left**). Fixed implant-supported mandibular prosthesis opposing a maxillary complete denture (**right**). Reprinted with permission from Rubenstein and Lang (2003). Copyright © 2003, Quintessence Publishing.

Materials used for these implants are primarily metals, stainless steel, Co–Cr alloy, and Ti alloy, due to their ease of fabrication in a conventional dental laboratory. Some advocate coating of metals with other inert materials, such as carbon and ceramics. It is suspected that these coatings will result in marginal improvement, as with endosseous dental implants.

Example 14-5

A bioengineer is asked why the roots of dental implants can be fixed by direct apposition of alveolar or maxillary bone but not in other implants such as the femoral prosthesis of a total hip joint replacement. What would be the major reason?

Answer

The tooth root is cone shaped and the load upon the tooth is predominantly compressive. These conditions favor stability. By contrast, in the hip joint there can be substantial bending. Even so, recent variants of hip implant designs have made use of a cone-shaped femoral component jammed into place. A further difference between dental and orthopedic implants is that the dental implant is percutaneous, so it can provide a pathway for pathogenic bacteria to enter the body. In addition, dental implants in bridge configuration can reduce forces in comparison with the case of a single tooth.

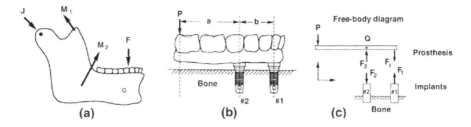

Figure 14-57. Simplified model of the jaw as a class 3 lever. The fulcrum is at the condyle (C), while the two major muscle forces (M_1 and M_2) act nearer to the fulcrum than the biting force F; J = joint reaction force. (**a**) A method for predicting the forces on two implants supporting a cantilever portion of a prosthesis. (**b**) A diagrammatic view of the situation in 2D. (**c**) Free-body diagrams of the prosthesis (**top**) and the implants (**bottom**). P = biting force; O = point at which summation of moments is being found. Reprinted with permission from Brunski (2003). Copyright © 2003, Quintessence Publishing.

14.4. INTERFACE PROBLEMS IN ORTHOPEDIC AND DENTAL IMPLANTS

The fixation of orthopedic and dental implants has been a most difficult and challenging problem. Fixation can be (1) active mechanical fixation by using screws, pins, wires, and bone cement, (2) passive mechanical fixation that usually entails relative motion between implant and tissue surfaces, (3) biological fixation by allowing tissues to grow into the interstices of pores or textured surface of implants, and (4) direct chemical bonding between implant and tissues.

This section is concerned with the pros and cons of the various fixation techniques. Also considered are newer techniques such as the use of electrical and pulsed electromagnetic field stimulation, chemical stimulation by using calcium phosphates, direct bonding with bone by glass-ceramics, and resorbable particle impregnated bone cement. Such techniques take advantages of both the immediate fixation of bone cement and long-term fixation of the tissue ingrowth.

One of the inherent problems of the orthopedic implantation is the fixation of the devices and the maintenance of the interface between the device and host tissue. Fixation can be classified into several categories as given below(see also Figure 14-58);

1. Direct interference or (passive)non-interference fit
2. Mechanical fixation using screws, bolts, nuts, wires, etc.
3. Bone cement interdigitation
4. Porous ingrowth (biological) fixation
5. Direct chemical bonding using adhesives or after coating with direct bonding layer
6. A combination of (3) and (4) by incorporating resorbable particles into bone cement
7. Porous ingrowth controlled by using electrical or electromagnetic stimulation.

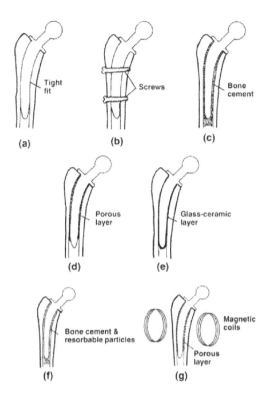

Figure 14-58. Schematic illustration of the various fixation methods: (**a**) direct interference or (passive) noninterference fit; (**b**) mechanical fixation using screws, bolts, nuts, wires, etc.; (**c**) bone cement or grouting; (**d**) porous ingrowth (biological) fixation; (**e**) direct chemical bonding using adhesives or after coating with direct bonding material layer; (**f**) bone cement with resorbable particles; (**g**) porous ingrowth controlled by using electrical or electromagnetic stimulation. Reprinted with permission from Park (1989). Copyright © 1989, Korean Orthopedic Association.

The major failure mode of modern orthopedic implants is loosening. Loosening is an interface problem; however, the underlying causes may include bone resorption around the implant, fatigue of the cement, biocompatibility problems, surgical technique, and others. We will review some of these fixation techniques and discuss some possible solutions related to the total joint and endosseous dental implants.

14.4.1. Bone Cement Fixation

The bone cement fixation creates two interfaces, i.e., (1) bone/cement and (2) cement/implant in the hip or knee joint fixation. According to an earlier study, the incidence of the problems (usually loosening) related to both interfaces for femoral prostheses were evenly divided about 10 and 11% for cement/implant and bone/cement interfaces, respectively. The cement/implant interface problems can be minimized by precoating the metal with bone cement or polymethylmethacrylate polymer, to which the new bone cement can adhere well. The precoating can achieve a good bonding between the "cement" and prostheses during the manufacturing process, as mentioned earlier.

Table 14-9. Mechanical Properties of Materials Used for Joint Prostheses and Bone

Materials	Young's modulus (GPa)	UTS* (MPa)	Elongation (%)	Density (g/cm³)
Metals				
316L S.S. (wrought)	200	1000	9	7.9
Co–Cr–Mo (cast)	230	660	8	8.3
Co Nic Cr Mo (wrought)	230	1800	8	9.2
Ti6A14V	110	900	10	4.5
Ceramics				
Alumina	400	260	nil	3.9
(Al₂O₃, polycrystalline)				
Glass-ceramic(Bioglass®)	200a	200	nil	2.5b
Hydroxyapatite	120	200	nil	3.2
Polymers				
PMMA (Solid)	3	65	5	1.18
PMMA Bone Cement	2	30	3	1.1
UHMW Polyethylene	1	30	200	0.94
Bone				
Femur(compact), long axis	17	130	3	2.0
Femur(compact), tangential	12	60	1	2.0
Spongy bone	0.1	2.	2.5	1.0

[a] UTS: ultimate tensile strength

[b] Estimated values.

The problem of the bone/cement interface cannot be so easily overcome since this problem is due to the intrinsic properties of the bone cement and of the bone, as well as extrinsic factors such as the cementing technique. The toxicity of the monomer, the inherent weakness of the cement as a material (see Table 14-9), and the inevitable inclusion of pores can contribute to the problem of loosening at the bone/cement interface.

The problem of the bone/cement interface may be solved by utilizing the concept of bone ingrowth. Specifically, the bone cement can be used for the initial fixation medium yet provide tissue ingrowth space later. This is done by incorporating resorbable particles such as inorganic bone, as shown in Figure 14-59. Studies indicate that the concept can be used effectively in experiments in rabbits and dogs. In one experiment the bone particle-impregnated bone cement was used to fix femoral stem prostheses in the femora of dogs (one group was experimental, the other control). After a predetermined time, the femora were harvested and sectioned into disks. Push-out tests were conducted to measure the interfacial strength for the bone/cement interface. The results (Figure 14-60) show that the experimental side showed increasing strength up to 5 months, while the control side showed a decrease in strength that stabilized after that time. The histology studies also disclosed integration of tissues into the spaces left by the dissolved particles. It was found that about 30% of bone particles can provide continuous porosity for the bone to grow yet provide a reasonable compromise to other parameters. If too high a concentration of bone particles was used, viscosity would be excessive during the polymerization/dough stage, and the strength of the cement might be reduced. At this time, the concept of bone particle-reinforced PMMA has not yet been demonstrated in human patients.

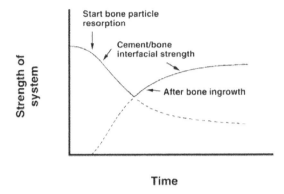

Figure 14-59. Basic concept of fixation by bone cement with resorbable particles. Immediate fixation is achieved as in ordinary bone cement. Particles are resorbed by the body, and bone grows into the resulting voids. Reprinted with permission from Liu et al. (1987). Copyright © 1987, Wiley.

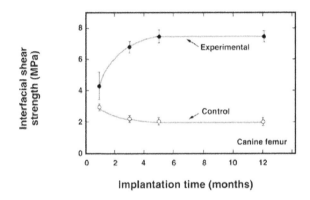

Figure 14-60. Maximum interfacial shear strength between bone and bone cement versus implant period. The femoral stems were implanted with ordinary bone cement and cement with bone particles. In both cases the interfacial strength stabilized after 5 months for this canine model. Reprinted with permission from Dai et al. (1991). Copyright © 1991, Wiley.

14.4.2. Porous Ingrowth (Biological) Fixation

The effort to develop a viable interface between tissue and implants has been underway over the last four decades ever since Smith (1963) tried to develop a bone substitute with porous aluminate ceramic impregnated with epoxy resin (called Cerocium®). Although the material showed good adherence to tissues, the pore size (average diameter, 18 μm) was too small to allow any bony tissue to grow into. Later, ceramics, metals, and polymers were used to test the ingrowth ideas. Basically, any inert material will allow bony tissues to grow into any spaces large enough to accommodate osteons. However, for the ingrown bone to be continuously *viable*, the pore space must be large enough (more than 75 μm in diameter for bony tissues), contiguous to the surface of the implant, and in contact with the bone. In addition, Wolff's law dictates that the ingrown tissues should be subjected to bodily loading in order to prevent resorption after the initial ingrowth has taken place.

Additional difficulties with the bony ingrowth concept include the fact that it is surgically unforgiving, a long immobilization time is required for the tissue to grow for initial fixation, the uncertainty of the time it will take until the patient can walk, and the difficulty in eradicating any infection that might occur. Moreover, once the interface is destroyed by accidental overloading, it cannot be reattached with certainty. Furthermore, the porous coating may weaken the underlying prosthesis itself. In metallic implants there is an increased danger of corrosion due to increased surface area. Due to these many problems and the poorer-than-expected clinical results of porous fixation, some have insisted that the cement technique is still the better choice at this time.

In order to alleviate these problems several modifications had been attempted:

1. *Precoating the metallic porous surface with ceramics or carbons.* This method has been tried with some limited success. The problems of coating deep in the pores and the thermal expansion difference between the metal and ceramic materials make a uniform and good adherent coating very difficult. Another attractive material for coating is hydroxyapatite ceramic, which is similar to bone mineral. It is not yet known whether this material is superior to other ceramics. Some preliminary studies in our laboratory indicate that in the early period of fixation the bioactive apatite coating may be more beneficial, but the effect may diminish later, as shown in Figure 14-61. In these studies, a simple cortical bone plug was replaced by an implant in the canine femur. The decrease in the interfacial strength at 12 weeks compared to controls is due to the increased concentration of calcium and phosphate ions in the pores. The rate of dissolution and absorption of the hydroxyapatite should be balanced for its optimal effect on the bony tissue formation and maintenance of the ingrown tissues.

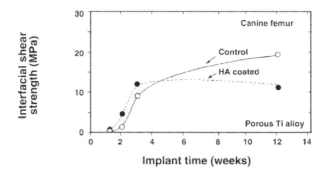

Figure 14-61. Maximum interfacial shear strength between bone and bioactive ceramic coated porous plug implants versus implant period. Plugs with and without (control) coating were implanted in the cortices of canine femurs. Reprinted with permission from Park (1988). Copyright © 1988, John Wiley & Sons.

2. *Precoating with porous polymeric materials on metal stem.* Theoretically this method is a better solution to the long-term implant problem since this method has two distinctive advantages over method (1) discussed above. First, the low-modulus polymeric material could transfer the load from the implant to the bone more gradually and evenly than metal. Moreover, it can prevent a stress-shielding effect on the ingrown tissues. Second, this method would reduce corrosion of the metal. One major problem with this technique is the weak interface strength between the polymer and metal stem especially in a dynamic loading condition in vivo.

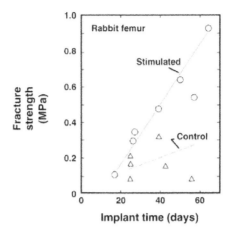

Figure 14-62. Maximum interfacial shear strength between bone and porous ceramic plug implants versus implant period with and without (control) direct current stimulation. Plugs were implanted in the cortices of canine femurs. Reprinted with permission from Park and Kenner (1975). Copyright © 1975, Marcel Dekker.

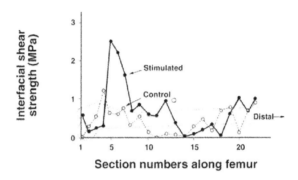

Figure 14-63. Maximum interfacial shear strength between bone and porous metallic intramedullary implant with and without (control) pulsed electromagnetic field stimulation. The porous implants were implanted in the medullary canals of canine femurs (one side was control, the other experimental). Unpublished data from J.B. Park, University of Iowa, 1988.

3. *Porous ingrowth with electrical or electromagnetic stimulation.* This technique combines porous ingrowth with the stimulating effect of electrical or electromagnetic fields. Direct current stimulation can indeed accelerate tissue ingrowth, as shown in Figure 14-62. The effect is strongest in the early stages of healing but diminishes later. The direct current stimulation has one distinctive problem, that is, the invasive nature of the implanted electrodes. The pulsed electromagnetic field stimulation method is better since stimulation can be carried out extracorporeally. A preliminary study using canine femur indicated that it can be effective, as shown in Figure 14-63. More studies are needed to verify this result.

4. *Porous ingrowth with use of filling materials.* There has been some effort to use filling materials around the porous implant since it is very difficult to prepare the tissue bed with the exact shape of the prosthesis, as is required to eliminate the micromovement of the prosthesis

after implantation. Bone matrix proteins and hydroxyapatite crystals can be used as filling materials. The success of this technique is not fully documented.

14.4.3. Direct Bonding between Bone and Implant

Some glass-ceramics showed direct bonding with the bone via a selective dissolution of the surface film. Glass-ceramics have not yet been successfully coated upon a metal surface; moreover, glass-ceramics are too brittle to use in implants as coating material.

14.4.4. Interference and Passive Fixation

These techniques have been applied in limited circumstances such as hip and finger joint prosthesis fixation. Due to the nature of the loading (mostly compressive) and (wedge) shape, an interference fit can be used to fix the femoral stem of a hip joint. This technique is largely used for ceramic (also some metallic) stems in view of the large size of the stem. One has to make the stem large due to the unpredictability of failure of brittle materials such as alumina ceramic. The stem can be seated, distributing the stress over a large area, and thus diminishing stress necrosis of the bone. Most likely, this fixation method will induce collagenous membrane formation at the interface between bone and implant unless relative motion between them can be eliminated. It is also conceivable that the sinking of the implant under load may continue throughout its life.

The passive fixation of finger joints such as Swanson's prosthesis is based on an entirely different concept of fixation. It depends on the development of a collagenous membrane between implant and bone in which the prosthesis glides in and out. This fixation can provide minimal rigidity of the joint, making it sufficient to hold a cup of coffee but not for putting screws into a wooden block.

In summary, one would like to have firm fixation of an implant for its usefulness and longevity. However, if the implant fails, a revision arthroplasty has to be performed and removal of the failed implant sometimes becomes a very difficult job. This is a rather catch-22 situation, since we need to make the interface strong in order to prevent failure of the implantation at least related to loosening of the implants.

Another problem is related to the longevity of implants. Due to the longer life of humans, we have to provide a long-lasting system, especially for young people who receive joint replacements for trauma more often than for arthritis. The original optimism regarding the porous ingrowth fixation method has been tempered somewhat, and we need to explore a better method of fixation. Certainly we should inform patients of the limitations of the implant and advise them accordingly so that they will not abuse it and use it properly. Finally, orthopedic doctors, biomedical engineers, and biologists should work together in order to solve the complicated problems of implant fixation.

14.4.5. Dental Implant Fixation

As mentioned earlier (§14.3), dental implants rely on the concept of direct bone apposition, termed osseointegration, on the implant surface to distribute mastication load. The exact mechanism of osseointegration is not known, although some ideas have been proposed, as illustrated in Figure 14-64. Osseointegration may fail to take place due to trauma during bone preparation, infections, or excess early functional loading during the initial healing phase. In addition, the interface may be destroyed by traumatic overload. Moreover, the dental implant is exposed to the "outside" of the body since the crown portion protrudes into the mouth. This makes it a transcutaneous or percutaneous implant, causing an added burden of sealing the implant to prevent invasion by foreign organisms.

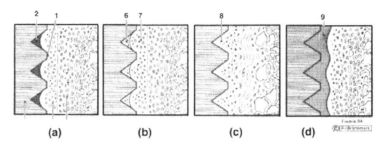

Figure 14-64. Schematic summary of the sequence of event of the osseointegration. (**A**) The screw-thread seats cannot initially be made to be congruent with the dental implant. The importance of the threaded implant is to create immediate stability after insertion and during the initial healing phase. (1) Contact between the fixture and bone (so-called immobilization). (2) Hematoma in a confined cavity, which is bordered by the fixture and bone. (3) Bone, despite careful preparation is thermally and mechanically damaged. (4) Unmolested bone tissue. (5) Fixture. (**B**) During the initial healing phase, the hematoma is transformed into new bone through in situ bone formation. (6) Damaged bone tissue heals through revascularization; demineralization and remineralization (7). (**C**) After the initial healing phase, vital bone is in direct contact with the surface of the fixture without any intermediary tissue. (8) Border zone is remodelled in response to functional loading. (**D**) In the case of osseointegration failure, nonmineralised connective tissue forms in the border zone in contact with the implants (9), which can be considered a form of pseudoarthrosis. Reprinted with permission from (Brånemark 1992). Copyright © 1992, Swedish Society of Medicine.

As for the potential for future development in this area, consider that early efforts to develop Brånemark style implants in the United States were met with initial success (Vitrident®, glassy carbon implant, Figure 14-49) in the late 1970s and early 80s. These implants failed commercially due to poor quality control and insufficient training by those who implanted them. The metal, Ti-alloy, used in the Brånemark-style implants has many favorable features, for example, it is easy to machine, has high strength and toughness, and is easily passivated. Some researchers have tried passivating the Ti alloy twice to reduce ion release and have claimed it yields better osseointegration. One may consider titanium oxide (TiO_2) or other oxide ceramics such as Al_2O_3 and ZrO_2 as tooth root materials.

PROBLEMS

14-1.   Why are the average maximum values of force at the hip and knee several times the body weight? Propose a method of measuring the forces transmitted through the joints in vivo.

14-2.   Describe the sequence of events by which the production of wear debris in an artificial joint can lead to joint pain and loosening of the prosthesis.

14-3.   A simple way to prevent the stem of a hip prosthesis from breaking through the cortex of a weak femur would be to use a metal band that circumscribed the weak area and distributed the load around the cortex. Describe the advantages and disadvantages of this approach.

14-4.   Assuming that you wished to attach a femoral head prosthesis by using a porous coating on a metallic stem, would you choose a ceramic, metallic or plastic coating material? Discuss all aspects of the problem including fabrication, compatibility, stress distribution, etc.

14-5.   What are some of the advantages and disadvantages of porous materials (for the attachment of joint replacements) when compared with PMMA bone cement fixation?

14-6.   It was reported that a maximum of only 15% of the available space in a porous implant was occupied by ingrown tissue for knee joints (especially the tibial component) retrieved after a few years. Discuss the consequences and possible reasons for such a small amount of tissue present.

14-7.   Explain how you would control the rate of resorbable particle dissolution for particle-reinforced bone cement. Consider the possibility of rapid early dissolution followed by slower dissolution over a period of several years.

14.8.   Discuss the idea of precoating radio-pacifying agents such as barium sulfate ($BaSO_4$) with PMMA to increase the strength of PMMA bone cement. Consider the fact that 10% by weight of barium sulfate inclusions decreases the strength of the cement by 10%.

14.9.   What are some of the advantages and disadvantages of porous materials (for the attachment of joint prostheses) in comparison with PMMA bone cement fixation?

14.10. Choose the most appropriate material for each question. Use letters.

A. Alumina (Al_2O_3) ceramic	F. 316L stainless steel
B. Polytetrafluoroethylene	G. Silicone rubber
C. Pyrolytic carbon with graphite	H. Ti6Al4V
D. Polymethyl methacrylate	I. Poly-L-lactic acid
E. UHMWPE	J. Catgut

   a. Used to make temporary orthopedic implants such as bone plates and screws.

   b. Used to make femoral stems of artificial hip joints.

   c. Used to make membrane of breast prosthesis.

   d. Used to make acetabular cup.

   e. Used to make synthetic absorbable sutures.

   f. Used to make hard contact lens.

   g. Used to make artificial teeth root.

   h. Used to make artificial heart valve discs.

   i. Used to make knitted artificial blood vessels.

   j. Used to make sutures less sticking to tissues.

14.11. The surface layers of metals are being oxidized as shown schematically in time sequence. The thickness of the oxide vary as follows (Kasemo and Lausmaa, 1985):

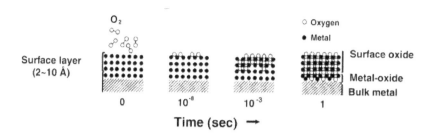

Metals	Thickness of oxide layer (Å)
Au, Pt, Pd	0~5
Fe, Ni, Al, Ti	15~50

a. Why are the Au, Pt, and Pd more oxidative or corrosion resistant than the Fe?

b. Why is Ti an excellent metal to be used as an implant while Al is not?

c. The time sequence of oxidation shows only up to 1 second. How many surface layers can the oxygen diffuse through the metal atoms? Assume a nominal diffusion coefficient of $3 \times 10^{-11} m^2/s$).

d. The molecular interaction between the oxide layer and biomolecules including tissues is an important avenue for understanding tissue compatibility. Suggest as much as you can for the interaction among them.

DEFINITIONS

Acetabulum: The socket portion of the hip joint.

Apatite: Mineral phase of bone usually found as hydroxyapatite, $Ca_{10}(PO_4)_6(OH)_2$.

Arthroplasty: Plastic repair of a joint.

Brånemark dental implant: Popular dental implant supposed to be integrated with bone of the root portion. It is made of Ti alloys such as Ti6Al4V.

C-FLEX® TPE: Polysiloxane modified styrene–ethylene/butylene–styrene block copolymer of thermoplastic elastomer.

c.p.-Ti: Commercially pure Ti having excellent corrosion resistance and biocompatibility.

CoCr alloy: Co-based alloys commonly used in orthopedic implants. Some are made castable (CoCrMo) or hot-forgable (CoNiCrMo). These alloys are stiff and strong and offer good corrosion resistance.

Dacron®: Polyethylene terephthalate polyester that is made into a fibrous form.

Edentulous: Without teeth.

Endosseous: In the bone, referring to dental implants fixed to the jawbone.

Finite-element analysis or model (FEA or FEM): study of a structure or structural member by subdividing it into small elements (mesh). A computer is used to analyze stresses and strains. The stresses/strains are then expressed in a graphic form.

Glass-ceramics: A ceramic made by controlled nucleation and crystallization for fine grains with capacity for tissue bonding by selected surface dissolution. Bioglass® is an example.

HDPE: High-density polyethylene, having a high m.w. and crystallinity, results in high-density and moderate m.w. (1/2 million amu). Before ultrahigh-m.w. polyethylene (UHMWPE) was available, HDPE polymers were used extensively in the early 1980s.

Hexsyn™: Hexene-based polyolefin rubber.

Hyaluronic acid: One of the polysaccharides commonly found in synovial fluid, the aortic wall, etc.

Hydrogel: Highly hydrated (over 30% by weight) polymer gel used to make soft contact lenses, such as acrylamide and poly-HEMA (hydroxyethylmethacrylate).

Krebs cycle: The sequence of reactions by which most living cells generate energy during the process of aerobic respiration. It takes place in the mitochondria, consuming oxygen, producing carbon dioxide and water as waste products, and converting ADP to energy-rich ATP.

Mersilene®: Type of woven polyethylene.

Nitinol®: Ni-Ti shape memory alloy originally experimented with at the Naval Ordinance Laboratory; hence the name.

Osseo(osteo)integration: Direct apposition of bone on the implant surface without intervening collagenous membrane formation. It is largely used in dentistry, not in orthopedics. Brånemark dental implant represents such a implant fixation technique in bone.

PGA: A biodegradable polymer made from glycolic acid.

Phusiline®: A polylactic acid.

PLA: A biodegradable polymer made form lactic acid.

Polymethylmethacylate (PMMA): An amorphous, hard plastic used to make bone cement and hard contact lens.

Proplast®: A composite material made of carbon and polytetrafluoroethylene. It has moderate porosity and strength.

Shape memory alloy (SMA): Alloys having "recovery of its shape" by thermomechanical transformation of its crystalline structure (see Nitinol®).

Silastic®: Polydimethyl siloxane silicone elastomer.

Stainless steel: Iron- (Fe) based alloys made to resist corrosion by adding Cr, Ni, and small amounts of Mo. By lowering C (0.8 to 0.3%), the corrosion resistance of the stainless steel can be enhanced (316 to 316L)

Subchondral: Bone lying under the cartilage of a joint.

Thixotropic gel: A gel that, when disturbed or shaken, will become fluid, and if left undisturbed will set into a gel; it is a reversible transformation that can occur under application of isothermal shearing stresses followed by rest. Ketchup is a typical example.

Ti6Al4V: One of the most widely used Ti alloys having a relatively low modulus for a metal (110 GPa) with high strength and corrosion resistance. It has about half the specific gravity of Co-based alloys and stainless steels. 6 w/o aluminum and 4 w/o vanadium are alloyed for preserving phases (β for V, α for Al), density, and process consideration.

Vitrident®: A defunct dental root implant made of glassy carbon.

UHMWPE: Ultrahigh-molecular-weight polyethylene, having m.w. = 2~4 million amu, semicrystalline, thermoplastic, linear polymer used in the acetabular cup of hips and tibia plateaus of joint fabrication.

Wollastonite: SiO_2–CaO ceramic used to make glass-ceramics with apatite.

BIBLIOGRAPHY

AAOS. 2002. *Wrist joint replacement (arthroplasty)*. http://orthoinfo.aaos.org.

Amstutz HC, Markolf KL, McNeices GM, Gruen TAW. 1976. Loosening of total hip components: cause and prevention. In *The Hip*, pp. 102–116. St. Louis: Mosby.

An HS, Juarez KK. 2006. *Artificial disc replacement*, www.spineuniverse.com/displayarticle.php/ article 1671.html.

An artificial intervertebral disc arthorplasty. 1995. Memphis, TN: Sofamor Danek Group.

BAK Interbody fusion system. 1996. Minneapolis, MN: SpineTech Inc.

Bao Q-B, Higham PA. 1991. *Hydrogel intervertebral disc nucleus*. US Patent No 5,047,055.

Bao Q-B, Higham PA. 1993. *Hydrogel bead intervertebral disc nucleus*. US Patent No 5,192,326.

Barb W, Park JB, Kenner GH, von Recum AF. 1982. Intramedullary fixation of artificial hip joints with bone cement-precoated implants, I: interfacial strengths. *J Biomed Mater Res* **16**:447–458.

Bechtol CO. 1976. *Bechtol patello-femoral joint replacement system*. Memphis, TN: Richards Manufacturing.

Beck JD, Pankow J, Tyroler HA, Offenbacher S. 1999. Dental infections and atherosclerosis. *Am Heart J* **138**(5 Pt 2):S528–533.

Black J, Dumbleton JH, eds. 1981. *Clinical biomechanics: a case history approach*. Edinburgh: Churchill Livingstone.

Blencke BA, Bromer H, Deutscher KK. 1978. Compatibility and long-term stability of glass-ceramic implants. *J Biomed Mater Res* **12**(3):307–316.

Boileau P, Walch G. 2006. Adaptability and modularity of shoulder prostheses. www.maitrise-orthop.com/corpusmaitri/orthopaedic/prothese.

Bragdon CR, Doherty AM, Rubash HE, Jasty M, Li XJ, Seeherman H, Harris W. 2003. The efficacy of BMP-2 to induce bone ingrowth in a total hip replacement model. *Clin Orthop Relat Res* **417**:50–61.

Brand RA, ed. 1987. *The hip: non-cemented hip implants*. St. Louis: Mosby.

Brånemark P-I. 1992. Osseointegration: a method of anchoring prostheses. *Swedish Soc Med* **1**:7–15.

Brånemark P-I, Hansson BO, Adell R, Breine U, Lidstrom J, Hallen O, Ohman A. 1977. *Osseous integrated implants in the treatment of the edentulous jaw, experience from a 10-year period*. Stockholm: Almqvist & Wiksell International.

Brantigan JW, Steffee AD, Geiger JM. 1991. A carbon fiber implant to aid interbody lumbar fusion: mechanical testing. *Spine* **16**(6 Suppl.):S277–S282.

Brunski JB. 2003. Biomechanics. In *Osseointegration in dentistry: an overview*, pp. 49–83. Ed BR Lang. Chicago: Quintessence Publishing.

Burton RI. 1973. Implant arthroplasty in the hand: an introduction. *Orthop Clin North Am* **4**:313–316.

Cales B, Blaise L, Villermaux F. 2000. Improved bearing system for THP using zirconia ceramic. In *Biomaterials and bioengineering handbook*, pp. 483–508. Ed DL Wise. NY: Marcel Dekker.

Charnley J. 1970. *Acrylic cement in orthopaedic surgery*. Edinburgh: Churchill/Livingstone.

Charnley J. 1979. *Low friction arthroplasty of the hip*. Berlin: Springer-Verlag.

Cranin AN, ed. 1970. *Oral implantology*. Springfield, IL: Thomas.

Dai KR, Liu YK, Park JB, Clark CR, Nishiyama K, Zheng ZK. 1991. Bone-particle-impregnated bone cement: An in vivo weight-bearing study. *J Biomed Mater Res* **25**(2):141–156.

Davis JR. 2003. *Handbook of materials for medical devices*. Materials Park, OH: ASM International.

Dumbleton JH. 1981. *Tribology of natural and artificial joints*. Amsterdam: Elsevier Scientific.

Eftekhar NS. 1978. *Principles of total hip arthroplasty*. St. Louis: Mosby.

Enker P, Steffee A, McMillin C, Keppler L, Biscup R, Miller S. 1993. Artificial disc replacement, preliminary report with 3-year minimum follow-up. *Spine* **18**(8):1061–1070.

Fenlin Jr JM. 1975. Total glenohumeral joint replacement. *Orthop Clin North Am* **6**:565–583.

Fernström U. 1972. The replacement of intervertebral discs with preservation of mobility. In *Die Wirgel-saule in Forschung und Proxis*, pp. 125–130. Ed H Junghanns. Stuttgart: Hippokrates.

Flatt AE, ed. 1972. *The care of minor hand injuries*. St. Louis: Mosby.

Frankel VH, Burstein AH. 1971. *Orthopedic biomechanics*. Philadelphia: Lea & Febiger.

Freeman MAR. 1980. The ICLH arthroplast of the knee joint. In *Total knee replacement*, pp. 59–82. Ed A Savastano. New York: Appleton-Century-Crofts.

Frost HM. 1973. *Orthopedic biomechanics*. Springfield, IL: Thomas.

Furnes O, Espehaug B, Lie SA, Vollset SE, Hallan G, Fenstad AM, Havelin L. 2005. *Prospective studies of hip and knee prostheses*. Norwegian Arthroplasty Register 1987–2004. Orthopedic Research Society Meeting, Washington, DC. Department of Orthopedic Surgery, Haukeland University Hospital.

Geesink RGT. 1993. Hydroxyapatite-coated hip implants: experimental studies. In *Hydroxyapatite coatings in orthopedic surgery*, pp. 151–170. Ed RGT Geesink, MT Manley. New York: Raven Press.

Ghista DN, Roaf R, eds. 1978. *Orthopedic mechanics: procedures and devices*. London: Academic Press.

Goel VK, Gilbertson L. 1997. The basic science of spinal instrumentation. *Clin Orthop Relat Res* **335**:10–31.

Goel VK, Hitchon PW, Grosland NM, Rogge TN, Sairyo K, Serhan HA. 1997. *Comparative kinematics of a collapsible and rigid anterior devices*. San Francisco: Orthopedic Research Society.

Goss CM, ed. 1975. *Gray's anatomy*. Philadelphia: Lea & Febiger.

Grau AJ, Becher H, Ziegler CM, Kaiser C, Lutz R, Bultmann S, Preusch M, Dorfer CE. 2004. Periodontal disease as a risk factor for ischemic stroke. *Stroke* **35**(2):496–501.

Greener EH, Harcourt JK, Lautenschlager EP. 1972. *Materials science in dentistry*. Baltimore: Williams & Wilkins.

Greenwald AS, Matejczyk MB. 1980. Knee joint mechanics and implant evaluation. In *Total Knee Replacement*, pp. 11–30. Ed AA Savastano. New York: Appleton-Century-Crofts.

Grenoble DE, Voss D. 1976. Materials and designs for implant dentistry. *Biomater Med Devices Artif Organs* **4**:133–169.

Griffith SL, Shelokov AP, Büttner-Janz K, LeMaire J-P, Zeegers WS. 1994. A multicenter retrospective study of the clinical results of the LINK® SB Charité intervertebral prosthesis: the initial european experience. *Spine* **19**(16):1842–1849.

Gschwend N. 1976. Design criteria, present indication, and implatation technuques for artificial knee joints. In *Advances in artificial hip and knee joint technology*, pp. 90–114. Ed M Schaldach, D Hohmann. Berlin: Springer-Verlag.

Haddad SL. 2006. *Total ankle arthroplasty*. www.orthopedictechreview.com.

Hench LL, Pachall HA. 1973. Direct chemical bond of bioactive glass-ceramic materials to bone and muscle. *J Biomed Mater Res* **7**(3):25–42.

Hirshhorn JS, McBeath AA, Dustoor MR. 1972. Porous titanium surgical implant materials. *J Biomed Mater Res Symp* **2**:49–69.

Homsy CA, Cain TE, Kessler FB, Anderson MS, King JW. 1972. Porous implant systems for prosthetic stabilization. *Clinic Orthop Relat Res* **89**:220–231.

In vitro testing of the RMC total knee. 1978. R&D Technical Monograph. Memphis, TN: Richards Manufacturing.

Johnston RC. 1987. The case for cemented hips. In *The Hip*, pp. 351–358. Ed RA Brand. St. Louis: Mosby.

Joint replacement in the upper limb. 1977. London and New York: Institution of Mechanical Engineers.

Kapandji IA. 1970. *Physiology of the joints*. Edinburgh: Livingstone.

Kapur KK. 1980. *Benefit and risk of blade implants: a critique*. NIH-Harvard Consensus Development Conference. Bethesda, MD: NIH Publication.

Kasemo B, Lausmaa J. 1985. Metal selection and surface characterization. In *Tissue-integrated prostheses, osseointegration in clinical dentistry*, pp. 99–116. Ed PI Brånemark, GA Zarb, T Albrektsson. Chicago: Quintessence Publishing.

Klawitter JJ, Hulbert SF. 1972. Application of porous ceramics for the attachment of load bearing internal orthopedic applications. *J Biomed Mater Res Symp* **2**:161–229.

Kostuik JP. 1992. The Kostuik artificial disc. In *Clinical efficacy and outcome in the diagnosis and treatment of low back pain*, pp. 259–270. Ed JN Weinstein. New York: Raven Press.

Liu YK, Park JB, Njus GO, Stienstra D. 1987. Bone-particle-impregnated bone cement: an in vitro study. *J Biomed Mater Res* **21**:247–261.

Malchau H, Herberts P, Garellick G, Soderman P, Eisler T. 2002. *Prognosis of total hip replacement*. Orthopedic Research Society Meeting, Dallas, Texas. Depatment of Orthopedics, Goteborg University.

Malchau H, Garellick G, Eisler T, Karrholm J, Herberts P. 2005. The Swedish hip registry, increasing the sensitivity by patient outcome data. *Clin Orthop Relat Res* **441**:19–29.

Marmor L. 1980. The Marmor type of knee replacement. In *Total knee replacement*, pp. 107–123. Ed AA Savastano. New York: Appleton-Century-Crofts.

Mathys-Medical 2006. *RM Finger prosthesis*. www.mathysmedical.com.

McAfee PC, Bohlman HH, Ducker T, Eismont FJ. 1986. Failure of stabilization of the spine with methylmethacrylate, a retrospective analysis of twenty-four cases. *J Bone Joint Surg Am* **68A**(8):1145–1157.

McKinney RVJ, Steflik DE, Koth DL. 1985. Evidence for a biological seal at the implant-tissue interface. In *The dental implant, clinical and biological response of oral tissues*, pp. 25–56. Ed RVJ McKinney, JE Lemons. Littleton, MA: PSG Publishing.

McMillin CR, Brantigan JW. 1993. *Design of CFRP lumbar/thoracic stackable spine fusion cages*. 38th International Society for the Advancement of Material and Process Engineering (SAMPE) Symposium, Anaheim, California.

Mears DC. 1979. *Materials and orthopaedic surgery*. Baltimore: Williams & Wilkins.

Meuli HCZ. 1975. Alloarthropstik des Handgelenks. *Z Orthop Ihre Grenzgeb* **113**:476–478.

Morscher E, ed. 1984. *The cementless fixation of hip endoprosthesis*. Berlin: Springer-Verlag.

Muratoglu OK, Kurtz SM. 2002. Alternative bearing surfaces in hip replacement. In *Hip replacement, current trend and controversies*, pp. 1–46. Ed RK Sinha. New York: Marcel Dekker.

Muratoglu OK, Bragdon CR, O'Connor DO, Jasty M, Harris WH, Gul R, McGarry F. 1999. Unified wear model for highly crosslinked ultra-high molecular weight polyethylene (UHMWPE). *Biomaterials* **20**:1463–1470.

Neer CS. 1974. Replacement arthroplasty for glenohumeral osteoarthritis. *J Bone Joint Surg Am* **56**(1):1–13.

Olson E, Hanely E, Rudet J, Baratz M. 1991. Vertebral collum allografts for the treatment of segmental spine defects: an experimental investigation in dogs. *Spine* **16**:1081–1088.

Olsson J, Stearns N. 2006. *Osseointegration of immrdiately loaded dental implants in the edenturous jaws: a study of the literature*. www.ki.se/odont/cariologi_endodonti/99B/ JohanOlsson_Nathon Stearns.pdf.

Pappas M, Buechel FF, DePalma AF. 1976. Cylindrical total ankle joint replacement: surgical and biomechanical rationale. *Clin Orthop Relat Res* **118**:82–92.

Park JB. 1983. Acrylic bone cement: In vitro and in vivo property–structure relationship: a selective review. *Ann Biomed Eng* **11**:297–312.

Park JB. 1988a. Biomaterials: an overview. In *encyclopedia of medical devices and instrumentation*, Vol. 1, pp. 328–350. Ed JG Webster. New York: John Wiley & Sons.

Park JB. 1988b. Unpublished data. University of Iowa.

Park JB. 1989. *Interface problems in orthopedic implants*. Invited Lecture, Korean Orthopedic Association, Pusan.

Park JB, Kenner GH. 1975. Effect of electrical stimulation on the tensile strength of the porous implant and bone interface. *Biomater Med Devices Artif Org* **3**:233–243.

Park JB, Choi WW, Liu YK, Haugen TW. 1986. Bone particle impregnated polymethylmethacrylate: in vitro and in vivo study. In *Tissue integration in oral and facial reconstruction*, pp. 118–124. Ed D Van Steenberghe. Amsterdam: Excerptu Media.

Park KD, Kang YH, Park JB. 1999. Interfacial strength between molded and UHMWPE-MMA monomer treated UHMWPE powder. *J Long-Term Effects Med Implants* **9**:303–318.

Paul JP. 1976. Loading on normal hip and knee joints and joint replacements. In *Advances in hip and knee joint technology*, pp. 53–70. Ed M Schaldach, D Hohmann. Berlin: Springer-Verlag.

Porous coated anatomic (PCA) total knee system. 1981. Rutherford, NJ: Orthopaedic Division, Howmedica.

Predecki P, Stephan JE, Auslander BE, Mooney VL, Kirkland K. 1972. Kinetics of bone growth into cylindrical channels in alumina oxide and titanium. *J Biomed Mater Res* **6**:375–400.

Raab S, Ahmed AM, Provan JW. 1982. Thin film PMMA precoating for improved implant bone-cement fixation. *J Biomed Mater Res* **16**(5):679–704.

Ray CD. 1991. Lumbar interbody threaded prosthesis. In *The artificial disc*, pp. 53–59. Ed M Brock, HM Mayer, K Weigel. Berlin: Springer-Verlag.

Ray CD. 1992. The artificial disc: introduction, history, and socioeconomics. In *Clinical efficacy and outcome in the diagnosis and treatment of low back pain*, pp. 205–225. Ed JN Weinstein. New York: Raven Press.

Regan JJ. 2006. *New device for cervical spine surgery*. www.spineuniverse.com.

Rubenstein JE, Lang BR. 2003. Prosdontic aspects of dental implnts. In *Osseointegration in dentistry: an overview*, pp. 85–124. Ed P Wortnington, JE Rubenstein, BR Lang. Chicago: Quintessence Publishing.

Sammarco GJ, Burstein AH, Frankel VH. 1973. Biomechanics of the ankle: a kinematic study. *Orthop Clin North Am* **4**:75–96.

Sasso RC. 2006. *Artificial cervical disc implant surgery: first of its kind in US*. SpineUniverse interview. http://www.spineuniverse.com/authorbio.php?authorID=59.

Savastano AA, ed. 1980. *Total knee replacement*. New York: Appleton-Century-Crofts.

Schueller DR, DeWald RL, Hammerberg KW, Mardjetko SM. 1994. *Anterior reconstruction using the Moss cage*. Ninth Annual Meeting, North American Spine Society (NASS), Minneapolis, Minnesota.

Scifert JL, Goel VK, Grobler LG, Sairyo K, Grosland NM, Chesmel KD. 1997. *Comparative load displacement behavior of rigid and hinged type pedicle-screw-rod fixation devices*. Minneapolis, MN: North American Spine Society.

Searson L, Gough M, Hemmings K. 2005. *Implantology in general dental practice*. London: Quintessence Publishing.

Shim HS. 1977. The strength of LTI carbon dental implants. *J Biomed Mater Res* **11**:435–445.

Shulman LB. 1991. Surgical considerations in implant dentistry. *J Dent Educ* **52**:713.

Shulman LB, Driskell TD, Block MS. 1997. Dental implants: a historic perspective. In *Implants in dentistry*, pp. 2–9. Ed MS Block, JN Kent, LR Guerra. Philadelphia: W.B. Saunders.

Small IA. 1980. *Benefit and risk of mandibular staple bone plates*. NIH-Harvard Consensus Development Conference. Bethesda, MD: NIH.

Smith L. 1963. Ceramic-plastic material as a bone substitute. *Arch Surg* **87**:653–661.

Solini A, Orsini G, Broggi S. 1989. Metal cementless prosthesis for vertebral body replacement of metastatic malignant disease of the cervical spine. *J Spinal Disord* **2**(4):254–262.

Sonstegard DA, Kaufer H, Matthews LS. 1977. The spherocentric knee: biomechanical testing and clinical trial. *J Bone Joint Surg Am* **59**(5):602–616.

Spector M. 1982. Bone ingrowth into porous polymers. *Biocompatibility of orthopedic implants*, Vol. 2. pp. 55–88. Ed DF Williams. Boca Raton, FL: CRC Press.

Spine solutions. 2006. http://www.spinesolutionsinc.com.

Stauffer RN. 1989. Current status of total ankle replacement. In *Joint surgery up to date*, pp. 83–93. Ed K Hirohata, M Kurosaka, TDV Cooke. Berlin: Springer-Verlag.

Steffee AD. 1992. The Steffee artificial disc. In *Clinical efficacy and outcome in the diagnosis and treatment of low back pain*, pp. 245–257. Ed JN Weinstein. New York: Raven Press.

Suggs JF, Li G, Park SE, Sultan PG, Rubash HE, Freiberg AA. 2006. Knee biomechanics after UKA and its relation to the ACL—a robotic investigation. *J Orthop Res* **24**(4):588–594.

Swanson AB. 1973. *Flexible implant resection arthroplasy in the hand and extremities*. St. Louis: Mosby.

Swanson SAV, Freeman MAR, eds. 1977. *The scientific basis of joint replacement*. New York: Wiley.

Taylor AR. 1970. *Endosseous dental implants*. London: Butterworths.

Taylor JK, Pope BJ. 2001. *The development of diamond as a bearing for total hip arthroplasty*. Scientific exhibit SE39, AAOS 2001. San Francisco.

Vasenius J, Vainionpää S, Vihtonen K, Mero M, Mikkola J, Rokkanen P, Törmälä P. 1989. Biodegradable self-reinforced polyglycolide (SR-PGA) composite rods coated with slowly biodegradable polymers for fracture fixation: strength and strength retention *in vitro* and *in vivo*. *Clin Mater* **4**(4):307–317.

Volz RG. 1977. Total wrist arthroplasty: a new approach to wrist disability. *Clin Orthop Relat Res* **128**:180–189.

Vuono-Hawkins M. 1991. *The design and evaluation of a thermoplastic elastomeric (tpe) lumbar intervertebral disc spacer.* New Brunswick, NJ: Rutgers University.

Waisbrod H. 1988. Treatment of metastatic disease of the spine with anterior resection and stabilization by means of a new cancellous metal construct: a preliminary report. *Arch Orthop Trauma Surg* **107**:222–225.

Walldius B. 1980. Arthroplasty of the knee: 27 years experience. In *Total knee replacement*, pp. 195–216. Ed AA Savastano. New York: Appleton-Century-Crofts.

Weinstein AM, Klawitter JJ, Cleveland TW, Amoss DC. 1976. Electrical stimulation of bone growth into porous Al_2O_3. *J Biomed Mater Res* **10**:231–247.

Weinstein JN. 2000. *The Dartmouth atlas of musculoskeletal health care.* Chicago, IL: AHA Press.

Wilde AH. 1989. Total condylar knee replacement: late results. In *Joint surgery up to date*, pp. 55–65. Ed K Hirohata, M Kurosaka, TDV Cooke. Berlin: Springer-Verlag.

Williams DF, Roaf R. 1973. *Implants in surgery.* Philadelphia: W.B. Saunders.

Wilson JN, Scales JT. 1970. Loosening of the total hip replacements with cement fixation. *Clin Orthop Relat Res* **72**:145–160.

Wright V, ed. 1969. *Lubrication and wear in joints.* Philadelphia: Lippincott.

Wroblewski BM. 1986. 15-21 year results of the Charnley low-friction arthroplasty. *Clin Orthop Relat Res* **211**:30–35.

Wynbrandt, J. 1998. *the excruciating history of dentistry: toothsome tales and oral oddities from babylon to braces.* New York, NY, St Martin's Press.

Yamamuro T, Shikata J, Okumura H, Kitsugi T, Kakutani Y, Matsui T, Kokubo T. 1990. Replacement of the lumbar vertebrae of sheep with cermaic prostheses. *J Bone Joint Surg Am* **72B**(5):889–893.

Youm Y, McMurty RY, Flatt AE, Gillespie TE. 1978. Kinematics of the wrist, I: an experimental study of radial ulnar devieation and flexion extension. *J Bone Joint Surg Am* **64A**:423–431.

Zimmer USA. 2006. *Elbow anatomy.* www.zimmer.com.

Zirconia femoral heads for total hip prostheses. 2000. Sixth World Biomaterials Congress Workshop, Kamuela (Big Island), Hawaii. Society For Biomaterials.

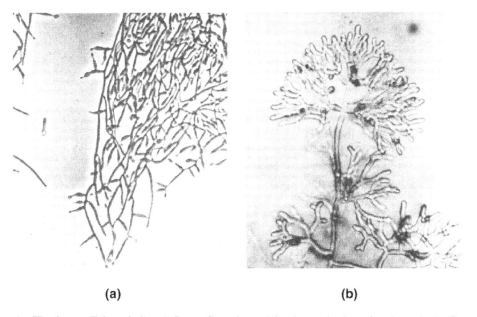

15

TRANSPLANTS

(a)   (b)

(**a**) The fungus *Tolypocladium inflatum* Gams is used for the production of cyclosporin A (Cy A), which has an antifungal effect on *Neurospora crassa*. (**b**) At the zone of growth inhibition, the fungus assumes the shape of a witch's broomstick (b). Reprinted with permission from Borkel (1982). Copyright © 1982, Elsevier.

As we have seen in the previous chapters, biomaterials have many uses in aiding healing, restoring a lost form or function, and in correction of deformity. The limitations of artificial materials become apparent when we realize that only the simplest mechanical, structural, optical, and chemical functions can be assumed by nonliving materials. Functions that can only be performed by living tissues can be restored either by transplanting a new tissue or a new organ or by regenerating the tissue or organ that has lost its function. In this chapter, transplants are considered.

15.1. OVERVIEW

Transplants may be classified according to the relationship between the donor and the recipient, as given in Table 15-1. Transplants are also classified according to their location in the recipient. Orthotopic transplants are placed in the same location as the original organ, while heterotopic transplants are placed in a different location.

Table 15-1. Types and Definitions of Transplants

Type	Definition
Autograft	Within one individual, from one part of the body to another
Isograft	Between genetically identical individuals, i.e., identical twins
Homograft, allograft	Between different individuals of the same species
Heterograft, xenograft	Between members of different species

The historical basis for transplantation can be traced back several thousand years to Greek myths of human–animal hybrids. In the Middle Ages there were anecdotal reports in stories of successful transplants of teeth, noses, and even whole limbs. At that time, noses were lost to sword cuts or from advanced syphilis. Instances were reported of autografts (transplants from one part of the body to another) of skin and flesh from the arm to the nose. The first such skin flap technique was reported by Tagliacozzi in 1596. Such techniques are currently used in modern plastic surgery. There were also reports of transplants of noses and teeth from slaves. It was thought, incorrectly, that the graft would survive only as long as the donor lived.

Modern transplantation is based on the technique of vascular anastomosis, developed in 1902, and the understanding and control of the immunological basis of rejection, which has evolved only in the past fifty years. Kidney transplants were tried in the early 1950s, and these failed due to rejection by the immune system. Immunosuppressive drugs were not available at that time. In 1954 a kidney transplant between two identical twins (isograft) was successful. Later isografts were successful, but allografts (transplants between two unrelated individuals of the same species) were generally unsuccessful despite the use of whole-body irradiation for immunosuppression. The period from 1962–1975 saw the development of immunosuppressive drugs, which substantially improved the success of transplants. Since the development of the immunosuppressive cyclosporin in 1975, further improvements have been realized, and organs such as liver and heart have been successfully transplanted. Transplant milestones are given in Table 15-2. More recently, tacrolimus, another immunosuppressive drug, has been used to enhance the survival of transplanted organs.

Table 15-2. Transplant Milestones in the United States and Canada

Organ	Date	Institution
Kidney*	1954	Brigham and Women's Hospital
Pancreas	1966	University of Minnesota
Liver*	1967	University of Colorado Health Sciences Center
Heart	1968	Stanford University Hospital
Heart-lung	1981	Stanford University Hospital
Lung-single	1983	Toronto General Hospital
Lung-double*	1986	Toronto General Hospital
Liver-living related	1989	University of Chicago Medical Center
Lung-living related	1990	Stanford University Medical Center

*Denotes first transplant of its kind in the world. Source: United Network for Organ Sharing. (2001). Transplant milestones in the United States and Canada.

Reprinted with permission from Cupples (2002). Copyright © 2002, Springer.

As for the donor, both living and cadaver donors are used in transplantation. Living donors are appropriate in the case of blood or bone marrow, which the body can replace, or in the case of the kidney, of which each person has two but can function adequately with one. The concept of a "cadaver" donor is based on the fact that the tissues and organs of the body do not die at the same rate. Obviously, an organ containing only dead cells is useless as a transplant, unless it is performing only a passive function. The requirement for a cadaver donor is that of *brain death*. Brain death involves (1) unreceptivity and unresponsivity, (2) absence of spontaneous respiration or movement, (3) absence of reflexes, and (4) a flat (zero signal) electroencephalograph (EEG). Patients with hypothermia or with severe endocrine or metabolic disturbances are excluded from consideration as brain-dead donors since they can at times be resuscitated from a very unresponsive state. The distinction between brain death and earlier concepts of death as the irreversible stoppage of heartbeat, blood circulation, and breathing is a relatively recent one. Until recently, the heart, not the brain, has been considered by most people to be the central organ of life and human existence. The brain stem controls breathing, and its death results in cessation of breathing; more than a few minutes of this results in permanent demise of the entire organism. The role of the brain stem is relevant since machinery to maintain respiration in the absence of brain stem viability has been available only in relatively recent years. Cadaver donors are usually those who have suffered a massive head injury or a cerebrovascular problem sufficient to kill the brain.

At body temperature, irreversible brain damage occurs about five minutes after cessation of blood circulation. The kidney, however, remains viable for about 1 to 2 hours without blood circulation. This amount of time permits transplantation if the donor and recipient are nearby and surgery can be done rapidly. Cooling of the kidney extends the viability time to 6–8 hours, with useful function retained at 24 hours; cooling combined with perfusion extends viability to 3 days. Cooling of the liver and heart extend viability to 5–8 hours. Organs may be preserved during cooling with agents such as lactobionic acid, raffinose, and hydroxyethyl starch. These prevent the cells from swelling during cold storage. These preservation procedures are useful in that they increase the availability of organs for transplantation. Table 15-3 gives the recommended times for each organ published by the UNOS (United Network for Organ Sharing).

Table 15-3. Organ Survival Time According to UNOS
(United Network for Organ Sharing)

Organ	Time (hours)
Lung	4–6
Liver	24
Pancreas	24
Kidney	48–72

Reprinted witih permission from Cupples (2002).
Copyright © 2002, Springer.

15.2. IMMUNOLOGICAL CONSIDERATIONS

Most of the early attempts at organ transplantation failed for reasons that became understood only recently. It is now known that the death of a transplanted tissue or organ is the result of an immunological response known as *rejection* by the recipient of the transplant. In animal experiments in the 1940s it was found that skin grafts between two different animals of the same species survive for about 7 days before being rejected. A second transplant between the same donor and recipient was found to be rejected twice as fast. This so called "second-set response" became an experimental model for study of the immunology of transplantation.

Rejection is currently prevented by the typing of tissues according to *histocompatibility*, and by the use of immunosuppressive drugs. Histocompatibility refers to the compatibility of different tissues in connection with immunological response. By contrast, *biocompatibility*, discussed in earlier chapters, refers to the compatibility of nonliving materials with living tissues and organisms. The destruction associated with rejection of a transplant occurs from the effects of antibody-activated macrophages and cytotoxic T cells. Hyperacute rejection, occurring within 24 hours following transplantation, is antibody mediated and is seen in recipients presensitized against donor antigens. Accelerated rejection, which occurs within 5 days, is considered a second-set reaction and may be mediated by antibodies or cells. Acute rejection, which occurs in the first few weeks after transplantation, is a T cell-mediated immune response of the recipient against the graft. Finally, chronic rejection is associated with a gradual decline in function of the graft and results from humoral factors and possibly a low-grade cellular attack. Rejection is viewed not as an all-or-none phenomenon but as a matter of degree. Incipient rejection episodes can be detected, and can be controlled with an increase of drug dosage or use of additional drugs. Drugs currently used include azathioprine (Imuran®) and corticosteroids, as well as more recently developed drugs such as cyclosporin and tacrolimus. Such drugs have revolutionized the transplantation of hearts and livers (see Figure 15-1).

Development of cyclosporin A (Cy A) was begun in 1970, and this drug was completely synthesized in 1980. It can be crystallized from seven-unit amino acids, as shown in Figure 15-2, and crystals of Cy A are shown in Figure 15-3. It has played a major role in improving survival rates, as illustrated in Figure 15-4 for kidney transplantation. Also, better survival rates resulted in more transplants being performed with all organs (see Figure 15-5). Table 15-4 gives data on the graft and patient survival rates at 1, 3, and 5 years.

Other immunosuppressive techniques that have been tried include the following. Splenectomy has been performed based on the fact that the spleen is the largest lymphoid organ. Removal of the spleen increases transplant survival in the first two years but reduces patient survival after that as a result of a higher risk of infection. The procedure, in conjunction

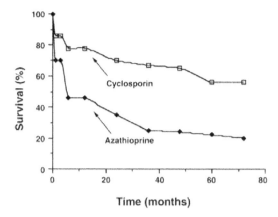

Figure 15-1. Comparison of survival of orthotopic liver transplant recipients under immuno-suppressive therapy with cyclosporin and azathioprine. Reprinted with permission from Cerilli (1988). Copyright © 1982, Lippincott, Williams & Wilkins.

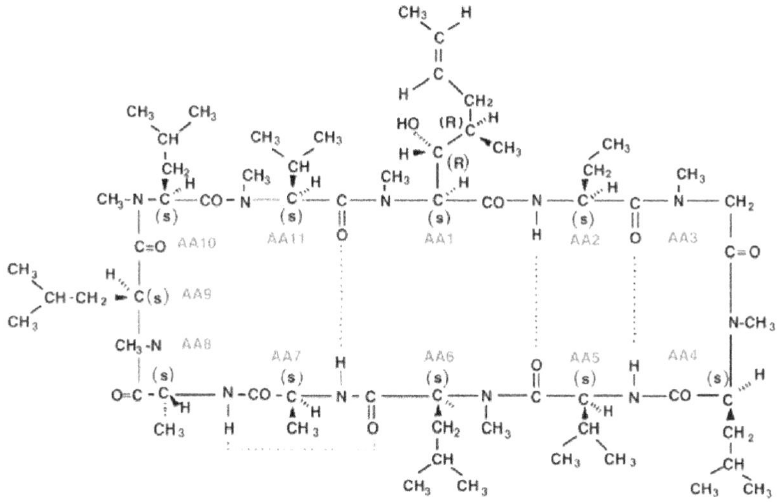

Figure 15-2. Molecular structure of cyclosporin ($C_6H_{111}O_{112}$; MW = 1202). Reprinted with permission from Borkel (1982). Copyright © 1982, Elsevier.

with antibiotic therapy, is still under investigation. Irradiation of the transplant has been tried; this procedure yields equivocal results and is being studied. Thoracic duct drainage is a short-term procedure in which lymph is removed from the main thoracic lymph duct, the lympho-cytes are extracted by centrifugation, and the remaining fluid returned to the body. This is no longer widely practiced. Whole-body irradiation was also used in early transplants; however, it is nonspecific and nontitratable, and has been supplanted by the use of drugs.

Figure 15-3. Crystals of cyclosporin A (crystallized from diisopropylether and photographed under red light). Reprinted with permission from Borkel (1982). Copyright © 1982, Elsevier.

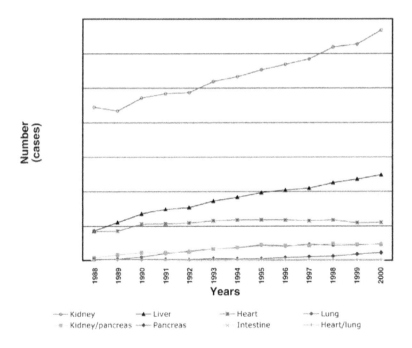

Figure 15-4. U.S. transplant volume by organ and year. Based on data from Cupples (2002).

Immunosuppressive therapy carries a variety of risks, including a susceptibility to infection in both the short term immediately after surgery as well as in the long term. Although antibiotic prophylaxis has improved, fungal infections present a significant problem in transplant recipients. Invasive fungal infections occur in 5 to 45% of recipients of solid transplant organs.

These infections are a major source of morbidity and mortality. There is also a significantly increased incidence of a variety of cancers, including Kaposi's sarcoma, other skin malignancies, and cancer of the lymphatic system and other organs. The incidence of cancer increases with time of follow-up; therefore, transplant patients are expected to be monitored for cancer development and other complications indefinitely. For example, the incidence of skin cancer in transplant recipients is about 4% at 1 year, to 44% at 7 years. Immunosuppressive drugs such as cyclosporin are also toxic to the kidney. Other side effects of immunosuppressive drugs include mood swings, sleep problems, gastrointestinal difficulties, fever, sexual dysfunction, cognitive dysfunction, tremor, poor coordination, headache, and hallucinations. These have the effect of reducing the quality of life of transplant recipients. The risks may be balanced by the fact that many recipients of vital organ transplants would have faced a more rapid death without them.

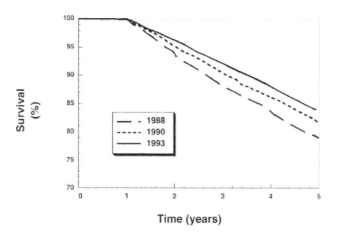

Figure 15-5. Comparison of survival of renal transplant recipients showing much improved graft survival during three time periods. Reprinted with permission from Hariharan et al. (2000). Copyright © 2000, Massachusetts Medical Society.

Histocompatibility is associated principally with the human lymphocyte antigen system. The ABO blood group antigens are considered to be important antigens in the context of transplantation. Almost all transplants are between ABO-compatible people; however, recently sporadic attempts have been made with additional immmunosuppression to cross the ABO blood group barrier in an effort to make more organs available. Testing for histocompatibility is currently done by serologic tests in which mixed lymphocyte cultures are incubated and examined for reactions. More recently, complement-dependent cytotoxicity (CDC) crossmatch and flow cytometry crossmatch (FCXM) are used prospectively for kidney transplants. Their use is being evaluated for other transplants Since liver failure and heart failure are invariably fatal, transplants of these organs are done even if full compatibility cannot be achieved.

Table 15-4. Graft and Patient Survival Rates at 1, 3, and 5 Years

Organ/ donor type	Graft surv. (G) Patient surv. (P)	Number of transplants 1997–98	1-year survival (%)	Number of transplants 1990–98	3-year survival (%)	5-year survival (%)
Cadaveric	G	13,235	89.4	56,352	76.3	64.7
donor kidney	P	13,235	94.8	56,352	88.9	81.8
Living donor	G	7,431	94.5	25,092	87.0	78.4
kidney	P	7,431	97.6	25,092	94.6	91.0
Liver	G	7.196	81.4	26,652	71.5	66.1
	P	7.196	87.9	26,652	79.2	74.2
Pancreas	G	324	76.2	824	49.9	41.6
	P	324	93.4	824	88.7	84.2
Kidney/pancreas	Kidney G	1,709	91.8	6,148	80.4	70.7
	Pancreas G	1,674	83.7	6,110	73.9	67.4
	P	1,716	94.4	6,156	88.0	82.7
Heart	G	4,316	85.1	19,269	76.0	68.5
	P	4,316	85.5	19,269	76.9	69.8
Lung	G	1,663	76.3	5,613	56.2	42.0
	P	1,663	77.0	5,613	58.1	44.2
Heart-lung	G	89	58.2	469	50.9	40.5
	P	89	60.0	469	51.4	41.7
Intestine	G	45	63.8	104	48.8	37.4
	P	45	78.9	104	62.0	49.6

Source: U.S. Department of Health and Human Services. (2000). Annual Report of the U.S. Scientific Registry for Transplant Recipients and the Organ Procurement and Transplantation Network: Transplant Data, 1990-1999.

Reprinted with permission from Cupples (2002). Copyright © 2002, Springer.

Example 15-1

Discuss transplantation of the stomach and of the internal ear.

Answer

As we have seen, avoidance of rejection is a challenge in transplantation surgery. Since it is possible to live reasonably well without a stomach by modification of the diet and by use of supplementary enzymes, there is probably not sufficient motivation to transplant the stomach. As for the internal ear, it would be desirable to cure deafness caused by inner ear disease or damage. However, establishment of the appropriate nerve connections is at present an insurmountable problem. Cochlear implants are not transplants; they are electronic devices that convert sound input to an electrical stimulus to the nerves in the ear.

15.3. BLOOD TRANSFUSIONS

The transfer of blood from one individual to another is a transplant of blood tissue. The task of the donor is made easy by the fact that blood is a liquid and is easily removed, and any lost blood is rapidly replenished by the body. Of great importance is the fact that only a limited number of histocompatibility antigens are relevant to the success of blood transfusions. These are the A, B, AB, and O blood groups, and the rh factor, positive or negative. Matching of blood groups is immunologically sufficient to ensure successful acceptance of the transfusion. Blood transfusion is commonly done during major surgery, and it can save lives in cases of severe bleeding from trauma or disease. An issue of current concern is the prevention of transmission of infectious disease such as acquired immune deficiency syndrome (AIDS) and hepatitis via transfused blood. This risk is minimized by the screening of donated blood for disease. In addition, some people oppose blood transfusion on religious grounds. Consequently, there is increasing interest in artificial blood substitutes. Perfluorocarbon liquids have been used as temporary artificial blood in some emergency patients for whom transfusions cannot be found. Oxygen is highly soluble in these materials. By contrast, in natural blood, oxygen is reversibly bound in the hemoglobin within red blood corpuscles. Another approach to the problem of disease transmission by blood transfusion is the use of autologous blood, i.e., blood from the same person who is to receive it. A person anticipating elective surgery can donate several pints of blood, which is refrigerated, and wait a month or so for the surgery, during which time the lost blood is replenished by the body. Obviously, such a procedure will not help in emergency cases.

15.4. INDIVIDUAL ORGANS

15.4.1. Kidney

Kidneys are transplanted more often than any other vital organ, as indicated in Figure 15-4, despite the availability of renal dialysis. Comparison of survival of renal transplant recipients showed much improved graft survival in the three time periods shown in Figure 15-5. As of 2001, about 12,000 kidney transplants, 2,100 heart transplants, 850 lung transplants, and 4,500 liver transplants were done each year in the United States, according to the United Network for Organ Sharing (see more yearly trends in Figure 15-4). The artificial kidney, described in Chapter 12, can sustain life in a patient lacking renal function; however, there are problems associated with physical and mental health, maintenance of the percutaneous device, and quality of life for the dialysis patient. Physical complications of dialysis include hypotension, cramps, malaise, headache, air embolism, aluminum toxicity, neuropathy, osteomalacia, dementia, and nausea. Peritoneal dialysis makes use of the diffusion properties of the natural abdominal membrane, which permits passage of larger molecules up to 30,000 daltons. Patients treated with this technique do not suffer severe complications discussed above such as neuropathy; however, there is a risk of infection.

In view of the above considerations, a transplanted natural kidney is a more desirable solution than artificial kidney dialysis. Both living and cadaver donors are used for kidney transplants; cadaveric sources account for 60–70% of kidney donors. Living donors are either blood relatives or spouses. In removing a kidney from a living donor, a diagonal incision is made paralleling the 11th rib, as shown in Figure 15-6. The rib is removed, the kidney is mobilized from the peritoneum, adrenal gland, and fat, and the renal vein and artery and the ureter are identified, clamped, and cut. The kidney is then removed and immediately perfused with a chilled saline solution containing procaine and heparin. As for cadaver donors, it is necessary to ascertain that the donor is free of communicable disease or cancer. Both kidneys as well as a segment of aorta and vena cava are removed from the cadaver, as shown in Figure 15-7, and are perfused and cooled as described above. The suturing procedure for reconnection of the vessels (anastomosis) is illustrated in Figure 15-8.

Immunosuppressive treatment for kidney transplant patients currently involves Imuran® and low-dose steroids in the case of a living human donor, and cyclosporin combined with Imuran® and low-dose steroids in the case of a cadaver donor. A rejection episode may occur in spite of the above precautions. Rejection is manifested in fever, tenderness of the kidney,

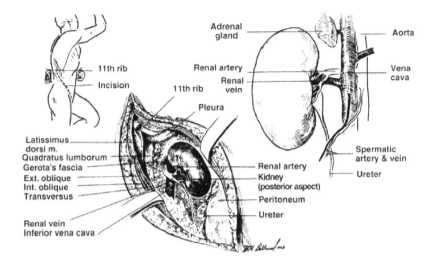

Figure 15-6. Surgical removal of kidney from a living donor. Reprinted with permission from Cerilli (1988). Reprinted with permission from Cerilli (1988). Copyright © 1982, Lippincott, Williams & Wilkins.

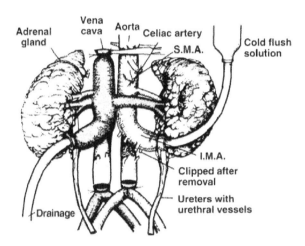

Figure 15-7. Surgical removal of kidney from a cadaver donor. Reprinted with permission from Cerilli (1988). Copyright © 1982, Lippincott, Williams & Wilkins.

weight gain, reduction in urinary output, and change in urinary composition. Rejection is treated with drugs; however, if the treatment fails during acute rejection, the transplanted kidney is removed. Kidney transplants are associated with a variety of complications, including osteoporosis and bone pain. Graft survival rates over periods of 1, 5, and 10 years are greater than 85, 60–70, and 40–50%, respectively. Patient survival is usually greater than graft survival since a patient who suffers transplant rejection may choose to undergo dialysis or possibly another transplant.

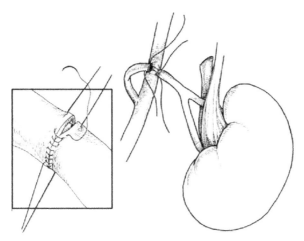

Figure 15-8. Anastomosis of blood vessels. Reprinted with permission from Cerilli (1988). Copyright © 1982, Lippincott, Williams & Wilkins.

15.4.2. Liver

The liver, as its name suggests, is essential to life, and its functions cannot be taken over, even temporarily, by any artificial device or material. The functions of the liver include active ones such as secreting bile, metabolism of carbohydrates, fats, and proteins, and detoxification of many substances. Patients with active chronic hepatitis, severe cirrhosis, inborn metabolic errors, and primary liver tumors are potential candidates for liver transplantation. The majority of end-stage liver failure is due to cirrhosis following alcohol abuse or chronic viral hepatitis.

Following the early success of kidney transplantation in the 1960s, clinical liver transplants were attempted. However, since none of the patients survived longer than 1 month, the procedure was abandoned for some time. Liver transplants were tried again as an experimental procedure with improved immunosuppressive drugs in the 1970s; however, more than half of these patients died within the first year. The introduction of cyclosporin as an immunosuppressive drug combined with improved surgical technique has led to dramatic improvements in the success of liver transplantation (Figure 15-1), and its acceptance as a clinical procedure.

Transplantation of the liver is rendered difficult by immunological considerations, and by the fact that the liver's blood supply is very complex, much more so than the kidney. Liver transplants involve interruption of the inferior vena cava, the main return to the heart. Heterotopic transplants (i.e., placement of the organ in a location other than the anatomically correct one) of the liver were tried initially to simplify the circulatory connections. Interruption of liver function is not tolerable, in contrast to the case of the kidney. Consequently, marginal viability problems and rejection episodes are very serious in liver transplants. Nevertheless, modern liver transplants have been successful, with 60–80% of recipients surviving 5 years. Living donor liver transplants to children have become accepted: a *piece* of a parent's liver is used for the transplant. The loss of a piece of liver is not a problem since the liver, like bone, is capable of true regeneration.

15.4.3. Heart and Lung

The first heart transplant performed by Dr. Christian Barnard in 1967 had a dramatic impact on society in view of the central role of the heart in maintaining life, and the psychological perception of the role of the heart. Dr. Barnard's patient survived 17 days, and heart transplants in the years immediately following did not last very long. Greater success in heart transplantation resulted from the introduction of cyclosporin, from increasing clinical experience with immunosuppressive drugs, and improved patient selection. Recipients of heart transplants tend to be relatively young people with end-stage cardiac disease. The survival rate of transplant recipients is now about 70 to 80% at 1 year after surgery. In addition to problems associated with immunosuppressive drugs, the transplanted heart tends to suffer vascular disease in about half of recipients who survive more than a year. A further problem in heart transplantation is that too few donors are available for the number of potential recipients. Attempts to circumvent this problem by using baboon hearts (in children) have not been successful. Since donated organs cannot be stored for long periods, a role is seen for cardiac assist devices and artificial hearts as a "bridge" to transplantation. Such devices can be used to support the patient for a period of days to weeks until an appropriate donor can be found. As described elsewhere in this book, the total artificial heart is experimental and has not been successful as a solution to end-stage heart disease.

Heart valve transplants, discussed elsewhere in this book, are routinely performed. These are based on porcine xenografts (transplants from another species) that have been fixed in glutaraldehyde. They do not contain any living cells. Consequently, rejection is not expected to be a problem.

Transplants of the lung or of the heart and lung are more difficult owing to the complexity of the surgical technique, and they are less commonly done than the heart alone. The first successful combined heart and lung transplant was done in 1981. Heart and lung transplant patients suffer a transient defect in lung function within the first month, and this is thought to result from lung denervation, interruption of the lymphatics, and surgical trauma. The nerve supply of the lung is important to its function. Transplanted lungs are always denervated, and the nerves do not regrow. Heart and lung differ immunologically, and rejection of each can occur independently. Survival of lung transplant recipients has recently been 85–90% over 1 year, but obliterative bronchiolitis occurs in 20 to 50% of long-term survivors and is a leading cause of morbidity and mortality in these people.

15.4.4. Bone, Bone Marrow, and Hands

Bone grafts, used widely in orthopedic surgery, serve the roles of structural support, of providing a framework for new bone tissue from the recipient to grow into, and a stimulus for new bone growth. The bone cells need not be living for a bone graft to succeed. Therefore, frozen bone tissue is suitable for this purpose. Autografts are commonly performed in which bone is removed from a location from which it can be spared, and used for repair. Bone from the rib, iliac crest, and fibula may be implanted in a viable form, by microsurgical anastomosis of the vascular supply. It is possible to walk relatively normally with a portion of the fibula removed. Allograft replacement of a large segment of bone removed for tumor surgery is also done, as shown in Figure 15-9.

Recently, successful allografts of human *hands* have been performed. These make use of advanced microsurgical technique.

Bone marrow is transplanted to correct severe immunological deficiency, aplastic anemia, and damage to the marrow due to radiation poisoning or chemotherapy for leukemia and other cancers. Histocompatibility matching is performed, and marrow is extracted from the iliac crest or sternum of a living donor under spinal or general anesthesia. The marrow is screened, heparinized, and injected intravenously into the recipient. There is significant morbidity associated with this procedure. For example, kidney failure occurs in about 20% of bone marrow recipients.

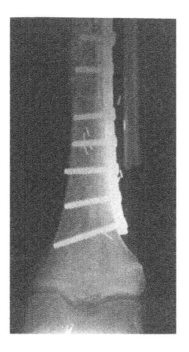

Figure 15-9. Large bone graft. Allograft replacement of tibial segment removed to treat tumor. Reprinted with permission from Cerilli (1988). Copyright © 1982, Lippincott, Williams & Wilkins.

15.4.5. Skin and Hair

Reconstructive surgery of the skin is performed to correct defects resulting from trauma, removal of skin tumors, or vascular problems such as pressure sores. It is sometimes desirable to provide "new" skin to replace the portion that was lost or damaged. The best source for such skin, if a relatively small amount is to be used, is elsewhere on the patient's body: an autograft. If the skin is simply rearranged in the vicinity of the defect, it is called a *flap*; if it is moved from elsewhere on the body, it is called a *graft*. Rearrangement of skin in a representative technique is shown in Figure 15-10. Grafts differ from flaps in that they are totally separated from their original blood supply and must develop a new supply from the location where they are placed. Consequently, the vascularity of the recipient bed is crucial to the success of a skin graft.

Hair transplants are a treatment for male pattern baldness. The procedure is done for cosmetic reasons. Hair transplants are always autografts, so rejection is not a problem. A local anesthetic is used to numb the scalp and saline is injected into the scalp to increase its turgor. Plugs of skin, including the hair follicles, are removed from the back part of the head, which resists balding. The plugs, about 4 mm in diameter, are harvested using a powered circular punch. Holes in the recipient area are cut using a slightly smaller punch and the plugs are re-implanted at the top and front of the head. More recent methods make use of smaller plugs that are less noticeable.

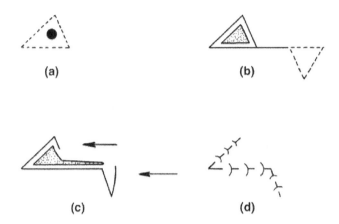

Figure 15-10. Surgical skin flap procedure: (**a**) tumor is identified and a triangular segment of skin marked; (**b**) segment with tumor is removed and incisions are made; (**c**) skin is moved; (**d**) wound is closed and sutured.

15.4.6. Nerve and Brain

The brain is an immunologically privileged site. The blood–brain barrier prevents cells of the immune system from reaching the brain, so that transplants into the brain are at low risk of rejection. Whole-brain transplants are not to be expected since (1) the recipient would presumably lose his/her memories with the brain, (2) volunteer living donors could not be found, (3) "cadaver donors," by definition brain dead, are unsuitable and nobody knows how to establish new nerve connections with the spinal cord. However, transplants of nerve tissue or neuroendocrine tissue into the brain have been considered for correction of Parkinson's disease and senile dementia. In the case of Parkinson's disease, autografts of tissue from the patient's own adrenal glands have been tried with some benefit, but not enough to justify further trials. Transplants of embryonic brain cells in rats will form new connections in the recipient rat, and improve the rat's brain function. As for humans, the use of human fetal tissue has dramatic ethical implications. Such transplants have been performed to treat Parkinson's disease. Many of the recipients had improvements in their ability to move, and better health with fewer medications; however, some patients suffered dyskinesia (abnormal involuntary movement and posture) following transplants.

15.4.7. Pancreas

In insulin-dependent diabetes, the endocrine function of the islets of Langerhans in regulating blood glucose is lost. Injections of insulin allow the patient to substitute an artificial or animal-based hormone for the natural one that has been lost. There remain many complications associated with insulin-treated diabetes. Many of these are related to the bolus character of the injected insulin and the associated fluctuations in blood sugar levels. Experimental treatment has been conducted with "artificial pancreas" devices consisting of a glucose sensor, a feedback amplifier, and a controlled insulin injector. Problems with these devices have resulted from a lack of biocompatibility of available glucose sensors, tissue growth around the sensor, and aggregation of the stored insulin.

Pancreatic transplantation is a possible solution; however, transplantation of the whole pancreas is made difficult by the corrosive nature of its exocrine secretion, a digestive enzyme; and by the vulnerability of pancreatic islet cells to rejection. Immunosuppressive therapy is therefore used to prevent rejection in patients who receive transplants of islet cells or the whole pancreas. At present, pancreas or islet cell transplants are done in patients with end-stage renal disease or in diabetics who already have a transplanted kidney and who are receiving immunosuppressive drugs. Many pancreas transplants are combined kidney–pancreas transplants. In a one-year follow-up, patient survival has exceeded 90% and graft survival more than 70%. Islet cell autografts into the portal vein have been done for patients undergoing removal of the pancreas for painful chronic pancreatitis, and some of these have been successful. An interesting and novel approach combining transplantation techniques with biomaterials is to implant only the islet cells in semipermeable artificial tubules or between semipermeable artificial membranes. Glucose and insulin can diffuse through the membranes, but immune cells and substances of large molecular weight cannot. Tubule-based islet cell transplants have been successful in controlling glucose levels in diabetes for about 12 months, after which the tubules become overgrown with recipient cells, slowing the diffusion process. Methods for pancreas support are still under investigation.

15.4.8. Cornea

Corneal transplantation (keratoplasty) is the most commonly done solid transplant procedure, as shown in Table 15-5. Conditions that impair the clarity of the cornea sufficiently to obscure vision include trauma, scarring following an infection, congenital abnormality, and degeneration. Success of corneal transplants is aided by the fact that the cornea is avascular. Diffusion to the blood system is very slow, as is the turnover of corneal proteins. Rejection, therefore, is not normally a problem. It occurs in a small number of cases and is treated by topical steroids. Donated corneas, owing to their slow metabolism, may be removed within 4 to 12 hours following death. They may be preserved for 96 hours to 3 weeks in a culture medium, or frozen in a cryoprotectant solution for as long as 18 months. The surgical procedure for the recipient involves great care in aligning the new cornea with surrounding structures. Suturing is done with 10-O or 11-O monofilament nylon suture under an operating microscope.

Example 15-2

Why and how would an "autograft" of a human leg be performed?

Answer

Autografts are usually performed with biological material that can be moved from one part of the body to another, as with bone and skin. This makes no sense in the case of a leg. However, a leg, arm, or finger that has been accidentally cut off can sometimes be reattached, a procedure called "replantation" rather than transplantation. For the procedure to be successful, most of the cells in the limb must remain alive. Therefore, the replantation surgery must be done quickly; it is very helpful to cool (but not freeze) the limb during the time prior to surgery.

Replantation of a limb involves joining the bone by the same techniques used for broken bones, e.g., bone plates. The muscles are joined individually by sutures. The blood vessels are also sutured together; connection of the smaller ones involves microsurgical techniques. The nerve fibers in the severed limb will die; however, if the nerve sheath is properly joined and aligned, new fibers can grow from the nerve cell bodies in the patient's body. In the case of a leg, this regrowth may take a year.

Table 15-5. Transplant Statistics, United States

| | Number of cases | | | |
Organ	1983	1984	Number waiting	Potential benefit
Cornea	21,500	23,500	3,500	250,000
Kidney	6,112	6,730	12,000	25,000
Heart	280	400	100	14,000
Liver	168	308	330	5,000
Pancreas	61	87	50	10,000
Heart-lung	13	17	50	1,000

Reprinted with permission from Cerilli (1988). Copyright © 1988, J.B. Lippincott.

15.5. REGENERATION

Transplantation of organs and tissues has proven successful in the treatment of many serious maladies. The lack of sufficient donors is likely to remain a major constraint on such procedures. An alternative approach is to consider the possibility of regenerating the diseased, damaged, or lost tissue or organ. The use of electrical stimulation to regrow bone across nonunions has been discussed elsewhere in this book. In a related development, attempts have been made to stimulate regeneration of amputated limbs. The background of this idea is the fact that animals of low order have remarkable regenerative capacity. For example, an earthworm cut in half can regrow into two new worms; some amphibians can regenerate a limb that has been amputated. In higher animals and in humans, regeneration is very limited, although young children can regrow an amputated fingertip. Regeneration in the lower animals is accompanied by a distinctive pattern of electrical potential in the injured limb, a potential that is opposite in polarity to that seen in a nonregenerating higher animal or human. Researchers have stimulated partial regeneration of limbs in amputated mammals such as rats or in a nonregenerating amphibian such as a frog, by applying an electrical potential pattern similar to that in a regenerating amphibian. These procedures are under investigation. While bone tissue has been regrown in humans in a clinical setting, whole limbs have not.

Another potential is seen in stem cell research to regenerate cells and tissues. Table 15-6 gives some potential uses of stem cell transplantation. Stem cells are found in bone marrow and in other parts of the body. Figure 15-11 illustrates the promising results of bone marrow transplantation over chemotherapy. Stem cell research in other areas such as nerve is illustrated in Figure 15-12. Stem cell therapy may offer benefits in patients with Parkinson's and Alzheimer's.

Table 15-6. Potential Future Uses for Stem Cells

- Facilitate gene therapy
- Harness graft-versus-solid tumor
- Microchimerism to induce tolerance to solid organ transplantation
- Repair and remodeling of other tissue after injury and scarring (cardiac, renal, liver, CNS)

Reprinted with permission from Mehta (2004). Copyright © 2004, Jones & Bartlett.

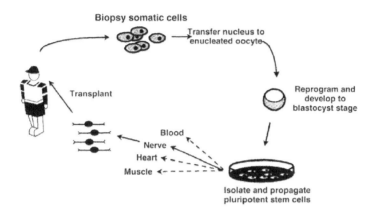

Figure 15-11. Stem cell therapy via somatic cell reprogramming. Reprinted with permission from Blakemore et al. (2000). Copyright © 2000, Wiley.

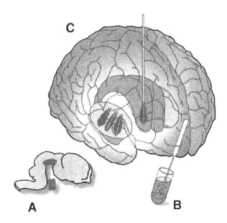

Figure 15-12. Schematic illustration of the most commonly used procedure for transplantation of human embryonic mesencephalic tissue into the striatum in patients with PD. Ventral mesencephalic DA-rich tissue (**A**) from human embryos aged 5.5–8 weeks postconception is dissociated (**B**), and then implanted unilaterally or bilaterally using stereotaxic surgery into the caudate nucleus or the putamen or both (**C**). Reprinted with permission from Lindvall (2000). Copyright © 2000, Wiley.

15.6. ETHICAL CONSIDERATIONS

Organ transplantation raises ethical, moral, and sociological issues not to be found in inert biomaterials. Issues that are of concern include the decision-making process as to which patients who desire transplants shall receive them, and the means by which donations are obtained. The high cost of transplant surgery tends to lead to patient selection favoring the wealthy, although medical insurance covers some procedures. On the donor side, we have described in §15.1 historical involuntary transplants that would be unacceptable in civilized societies today. Fiction author Larry Niven has created a hypothetical future society in which technical obstacles to transplantation of all organs have been overcome. This fictional society used capital punishment subjects as forced transplant donors. To extend their lives by a ready supply of transplants, individuals in this society then voted progressively more minor crimes, including traffic citations, to be capital offenses.

In the modern world, there is evidence over the last fifteen years that organs are being sold in countries outside the United States. This practice is problematical in that it is viewed as exploiting poor people who are coerced into donation by economic forces. Although the voluntary donation of an organ to a family member can be a very positive altruistic experience, there have been cases of family members of sick people who have been pressured into donating organs. In view of these issues, the American Society of Transplant Surgeons has prepared a set of guidelines for the procurement and use of transplant organs in the United States. An even more troubling development is the use of organs from executed convicts or patients under anesthesia in some less-developed countries, particularly in Asia. Many prisoners who become unwitting donors are convicted after hurried trials based on confessions extracted under torture; human rights groups say they may be innocent. Execution is by a gunshot to the back of the head. Organs have been removed from prisoners still breathing and moving.

PROBLEMS

15.1. Although transplants perform functions that cannot be carried out by inert biomaterials, various biomaterials are used in transplantation surgery. What are these biomaterials? Is their role any different in the case of transplants than in applications not involving transplants?

15.2. Discuss transplantation of (a) the external ear, (b) the entire brain, (c) tonsils (Jingade and Sangur, 2006), (d) entire arms, (e) the whole eye, and (f) whole teeth. Include in your discussion the questions of need, technical feasibility, ethics, economic, and social aspects.

15.3. Discuss the relative merits of heart valve replacement by porcine xenograft, by a tilting disk valve design using pyrolytic carbon, and by a silicone ball in cage valve design. Include in your discussion the issues of hemodynamic flow, likely failure mechanisms, use of anticoagulant drugs, and expected performance.

15.4. Discuss the differences between replacement heart valves based on porcine xenograft valves and valves made from bovine pericardium (see Chapter 12). What are their pros and cons?

15.5. Many people object to the use of animals for testing biomaterials. Under what circumstances are in-vivo tests required for evaluating a particular device such as a heart valve? Can you suggest ways of avoiding animal tests entirely? Keep in mind the fact that testing with human patients is in vivo but does not involve animals. Under what circumstances do you think it is ethical to test an unproven biomaterial in humans?

15.6. From the following list select the most appropriate *two* materials for each application:

A. Collagen   B. Polypropylene   C. Polyester(PET)
D. Polytetrafluoroethylene (PTFE)   E. Cotton
F. Polylactic acid (PLA)   G. Stainless steel

    a. Absorbable sutures   b. Nonabsorbable sutures
    c. Synthetic sutures   d. Natural sutures
    e. Highest density   f. Lowest density

15.7. From the following list select the most appropriate one.

A. Ligamentum nuchae   B. Cartilage   C. Teeth enamel   D. Skin
E. Tendon   F. Artery   G. Platelet   H. Bone

    a. Low coefficient of friction   b. Laplace equation
    c. Elastin   d. Langer's line
    e. Tensile force transmission   f. $Ca_5(PO_4)_3(OH)$
    g. Emboli   h. Wolff's law

15.8. Choose the most appropriate material for each question. Use alphabets.

A. Alumina (Al_2O_3) ceramic   B. 316L stainless steel
C. Polyester   D. Silicone rubber
E. Pyrolytic carbon with graphite   F. CoNiCrMo (MP35N)
G. Polymethyl methacrylate   H. Poly-L-lactic acid
I. UHMWPE

a. Used to make temporary orthopedic implants such as bone plates and screws.
b. Used to make femoral stems of artificial hip joints.
c. Used to make Swanson type (integral hinge) finger joint prosthesis.
d. Used to make acetabular cup.
e. Used to make synthetic absorbable sutures.
f. Used to make bone cement powder.
g. Used to make artificial teeth.
h. Used to make artificial heart valve discs.
i. Used to make knitted artificial blood vessels.
j. Used to make sutures.

15.9. Select *one material* from the materials listed:

A. Graphite   B. PMMA   C. Bone   D. Stainless steel
E. Copper   F. $BaTiO_3$ G. Single-crystal alumina

a. Highest electrical resistivity   b. Lowest density
c. Lowest electrical resistivity   d. Lowest acoustic impedance (Z)
e. Piezoelectric ceramic.   f. Highest permeability of O_2 gas
g. Hard contact lens material   h. Composite material
i. Lowest friction coefficient   j. Highest refractive index.

15.10.   Choose the most appropriate material for each question. Use alphabets.

A. Alumina (Al_2O_3) single crystal   I. 316L stainless steel

B. PTFE(Teflon®)   J. Silicone rubber

C. Pyrolytic carbon with graphite   K. Ti6Al4V

D. PMMA   L. PLA

E. UHMWPE   M. HDPE

F. Hydroxyapatite   N. CoCrMo alloy

G. Poly(HEMA)   O. Bioglass®

H. Pt   P. Catgut

a.   Used to make temporary orthopedic implants such as bone plates and screws.
b.   Used to make femoral stems of artificial hip joints.
c.   Used to make membrane of breast prosthesis.
d.   Used to make acetabular cup.
e.   Used to make synthetic absorbable sutures.
f.   Used to make hard contact lens.
g.   Used to make artificial teeth root.
h.   Used to make artificial heart valve discs.
i.   Used to make knitted artificial blood vessels.
j.   Used to make sutures less sticking to tissues.
k.   Used to coat the surface of femoral stem for better tissue ingrowth.
l.   Used to make oxygenator membrane.
m.   Used to make soft contact lenses.
n.   Used to make electrode for cochlear implants.
o.   Used to make acetabular cup by Charnley in earlier times (before 80s).

15.11.   Select the most appropriate one from the lists. One for each.

A. DLC	B. CoCrMo alloy	C. CoNiCrMo alloy
D. Hg amalgam	E. Nitinol®	F. Al_2O_3
G. ZrO_2	H. Bioglass®	I. Hydroxyapatite
J. LTI	K. PolyHEMA	L. Kevlar®

a.   Usually cast
b.   Sapphire or ruby
c.   Hot-forged
d.   Aromatic polyamide
e.   Hydrogel
f.   Shape memory alloy
g.   Fake diamond
h.   CaP ceramic, bone bonding
i.   Isotropic carbon
j.   SiNaCaP glass, bone bonding
k.   Amorphous diamond
l.   Ag-Sn alloy is used to make solid solution

A. Galvanic corrosion   B. Passivation   C. Pourbaix diagram
D. Tacticity   E. Vulcanization   F. Branching   G. Weibull modulus

   a.   Crosslinking of rubbers
   b.   Plot of electric potential versus pH
   c.   Deliberate oxidation of surface layer
   d.   Arrangement of side groups of linear chain
   e.   Macroscopic differences in electrochemical potential
   f.   Chains grown from the sides of the main backbone chains
   g.   Measure of strength distribution, large for metals, small for ceramics

15.12.   Select the most related one.

A. Ceramics

a. Al_2O_3	1. LTI	( )
b. $Ca_5(PO_4)_3(OH)$	2. Resorbable ceramic	( )
c. Pyrolytic carbon	3. Ruby	( )
d. TCP	4. Mineral phase of bone	( )
e. ZrO_2	5. Radio-opacifier for x-ray	( )
f. Barium sulfate	6. SiO_2–CaO–Na_2O–P_2O_5	( )
g. Glass-ceramic	7. Y_2O_3 stabilized	( )

B. Metals

h. Ti	1. Castable alloy	( )
i. CoNiCrMo	2. SME alloy	( )
j. Platinum	3. Austenitic phase	( )
k. 316L stainless steel	4. Wrought alloy	( )
i. CoCrMo	5. Allotropic	( )
l. Ti6Al4V	6. Precipitation hardening	( )
m. NiTi	7. Highly cathodic	( )

C. Polymers

n. PMMA	1. Delrin®	( )
o. Aromatic polyamide	2. Styrene/butadiene rubber	( )
p. PolyHEMA	3. Neoprene rubber	( )
q. Polyacetal	4. Synovial fluid	( )
r. Polysaccharide	5. Kevlar®	( )
s. SBR	6. Hydrogel	( )
t. Polychloroprene	7. IOLs	( )

15.13.   A bioengineer is asked by a supervisor to construct a bone fracture plate from the following materials:

A. CoCrMo   B. ZrO_2 C. PLA   D. 316L SS   E. Ti6Al4V

   a. Which one would be chosen for its flexibility?
   b. Which one would be chosen for its tensile strength?
   c. Which one would be chosen for its specific tensile strength [s/r]?
   d. Which one should be chosen for its resorption capacity?
   e. Choose one material and give three reasons(advantages) for using
      it for making a fracture plate.

DEFINITIONS

Allograft: Transplant between different individuals of the same species.

Anastomosis: Joining two blood vessels or other tubular structures in the body.

Antigen: A foreign protein that elicits an immune system response.

Autograft: Transplant within one individual, from one part of the body to another.

Biocompatibility: Compatibility of nonliving implant materials with the body.

Brain death: Death of the brain including the brain stem resulting in unreceptivity and unresponsivity, absence of spontaneous respiration or movement, absence of reflexes, and a flat EEG.

Cyclosporin: An immunosuppressive drug.

EEG: Electroencephalogram, which is a plot of the electrical activity of the brain as measured via electrodes on the scalp.

Histocompatibility: Compatibility of transplant tissue with the recipient's body in relation to immune response.

Iliac: Referring to the ilium, the large flat bone of the pelvis.

Island of Langerhans: A cluster of endocrine cells found in the pancreas; they secrete insulin.

Isograft: Transplant between genetically identical individuals, i.e., identical twins.

Graft: A transplant.

Homograft, allograft: Transplant between different individuals of the same species.

Heterograft, xenograft: Transplant between members of different species.

Heterotopic: Transplanted in a location different from the anatomically correct one.

Hypothermia: Abnormally low body temperature. It can cause an unresponsive condition that can mimic brain death, so transplants are not taken from hypothermic patients.

Kaposi's sarcoma: A type of cancer, once rare, now largely associated with immune system deficiencies such as those produced in AIDS or by immunosuppressive drugs used in transplant procedures. This disease produces bluish skin lesions.

Keratoplasty: Replacement of the cornea.

Orthotopic: Transplanted in the anatomically correct location.

Parkinson's disease: A neurological disorder characterized by tremor and paralysis. Attempts have been made to cure it by transplantation of brain or adrenal cells into the brain.

Porcine: From a pig.

Replantation: Surgical reattachment of a severed body part.

T cells: Immunologically competent cells derived from the thymus.

Transplantation: Transfer of a tissue or organ from one body to another, or from one location in a body to another.

Tubule: A small tube.

BIBLIOGRAPHY

Bach FH, Soares M, Lin Y, Ferran C. 1999. Barriers to xenotransplantation. *Transpl Proc* **31**(4):1819–1820.

Barbucci R. 2002. *Integrated biomaterials science*. New York: Kluwer Academic/Plenum.

Becker RO. 1972. Stimulation of partial limb regeneration in rats. *Nature* **235**:109–111.

Blakemore WF, Smith PM, Franklin RJM. 2000. Remyelinating the remyelinated CNS. In *Neural transplantation in neurodegenerative disease: current status and new directions*, pp. 289–301. Ed DJ Chadwick, JA Goode. Chichester: Wiley.

Bohinjec M, Rozman P. 2002. Immunogenetics in transfusion medicine: molecular and cellular basis of immunity and alloimmunity. Amsterdam: Elsevier.

Borkel JF. 1982. The history of cyclosporin A and its significance. In *Cyclosporin A*, pp. 5–34. Ed DJG White. Amsterdam: Elsevier Biomedical Press.

Bowdish ME, Barr ML. 2004. Living lobar lung transplantation. *Respir Care Clin North Am* **10**(4):563–579.

Bretzel RG, Hering BJ, Federlin KF. 1995. Islet cell transplantation in diabetes mellitus: from bench to bedside. *Exp Clin Endocrinol Diabetes* **103**(Suppl 2):143–59.

Calafiore R. 1998. Actual perspectives in biohybrid artificial pancreas for the therapy of type 1, insulin-dependent diabetes mellitus. *Diabetes Metab Rev* **14**(4):315–324.

Cascalho M, Platt JL. 2005. New technologies for organ replacement and augmentation. *Mayo Clin Proc* **80**(3):370–378.

Cerilli GJ. 1988. *Organ transplantation and replacement*. Philadelphia: Lippincott.

Chick WL, Like AA, Lauris V. 1975. Beta cell culture on synthetic capillaries: an artificial endocrine pancreas. *Science* **87**:847–849.

Chkhotua A, Shohat M, Tobar A, Magal N, Kaganovski E, Shapira Z, Yussim A. 2002. Replicative senescence in organ transplantation-mechanisms and significance. *Transpl Immunol* **9**(2–4):165–171.

Clarkson ED. 2001. Fetal tissue transplantation for patients with Parkinson's disease: a database of published clinical results. *Drugs Aging* **18**(10):773–785.

Cohen EP, Lawton CA, Moulder JE. 1995. Bone marrow transplant nephropathy: radiation nephritis revisited. *Nephron* **70**(2):217–222.

Compston J, Shane E. 2005. *Bone disease of organ transplantation*. Burlington, MA: Elsevier Academic.

Compton CN, Raaf JH. 2005. Vascular access devices. In *The bionic human*, pp. 561–587. Ed FE Johnson, KS Virgo. Totowa, NJ: Humana Press.

Crawford SW. 1995. Respiratory infections following organ transplantation. *Curr Opin Pulm Med* **1**(3):209–215.

Cupples SA. 2002. Overview of organ transportation. *Solid organ transplantation: a handbook for primary health care providers*, pp. 1–15. Ed SA Cupples, L Ohler. New York: Springer.

Delloye C, Cnockaert N, Cornu O. 2003. Bone substitutes in 2003: an overview. *Acta Orthop Belg* **69**(1):1–8.

Deschamps JY, Roux FA, Sai P, Gouin E. 2005. History of xenotransplantation. *Xenotransplantation* **12**(2):91–109.

Diamond SG, Markham CH, Rand RW, Treciokas LJ. 1994. Four-year follow-up of adrenal-to-brain transplants in Parkinson's disease. *Arch Neurol* **51**(6):559–563.

Ellingsen JE, Lyngstadaas SP. 2003. *Bio-implant interface: improving biomaterials and tissue reactions*. Boca Raton, FL: CRC Press.

Epstein S, Shane E, Bilezikian JP. 1995. Organ transplantation and osteoporosis. *Curr Opin Rheumatol* **7**(3):255–261.

Fine RN, Kelly DA, Webber SA. 2003. *Pediatric solid organ transplantation*. Philadelphia: W.B. Saunders.

Francois CG, Breidenbach WC, Maldonado C, Kakoulidis TP, Hodges A, Dubernard JM, Owen E, Pei G, Ren X, Barker JH. 2000. Hand transplantation: comparisons and observations of the first four clinical cases. *Microsurgery* **20**(8):360–371.

Gage NH, Dunnett SB, Stenevi U, Bjorklund A. 1983. Aged rats: recovery of motor impairments by interstitial nigral grafts. *Science* **221**(4614):966–969.

Gourishankar S, Halloran PF. 2002. Late deterioration of organ transplants:a problem in injury and homeostasis. *Curr Opin Immunol* **14**(5):576–583.

Greenstein JL, White-Scharf M. 1998. Xenotransplantation: a review of current issues. *J Biolaw Bus* **2**(1):46–50.

Hadley S, Karchmer AW. 1995. Fungal infections in solid organ transplant recipients. *Infect Dis Clin North Am* **9**(4):1045–1074.

Hagell P, Piccini P, Bjorklund A, Brundin P, Rehncrona S, Widner H, Crabb L, Pavese N, Oertel WH, Quinn N, Brooks DJ, Lindvall O. 2002. Dyskinesia following neural transplantation in Parkinson's disease. *Nat Neurosci* **5**(7):627–628.

Hariharan S, Johnson CP, Bresnahan BA, Taranto SE, McIntosh MJ, Stablein D. 2000. Improved graft survival after renal transplantation in the United States, 1988 to 1996. *N Engl J Med* **342**(9):605–612.

Harlan D, Kirk A, Kenyon N. 1999. *Organ transplantation*. International Information Resources Inc., CenterNet, Association of Academic Health Centers (US). Bethesda, MD: NIH.

Jingade RRK, Sangur RR. 2005. Biomechanics of dental implants: an FEM study. *J Indian Prosthodont Soc* **5**(1):18–22.

Khalpey Z, Koch CA, Platt JL. 2004. Xenograft transplantation. *Anesthesiol Clin North Am* **22**(4):871–885.

Kirklin JK, Young JB, McGiffin D. 2002. *Heart transplantation*. New York: Churchill Livingstone.

Kolata G. 1982. Grafts correct brain damage. *Science* **217**:342–217.

Korossis SA, Fisher J, Ingham E. 2000. Cardiac valve replacement: a bioengineering approach. *Biomed Mater Eng* **10**(2):83–124.

Land WG. 2004. Aging and immunosuppression in kidney transplantation. *Exp Clin Transpl* **2**(2):229–237.

Lieberman JR, Friedlander GE. 2005. *Bone regeneration and repair: biology and clinical applications*. Totowa, NJ: Humana Press.

Lindvall O. 2000. Neural transplantation in Parkinson's disease. In *Neural transplantation in neurodegenerative disease: current status and new directions*, pp. 110–128. Ed DJ Chadwick, JA Goode. Chichester: Wiley.

Maddrey WC, Schiff ER, Sorrell MF. 2001. *Transplantation of the liver*. Philadelphia: Lippincott Williams & Wilkins.

Mehta P. 2004. The scientific basis of pediatric bone marrow transplantation. In *Pediatric stem cell transplantation*, pp. 4–12. Ed P Mehta. Sudberry, MA: Jones & Bartlett.

Morozumi K, Yamaguchi Y. 2000. In Proceedings from conference on transplant pathology: July 10, 1999, Tokyo, Japan. Copenhagen: Munksgaard.

Morris J. 2004. Neuroblastoma. In *Pediatric stem cell transplantation*, pp. 131–155. Ed P Mehta. Sudberry, MA: Jones & Bartlett.

Morris PJ. 2001. *Kidney transplantation: principles and practice*. Philadelphia: W.B. Saunders.

Niven L. 1974. *The long arm of Gil Hamilton*. New York: Ballantine.

Nógrádi A. 2005. *Transplantation of neural tissue into the spinal cord*. Georgetown, TX: Landes Bioscience (Eurekah.com).

Norwood OT, Shiell RC. 1984. *Hair transplant surgery*. Springfield, IL: Thomas.

Ochi M, Uchio Y, Kawasaki K, Wakitani S, Iwasa J. 2002. Transplantation of cartilage-like tissue made by tissue engineering in the treatment of cartilage defects of the knee. *J Bone Joint Surg Br* **84**(4):571–578.

Oellerich M. 2001. In Proceedings of the Analytica Conference 2000 symposium on laboratory evaluation and monitoring of transplant recipients: Munich, Germany, April 11, 2000. New York: Pergamon Press.

Olbrisch ME, Benedict SM, Ashe SM, Levinson JL. 2002. Psychosocial assessment of organ transplant candidates. *J Consult Clin Psychol* **70**:771–783.

Pietra BA. 2003. Transplantation immunology 2003: simplified approach. *Pediatr Clin North Am* **50**(6):1233–1259.

Prokop A, Hunkeler D, Cherrington A. 1997. *Bioartificial organs: science, medicine, and technology*. New York: New York Academy of Sciences.

Qian Y, Dana MR. 2001. Molecular mechanisms of immunity in corneal allotransplantation and xenotransplantation. *Expert Rev Mol Med* **2001**:1–21.

Racusen LC, Solez K, Burdick JF, NetLibrary Inc. 1998. *Kidney transplant rejection diagnosis and treatment*. New York: Marcel Dekker.

Rapaport FT, Dausset J. 1968. *Human transplantation.* New York: Grune and Stratton.

Remuzzi G, Perico N. 1995. Cyclosporine-induced renal dysfunction in experimental animals and humans. *Kidney Int Suppl* **52**:S70–74.

Rigg KM. 1995. Renal transplantation: current status, complications and prevention. *J Antimicrob Chemother* **36**(Suppl B):51–57.

Rose ML. 2001. *Transplant-associated coronary artery vasculopathy.* Austin, TX: Landes Bioscience.

Rubin RH, Young LS. 2002. *Clinical approach to infection in the compromised host.* New York: Kluwer Academic/Plenum.

Schaffner A, ed. 2001. Immunocompromised host society consensus conference on epidemiology, prevention, diagnosis, and management of infections in solid-organ transplant patients: Davos, Switzerland, 23 June 1998, fully updated summer 2000. Chicago: University of Chicago Press.

Shapiro R, Jordan ML, Scantlebury VP, Vivas C, Gritsch HA, Corry RJ, Egidi F, McCauley J, Ellis D, Gilboa N. 1995. The superiority of tacrolimus in renal transplant recipients: the Pittsburgh experience. *Clin Transpl* **1995**:199–205.

Smith CS. 2001. On death row, China's source of transplants. *The New York Times.* 18 October, p. 1.

Solez K, Racusen LC, Billingham ME. 1996. *Solid organ transplant rejection mechanisms, pathology, and diagnosis.* New York: Marcel Dekker.

Stevens JM, Hilson AJ, Sweny P. 1995. Post-renal transplant distal limb bone pain: an under-recognized complication of transplantation distinct from avascular necrosis of bone? *Transplantation* **60**(3):305–307.

Stratta RJ, Taylor RJ, Larsen JL, Cushing K. 1995. Pancreas transplantation. *Ren Fail* **17**(4):323–337.

Taylor JK, Pope BJ. 2001. *The development of diamond as a bearing for total hip arthroplasty.* Scientific exhibit SE39, AAOS 2001. San Francisco.

Thiru S, Waldmann H. 2001. *Pathology and immunology of transplantation and rejection.* Oxford: Blackwell Science.

Toledo-Pereyra LH, Lopez-Neblina F. 2003. Xenotransplantation: a view to the past and an unrealized promise to the future. *Exp Clin Transpl* **1**(1):1–7.

Tramovich TA, Stegman SJ, Glogau RG. 1989. *Flaps and grafts in dermatologic surgery.* Chicago: Year Book Medical Publishers.

Tseng YL, Sachs DH, Cooper DK. 2005. Porcine hematopoietic progenitor cell transplantation in nonhuman primates: a review of progress. *Transplantation* **79**(1):1–9.

Valantine H, Zuckermann A. 2005. From clinical trials to clinical practice: an overview of Certican (everolimus) in heart transplantation. *J Heart Lung Transpl* **24**(4 Suppl):S185–190; discussion S210–211.

van Zuijlen PP, Vloemans JF, van Trier AJ, Suijker MH, van Unen E, Groenevelt F, Kreis RW, Middelkoop E. 2001. Dermal substitution in acute burns and reconstructive surgery: a subjective and objective long-term follow-up. *Plast Reconstr Surg* **108**(7):1938–1946.

Wang-Rodriguez J, Rearden A. 1995. Effect of crossmatching on outcome in organ transplantation. *Crit Rev Clin Lab Sci* **32**(4):345–376.

Wilkes DS, Burlingham WJ. 2004. *Immunology of organ transplant.* Dordrecht: Kluwer Academic/Plenum.

Williamson SL. 2001. Immunology for kidney transplant recipients: creation of an educational CD-ROM using vector animation software. PhD dissertation, Johns Hopkins University.

16

TISSUE ENGINEERING MATERIALS
AND REGENERATION

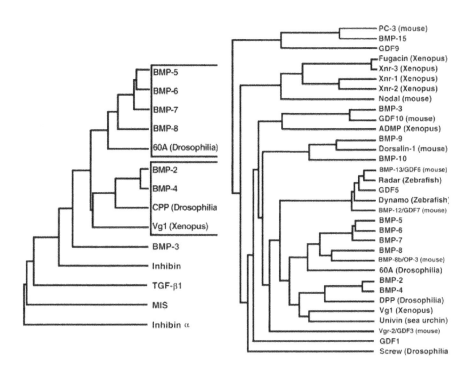

Left: Amino acid sequence relationship between bone morphogenetic protein (BMP) molecules identified in osteoinductive extracts derived from bone, as well as representative other members of the TGF superfamily. This figure was generated using the Genetics Computer Group software program PileUp, utilizing the amino acid sequences from the first conserved cysteine residue in the mature part of the human molecules to the carboxy terminus. The length of the horizontal lines corresponds to the number of differences between the proteins or groups of proteins, i.e., shorter lines indicate that the molecules are more closely related in amino acid sequence. For example, BMP-2 and BMP-4 have very few amino acid differences, whereas BMP-2 and TGF-β have many. Reprinted with permission from Wozney (1999). Copyright © 1999, Quintessence Publishers. **Right**: Amino acid sequence relationships between members of the TGF superfamily. All sequences are human unless otherwise indicated. This figure was generated as described above. BMP = bone morphogenetic protein; GDF = growth and differentiation. Reprinted with permission from Wozney (1999). Copyright © 1999, Quintessence Publishers.

16.1. OVERVIEW

In prior chapters we have seen that some functions of diseased or injured organs and tissues can be assumed with the aid of synthetic materials. Synthetic materials, being nonliving, at best assume passive structural (mechanical), passive optical (refractive), or passive chemical (diffusion) functions. Transplanted tissues and organs can perform functions that cannot be achieved by nonliving materials; however, transplants are limited by availability of donors and the need to control rejection. A superior approach, if it can be achieved, is to consider the possibility of *regenerating* the diseased, damaged, or lost tissue or organ.

Tissue engineering is intended to achieve regeneration by growing a viable new tissue or organ. Tissue engineering entails harvesting donor tissue that is then dissociated into individual cells. Donor cells may be adult or embryonic. If they are adult cells, they may be differentiated cells or stem cells (see below). These cells may be directly implanted, or encouraged to proliferate in an organized manner in tissue culture (referred to as ex vivo). The organization of cultured cells can be guided by growing them on a substrate known as a scaffold. The cultured tissue is then implanted. The knowledge base required for tissue engineering is interdisciplinary, since one must understand the cell biology of proliferation and differentiation as well as the materials science of substrate or scaffold materials used in tissue or organ culture. It is the latter aspect that is emphasized in this chapter. Clinical case histories are also given; the reader should appreciate that this is a rapidly evolving field and that new applications are likely to emerge rapidly.

Table 16.1. Some Applications of Tissue Engineering

Applications	Examples
Cell production in vitro	Bone marrow cell production
Extra corporeal devices	Artificial liver
Tissue growth and repair in situ	Nerve regeneration Artificial skin Bone and cartilage Blood vessel
Implantable devices	Endotherialized vascular grafts Bone and cartilage implants Artificial pancreatic islets Skin regeneration template

Adapted with permission from Engelberg and Kohn (1991). Copyright © 1991, Butterworth-Heinemann and the Biological Engineering Society.

Tissue engineering involves studies from cells to organ biology and physiology as well as the interaction with substrates (biomaterials), as illustrated in Figure 16-1. The role of scaffolds (temporary materials structured in such a way as to guide growth of cells in vitro or in vivo) in tissue engineering is depicted in Figure 16-2. Another approach to the problem of replacing tissues or organs is reproduction of the whole organ or body by cloning from somatic (body) cells rather than from reproductive cells. This has been done experimentally in animals. Dolly was a sheep reproduced in this way. The spare body part concept may not be acceptable in humans due to ethical as well as technical problems. Table 16-1 shows some examples of the applications of tissue engineering. The growth of cells in a scaffold can be facilitated by other biological entities such as growth factors, as given in Table 16-2. The scaffold (matrix)

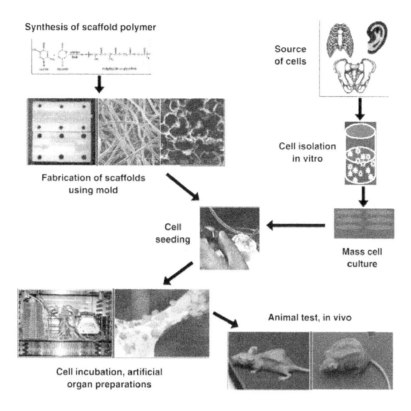

Figure 16-1. Schematic representation of tissue engineering. The final products (in this case, ear and nose) could be used as implants if the immunological reaction could be controlled. Courtesy of Drs. Gilson Khang and Hae Bang Lee, Korea Institute of Chemical Technology, Taejun, Korea, 2006.

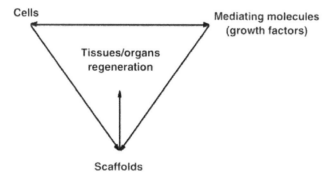

Figure 16-2. Schematic representation of the role of scaffolds in tissue engineering.

Table 16-2. Examples of Matrices, Cells, and Regulators
Employed in Tissue Engineering

A. Matrices (porous structures)
    a. Absorbable
        1. Natural polymers
            Collagen (types I, II, III, IV)
            Collagen–glycosaminoglycan copolymer
            Fibrin
            Poly(hydroxybutylate), PHB
            Poly(hydroxyvaleric acid), PHV
            Sodium alginate
            Chitin and Chitosan

        2. Synthetic polymers
            Polylactic acid
            Polyglycolic acid
            Poly(ε-caprolactone)
            Polyanhydrides
            Poly(ortho esters)

        3. Composites
            Bone particles/natural or synthetic polymers

        4. Natural mineral
            Anorganic bone(human and bovine bone)
            Reprocessed whole bone
            Anorganic mineral
            Hydroxyapatite*

    b. Nonresorbable
        1. Synthetic polymer
            Polytetrafluoroethylene

        2. Synthetic ceramics
            Hydroxyapatite*
            Calcium phosphate (mono-, di-, tri-, and tetra CaP)
            Glass-ceramic (Bioglass®)
            Calcium sulfate (Plaster of Paris)

B. Cells
    a. Autologous parenchymal cells
    b. Allogeneic parenchymal cells
    c. Marrow stromal stem cells

C. Soluble Regulators
    a. Growth factors (polypeptide mitogens)
    b. Differentiation factors (e.g., bone morphogenetic protein)

* Hydroxy apatite can be obtained from natural source as well as synthesized

Modified with permission from Spector (1999). Copyright © 1999, Wiley.

materials, cells, and soluble cell regulators have to work together to get optimal regeneration of tissues and organs. Some writers have considered tissue engineering to be a paradigm shift in medicine. However, successful clinical implementations to date have been few. Little is known about the processes by which cells proliferate and differentiate in artificial scaffolds. It is not known for the scaffold what is the optimal "open" pore size, pore shape, pore density, pore distribution, and substrate, despite much effort.

Natural tissue matrix materials have been used as scaffolds. Although the concept seems reasonable, success has not yet been achieved. A related concept, that of a "replaneform," makes use of a natural biological structure such as a sea urchin tentacle as a mold to cast a replica in an artificial biomaterial. A porous (bone) tissue ingrowth substrate was made with metal alloys and arterial graft material was made with silicone rubber. Although these processes produced "good" pore structures for the cells/tissues to grow into, no definitive advantages were found compared to fully synthetic porous implants. In the following sections the fundamental properties and design of scaffolds for tissue engineering are discussed.

16.2. Substrate Scaffold Materials

Biodegradable and bioabsorbable materials used in tissue engineering to guide and support the growth of new tissues. Polymeric materials may be used in prefabricated biomaterial-cell constructs to deliver cells to the patient for the purpose of regeneration. These materials are readily decomposed, and the decomposition products are metabolized and excreted from the body with minimum harmful effect on tissues or organs. Most studies on biodegradable materials are centered around polymers due to their similarity in molecular structure to that of tissue components. Moreover, it is easy to alter polymer molecules. Many polymers are subject to degradation by enzymes, a fact that is beneficial in the context of absorbable materials. The enzymatically degradable materials usually come from natural tissues or their constituents, such as collagen, while the nonenzymatically degradable ones are made from synthetic materials. The biodegradable materials could be utilized to deliver drugs, to close internal organs after surgery, and to reduce bone fracture. They can serve as a matrix or substrate material for tissue engineering. It is also important to consider the various factors involved with biodegradation in terms of *rate of tissue healing* in various parts of the body, *rate of decomposition* of the material, and the *rate of removal* of the decomposition products. We will focus more on tissue engineering than drug delivery since the latter has traditionally been treated in the pharmacological industry rather than in the context of implants or tissue engineering. However, there is little difference between the two, except the rate and environment of degradation and the encapsulating material is cells in tissue engineering and drugs in drug delivery.

The cell delivery medium is a very important aspect of tissue engineering since the cells must grow and multiply during the culturing period as well as function in vivo. In addition, the medium should not remain in the body after it serves its purpose. The scaffold or delivery material usually must be dissolved and absorbed by the body as the new cells take over the biological function. This dissolution should occur without harmful effects on the surrounding tissues. Therefore, it is only natural that absorbable materials such as collagen have been utilized. The main focus of the biodegradable polymer research has been their loss of strength and mass, interaction with cells (biocompatibility), the rate of drug release, etc.

Tissue engineering substrate materials should have similar biocompatibility as other biomaterials. Biocompatibility includes chemical, mechanical, pharmacological, and, where applicable, antithrombogenic or optical compatibility. The substrate materials should have affinity to the cells on the molecular scale (nanometer scale or below). In addition, they should be resorbed or absorbed by the body without affecting the function of the cells and tissues grown onto the substrates as well as the organs where the new tissues reside. Most polymers will degrade in vivo, as studied in a previous chapter, but the performance of a polymer in the present context is a function of its *rate of degradation*. Table 16-3 lists some structural factors of polymers that could influence their degradation rate.

Table 16-3. Structural Factors to Control Polymer Biodegradability

Factors		Control
Molecular structure groups,		Chemical main chain bonds, side and functional
Crystalline/aggregate state		Polymer blend, processing, copolymerization
Bulk state		Fiber, film, composite
Surface state	Area	Pore size, porosity, pore size distribution
	Hydrophilic/hydrophobic balance	Copolymerization, introduction of functional groups

Adapted with permission from Engelberg and Kohn (1991). Copyright © 1991, Elsevier.

The ratio between the characteristic time constant for biodegradation of the implant substrate at the tissue site, t_b and the time constant for healing by synthesis of new tissue inside the implant, t_h, is denoted as

$$O(1) = \frac{t_b}{t_h}, \qquad (16\text{-}1)$$

If this ratio is close to one, the substrate material may serve its purpose properly.

Another important requirement of substrate materials is the depth in a material at which cells can receive adequate nutrition by diffusion. This can be quantified in terms of the critical path length, l_c, beyond which cell migrations beyond a source of nutrition cannot take place. This calculation can be done with the aid of a dimensionless cell life-line number S, which is a dimensionless quantity that expresses the relative importance of reactions that consume nutrients and diffusion of nutrients. The diffusion length, l, is the distance over which nutrients diffuse in the tissue, and the rate of consumption of nutrients is r (moles/cm^3/s). The diffusivity of nutrient in the medium of implant is D (moles/cm^2/s), and the nutrient concentration at or near the surface of the implant is c_0 (moles/cm^3):

$$S = rl^2 / Dc_0. \qquad (16\text{-}2)$$

When S is on the order of 1, l is approximately l_c. Under these conditions, cells can migrate from host tissue into the implanted template without the requirement of the nutrient concentration in excess of that supplied by the diffusion. Therefore, Eq. (16-2) can be used to calculate the thickness (for half maximum concentration of nutrient) of the implant beyond which the cells require the presence of capillaries for adequate nutrient transport. Although these equations were developed to apply for the regeneration templates of skin, similar analysis could be made in other examples of cell migration and growth template development, such as for nerve and for knee meniscus.

There are three basic mechanisms of (bio)polymer chemical degradation, as shown in Figure 16-3. All mechanisms involve conversion of water-insoluble constituents to the water-soluble state by cleavage of crosslinks between water-soluble polymer chains (mechanism I), transformation or cleavage of side groups leading to formation of polar or charged side groups (mechanism II), and cleavage of backbone chain linkages among repeating units.

Physical degradation of polymers involves surface or bulk processes, or sometimes a combination of the two. Bulk degradation results from the fact that the rate of uptake of water is faster than the rate of conversion of polymer into water-soluble materials. Since bulk degradation occurs throughout the volume of the material, the final collapse of the material can appear abruptly. Hydrophilic polymers often exhibit this behavior. Hydrophobic polymers may degrade the surface first, leaving the inner structure intact, as shown in Figure 16-4. The surface-degrading polymers offer easier control of the rate of degradation.

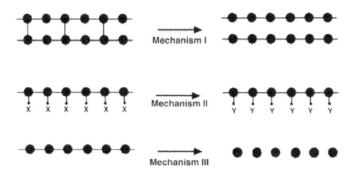

Figure 16-3. Mechanism of chemical degradation of polymers. Reprinted with permission from Kohn and Langer (1996). Copyright © 1996, Academic Press.

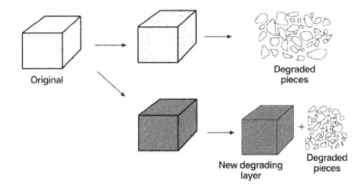

Figure 16-4. Schematic representation of two types of polymer degradation: bulk (top) and surface.

There are two types of polymeric substrate materials. One is natural, the other synthetic, as shown in Table 16-2. Polymers are also classified according to the degradability by enzyme, as given in Table 16-4, and nondegradability by enzyme in Table 16-5. Natural polymers like collagen are enzymatically degradable. Collagen can have some drawbacks as an implantable material. Structure and properties of natural materials such as collagen are species specific and tissue specific, making it difficult to obtain uniform raw materials. As for synthetic materials, since immunogenic activities often remain even after extensive processing, it can be difficult to adequately process them to obtain uniform raw materials. Processing methods include melt extrusion, which requires elevated temperatures. Some biocompatible polymers are structurally sensitive to such elevated temperature. By contrast, synthetic polymers such as poly(glycolic acids) have the opposite characteristics. Their main drawback is that their degrading products are not as compatible as those of the natural polymers, although there are some exceptions.

Table 16-4. List of Enzymatically Degradable Polymers, Structures,l
Examples, Immunogenicity, and Degraded Products

Polymers	Structure	Examples		Immuno-genicity	Enzymes	Degraded products
Proteins						
Polypeptides		Albumin	R: the residues of 20	–	Peptidase	α-Amino acids
		Fibrinogen	kinds of α-amino acids	±	Chymotrypsin	
		Collagen	(Gly-Pro-X)$_n$	±	Pepsin	
		Gelatin	X: other amino acids	–	Papain, etc	
Polyaminoacids		Poly-L-glutamic acid	(R = (CH$_2$)$_2$COOH)	+	α-Amino acids	
		Poly-L-leucine	(R = CH$_2$CH$_2$CH$_2$CH$_3$)			
		Poly-L-lysine	(R = (CH$_2$)$_3$NH$_2$)			
Polysaccharides		Amylose	(R = H)	–	Amylase	Glucose
		Hydroxyethylstarch	(R = CH$_2$CH$_2$OH)			
		Dextran	(R = H)	–	Amylase	Glucose
		Alginic acids		±		D-Mannuronic and L-Gluronic acids
		Chitin	(R = COCH$_3$)	±	Lysozyme	N-Acetyl-glucosamine
Polyesters		Poly(β-hydroxy-alkanoate)	(R = Me, Et, etc.)	–	Esterase	β-Hydroxybutyric acid, etc.
Nucleic acids		Synthetic DNA	(B: thymine, etc.)		Nuclease	Nucleosides Phosphoric acid

Adapted with permission from Kimura (1993). Copyright © 1993, CRC Press.

16.2.1. Natural Polymers

16.2.1.1. Collagen

The basic structure and morphology of collagen are presented in Chapter 10. Collagen is a family of fibrous insoluble proteins having a triple helical conformation extending over a major part of the molecule. Collagen has been used in medicine for many years as absorbable suture (cat gut) material, derived from sheep intestine. Collagen is also of use in drug delivery and tissue engineering. Applications include shields in ophthalmology, injectable dispersions for local tumor treatment, sponges carrying antibiotics, and small pellets impregnated with protein based drugs. Table 16-6 shows some of the applications of collagen.

Table 16-5. List of Non-Enzymatically Degradable Polymers,
Structures, Examples, and Degraded Products

Polymers	Structure	Examples	Degraded products
Polyesters			
Poly(α-hydroxy acids)		Polyglycolide (R = H) Polylactide (R = Me) Polyglactin (glycolic/lactic copolymer) Poly(α-malic acid) (R = CH$_2$COOH) etc.	Glycolic acid Lactic acid Malic acid
Poly(ω-hydroxyl acids)		Poly-ε-caprolactone (x = 5) Poly(β-hydroxyalkanoate) (R = Me, Et)*	5-Hydroxyhexanoic acid 3-Hydroxybutyric acid
Poly(ester-ether)		Poly(1,4-dioxan-2-one) (x = 1) Poly(1,4-dioxepan-7-one) (x = 2)	(2-Hydroxyethoxy) acetic acid
Poly(ester-carbonate)		Poly(glycolide-co-1,3-dioxan-2-one)	Glycolic acid
Polyanhydride		Poly(sebacic anhydride) (x = 6)	Sebacic acid
Polyorthoester			Alcohols
Polycarbonate		Poly(1,3-dioxan-2-one) (x = 3)	Trimethylene glycol
Poly(amide ester)		Polydepsipeptide (R = R' = Me, x = 1)**	Amino acid, hydroxy acid
Poly(α-cyanoacrylate)		Poly(ethyl-α-cyanoacrylate) (R = Et)	Formalin Ethyl cyanoacetate
Inorganic polymers		Polyphosphazene (R = imidazoyl, p-cresyl)	Phosphoric acid, ammonia, etc.
Non-polymer	Ca$_{10}$(PO$_4$)$_6$(OH)$_2$	Hydroxyapatite	Calcium phosphate

*The biopolymers can be hydrolyzed very slowly by non-enzymatic processes.
**Accelerated by esterase and peptidase.
Adapted with permission from Kimura (1993). Copyright © 1993, CRC Press.

a. *Physical modification of collagen structure.* Collagen structure is temperature sensitive. Most natural collagen is insoluble. To prepare reconstituted collagen for use in biomaterials, the collagen may be heated in dilute acid to about 40°C, or treated with KSCN to dissolve it. Collagen that has been so denatured has a random arrangement of its macromolecules. Such collagen is called gelatin.

Table 16-6. Some Applications of Collagen-Based Materials

Applications	Physical state
Sutures	Extruded tape
Hemostasis	Sponge, fiber, felt
Arterial graft	Extruded tube, reprocessed human/animal artery
Heart valves	Processed porcine heart valve
Tendon, ligaments	Processed tendon
Burn treatment	Porous collagen-glycosaminoglycan (GAG) polymer
Nerve regeneration	Porous collagen-glycosaminoglycan (GAG) polymer
Meniscus regeneration	Porous collagen-glycosaminoglycan (GAG) polymer
Intradermal cosmetic augmentation	Injectable collagen particles
Gynecological application	Sponge
Drug delivery	Various forms

Adapted with permission from Yannas (1996). Copyright © 1996, Academic Press; and Li (2000). Copyright © 2000, CRC Press.

The characteristic banding of collagen fibrils with a periodicity of 64 nm may exist at low pH (< 4.25) (see Figure 9-2). The transition does not affect the triple helix coil structure. The thrombogenic properties of the collagen can be changed by eliminating the banded structure by heat or chemical treatment.

Porous collagen can be obtained by freezing the collagen fiber solution and then evaporating the frozen ice in a vacuum. The resulting structure is a negative image of the network of dendritic ice crystals produced during the freezing process. The size and direction of pores can be controlled by the freezing temperature. Lower temperature gives smaller pores. Control of porosity can also be achieved by control of heat flux during freezing in vacuum. A pore size of 10 μm or higher is needed for ingress of soft tissue cells, while 50 μm or more is necessary for hard tissue (bone) cells. It is also very important that the pores are interconnected through and through to allow cellular activities in the pores.

b. *Chemical modification of collagen.* The main method of changing the rate of degradation or resorption of collagen is crosslinking via dehydration (to < 1% water content) or by a chemical reagent. The dehydration causes formation of interchain peptide bonds. The average molecular weight between crosslinks could reach about 70,000 g/mol after exposure for a few hours above 105°C. Dialdehyde and glutaraldehyde have been used to crosslink collagen. Two molecules of glutaraldehyde are crosslinked via two lysine side groups forming an anabilysine structure:

$$HO_2C-CH(NH_2)-[CH_2]_4-N$$

(16-3)

Glutaraldehyde-treated collagen showed a lesser immunogenic reaction when implanted than the dialdehyde-treated one. However, immunogenicity is not a serious concern for most applications using collagen.

c. *Properties of collagen.* As with other biological materials, properties of collagen depend on many factors such as the source of raw material, wet or dry, test condition such as rate of loading, temperature, and direction of loading with respect to the natural orientation of collagen fibers. Results of typical uniaxial testing of a collagen rich tissue are given in Figure 9-17 for rat tail tendon. Collagen in tendons and ligaments is highly oriented, and so it is anisotropic. The anisotropic nature of a collagenous tissue is exhibited in Figure 9-18 for abdominal human skin.

The wet collagen fibers show a typical mechanical behavior of soft tissues, exhibiting a sigmoid shape (S-type) stress–strain curve. The initial portion has an extremely low resistance to stress, and stiffening takes place with the orientation of collagen fibers at higher strain in the second portion of the curve. At very high strain (the third portion), the fibers begin to slip past each other while some very taut fibers begin to break, giving rise to a reduction in stiffness prior to gross failure. The mechanical properties of collagen fibers are given in Table 9-9. Other soft collagenous tissues — such as knee meniscus, spinal disc, joint cartilage, etc. — will have much lower modulus and strength than the oriented fiber tissues, as discussed. These tissues are actually natural composites with collagen as one constituent. Bone and dentin are hard tissues that contain collagen and a stiffer component, a calcium phosphate mineral considered as hydroxyapatite.

16.2.1.2. Poly(hydroxybutylate) (PHB) and Poly(hydroxyvaleric acid) (PHV)

These are natural polymers made by microorganisms. Poly(β-hydroxybutylate) (PHB) and poly(hydroxyvaleric acid) (PHV) repeating units in a copolymer can be represented as follows:

$$
\left[\begin{array}{c} O \\ \| \\ C-CH_2-CH-O \\ | \\ CH_3 \end{array}\right]_X
\left[\begin{array}{c} O \\ \| \\ C-CH_2-CH-O \\ | \\ CH_2CH_2 \end{array}\right]_Y
\tag{16-4}
$$

Hydroxybutyric acid (HB)   Hydroxyvaleric acid (HV)

PHB and PHV are polyesters, made by solid-phase peptide chemistry and genetically engineered microorganisms. The amino acid polymers are formed into films, and then crosslinked. The PHB degrades into D-3-hydroxybutiric acids in vivo, which is normally present in blood, making it very biocompatible. The thermal and mechanical properties of PHB are similar to PLA due to the similar repeating units. However, PHB is highly crystalline and brittle before being copolymerized with up to 30% PHV, making the copolymer more flexible but less crystalline. The copolymers are used for drug delivery matrices, sutures, and as components of artificial skin.

16.2.1.3. Chitin and Chitosan

Chitin is a natural polymer that can be obtained from crustacean exoskeletons or via fungal fermentation processes. Chitosan is an N-deacetylated derivative of chitin that is prepared by treating the chitin at 110–120°C for 2–4 hours in a 40–50% NaOH solution. Chitin is a polysaccharide and can be digested by lysozyme. Both chitin and chitosan can be made into films, fibers, and gels from dissolved solutions. Many derivatives from chitosan can be obtained, as illustrated in Figure 16-5.

Wound dressing made of chitin promotes the healing of skin wounds and regeneration of connective tissues. Also, chitin sutures can be made stronger than other resorbable sutures such as catgut. It is interesting that, while the amount of fibroblast cell attachment on the surface of a chitosan–collagen blend increases with increasing amounts of chitosan, the relative amount of cell growth showed the opposite, as shown in Figure 16-6.

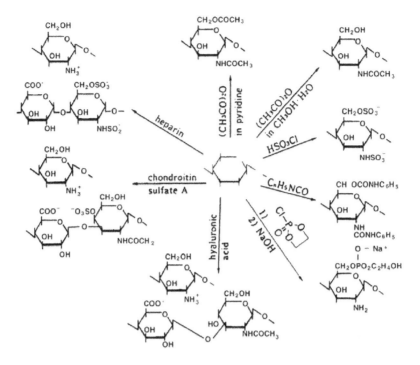

Figure 16-5. Possible derivatives from chitosan. Reprinted with permission from Aiba et al. (1987). Copyright © 1987, Butterworths-Heinemann.

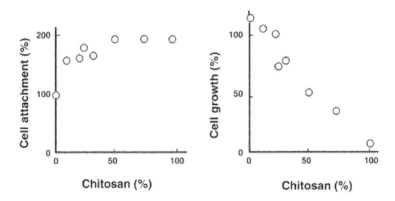

Figure 16-6. Rate of cell attachment and cell growth in chitosan–collagen blend. Reprinted with permission from Popowicz et al. (1985). Copyright © 1985, Akademie Verlag.

16.2.2. Synthetic Polymers

16.2.2.1. Poly(α-hydroxyesters): poly(glycolic acid) (PGA) and poly(lactic acid) (PLA)

The poly(α-hydroxyesters) have been synthesized to make absorbable polymers that can be hydrolyzed in vivo. The general chemical repeating unit can be represented as

$$\left[O - \underset{R}{\overset{H}{\underset{|}{C}}} - \overset{O}{\overset{\parallel}{C}} \right] \tag{16-5}$$

Depending on the type of side groups, the polymer can be glycolic (R=H), lactic (R=HC$_3$), and malic (R=CH$_2$COOH) acid. Sometimes the material is copolymerized to modify its properties, such as in polyglactin, which is a copolymer of poly(glycolic) and polyl(actic) acid.

The PGA and copolymers of PGA and PLA have been used as suture materials commercially known as Dexon® (Davis and Geck, Danbury, CT) and Vicryl® (Ethicon, Somerville, NJ), respectively, since 1970. Also, efforts have been made to produce bioabsorbable pins, screws, and plates for oral, maxillofacial, and orthopedic surgery from these polymers.

Copolymers, poly (lactic-co-glycolic acid), abbreviated PLGA, have been made as foam scaffolds to serve as templates for cell culture in tissue engineering. Foams have been fabricated as follows. PLGA is dissolved in chloroform and particles of a porogen (which generates pores) such as sodium chloride or gelatin are added to the solution. The suspension is cast as a layer and the solvent is evaporated. The porogen particles are then dissolved, leaving a foam.

Table 16.7. Properties of Synthetic Absorbable Sutures (diameter 0.209–0.239 mm)

Polymer	Trade name	Knot strength (N)	Tensile elongation (%)	Fiber form
PGA	Dexon (Davis & Geck), Opepolyx (Nippon Shoji), etc.	15.2	20	Multifilament
Polyglactin	Vicryl (Ethicon)	12.1	14	Multifilament
Polyglactin	Vicryl (Ethicon)	12.1	14	Multifilament
Polydioxanone	PDS (Ethicon)	12.3	20	Monofilament
Polyglyconate	Maxon (Davis & Geck)	10.6	31	Monofilament

Adapted with permission from Kimura (1993). Copyright © 1993, CRC Press.

The PGA loses its strength 2–4 weeks after initial implantation due to the hydrophilic nature of PGA; the Dexon® sutures lose their mechanical strength more rapidly. Slower-degrading hydrophobic PLA was copolymerized to adapt the material properties of PGA to a wider range of possible applications. Table 16-7 gives some properties of synthetic absorbable sutures, and Figure 16-7 shows the strength change with time when implanted in rabbit muscle. The changes in molecular weight, strength, and mass can be generalized as shown in Figure 16-8. Also, the chemical, physical, and surface nature of the polymer as well as the environment will greatly influence the degradation process.

Fibers of PGA or PLA can be sintered together with PLA and PGA powders with a self-reinforcing (SR) technique, at high temperature and pressure, resulting in materials with significantly higher strength and modulus that are suitable for use as fracture fixation devices such as bone screws and pins as mentioned earlier. The PGA is a linear aliphatic polyester and highly crystalline. Therefore, it has a high melting temperature and high solvent resistance, as given in Table 16-8.

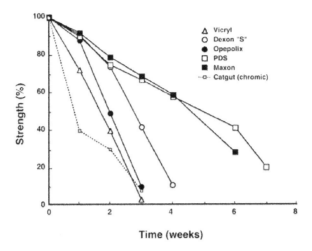

Figure 16-7. Strength change with implantation time of various sutures in muscles of rabbit. Reprinted with permission from Kimura (1993). Copyright © 1993, Chemical Rubber Company.

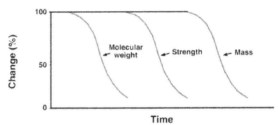

Figure 16-8. Schematic illustration of polymer degradation with respect to molecular weight, strength, and mass. Adapted with permission from E. Chiellini, *Tutorial course: Tissue engineering of bone and cartilage: current trend and future perspectives* (2005). Copyright © 2005, E. Chiellini.

Both PGA and PLA polymers are degraded to glycolide and lactide, respectively, mainly by hydrolysis, but enzymes in the living surroundings also contribute to their absorption in vivo, as shown schematically in Figure 16-9. In aqueous solution, the polyglycolide causes a more acidifying effect than the polylactide. Figures 16-9 present scanning electron microscope photographs of PGA suture before and after implantation.

The PLA has two distinctive stereoregular, isomeric molecular strucures — D-PLA and L-PLA — and the racemic form, D,L-PLA.

$$ \begin{array}{ccccc} \text{PDLA} & \longleftarrow & \text{D form} & \vdots & \text{L form} & \longrightarrow & \text{PLLA} \end{array} \qquad (16\text{-}6) $$

Table 16-8. Mechanical Properties and Transition Temperatures of Some Resorbable Synthetic Polymers

Polymer (kg/mol)	Tensile strength (MPa)	Tensile modulus (MPa)	Yield strain (%)	Breaking strain (%)	T_m (°C)	T_g (°C)
PGA (50)	–	–	–	–	210	35
L-PLA (50)	28	1200	3.7	6.0	170	54
L-PLA (100)	50	2700	2.6	3.3	159	58
L-PLA (300)	48	3000	1.8	2.0	178	59
D,L-PLA (107)	29	1900	4.0	6.0	–	51
D,L-PLA (550)	35	2400	3.5	5.0	–	53

Adapted with permission from Engelberg and Kohn (1991). Copyright © 1991, Elsevier.

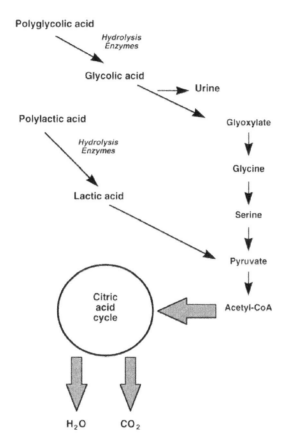

Figure 16-9. Biodegradation of polyglycolide and polylactide. Reprinted with permission from Partio (1992). Copyright © 1992, E.K. Partio.

The L-PLA is used most frequently since the byproducts of hydrolysis are similar to naturally occurring L(+) lactic acid. The semicrystalline L-PLA has higher mechanical strength than the noncrystalline D,L-PLA; therefore, the former is used for orthopedic and dental implants such as pins and screws, while the latter is used as a drug-delivery medium. Also, the PGA degrades faster than the PLA, as shown in Figure 16-10. It is, however, interesting that when PGA and PLA is copolymerized the PGA loses crystallinity faster than the rate of hydrolysis, resulting in a faster rate of degradation in 50:50 copolymer than in either pure PGA or PLA, as shown in Figure 16-11.

Figure 16-10. SEM picture of PGA sutures before and after implantation. Reprinted with permission from Ruderman (1973). Copyright © 1973, Wiley.

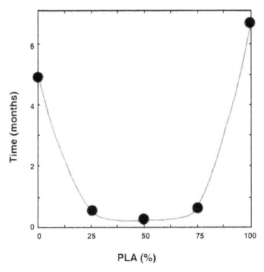

Figure 16-11. Times for 50% weight loss of PGA as a function of its unit composition implanted under the dorsal skin of a rat. Reprinted with permission from Miller et al. (1977). Copyright © 1977, Wiley.

Example 16.1

PLGA is mixed with an equal amount (by weight) of calcium phosphate, hydroxyapatite (Ha et al., 1993) to make a bone regeneration scaffold. If the scaffold contains 90% porosity with a pore size of 150 μm diameter,

    a.   Calculate its density.

    b.   Calculate its strength and modulus. Assume that a Voigt model can be applied (see Table 8-1)

    c.   What would be its density if the pores are filled with body fluid?

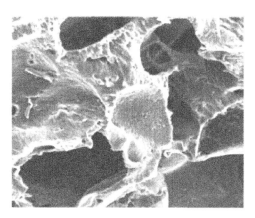

SEM picture of PLGA scaffold. This is about 90% porosity with a 150-μm pore diameter. Reprinted with permission from Chiellini (2005). Copyright © 2005, E. Chiellini.

Answer

    a.   The densities of PLGA and HA are approximately 1 and 3 g/cm³, respectively; therefore, based on 100 g, the volume fraction of each component (they are given the name of the material) would be

$$PLGA = 50 \text{ g}/1\text{g/cm}^3 = 50 \text{ cm}^3,$$

$$HA = 50\text{g}/3\text{g/ cm}^3 = 16.67 \text{ cm}^3,$$

$$\text{Porosity} = 0.9, \ PLGA = 50/(50 + 16.67) \times 0.1 = 0.075,$$

$$HA = 0.1 - 0.075 = 0.025,$$

And, therefore,

$$\text{Density of the scaffold} = 1 \times 0.075 + 3 \times 0.025 = \underline{0.15 \text{ g/cm}^3}.$$

    b.   The strength and modulus of each component are:

Component	Strength (MPa)	Modulus (GPa)
PLGA	50	2
HA	100	50

$$\text{Strength} = 2\text{GPa} \times (100 \text{ MPa}/50 \text{ GPa}) \times 0.075 + 50 \text{ GPa}(100 \text{ MPa}/50 \text{ GPa}) \times 0.025$$
$$= \underline{2.8 \text{ MPa}}.$$

$$\text{Modulus} = 2 \text{ GPa} \times 0.075 + 50 \text{ GPa} \times 0.025 = \underline{1.4 \text{ GPa}}.$$

c. Assuming the density is $1\,g/cm^3$ for body fluid, then

$$\text{Density} = 1 \times 0.975 + 3 \times 0.025 = \underline{1.05\ g/cm^3}.$$

Note that the pore size does not affect the calculations. It is also important that the pores are interconnected, open pores for cells to grow.

16.2.2.2. Poly(ε-caprolactone)

The poly(ε-caprolactone) has the following repeating unit:

$$\left[O-(CH_2)_5-\overset{\overset{\textstyle O}{\textstyle \|}}{C} \right]_n \tag{16-7}$$

The polymer contains five $(CH)_2$ units in the repeating unit, making the chains much more flexible than PGA, which has one. Therefore, thermal and mechanical properties decreased considerably compared to PGA. However, the rate of biodegradation is slower than even PLA, making it better suited to slow drug release applications such as one-year implantable contraceptives, biodegradable wound closure staples, etc.

Example 16.2

The solution or melt crystallization has been used to purify materials (zone melt and solidification in metals and ceramics). The original study on the DNA was done by crystallization and an x-ray diffraction technique by Crick and Watson. The poly(ε-caprolactone) was solutionized in 66% propanol and crystallized. The electron microscopic picture shows lamellar structure, as shown below. The x-ray (electron) diffraction study showed orthorhombic crystal structure with dimensions $a = 7.496 \pm 0.002$, $b = 4.974 \pm 0.001$, and $c = 17.297 \pm 0.0023$ Å (fiber axis), as shown in the following figure:

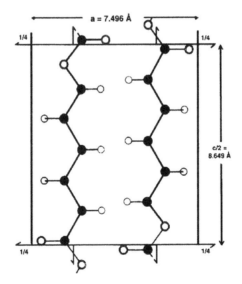

Unit cell and chain arrangement of poly(ε-caprolactone) in the ac projection. Only half $(c/2)$ of the unit cell is shown in the c-direction.

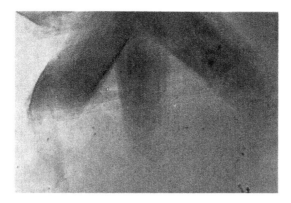

Single crystal from toluene–66% propanol solution, shadowed with chromium.

a. Calculate the density and compare with the value given by the authors of 1.146 g/cm^3.
b. If the chains are folded themselves, how many folds would be present in one chain with a molecular weight of 500,000 g/mol. Assume the fold length (~10 nm, see Figure 2-16) is the same as the thickness of the single lamellar. Disregard fold thickness at each fold end.

Answer

a. There are four repeating units (see Eq. (16-7)) per unit cell:

$$\text{Density} = \frac{\text{mass}}{\text{volume}} = \frac{4 \text{ repeating unit/u.c.}}{7.496 \text{ Å} \times 4.974 \text{ Å} \times 17.297 \text{ Å} \times 6.02 \times 10^{23} \text{atoms/mol}}$$
$$= \frac{4 \times (2 \times 16 + 6 \times 12 + 10 \times 1) \text{ g/u.c.}}{645 \text{ Å}^3 \times 6.02 \times 10^{23} \text{atoms/mol}} = \underline{1.174 \text{ g/cm}^3}$$

The (theoretical) value is somewhat higher than the reported value of 1.146 g/cm^3.

b. The number of folds (n) per chain length, L, would be

$$n - 1 = L/l,$$

where l is the fold length, disregarding fold length at each fold end. Therefore,

$$n - 1 = \frac{500,000 \text{ g/mol}}{10 \text{ nm}/17.297 \text{ Å} \times 114 \text{ g/mol}} = 759$$
$$n = \underline{760}$$

Obviously, the number of folds of each chain depends on the molecular weight. Although thermodynamically favorable for chain folding, in early development of chain folding theory transmission electron microscopy and diffraction studies were required to vertify folds. The exact mechanism of chain folding is not fully verified, although some models are available.

16.2.2.3. Polyanhydrides

The repeating unit of poly(sebacic acid–hexadecadioic acid) with a one-to-one ratio of the co-polymer is

$$\left[O-\overset{\overset{O}{\|}}{C}-(CH_2)_8-\overset{\overset{O}{\|}}{C} \right]_1 \left[O-\overset{\overset{O}{\|}}{C}-(CH_2)_{14}-\overset{\overset{O}{\|}}{C} \right]_1 \tag{16-8}$$

Sebacic acid (SA)   Hexadecandioic acid (HDA)

The polymer is hydrolyzed fast in vivo, especially the aliphatic polyanhydride; therefore, aliphatic–aromatic copolymers are made to control the rate of degradation. This polymer has an excellent biocompatibility in vivo but will react with drugs containing free amino groups, limiting their use as a drug-delivery matrix. The thermal and mechanical properties are not as good as polycaprolactone since it contains many more $(CH)_2$ groups in the main chain.

16.2.2.4. Poly(ortho esters)

One of the poly(ortho esters) repeating unit is

$$\left[(CH_2)_6 \right]_{65} O \left[\cdots \right]_{100} O \left[CH_2-\bigcirc-CH_2 \right] \tag{16-9}$$

1,6-HD   DETOSU   t-CDM

The poly(ortho esters) undergo surface degradation, making them ideal as a drug-delivery vehicle. The rate can be controlled by incorporating acidic or basic excipients into the polymer matrix since the orthoester link is less stable in an acid than in a base.

16.2.3. Composites for Bony Ingrowth

Since bone is a natural composite of collagen, a natural (bio)polymer, and mineral, hydroxyapatite, one could make (bone) scaffolds as porous synthetic composites of a resorbable polymer and apatite mineral. The most important parameter in achieving ingrowth in such materials is the porosity in which the cells can reside and grow. As with polymeric scaffolds, the rate of degradation of polymer–apatite composites depends on many aspects, including pore size, size distribution, shape, "open or closed pore structure," volume, relative amount of each constituent material, pH of surroundings during degradation, to name a few. As with other scaffolds, growth factors can be incorporated during scaffold formation. The main difference between the scaffold and the implant is the seeding of cells into the scaffold structure. If one takes a bone substitute (see §13.4.3) and seeds it with cells, one could implant it. Of course, one could also incorporate biological mediators, and growth factors as well. It is, however, essential to know many basic facts, as given in Table 16-2.

16.3. STERILIZATION OF SCAFFOLDS

As mentioned in the previous chapter, all implants must be sterilized before implantation. The biodegradable polymers pose some challenge to sterilization due to their low thermal transition temperatures (T_g and T_m), and their tendency to hydrate. Moreover, some polymers are made porous for cell seeding and controlled diffusion rate. This porosity can be filled with chemical sterilizing agents if they are used; the agents must then be flushed out. The most commonly used sterilization techniques for biodegradable polymers utilize chemicals, heat, steam, radiation, or a combination thereof. A summary of the sterilization methods used for biodegradable polymers and associated advantages and disadvantages is given in Table 16-9.

Table 16-9. Standard Sterilization Techniques Applied to Biodegradable Polymers

Sterilization technique	Advantages	Disadvantages
Steam sterilization (high steam pressure, 120–135°C)	No toxic residues	Deformation and degradation due to hydrolysis, limited usage
Dry heat sterilization (160–190°C)	No toxic residues	Melting and softening of polymer, not usable
Radiation (ionizing or γ)	High penetration, low chemical reactivity, efficient	Deterioration by crosslinking or breakage of polymer chains, may decrease M_w by up to 40%
Gas sterilization (ethylene/propylene oxide)	Low temperature residues are toxic, may alter properties	Lengthy process due to degassing,
Filtration, usually microporous glass or ceramic at high pressure	No damage to materials	Can only be used on liquid materials

Adapted with permission from Rozema et al. (1991). Copyright © 1991, Wiley; and Konig (1997). Copyright © 1997, Wiley.

Some sterilization methods can significantly affect the mechanical and physical properties of biodegradable polymers due to depolymerization by chain scission, hydrolysis by water molecules, and shape changes by heat. In addition, sterilization procedures can leave harmful residues, causing less than optimal performance of the biodegradable materials in vivo and in vitro. The specific effects of different sterilization techniques are determined by the sterilization parameters, the method used for fabrication, and the material itself. Therefore, it is important that choice of a particular sterilization regimen be made after careful studies. A new standard for sterilizing devices made of biodegradable polymer should be established. The ASTM does not have a general standard at this time. There is only one standard relating to the biodegradable polymer (F1635-95, in-vitro degradation testing of poly (L-lactic acid) resin, and fabricated form for surgical implants) listed at this time.

16.4. REGENERATION STIMULATED ELECTRICALLY

The objective of electrical stimulation in this setting is to accelerate healing or the regenerative process that normally occurs (see §13.4.1 for bone stimulation), or to induce the body to initiate a regenerative response that should not ordinarily occur. The use of electrical stimulation to regrow bone across nonunions has been discussed elsewhere in this book. Experiments have been conducted to stimulate regeneration of amputated limbs. The background of this idea is the fact that animals of low order have remarkable regenerative capacity. For example, an earthworm cut in half can regrow into two new worms; some amphibians can regenerate a limb that has been amputated. In higher animals and in humans, this sort of regeneration is very limited, although young children can regrow an amputated fingertip. Regeneration in the lower animals is accompanied by a distinctive pattern of electrical potential in the injured limb, a potential opposite in polarity to that seen in a nonregenerating higher animal or human. For example, electrical signals in amphibians, which can naturally regenerate lost limbs, differ from those in mammals, which ordinarily do not regenerate lost limbs. Researchers have stimulated partial regeneration of limbs in amputated mammals such as rats or in a nonregenerating amphibian such as a frog, by applying an electrical potential pattern similar to that in a regenerating amphibian. While bone tissue has been regrown by electrical stimulation in humans in a clinical setting, whole limbs have not.

16.5. CELLULAR ASPECTS, VIABILITY, STEM CELLS

Tissues prepared by tissue engineering should be viable after they are implanted in order to be functional. Viability requires nutrition for the cells. This can be supplied by diffusion if the tissue layer is thin, on the order of a millimeter or so. Otherwise, provision must be made for vascular infiltration. Therefore, tissue engineering scaffolds are designed to be porous so that capillaries from the recipient can penetrate the implanted tissue. The porosity, distribution of pore sizes, and geometrical continuity will govern how the biomaterial and tissue will interact when implanted.

Stem cells are undifferentiated cells capable of proliferation, of self-renewal, and of differentiation into at least one type of specialized cells. The controlled growth of human embryonic stem cells generated widespread interest in them as a source of cells for tissue engineering and other approaches to regeneration. Stem cells are of interest since they can in principle provide an unlimited supply of cells capable of giving rise to some or all tissues, including nerve. Embryonic stem cells would generate the same sort of immune response as a transplanted organ; however, since one has access to them before differentiation, it may be possible to modify them genetically to reduce or eliminate immune incompatibility.

There are ethical aspects to the use of human embryonic stem cells. Thus far, such cells have been derived from early-stage embryos produced for in-vitro fertilization. More embryos are produced in this process than are used to assist in reproduction. Use of them for experiments in regeneration has been considered superior to storing them indefinitely frozen, or discarding them. Donors have provided informed consent during this process.

Stem cells are not only found in the embryo: adults also carry stem cells in tissue niches such as bone marrow, brain, liver, and skin. The promise of these stem cells in tissue engineering is that they may be harvested, induced to proliferate, guided to differentiate into a desired tissue or organ, and then put back into the same person's body. Since the tissue would be derived from a person's own cells, there is no risk of rejection or concern about histocompatibility. A challenge is that adult-derived stem cells are often present in low concentrations, for example, 1 in 100,000 cells in bone marrow.

16.6. BLADDER REGENERATION

The study of bladder regeneration is motivated by the fact that many people who suffer severe disease of the bladder choose to have part or all of this organ removed. As of this writing, segments of the gastrointestinal tract are used as autologous transplants to replace or repair lost tissue. This can cause complications such as infection, metabolic disturbance, urolithiasis, perforation, and cancer. Growth of a new bladder or bladder segment from natural urothelial cells would solve this problem. Until recently, it was difficult to achieve sufficient proliferation of these cells in tissue culture, but it is currently possible to expand the area of a one-square-centimeter specimen by more than a million within two months. Following this proliferation of urothelial and muscle cells, the cells are seeded onto a porous scaffold and allowed to attach and form sheets of cells. The sheet, consisting of cells and scaffold, may then be implanted. Such implants have been successful in animal trials.

16.7. CARTILAGE REGENERATION

Cartilage in the joints does not have good healing capacity since it lacks blood vessels and nerves. Moreover, chondrocytes surrounded by an extracellular matrix cannot migrate from healthy tissue to a nearby injury, in contrast to most other tissue. For that reason, joint replacement by nonliving materials has been developed, as discussed elsewhere in this book. Such joint replacements involve major surgery and themselves have no self-repair capacity whatever.

Defects in the cartilage of the knee have been repaired in humans with tissue derived from tissue engineering methods (Ochi et al., 2002). In these patients, defects in the load-bearing cartilage of a femoral condyle or patellar facet occurred due to trauma (in most of the cases) or disease (in some cases). Symptoms included pain, swelling, joint locking, and crepitus. To repair the defects, autologous chondrocytes were taken from cartilage segments 500 mg or smaller from detached cartilage fragments or an unloaded portion of the joint. The cells were isolated from the cartilage by digestion with trypsin, and then collagenase. The cells were incubated on a gel medium to encourage proliferation. The gel medium was transplanted into patients' cartilage defect after 21 to 26 days. The transplanted cartilage, after 6 months to 2 years, had similar appearance and hardness in comparison with the normal cartilage. All of the patients experienced significant improvement, and most had excellent results. As of this writing, tissue engineering has not yet progressed to the point of regenerating an entire joint surface.

16.8. SKIN REGENERATION

Regeneration of skin is desirable when areas of skin have been lost due to disease (such as cancer, diabetes, or ulceration) or trauma (principally, severe burns). Synthetic artificial skin can save the life of a burned patient by preventing loss of electrolytes and preventing infection, as discussed in Chapter 11, but the healing process produces scar tissue, not normal skin. Autologous split-thickness grafts are considered the mainstay of treatment for large, full thickness skin wounds. Scarring also occurs after this treatment. This scar tissue does not have the flexibility of normal skin; it may experience contracture, and its cosmetic appearance is inferior to that of normal skin. Deep lesions are problematical, particularly in diabetics, who have impaired healing capacity.

Tissue engineering has been used clinically in the treatment of a severe and deep lesions in a diabetic. Autologous fibroblasts and keratinocytes were harvested and grown on scaffolds containing an ester derivative of hyaluronic acid. Cultured tissue was placed on the wound at 7-day intervals. The ulcer healed completely 60 days after the first graft. After 16 months no lesions recurred, and the patient was able to walk without support.

As for burn patients, an autograft was compared with a reconstruction based on an autograft-based tissue engineering approach. In the latter, a porous, 1-mm thick membrane was used. It was composed of bovine collagen fiber template in which the fibers were coated with elastin hydrolysate derived from bovine ligamentum nuchae. The autograft was placed on top of this membrane during the reconstruction. The membrane approach showed no statistically significant long-term improvement compared with conventional autografts, though it was beneficial in the short term and for small wounds.

16.9. BONE REGENERATION

The healing of a bone fracture is regenerative in that the new bone is equivalent to the preexisting bone and no scar is formed. If a large segment of bone is lost due to disease or injury, natural healing will not bridge the gap. Bone grafts are currently used, but if one uses autologous bone there is a limited supply. Therefore, osteoblasts have been cultured on porous PLGA polymer and tested in animal experiments. It would be logical to use hydroxyapatite or glass-ceramics (§§6.3 and 6.4) to induce bone regeneration. However, this has not been attempted as of yet.

16.10. REGENERATION IN THE CARDIOVASCULAR SYSTEM

The limitations of available heart valve substitutes include the use of long-term anticoagulation drugs to prevent clotting in these implants, their inability to repair damage, and susceptibility to infection. Therefore, research has been conducted on the prospects for growing heart valves via tissue engineering

As for arteries, synthetic materials have been used with success in the aorta and relatively large vessels 6 to 10 mm in diameter. For small vessels 3-4 mm in diameter, such as those in the coronary circulation, synthetic vessels do not work: they fill with clot. An initial step in tissue engineering of blood vessels is to grow a layer of autologous endothelial cells upon a graft of Dacron or expanded polytetrafluoroethylene (ePTFE). The cells can be harvested from peripheral veins or small blood vessels in the patient's fat. This approach has been used clinically. Such grafts cannot remodel as would a normal vessel, and they cannot actively change diameter to control blood flow. Functional blood vessel replacements are currently under study to solve these problems. Vascular cells have been embedded in a reconstituted animal collagen gel matrix. Thus far, collagen-based tissue-engineered vessels are not strong enough for use. To achieve properties comparable to a natural blood vessel, the cultured vessel must be subjected to pulsatile pressure as it is grown; otherwise, it bursts at a relatively low pressure. Vessels grown from rolls of layers of cultured cells show ten times better burst strength than collagen-based ones. Replacement pulmonary arteries for potential use in congenital defects in babies have been grown in culture from umbilical cord cells.

16.11. SUMMARY

A variety of polymeric materials has been discussed in the context of biodegradable materials, mostly polymers, for use in tissue engineering. These materials may be of natural origin or purely synthetic. Ceramics and glasses are not discussed in this chapter, although some, such as calcium phosphate compounds, could be used as an excellent scaffold. Tricalcium phosphate or hydroxyapatite are such materials, and they are discussed in Chapter 6. They may serve as a scaffold for the growth of cultured cells to form an organized tissue. Several case studies have also been presented in this rapidly evolving field.

PROBLEMS

16-1. Pores are important for the cells to enter and grow, as well as in capillary suction of tissue fluid and blood. They also provide a very large surface area for invading cells or for interaction with presiding cells. Calculate the surface area of a collagen sponge that is made of spherical pores (average 50 μm in diameter) that touch each other in one cubic centimeter. First calculate the number of spheres in 1 cm^3, and then the surface area of the spheres. Assume the spheres (pores) are packed most efficiently.

16-2. A bioengineer is trying to study the effect of degradation of collagen scaffolds using a mechanical test. Design an experiment to accomplish the goal. The bioengineer can use a uniaxial-material testing machine, a temperature-controlled wet chamber for the specimen in which it can be loaded, and one also could use enzymes such as collagenase in the chamber (Hoffman et al., 1972). What are the parameters — i.e., stiffness, strength, half-time of complete degradation — that can be obtained? Can you measure the amount of degradation products with time? How would you accomplish this? Be specific.

16-3. Biodegradable materials most likely go back to their original molecular state (monomer) before they are assembled into final products (Frisch et al., 1994). Some monomers (lactide/glycolide) can be toxic after becoming acid in solution or body fluid. They become toxic for tissues if they are concentrated in one location. Although the body may tolerate this in small amounts for cases such as sutures, it would be quite detrimental to cells in the case of scaffolds for tissue engineering. Propose one or more ways of neutralizing the acid as biodegradable polymers degrade.

16-4. Fifteen milligrams of PLGA were degraded in a 15-ml flask and 15 mg of phosphate buffer were added. The pH was monitored with time and the results are depicted as follows (figure courtesy of Dr. J.W. Lee, University of Iowa, Iowa City, 2006):

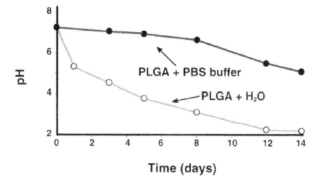

a. Calculate the concentration of acid if the pH dropped from 7.0 to 3 for a control specimen without buffer.

b. Calculate the amount of buffer in milligrams needed to neutralize 1 mg of PLGA.

c. Propose a mechanism of how the acids are being neutralized.

16-5. The idea of making the same cellular structure for better ingrowth of cells and tissues was tried in earlier times as "replaneform," but it did not prove to be any better than a "porous" implant if the pores were open and interconnected, and larger than the cell size. Speculate the reasons for the less than optimal condition for tissue/cell (in)growth. Can you suggest better methods?

16-6. One would like to use "nanotechnology" for making scaffolds.

a. Define nanotechnology.

b. Discuss means of making "nanomaterials" from the scaffold materials listed in Table 16-2.

c. Discuss the advantages and disadvantages of making "nanomaterials" for application in scaffolds for tissue engineering.

16-7. One would like to use "nanotechnology" for drug delivery.

a. Discuss the difference between conventional drug delivery by incorporating drugs into excipients and making different-sized microspheres (MS) that will release drugs at a different rate due to surface area and surface energy.

b. Discuss the advantages and disadvantages of making "nanospheres" (NS) for application in drug delivery.

16-8. One would like to deliver cancer drugs to a precise location by using NS (nanotechnology) and "magnetic particles." The magnetic particles can be guided by an MRI- (magnetic resonance) type device.

a. Discuss the advantages and disadvantages of making such a sphere (drug and magnetic particles) and delivering locally to the cancer.

b. Magnetic spheres can also be utilized for heating via "magnetic induction" by radio frequency (rf). Discuss the advantages and disadvantages of such applications in brain cancer.

16-9. One would like to deliver genes to a precise location by using NS (nanotechnology) and "magnetic particles." The magnetic particles can be guided by an MRI-type device.

a. Discuss the advantages and disadvantages of making such spheres (genes and magnetic particles) and delivering into local genes.

b. Discuss the experimental protocols of incorporating genes to magnetic particles, testing the viability of such particles, and separating the magnetic particles from the genes after the genes are in place.

SYMBOLS/DEFINITIONS

Symbols

l_c: Critical cell path length in porous scaffold or depth of diffusion of cells.

l: Diffusion path length.

r: Rate of consumption of the nutrient by the cells (moles/cm^3/s).

D: Diffusivity of nutrient in the medium of implant (moles/cm^2/s).

C_0:   Nutrient concentration at or near the surface of an implant (mol/cm^3).

S: Cell lifeline number.

t_b: Time constant for biodegradation of the implant substrate at the tissue site.

t_h: Time constant for healing by synthesis of new tissue inside the implant.

$O(l)$: Ratio of t_b/t_h.

Definitions

Aldehyde: Any one of a group of organic chemical compounds having the radical CHO, derived from the primary alcohols by oxidation, and yielding acids on further oxidation. Formaldehyde is an aldehyde produced by the oxidation of methanol (CH$_3$CHO).

Autograft: A transplant from a part of one's body to another part of the body.

Autologous: Derived from a person's own tissue.

Bioabsorption: Being absorbed by blood or tissues of the decomposed polymer by products in vivo.

Biodegradation: Being decomposed by bacterial or enzymatic action in vivo.

Chemical degradation: Erosion of polymers by breakage main chains or crosslinks.

Chitin: A horny substance forming the hard outer covering of lobsters, crabs, beetles, crickets, and some fungi; closely related to cellulose.

Chitosan: A derivative of chitin by N-deacetylation.

Collagen: Protein substance in the fibers of connective tissues, bone, and cartilage of vertebrates. Boiling with water converts collagen to gelatin.

Collagen-glycosaminoglycan copolymer: Two natural polymers made into a copolymer, one being protein (collagen) and another polysaccharide, would give good mechanical, chemical, and biological properties for such applications as scaffold material.

Dexon®: Suture made of copolymer of PGA and PLA (Davis and Geck, Danbury, CT).

Fibrin: An insoluble protein formed from fibrinogen during the clotting of blood. It forms a fibrous mesh that impedes the flow of blood.

Gelatin: Denatured collagen obtained by boiling bones and cartilage.

Glutaraldehyde (C$_5$H$_8$O$_2$): Chemical made from cyclohexene in atmospheric reaction, used to preserve tissues and sterilization.

Immunogenicity: The property of causing immunity (antigenicity) to a disease.

PGA: Poly(glycolic acid), one of the synthetic polymers, poly(α•-hydroxyesters) used to make biodegradable scaffolds.

PHB: Poly(hydroxybutylate), a natural polymer usually copolymerized with PHV to make biodegradable polymer scaffolds.

PHV: Poly(hydroxyvaleric acid), a natural polymer usually copolymerized with PHB to make biodegradable polymer scaffolds.

Physical degradation: Erosion of polymers at the surface or bulk.

PLA: Poly(lactic acid), one of the synthetic polymers, poly(α-hydroxyesters) used to make biodegradable scaffolds.

Polyanhydride: Step-reaction polymerized polymer from $HOOC(CH_2)_xCOOH$ resulting in $HO[-CO(CH_2)_xCOO-]_yH + H_2O$.

Poly(ϵ-caprolactone) (PCL): A biodegradable, biocompatible, thermoplastic homopolymer used to coat, fabricate scaffold, films, etc. Can be copolymerized to modify its properties.

Poly(ortho esters) (POE): A polymer produced from 3,9-bis(ethylidene 2,4,8,10-tetraoxaspirol [5,5]undecane) (DETOSU) condensed with di-alcohols. This generation of POEs released nonacidic byproducts and eliminated the autocatalytic process. Combined with certain additives, POEs were found to degrade by surface erosion, making DETOSU-based POEs well suited for controlled drug delivery.

Racemic: Consisting of an optically inactive, equimolar mixture of the dextrorotatory and levorotatory forms of certain substances.

Resorption: A material, typically a polymer, being absorbed by the body in vivo.

Sodium alginate $(C_6H_7NaO_6)_n$: Sodium salt of alginic acid, with 10~600 k m.w. dissolves slowly in water, forming a viscous solution; used as stabilizer, thickener, gelling agent, and emulsifier.

Stem cells: Undifferentiated cells capable of proliferation and of self-renewal and of differentiation into at least one type of specialized cells.

Tissue engineering: An interdisciplinary field involving engineering and life sciences for the development of biological substitutes containing living cells, with the aim to restore, maintain, or improve cells, tissues, and organs in the body.

Vicryl®: Suture made of copolymer of PGA and PLA (Ethicon, Somerville, NJ).

BIBLIOGRAPHY

Aiba S, Minoura N, Tagachi K, Fujiwara Y. 1987. Covalent immobilization of chitosan derivatives onto polymeric film surfaces with the use of a photosensitive heterobifunctional crosslinking reagent. *Biomaterials* **8**:481.

Atala A. 2001. Bladder regeneration by tissue engineering. *BJU Int* **88**(7):765–770.

Atala A, Lanza R. 2001. *Methods of tissue engineering*. New York: Academic Press.

Barbucci R. 2002. *Integrated biomaterials science*. New York: Kluwer Academic/Plenum.

Bassett P, DiClemente SC. 1999. *Emerging markets in tissue engineering: angiogenesis, soft and hard tissue regeneration, xenotransplant, wound healing, biomaterials and cell therapy*. Southborough, MA: D&MD Reports.

Becker RO. 1961. Search for evidence of axial current flow in peripheral nerves of salamander. *Science* **134**:101–102.

Bishop AE, Buttery LDK, Polak JM. 2002. Embryonic stem cells. *J Pathol* **197**(4):424–429.

Bock G, Goode J. 2003. *Tissue engineering of cartilage and bone*. CIBA Foundation Symposia Series, no. 249. New York: Wiley.

Bronzino JD, ed. 2006. *Tissue engineering and artificial organs (the biomedical engineering handbook)*. Boca Raton, FL: CRC Press.

Chiellini E. 2005. Tutorial course: Tissue engineering of bone and cartilage: current trend and future perspectives. October 18.

Dalla PL, Cogo A, Deanesi W, Stocchiero C, Colletta VC. 2002. Using hyaluronic acid derivatives and cultured autologous fibroblasts and keratinocytes in a lower limb wound in a patient with diabetes: a case report. *Ostomy Wound Manage* **48**(9):46–49.

Deutsch M, Meinhart J, Fischlein T, Preiss P, Zilla P. 1999. Clinical autologous in vitro endothelialization of infrainguinal ePTFE grafts in 100 patients: a 9-year experience. *Surgery* **126**(5):847–855.

Engelberg I, Kohn J. 1991. Physicomechanical properties of degradable polymers used in medical applications: a comparative study. *Biomaterials* **12**:292–304.

Folkman J, Hochberg M. 1973. Self-regulation of growth in three dimensions. *J Exp Med* **138**(4):745–753.

Friess W. 1998. Collagen: biomaterial for drug delivery. *Eur J Pharmacol Biopharmacol* **45**(2):113–136.

Frisch KC, Eldred EW, eds. 1994. *Proceedings of the 25th anniversary symposium of the polymer institute: service to industry, government and education*. Lancaster, PA: Technomics Publishers.

Ha S-W, Mayer J, Wintermantel E. 1993. Micro-mechanical testing of hydroxyapatite coatings on carbon fiber reinforced thermoplastics. In *Bioceramics*, Vol. 6, pp. 489–493. Ed P Ducheyne, D Christiansen. Oxford: Pergamon.

Hasirci N, Hasirci V. 2004. *Biomaterials: from molecules to engineered tissues*. New York: Kluwer Academic/Plenum.

Heijkants RG, van Calck RV, De Groot JH, Pennings AJ, Schouten AJ, van Tienen TG, Ramrattan N, Buma P, Veth PR. 2004. Design, synthesis and properties of a degradable polyurethane scaffold for meniscus regeneration. *J Mater Sci Mater Med* **15**(4):423–427.

Helmus MN. 2003. *Biomaterials in the design and reliability of medical devices*. Tissue Engineering Intelligence Unit 5. Georgetown, TX: Landes Bioscience (Eurekah.com).

Hench LL, Jones JR, eds. 2005. *Biomaterials, artificial organs and tissue engineering*. Cambridge: Woodhead.

Herndon DN, ed. 1996. *Total burn care*. London: Saunders.

Hoerstrup SP, Kadner A, Breymann C, Maurus CF, Guenter CI, Sodian R, Visjager JF, Zund G, Turina MI. 2002. Living, autologous pulmonary artery conduits tissue engineered from human umbilical cord cells. *Ann Thorac Surg* **74**(1):46–52; discussion 52.

Hoffman AS, Grande LA, Gibson P, Park JB, Daly CH, Ross R. 1972. Preliminary studies on mechano-chemical-structure relationships in connective tissues using enzymolysis techniques. In *Perspectives of biomedical engineering*, pp. 173–176. Ed RM Kenedi. Baltimore, MD: University Park Press.

Kimura Y. 1993. Biodegradable polymers. In *Biomedical applications of polymeric materials*, pp. 163–189. Ed T Tsuruta, T Hayashi, K Kataoka, K Ishihara, Y Kimura. Boca Raton, FL: CRC Press.

Kohn J, Langer R. 1996. Biosorbable and bioerodible materials. In *Biomaterials science: an introduction to materials in medicine*, pp. 64–73. Ed BD Ratner, AS Hoffman, FJ Schoen, JE Lemons. San Diego: Academic Press.

Konig C, Ruffieux K, Wintermantel E, Blaser J. 1997. Autosterilization of biodegradable implants by injection molding process. *J Biomed Mater Res* **38**:115–119.

Lanza RP, Langer RS, Vacanti J, eds. 2000. *Principles of tissue engineering*. San Diego: Academic Press.

Li S-T. 2000. Biologic materials: tissue-derived biomaterials (collagen). In *the biomedical engineering handbook*, 2d ed., pp. 42/1–23. Ed JD Bronzino. Boca Raton, FL: CRC Press.

Lu HH, El-Amin SF, Scott KD, Laurencin CT. 2003. Three-dimensional, bioactive, biodegradable, polymer-bioactive glass composite scaffolds with improved mechanical properties support collagen synthesis and mineralization of human osteoblast-like cells in vitro. *J Biomed Mater Res A* **64**(3): 465–474.

Lu HH, Tang A, Oh C, Spalazzi JP, Dionisio K. 2005. Compositional effects on the formation of a calcium phosphate layer and the response of osteoblast-like cells on polymer-bioactive glass composites. *Biomaterials* **26**:6323–6334.

McLaren A. 2001. Ethical and social considerations of stem cell research. *Nature* **414**(6859):129–131.

Miller PA, Brady JM, Cutright DE. 1977. Degradation rates of oral resorbable implants (polylactates and polyglycolates): rate modification with changes in PLA/PGA copolymer ratios. *J Biomed Mater Res* **11**(5):711–719.

Moss SC. 2002. Advanced biomaterials: characterization, tissue engineering, and complexity. Paper presented at Materials Research Society Symposium, November 26–29. Boston: Materials Research Society.

Murphy WL, Peters MC, Kohn DH, Mooney DJ. 2000. Sustained release of vascular endothelial growth factor from mineralized poly(lactide-co-glycolide) scaffolds for tissue engineering. *Biomaterials* **21**(24):2521–2527.

Nerem RM, Seliktar D. 2001. Vascular tissue engineering. *Annu Rev Biomed Eng* **3**:225–243.

Ochi M, Uchio Y, Kawasaki K, Wakitani S, Iwasa J. 2002. Transplantation of cartilage-like tissue made by tissue engineering in the treatment of cartilage defects of the knee. *J Bone Joint Surg Br* **84**(4): 571–578.

Palsson B. 2003. *Tissue engineering*. Boca Raton, FL: CRC Press.

Partio EK. 1992. *Absorbable screws in the fixation of cancellous bone fractures and arthrodeses: a clinical study of 318 patients*. Department of Orthopaedics and Traumatology, Helsinki University.

Peter SJ, Miller MJ, Yasko AW, Yaszemski MJ, Mikos AG. 1998. Polymer concepts in tissue engineering. *J Biomed Mater Res* **43**(4):422–427.

Petite H, Quarto R. 2005. *Engineered bone*. Tissue Engineering Intelligence Unit. Georgetown, TX: Landes Bioscience (Eurekah.com).

Popowicz P, Kurzyca J, Dolinska B, Popowicz J. 1985. Cultivation of MDCK epithelial cells on chitosan membranes. *Biomed Biochim Acta* **44**:1329.

Prokop A, Hunkeler D, Cherrington AD, eds. 1998. *Bioartificial organs*. Papers presented at a conference entitled Bioartificial Organs: Science and Technology held July 21–25, 1996, Nashville, Tennessee by the Engineering Foundation. New York: New York Academy of Sciences.

Ratner BD, Bryant SJ. 2004. Biomaterials: where we have been and where we are going. *Annu Rev Biomed Eng* **6**:41–75.

Roy TD, Simon JL, Ricci JL, Rekow ED, Thompson VP, Parsons JR. 2003. Performance of degradable composite bone repair products made via three-dimensional fabrication techniques. *J Biomed Mater Res A* **66**(2):283–291.

Rozema FR, Boering G, van Asten JAAM, Nijenhuis AJ, Penning AJ. 1991. The effects of different steam-sterilization programs on material properties of poly(L-lactide). *J Appl Biomater* 24:23–28.

Ruderman RJ, Bernstein E, Kairinen E, Hegyeli AF. 1973. Scanning electron microscopic study on surface changes on biodegradable sutures. *J Biomed Mater Res* **7**:215–229.

Saltzman WM. 2004. *Tissue engineering: engineering principles for the design of replacement organs and tissues*. Oxford: Oxford UP.

Shi D. 2004. *Biomaterials and tissue engineering*. Berlin: Springer-Verlag.

Shoichet MS, Hubbell JA, eds. 1998. *Polymers for tissue engineering*. Boston: Brill Academic.

Shoichet MS, Schmidt C. 2001. *Neural tissue engineering*. Oxford: Elsevier.

Spector M. 1999. Basic principles of tissue engineering. In *Tissue engineering: application in maxillofacial surgery and periodontics*, pp. 3–16. Ed SE Lynch, RJ Genco, RE Marx. Chicago: Quintessence Publishers.

Stock UA, Vacanti JP, Mayer Jr JE, Wahlers T. 2002. Tissue engineering of heart valves: current aspects. *Thorac Cardiovasc Surg* **50**(3):184–193.

Teoh SH, ed. 2002. Symposium B: biomaterials and tissue engineering, international conference on materials for advanced technologies (ICMAT 2001), July 1–6, 2001. International Conference on Materials for Advanced Technologies. Amsterdam: Elsevier.

Thomson JA, Itskovitz-Eldor J, Shapiro SS, Waknitz MA, Swiergiel JJ, Marshall VS, Jones JM. 1998. Embryonic stem cell lines derived from human blastocysts. *Science* **282**(5391):1145–1147.

Thomson RC, ed. 1998. Biomaterials regulating cell function and tissue development: symposium held April 13–14, 1998, San Francisco, California, U.S.A. Warrendale, PA: Materials Research Society.

van Zuijlen PP, Vloemans JF, van Trier AJ, Suijker MH, van Unen E, Groenevelt F, Kreis RW, Middelkoop E. 2001. Dermal substitution in acute burns and reconstructive surgery: a subjective and objective long-term follow-up. *Plast Reconstr Surg* **108**(7):1938–1946.

Wong JY, ed. 2004. *Architecture and application of biomaterials and biomolecular materials*. Materials Research Society Symposium Proceedings. Boston: Materials Research Society.

Wozney JM. 1999. Biological and clinical applications of rhBMP-2. In *Tissue engineering applications in maxillofacial surgery and periodontics*, pp. 103–123. Ed SE Lynch, RJ Genco, RE Marx. Chicago: Quintessence Publishing.

Yannas IV. 1996. Natural materials. In *Biomaterials science: an introduction to materials in medicine*, pp. 84–94. Ed BD Ratner, AS Hoffman, FJ Schoen, JE Lemons. San Diego: Academic Press.

Yaszemski MJ. 2004. *Tissue engineering and novel delivery systems*. New York: Marcel Dekker.

APPENDICES

Appendix I: Physical Constants and Conversions

Physical Constants

Name	Symbol	Units
Atomic mass unit	amu	1.66×10^{-24} g
Avogadro's number	N	6.023×10^{23}/mol (or molecules/g mol)
Boltzmann's constant	k	1.381×10^{-23} J/K or 8.63×10^{-5} eV/K
Permittivity (vacuum)	ε	8.85×10^{-12} C/V·m
Electron charge	q	1.602×10^{-19} C
Electron volt	eV	1.60×10^{-19} J
Faraday's constant	F	0.649×10^{4} C/mol
Gas constant	R	8.314 J/mol K or 1.987 cal/mol K
Planck's constant	h	6.62×10^{-34} J·s
Standard gravity	g	9.807 m/s^2
Velocity of light	c	3×0^{6} m/s

Conversions

1 Angstrom (Å)	=	10^{-10} m
1 Ampere (Å)	=	1 Coul/s
1 Inch (in)	=	0.0254 m
1 Calorie (Cal)	=	4.1868 J
1 Erg	=	10^{-7} J
1 Dyne	=	10^{-5} N
1 Joule (J)	=	1 N·m
1 kg force (kgf)	=	9.807 N
1 Pound (force) (lb)	=	4.448 N
1 Pound (mass)	=	0.4536 kg
1 Atmosphere (standard)	=	0.1 MPa
1 mm Hg (60°F)	=	1.329×10^{2} Pa
1 Inch of Hg (60°F)	=	3.37685×10^{3} Pa
1 Dyne/cm^2	=	0.1 Pa
1 kgf/mm^2	=	9.807×10^{6} Pa
1 lb/in^2 (psi)	=	6.895×10^{3} Pa
1 MPa	=	145 psi
1 Poise	=	0.1 Pa·s
1 Rad	=	0.01 Gy

Appendix II: SI Units

The International System of Units of SI (Le Système International d'Unités) defines the *base units* as follows:

Base Units

Quantity	Unit	Symbol
Length	Metre	m
Mass	Kilogram	kg
Time	Second	s
Electric current	Ampere	A
Temperature	Kelvin	K
Amount of substance	Mole	mol

Derived Units

Quantity	Unit	Symbol	Formula
Frequency	Hertz	Hz	$1/s$
Force	Newton	N	$kg \cdot m/s^2$
Pressure, stress	Pascal	Pa	N/m^2
Energy, work, quantity of heat	Joule	J	$N \cdot m$
Power	Watt	W	J/s
Absorbed dose	Gray	Gy	J/kg

Appendix III: Common Prefixes

Multiplication factor	Prefix	Symbol
10^{18}	peta	P
10^{12}	tera	T
10^9	giga	G
10^6	mega	M
10^3	kilo	k
10^{-2}	centi	c
10^{-3}	milli	m
10^{-6}	micro	μ
10^{-9}	nano	n
10^{-12}	pico	p
10^{-15}	femto	f
10^{-18}	atto	a

Appendix IV: Properties of Selected Elements

Element	Symbol	Atomic number	Atomic weight (amu)	T_m (°C)	Solid density ρ (g/cm³)	Crystal Structure*	Atomic radius (nm)
Hydrogen	H	1	1.008	−259.14	–	–	0.046
Lithium	Li	3	6.94	180	0.534	bcc	0.1519
Beryllium	Be	4	9.01	1289	1.85	hcp	0.114
Boron	B	5	10.81	2103	2.34	–	0.046
Carbon	C	6	12.011	>3500	2.25	hex	0.077
Nitrogen	N	7	14.007	−210	–	–	0.071
Oxygen	O	8	15.999	−218.4	–	–	0.060
Fluorine	F	9	19.0	−220	–	–	0.06
Neon	Ne	10	20.18	−248.7	–	fcc	0.160
Sodium	Na	11	22.99	97.8	0.97	bcc	0.1857
Magnesium	Mg	12	24.31	649	1.74	hcp	0.161
Aluminum	Al	11	26.98	660.4	2.70	fcc	0.14315
Silicon	Si	14	28.09	1414	2.33	diamond cubic	0.1176
Phosphorus	P	15	30.97	44	1.8	–	0.11
Sulfur	S	16	32.06	112.8	2.07	–	0.106
Chlorine	Cl	17	35.45	−101	–	–	0.0905
Argon	Ar	18	39.95	−189.2	–	fcc	0.192
Potassium	K	19	39.10	63	0.86	bcc	0.2312
Calcium	Ca	20	40.08	840	1.54	fcc	0.1969
Titanium	Ti	22	47.90	1672	4.51	hcp	0.146
Vanadium	V	23	50.94	1910	6.09	bcc	0.132
Chromium	Cr	24	52.00	1863	7.20	bcc	0.1249
Manganese	Mn	25	54.94	1246	7.2	–	0.112
Iron	Fe	26	55.85	1538	7.88	bcc	0.1241
Cobalt	Co	27	58.93	1494	8.9	hcp	0.125
Nickel	Ni	28	58.71	1455	8.90	fcc	0.1246
Copper	Cu	29	63.54	1084.5	8.92	fcc	0.1278
Zinc	Zn	30	65.37	419.6	7.14	hcp	0.139
Yttrium	Y	39	88.9	1522	4.47	hcp	–
Zirconium	Zr	40	91.22	1852	6.51	hcp	0.159
Molybdenum	Mo	42	95.94	2617	10.22	bcc	0.136
Silver	Ag	47	107.87	961.9	10.5	fcc	0.1444
Tin	Sn	50	118.69	232	7.3	bct	0.1509
Antimony	Sb	51	121.75	630.7	6.7	–	0.1452
Cesium	Cs	55	132.9	28.4	1.9	bcc	0.262
Tungsten	W	74	183.9	3387	19.4	bcc	0.1367
Gold	Au	79	197.0	1064.4	19.32	fcc	0.1441
Mercury	Hg	80	200.6	−38.86	–	–	0.155
Lead	Pb	82	207.2	327.5	11.34	fcc	0.1750

* Solid density and crystal structure are for room temperature.

Appendix V: Properties of Selected Engineering Materials (20°C)

Material	Density (g/cm^3)	Thermal conductivity $(W/m \cdot °C)$	Linear expansion $(°C^{-1})$	Electrical resistivity, ρ (ohm·m)	Average modulus of elasticity, E	
					MPa	psi
Metals						
Aluminum (99.9+)	2.7	0.22	22.5×10^{-6}	29×10^{-9}	70,000	10×10^6
Copper (99.9+)	8.9	0.40	17×10^{-6}	17×10^{-9}	110,000	16×10^6
Iron (99.9+)	7.88	0.072	11.7×10^{-6}	98×10^{-9}	205,000	30×10^6
Monel (70 Ni–30 Cu)	8.8	0.025	15×10^{-6}	482×10^{-9}	180,000	26×10^6
Silver (sterling)	10.4	0.41	18×10^{-6}	18×10^{-9}	75,000	11×10^6
Steel (1020)	7.86	0.050	11.7×10^{-6}	169×10^{-9}	205,000	30×10^6
Steel (1040)	7.85	0.048	11.3×10^{-6}	$171 \times 10{-9}$	205,000	30×10^6
Steel (1080)	7.84	0.046	10.8×10^{-6}	180×10^{-9}	205,000	30×10^6
Steel (18Cr–8Ni stainless)	7.93	0.015	9×10^{-6}	700×10^{-9}	205,000	30×10^6
Ceramics						
Al_2O_3	3.8	0.029	9×10^{-6}	$>10^{12}$	350,000	50×10^6
Graphite (bulk)	1.9	–	5×10^{-6}	10^{-5}	7,000	1×10^6
MgO	3.6	–	9×10^{-6}	10^3 (1100°C)	205,000	30×10^6
Quartz (SiO_2)	2.65	0.012	–	10^{12}	310,000	45×10^6
SiC	3.17	0.012	4.5×10^{-6}	0.025 (1100°C)	–	
TiC	4.5	0.030	7×10^{-6}	50×10^{-8}	350,000	50×10^6
Glass						
Plate	2.5	0.00075	9×10^{-6}	10^{12}	70,000	10×10^6
Borosilicate	2.4	0.0010	2.7×10^{-6}	$>10^{15}$	70,000	10×10^6
Silica	2.2	0.0012	0.5×10^{-6}	10^{18}	70,000	10×10^6
Polymers						
Melamine-formaldehyde	1.5	0.00030	27×10^{-6}	10^{11}	9,000	1.3×10^6
Phenol-formaldehyde	1.3	0.00016	72×10^{-6}	10^{10}	3,500	0.5×10^6
Urea-formaldehyde	1.5	0.00030	27×10^{-6}	10^{10}	10,300	1.5×10^6
Rubbers (synthetic)	1.5	0.00012	–	–	4–75	600–11,000
Rubber (vulcanized)	1.2	0.00012	81×10^{-6}	10^{12}	3,500	0.5×10^6
Polyethylene (L.D.)	0.92	0.00034	180×10^{-6}	10^{13}–10^{16}	100–350	14,000–50,000
Polyethylene (H.D.)	0.96	0.00052	120×10^{-6}	10^{12}–10^{16}	350–1250	50,000–180,000
Polystyrene	1.05	0.00008	63×10^{-6}	10^{16}	2,800	0.4×10^6
Polyvinylidene chloride	1.7	0.00012	190×10^{-6}	10^{11}	350	0.05×10^6
Polytetrafluoroethylene	2.2	0.00020	100×10^{-6}	10^{14}	350–700	50,000–100,000
Polymethyl methacrylate	1.2	0.00020	90×10^{-6}	10^{14}	3,500	0.5×10^6
Nylon	1.15	0.00025	100×10^{-6}	10^{12}	2,800	0.4×10^6

NAME INDEX

SUBJECT INDEX

Lightning Source UK Ltd.
Milton Keynes UK
UKOW05f0610010915

257856UK00001B/13/P